Food Safety

This book is an updated reference source on food safety best practices. The chapters discuss analytical approaches to measuring food contaminants, quality control and risk assessment of food storage, food irradiation, etc. The contributors discuss how quality control and management help to establish sustainable and secure food systems globally. The book covers topics such as techniques to measure food contaminants, toxins, heavy metals and pesticide content in food.

FEATURES

- Examines the role of food safety approaches in global food supply chains
- Describes various detection techniques for food contaminants and toxins
- Discusses the application of nanotechnology and other innovations in food safety and risk assessment
- Reviews the international regulations for management of food hazards
- Includes the hazard analysis critical control points (HACCP) principles

This book is an essential resource to help students, researchers, and industry professionals understand and address day-to-day problems regarding food contamination and safety and their impact on human health.

Food Safety
Quality Control and Management

Edited by
Mohammed Kuddus
Syed Amir Ashraf
Pattanathu Rahman

CRC Press
Taylor & Francis Group
Boca Raton London New York

CRC Press is an imprint of the
Taylor & Francis Group, an **informa** business

Designed cover image: Shutterstock

First edition published 2024
by CRC Press
2385 NW Executive Center Drive, Suite 320, Boca Raton FL 33431

and by CRC Press
4 Park Square, Milton Park, Abingdon, Oxon, OX14 4RN

CRC Press is an imprint of Taylor & Francis Group, LLC

© 2024 selection and editorial matter, Mohammed Kuddus, Syed Amir Ashraf and Pattanathu Rahman; individual chapters, the contributors

ISBN: 978-1-032-36999-0 (hbk)
ISBN: 978-1-032-37000-2 (pbk)
ISBN: 978-1-003-33485-9 (ebk)

DOI: 10.1201/9781003334859

Typeset in Palatino
by MPS Limited, Dehradun

Contents

Preface

Welcome to this comprehensive compilation exploring various aspects of food safety, quality control, and the analysis of contaminants in food products. In today's globalized world, ensuring the safety and quality of the food we consume is of paramount importance. This collection of topics aims to shed light on the analytical approaches, techniques, and advancements in the field of food safety and quality assurance. There are 16 manuscripts accepted for publication in this book contributed by 56 authors from India, China, Italy, Brazil, West Indies, Mexico and Bangladesh. The contributors' co-operation and timely responses to complete this book is highly commendable.

Chapter 1: In this chapter, authors delve into the various analytical approaches employed for the measurement and identification of food contaminants. They explore the latest advancements in instrumental techniques, sample preparation methodologies and data analysis methods that aid in accurate and reliable detection of contaminants in food.

Chapter 2: This chapter focuses on quality control measures and risk assessment strategies implemented in the food industry to mitigate potential hazards associated with storage and packaging processes.

Chapter 3: This chapter explores the applications of this powerful analytical tool in identifying and quantifying contaminants in various food matrices, enhancing food safety standards.

Chapter 4: In this chapter, the authors examine the role of FTIR analysis in verifying the authenticity of food products, detecting adulteration and assessing their overall quality.

Chapter 5: This chapter focuses on the analysis of food preservatives using chromatographic techniques, highlighting the importance of accurate quantification and monitoring of these additives for consumer safety.

Chapter 6: This chapter explores the methods and techniques employed for the detection and quantification of aflatoxins in dairy products, aiming to minimize their presence and protect public health.

Chapter 7: This chapter investigates the analytical methods used to analyze heavy metal contamination in seafood, enabling effective monitoring and control measures to ensure the safety of these highly consumed food items.

Chapter 8: In this chapter, authors explore the application of ELISA and PCR techniques for the rapid and accurate assessment of biological contaminants in food, facilitating timely intervention and prevention.

Chapter 9: This chapter discusses intelligent point-of-care testing methods specifically designed for the detection of mycotoxins, ensuring prompt identification and prevention of these harmful toxins.

Chapter 10: This chapter examines the various methods employed for mycotoxin degradation in food, including physical, chemical and biological approaches, aiming to reduce the health risks associated with mycotoxin contamination.

Chapter 11: This chapter provides an in-depth exploration of food irradiation technology, its applications, benefits and regulatory considerations in ensuring food safety.

Chapter 12: This chapter investigates the utilization of inorganic nanoparticles for the detection, elimination and monitoring of contaminants in food, highlighting their potential for sustainable food safety practices.

Chapter 13: This chapter explores the current applications of nanozymes in food safety, addressing their challenges and potential future advancements in ensuring the quality and integrity of food products.

Chapter 14: This chapter showcases recent innovations, including biosensors, smart packaging, and blockchain, and their significant contributions to improving food safety systems, supply chain transparency and consumer trust.

Chapter 15: This chapter focuses on the management of food allergens, including allergen detection techniques, labeling regulations and best practices in food production and handling, aiming to mitigate the potential health hazards associated with allergenic foods.

Chapter 16: This chapter explores the assessment of food contaminants specifically in meat products, emphasizing the importance of reliable analysis and quality control measures to safeguard consumers' health.

We hope this compilation serves as a valuable resource for students, researchers, food industry professionals, and policymakers, providing insights into the analytical approaches, techniques, and technologies employed in the assessment and management of food safety and quality. The contributors' expertise and the breadth of topics covered ensure a comprehensive understanding of the challenges and advancements in this critical field. We are grateful to the CRC Press for their determined effort to publish the book on schedule.

Mohammed Kuddus
Syed Amir Ashraf
Pattanathu Rahman
August 2023

Editors

Mohammed Kuddus, PhD, is the head of the Department of Biochemistry, College of Medicine, University of Hail, Hail, Kingdom of Saudi Arabia. He earned a PhD in biochemistry and enzyme biotechnology at Sam Higginbottom University of Agriculture, Technology and Sciences (SHUATS), Allahabad, India. Prof. Kuddus has more than 95 publications, 10 books, 22 book chapters, 10 research grants and a patent to his credit. He is included in the Stanford University List of World's Top 2% Scientists (2023). He is a recipient of Young Scientist Project from DST, India and IFS, Sweden.

Syed Amir Ashraf, PhD, is a faculty member at the Department of Clinical Nutrition, College of Applied Medical Sciences, University of Hail, Kingdom of Saudi Arabia. His primary research focuses on food bioactive characterization, nutraceutical and functional foods, food safety and management, metabolic disorders, cancer and bioinformatics. Dr. Ashraf has published more than 70 research articles in peer-reviewed journals along with five book chapters. He is an editorial board member and reviewer of various reputed international journals.

Pattanathu Rahman, PhD, is a senior academic member at Liverpool John Moores University, Liverpool, England. He has 25 years of academic experience in research, innovation, management and commercialization. He served as an associate professor at the University of Portsmouth and Teesside University and founded TeeGene Biotech and Tara Biologics. He has discovered novel biosurfactant-producing bacteria at Teesside, and Dr. Rahman is the author of 70+ peer-reviewed journal articles with 8,600 citations, an h-index of 36, and the editor of three books on biosurfactants. His vast technical and management experience in biotechnology has led to various cutting-edge technologies, product discovery, scale-up, tech transfer and commercialization. He is included in the Stanford University List of World's Top 2% Scientists (2021, 2022, 2023). He is a visiting professor at SOA University, Bhubaneswar, India.

Contributors

Shafi Ahmed
Department of Agro Product Processing
    Technology
Jashore University of Science and Technology
Jashore, Bangladesh

Md. Akhtaruzzaman
Department of Agro Product Processing
    Technology
Jashore University of Science and Technology
Jashore, Bangladesh

Md. Shofiul Azam
Laboratory of Food Microbiology and Safety
Department of Food Engineering
Dhaka University of Engineering and
    Technology, Gazipur
Dhaka, Bangladesh

Eliana Badiale-Furlong
Postgraduate Program in Engineering and
    Food Science
Chemistry and Food School
Federal University of Rio Grande
Porto Alegre, Brazil

Neela Badrie
Department of Food Production
Faculty of Food and Agriculture
The University of the West Indies
St. Augustine, Trinidad and Tobago

Pooja Nivrutti Bhagat
Dairy Engineering Division
Indian Council of Agricultural Research
    (ICAR)-National Dairy Research Institute
    (NDRI)
Haryana, India

Anikesh Bhardwaj
Department of Chemistry
Chandigarh University
Mohali, India

Deepshikha Buragohain
Department of Life Science and Bioinformatics
Assam University
Assam, India

A.E. Cedillo-Olivos
Departamento de Ingeniería Bioquímica
Escuela Nacional de Ciencias Biológicas
Instituto Politécnico Nacional
Mexico City, Mexico

R. B. Colorado
Departamento de Ingeniería Bioquímica
Escuela Nacional de Ciencias Biológicas
Instituto Politécnico Nacional
Mexico City, Mexico

Carlos Adam Conte Jr.
Center for Food Analysis (NAL)
Technological Development Support
    Laboratory (LADETEC)
Federal University of Rio de Janeiro
Rio de Janeiro, Brazil

Ayantika Das
Dairy Microbiology Division
Indian Council of Agricultural Research
    (ICAR)-National Dairy Research Institute
    (NDRI)
Haryana, India

Vaishali Lekchand Dasriya
Dairy Microbiology Division
Indian Council of Agricultural Research
    (ICAR)-National Dairy Research Institute
    (NDRI)
Haryana, India

Juliane Lima da Silva
Postgraduate Program in Technological and
    Environmental Chemistry
Chemistry and Food School
Federal University of Rio Grande
Porto Alegre, Brazil

Francine Kerstner de Oliveira
Postgraduate Program in Engineering and
    Food Science
Chemistry and Food School
Federal University of Rio Grande
Porto Alegre, Brazil

Tejpal Dhewa
Department of Nutrition Biology
Central University of Haryana
Haryana, India

Harmeet Singh Dhillon
Dairy Microbiology Division
Indian Council of Agricultural Research
    (ICAR)-National Dairy Research Institute
    (NDRI)
Haryana, India

Mariya Divanshi
Dairy Microbiology Division
Indian Council of Agricultural Research (ICAR) –
    National Dairy Research Institute (NDRI)
Haryana, India

Flavio Dias Ferreira
Academic Department of Food (DAALM)
Federal Technological University of Paraná
    (UTFPR)
Curitiba, Brazil

Yu Gao
College of Plant Protection
Jilin Agricultural University
Changchun, China

Jaqueline Garda-Buffon
Postgraduate Program in Engineering and
    Food Science
Chemistry and Food School
Federal University of Rio Grande
Porto Alegre, Brazil

S. A. González
Departamento de Ingeniería Bioquímica
Escuela Nacional de Ciencias Biológicas
Instituto Politécnico Nacional
Mexico City, Mexico

Md. Abir Hossain
Laboratory of Food Microbiology and Safety
Department of Food Engineering
Dhaka University of Engineering and
    Technology, Gazipur
Dhaka, Bangladesh

Xiaofeng Hu
Key Laboratory of Detection for Mycotoxins
Ministry of Agriculture and Rural Affairs
Oil Crops Research Institute
Chinese Academy of Agricultural Sciences
Beijing, China

Marco Iammarino
Chemistry Department
Istituto Zooprofilattico Sperimentale della
    Puglia e della Basilicata
Foggia, Italy

Rumana A. Jahan
Centre for Advanced Research in Sciences
University of Dhaka
Dhaka, Bangladesh

C. Jiménez-Martínez
Departamento de Ingeniería Bioquímica
Escuela Nacional de Ciencias Biológicas
Instituto Politécnico Nacional
Mexico City, Mexico

Wenjing Kang
School of Food Science and Engineering
Key Laboratory of Tropical and Vegetables
    Quality and Safety for State Market
    Regulation
Hainan University
Haikou, China

Sanjeev Kumar
Department of Life Sciences and Bioinformatics
Assam (Central) University
Assam, India

Ankita Kumari
Animal Biochemistry Division
National Dairy Research Institute
Haryana, India

Peiwu Li
Key Laboratory of Detection for Mycotoxins
Ministry of Agriculture and Rural Affairs
Oil Crops Research Institute
Chinese Academy of Agricultural Sciences
Beijing, China

Yin Liu
School of Food Science and Engineering
Key Laboratory of Tropical and Vegetables
    Quality and Safety for State Market
    Regulation
Hainan University
Haikou, China

Md. Hasan Tarek Mondal
Department of Food Engineering
Dhaka University of Engineering and
    Technology, Gazipur
Dhaka, Bangladesh

Maria Lúcia Guerra Monteiro
Center for Food Analysis (NAL)
Technological Development Support
    Laboratory (LADETEC)
Federal University of Rio de Janeiro
Rio de Janeiro, Brazil

Priyakshi Nath
Department of Life Sciences and Bioinformatics
Assam (Central) University
Assam, India

Luiz Torres Neto
Center for Food Analysis (NAL)
Technological Development Support
 Laboratory (LADETEC)
Federal University of Rio de Janeiro
Rio de Janeiro, Brazil

Wesclen Vilar Nogueira
Postgraduate Program in Engineering and Food
 Science
Chemistry and Food School
Federal University of Rio Grande
Porto Alegre, Brazil

Raghu H.V.
Dairy Microbiology Division
Indian Council of Agricultural Research (ICAR)-
 National Dairy Research Institute (NDRI)
Haryana, India

Soniya Ashok Ranveer
Dairy Microbiology Division
Indian Council of Agricultural Research (ICAR)-
 National Dairy Research Institute (NDRI)
Haryana, India

Mrinal Samtiya
Department of Nutrition Biology
Central University of Haryana
Haryana, India

Xuemei Tang
School of Food Science and Engineering
Key Laboratory of Tropical and Vegetables
 Quality and Safety for State Market Regulation
Hainan University
Haikou, China

Aurelia Di Taranto
Chemistry Department
Istituto Zooprofilattico Sperimentale della
 Puglia e della Basilicata
Foggia, Italy

Md. Nazim Uddin
Bangladesh Agricultural Research Institute
Gazipur City, Bangladesh

Md. Wahiduzzaman
Bio-Med Big Data Center
CAS Key Laboratory of Computational Biology
CAS-MPG Partner Institute for Computational
 Biology
Shanghai Institute of Nutrition and Health
University of Chinese Academy of Sciences
Chinese Academy of Sciences
Shanghai, China

Shenling Wang
Key Laboratory of Detection for Mycotoxins
Ministry of Agriculture and Rural Affairs
Oil Crops Research Institute
Chinese Academy of Agricultural Sciences
Beijing, China

Long Wu
School of Food Science and Engineering
Key Laboratory of Tropical and Vegetables
 Quality and Safety for State Market
 Regulation
Hainan University
Haikou, China

Ting Wu
School of Food Science and Engineering
Key Laboratory of Tropical and Vegetables
 Quality and Safety for State Market
 Regulation
Hainan University
Haikou, China

Meng-Lei Xu
Jilin Provincial Key Laboratory of Nutrition and
 Functional Food
College of Food Science and Engineering
Jilin University
Changchun, China

Miaomiao Yang
School of Food Science and Engineering
Key Laboratory of Tropical and Vegetables
 Quality and Safety for State Market
 Regulation
Hainan University
Haikou, China

Wei Zeng
School of Food Science and Engineering
Key Laboratory of Tropical and Vegetables
 Quality and Safety for State Market
 Regulation
Hainan University
Haikou, China

Zhaowei Zhang
Key Laboratory of Detection for Mycotoxins,
 Ministry of Agriculture and Rural Affairs,
 Oil Crops Research Institute
Chinese Academy of Agricultural Sciences
Beijing, China

1 Analytical Approaches for Measurement of Food Contaminants

Soniya Ashok Ranveer, Vaishali Lekchand Dasriya, Pooja Nivrutti Bhagat, Harmeet Singh Dhillon, and H.V. Raghu

1.1 INTRODUCTION

There are many connection between the production and delivery of food from farmers to consumers (planting, breeding, harvesting, packing, transferring to retail markets, storing, shipping, importing, processing, and shelf storage) (Keding *et al.*, 2013). There are many possibilities for involvement these steps in the food chain, including the use of "pesticides, agricultural bioengineering, veterinary drug administration, storage and handling conditions, processing applications, economic gain strategies, food additives, packaging material", etc. As a result of the potential for contamination or the introduction (both purposeful and unintentional) of hazardous substances or constituents, each of these procedures can have a significant impact on food quality and safety. Food safety and quality laws and regulations have been enacted and are being improved to safeguard farmers, consumers, and the food manufacturing sector (Pou *et al.*, 2022). When there is a risk of illness from eating something, the dosage usually determines how bad the sickness will be. There is often a tolerance limit below which no negative effects are seen. Tolerance levels are set by the EPA and enforced by the FDA and USDA in the United States (Trichilo and Schmitt, 1989). However, events related to the microbiological and chemical safety of food continue to happen. Pesticides, mycotoxins, veterinary medicine residues, some food additives, food adulterants, packaging hazardous compounds, and environmental pollutants all raise serious safety concerns (Sabui *et al.*, 2022). Allergens (e.g., sulfite and histamine), heavy metals (e.g., mercury, cadmium, and lead), pesticides (e.g., dimethoate, omethoate, and isophenfos-methyl), and veterinary drugs (e.g., lactam, sulfonamide, chloramphenicol, and nitrofurans) are among the most commonly reported hazards within the chemical category (SOEST, and Fritschi, 2004). Contamination from microorganisms, such as fungi, viruses and bacteria are associated with foodborne diseases in humans (Bari and Yeasmin, 2018). GMOs and the products they produce are other groups that can cause alarm. The introduction and use of GMOs in food items led to the creation of safety labeling standards mandated by law (Yang and Chen, 2016). To guarantee food quality, safety, and equitable commerce, there must be effective and trustworthy detection and analysis procedures.

1.2 ANALYTICAL APPROACH

Analyzing potential food hazards can be done using any number of techniques and procedures. The purposes, cost, energy, and time, dependability of the procedure, the sophistication of the food matrix, distrusted amount of contamination, and availability of analyzers are only a few of the aspects that play a role in deciding which approach to take (Stroka and Anklam, 2002). The goals may be as simple as identifying a family of pollutants as a whole, or as sophisticated as pinpointing the concentration of a single contaminant or even detecting previously unreported adulterants. Cost-effective and time-efficient screening procedures are increasingly being adopted by regulatory bodies and the business sector as a whole. However, quantitative procedures that call for specialized hardware may be required for the analysis to reach its conclusions (Higueras *et al.*, 2012). When this occurs, businesses may elect to have their samples analyzed by independent research facilities. After settling on an approach, it's time to think about how to collect and analyze samples.

1.2.1 Qualitative or Semiquantitative Methods

Semiquantitative techniques can estimate the concentration of a detected pollutant or residue, whereas qualitative techniques can only identify its presence (Van Aken *et al.*, 2006). The primary advantages of these approaches are their low price, quickness, and ease of use. Some examples of such methods include thin-layer chromatography (TLC), enzyme inhibition, and immunoassay (Table 1.1) (Chen and Schwack, 2014).

1.2.2 Quantitative Methods

Gas chromatography and high-performance liquid chromatography are the two most common analytical techniques used for the quantitative study of chemical pollutants and residues in food (Liu *et al.*, 2022). GC has historically been paired with even more accurate detectors for the study of

DOI: 10.1201/9781003334859-1

Table 1.1: Antibiotic, Mycotoxin, and Pesticide Analyses in Food: A Summary

Contaminant	Semiquantitative or Qualitative (Screening Methods)	Quantitative	
Pesticides	TLC, HPLC, HPLC Enzyme inhibition Immunoassay	Multiresidue (MRMs) Single-residue (SRMs) GC (mostly)	Single-residue (SRMs) GC (mostly) GC (mostly)
Mycotoxins	TLC, GC, Immunoassay Capillary electrophoresis Immunoassays	HPLC (mostly)	
Antibiotics	Enzyme substrate assays and Immunoassays	HPLC (mostly), Microbial growth inhibition GC Receptor assays Immunoassays	

food pollutants and residues due to its superior separation efficiency compared to HPLC (McCalley, 2002). Even though GC analysis requires derivatization of polar analytes, it is preferred for multicomponent contaminant and residue analysis due to the combination of GC with mass spectrometry (MS) and the availability of high-cost benchtop GCMS devices (Loos et al., 2016). Mycotoxins, polar insecticides, and the vast majority of veterinary medication residues are examples of thermally labile and/or massive analytes that currently require HPLC for analysis (Sorbo et al., 2022). HPLC-mass spectrometry (LC-MS) is a relatively new technique that allows for the direct, selective, and sensitive analysis of polar analytes (Kang, 2012). As we move away from pesticides that are persistent and less polar and toward those that are more rapidly degradable, more polar, and thermolabile, LC-MS is being used for multiclass, multi-residue analysis. Using GC-MS or LC-MS allows for the potential spectroscopic resolution of coeluting peaks, as well as the simultaneous quantification and structural identification of a broad range of chemicals (Table 1.1) (Kadokami et al., 2022). Additionally, immunoassay analytical methods are employed for the identification and quantification of single and multiple pollutants or residues. Immunoassays allow for the quick and accurate detection of a wide range of pollutants, including drugs, pesticides, and mycotoxins. Enzyme-linked immunosorbent assays (ELISAs) and immunosensor methods are two of the most sensitive immunoassay techniques for detecting harmful analytes, whereas immu-noaffinity chromatography is used to concentrate and purify the analyte of interest (Alnassrallah et al., 2022). Immunoassays are not only highly sensitive, but also easy, rapid, and cost-effective compared to other methods used for the quantification of pollutants and residues (Lei et al., 2022). It's possible, though, that the test antibodies will show cross-reactivity (affinity) with other substances that share a similar chemical structure.

1.3 PESTICIDE RESIDUE ANALYSIS

Acaricides, molluscicides, pheromones, repellents, nematicides, plant growth regulators, and rodenticides are all examples of pesticides (Yadav and Devi, 2017). The FDA has released a chemical nature of pesticides compound (including 14 terms). To this day, pesticides are still essential in preventing plant disease and ensuring the world's population has enough to eat (Tyczewska et al., 2018). It is projected that 30% of crop yield would be lost if pesticides were not applied. Acute neurologic toxicity, cancer, chronic neurodevelopment impairment, and immuno-logical, reproductive, endocrine system dysfunction are a few of the potential risks associated with pesticide exposure (Upadhayay et al., 2020). Due to the potential toxicity of pesticides to animal health and the environment, there are already stringent rules for the registration (including the establishment of tolerance thresholds) and usage of pesticides around the world. The Environmental Protection Agency (EPA) mandates that scientific studies be conducted to ascertain the character and magnitude of harmful effects, as well as to identify the threshold below which these effects are no longer observable (no observed adverse effect level, NOAEL). With the help of a safety factor, often set at 100 (i.e., ADI = NOAEL/100), the NOAEL can be converted into the appropriate daily intake (ADI) for humans (Shah et al., 2017). To properly conduct risk assessment studies, researchers need to take into account not only the dangers associated with exposure to pesticides but also the overall aggregate exposure of different subpopulations (Lentz et al., 2015).

Pesticide residue in or on foods and how often those foods are eaten both contribute to dietary exposure. Chronic exposure risks can be estimated by comparing the estimated number of pesticides consumed over a long period with the ADI (Alla *et al.*, 2015). As a result of decreasing tolerance levels, more precise and sensitive analytical procedures have been developed to ensure that food continues to suit consumers' needs.

1.3.1 Different Analytical Techniques

Analyzing pesticide traces in food is due to many different reasons. The food matrix is complicated; there are more matrix components than target pesticides, pesticide doses may be in the picogram or femtogram range, and there are major discrepancies between the physical and chemical properties of pesticides (Sannino, 2008). Pesticide analysis techniques can be broken down into two broad groups: single residue methods (SRM) and multiple residue methods (MRM) (Salvador *et al.*, 2020). The goal of SRM development was to create a method for measuring a single analyte and, in many cases, its potentially harmful metabolites and transformation products. Pesticides can have a wide range of properties, including pH, polarity, non-polarity, volatility, and acidity (Prieto *et al.*, 2010). Inside this light, a MRM that can detect many pesticides in a single analysis is the most time- and cost-effective method. Additionally, AOAC International has created an MRM for pesticide residues called the "AOAC Pesticide Screen (970.52)" (Rejczak and Tuzimski, 2015). Current MRMs utilized by the FDA and USDA rely on GC and HPLC analysis for identification and quantification (Botitsi *et al.*, 2011). Before chromatographic analysis, MRMs perform optimal sampling, extraction, and fractionation/cleanup procedures to ensure that the vast majority of the pesticide residues present in the sample matrix are successfully transferred to the organic phase (Ismail et al., 2010). With the use of a water-miscible solvent, pesticide residues can be efficiently partitioned from the sample matrix into the organic phase. After that, the mixture is separated using a nonpolar solvent that is miscible with the polar solvent but not water. Cleaning up the extracted material to reduce matrix effects and improve sensitivity and selectivity follows the extraction process (Dasriya *et al.*, 2021). Low-polarity solvent mixes are eluted from adsorption columns filled with Florisil, alumina, or silica gels during the cleanup process (Tekel and Hatrík, 1996).

1.3.2 Analytical Techniques

There is a large variety of analytical methods that can be employed in pesticide analysis for purposes of detection, identification, and/or quantification. This discussion includes both established and emerging analytical methods for tracing the origins of pesticide residues. When something is detected, identified, and/or quantified, the term "detection" is used. Pesticide analysis can make use of several different types of analytical methods. Some of the most cutting-edge breakthroughs and current methods for analysing samples for pesticide residues are discussed here.

1.3.3 Biochemical Techniques

Several different biochemical methods can be used to detect pesticides, including enzyme inhibition tests and immunoassays. Enzyme inhibition tests are a part of many of today's commercially accessible test kits (Dasriya *et al.*, 2021). These assays work on the concept that pesticides present in the sample will inhibit an enzyme necessary for key processes in insects. If there are no pesticides around, the enzyme will react with the substrate and induce a color change. If there is no visible color shift, the test is positive; more advanced analysis, such as HPLC and GC (Dasriya *et al.*, 2021), will be required to confirm the presence and concentration of the target pesticides. Immunoassays can be modified to serve a variety of applications, from rapid screening [field-portable] to more in-depth quantitative analysis in a lab setting, whereas enzyme inhibition assays are mostly utilized as screening procedures despite their low sensitivity and selectivity (Ju *et al.*, 2016). When compared to more traditional approaches, immunoassays are advantageous because of their low complexity, high sensitivity, and high throughput. Furthermore, unless cross-reactivity exists, there is no need for substantial cleansing of extracts. As a result, with a large enough sample size, this method can be quite beneficial for program monitoring. There are a wide variety of uses for class- and compound-specific immunoassays. ELISAs make up the vast majority of the immunoassays utilized in the analysis of pesticide residue. Antibodies used in cyclodiene insecticide and triazine herbicide ELISA testing are both specific to their target classes and highly reactive, allowing for the detection of structurally similar chemicals (Ismail *et al.*, 2010).

1.3.4 Chromatographic Techniques

1.3.4.1 Thin Layer Chromatography

To speed up the examination process, pesticides can be screened using thin-layer chromatography (TLC). It is not employed as a quantitative approach due to its low-resolution capacity, low accuracy, and inadequate recognition in comparison to HPLC and GC (Nestola and Schmidt, 2016). TLC, on the other hand, can be employed as a semiquantitative approach before additional precise detection and quantification. Insect enzyme inhibitors such as cholinesterases can be detected and estimated as one possible use case. Insecticides belonging to the OP and carbamate classes are among those that can block the activity of these enzymes (Songa and Okonkwo 2016). After a crude extract has been separated using TLC, the enzyme (or enzymes) and a substrate (that will be hydrolyzed into a colored product) is then sprayed onto the plate. Pesticide residues prevent enzymes from producing a color change, and the size of the resulting "zone of inhibition" depends on how much pesticide is present, specifically (Gavahian *et al.*, 2021).

1.3.4.2 Gas Chromatography

In recent years, the invention of fused silica capillary columns has made it possible to separate and detect a wide variety of pesticides that share similar physical and chemical properties. The OC and OP families of pesticides are both volatile and thermally stable, making GC the method of choice for their determination. The composition of the pesticides informs the selection of columns and detectors (Sannino, 2008). Columns made of diphenyl and dimethylpolysiloxane (5%, 95%) are frequently used in MRMs. Many heteroatoms, including O, S, N, Cl, Br, and F, can be found in a single molecule of a pesticide. So, element-selective detectors like a flame photometric detector (FPD) are commonly utilized for the detection of P-containing substances (Seiber *et al.*, 2021). The FPD is commonly used to identify OP pesticides in a wide variety of crops without resorting to laborious cleanup procedures. The great sensitivity of the electron capture detector to organic halogen compounds makes it ideal for the measurement of OC. Conventional GC analysis is constrained by the need for multiple injections due to the MRM strategy for multiclass detection with these selective detectors (Botitsi *et al.*, 2011).

1.3.4.3 High-Performance Liquid Chromatography

The proliferation of low-volatility, high-polarity, and heat-labile pesticides necessitated the improvement of high-performance liquid chromatography (HPLC) study for their separation and detection. Many classes of pesticides, including "N-methyl carbamate (NMC), urea herbicides, benzoylurea insecticides, and benzimidazole fungicides", are studied using high-performance liquid chromatography (HPLC). Reversed-phase chromatography using C18 or C8 columns and an aqueous mobile phase is commonly used to evaluate these substances, and the results are then detected using either UV absorption, a UV diode array, fluorescence, or mass spectrometry (McCalley, 2002). Purified HPLC separation and UV detection at 254 nm allow for quantitative evaluation of phenylurea herbicide concentrations with high selectivity and sensitivity. Compared to ultraviolet light, fluorescence is a more sensitive method for detecting benzimidazole fungicides. Post-column derivatization is used in situations where the sensitivity of UV and fluorescence detection is low. The equipment needed for post-column derivatization, such as a mixing chamber and a reactor, is not always readily available. Another important drawback is competition with other chemicals that exhibit fluorescence (Baeyens, *et al.*, 1998). Using fluorescence or UV after standard HPLC analysis of pesticides in complicated systems is often insufficient. Since spectrum variations are sometimes too weak to be resolved, even diode array detection may not be an option (Wilson and Brinkman 2003). The use of MS detection has broadened the applications for HPLC analysis of pesticides. One of the most effective methods for analyzing polar, ionic, and thermally labile pesticides is liquid chromatography-mass spectrometry (LC-MS) (Ingelse *et al.*, 2001).

1.4 MYCOTOXIN ANALYSIS

Filamentous fungi, commonly referred to as molds, can grow on food products and develop mycotoxins, a class of chemical poisons. Aspergillus, Fusarium, and Penicillin are three of the most common fungus genera that yield mycotoxins (Bhatnagar *et al.*, 2002). Environmental factors (e.g., temperature, humidity, weather variations, damage to kernels, and a pest infestation) can all cause crops to get contaminated with fungal growth, which can then lead to mycotoxin contamination (Bruns, 2003). Fungal growth can also be triggered by environmental factors such as excessive soil dryness or an imbalance in the plant's ability to absorb nutrients. Toxins produced by molds, known as mycotoxins, can contaminate foods at any point in the chain.

1.4.1 Rapid Methods of Detection

1.4.1.1 Thin Layer Chromatography

The Association for the Analysis of Agricultural Chemicals (AOAC) International has approved a plethora of thin layer chromatography (TLC) methods for the analysis of mycotoxins. Toxins like "dioxins and nitrates (DON)" in grains like wheat and barley, "aflatoxin" in peanuts and corn, "aflatoxin M1" in dairy products like milk and cheese, "OTA" in grains like barley and green coffee, and "zea" in corn are all analyzed (Alshannaq and Yu, 2017). Screening typically makes use of traditional TLC methods, with detection limits as low as 2 ng/g. In the event of a positive finding, additional, more precise quantitative testing will be performed (Singh and Mehta, 2020). When TLC is utilized in conjunction with IAC, the results from mycotoxin analysis are more accurate (Turner *et al.*, 2009).

1.4.1.2 Immunoassays

Radioimmunoassay (RIA), enzyme-linked immunosorbent assay (ELISA), and fluorescence polarization immunoassay are the three main immunoassays used for the measurement of mycotoxins (FPIA) (Tian *et al.*, 2018). Mycotoxin (such as aflatoxin) radiolabeling in RIA has been mostly displaced by ELISA (Candlish, 1991). Due to the low molecular weight (MW) of mycotoxins, it is possible to make a qualitative assessment of the presence of "aflatoxins B1, B2, G1, and G2" possible in as short as 5 minutes (de Oliveira et al., 2014). On the other hand, a reference procedure like HPLC must confirm any positive results. On the other hand, biosensors are small analytical devices that combine a transduction system and biological components (e.g., nucleic acids, antibodies, cells, or enzymes). The signal generated by the target molecule's interaction with the biological component is processed by the transduction system. Biosensors are becoming increasingly popular for the detection of mycotoxins, which are produced by the fungus (Table 1.2) (Dey *et al.*, 2022).

Table 1.2: Examples of Commercial Mycotoxin Residue Analysis Test Kits

Type of Test	Commercial Supplier	Type of Matrices	Example of Mycotoxin
Lateral flow receptor	Charm ROSAR series (Charm Sciences, Inc., 659 Andover Street, Lawrence, MA)	Milk, grain, feedstuffs, wine, and grape juice	Aflatoxin M1 and M2, DON, fumonisin, ochratoxin, T-2 toxin, and zearalenone
Immunoaffinity columns with HPLC or fluorometry	AflaTestR WB and sister products (VICAM, 34 Maple St. Milford, MA)	Coffee, milk, milk products, nuts, wheat and grains, and other foods	Aflatoxins M1, B1, B2, G1, and G2, citrinin, DON, fumonisin, ochratoxin, T-2 and HT-2 toxins, and zearalenone
Cleanup columns for HPLC, GC, or TLC analysis	MycosepR and sister products (Romer Labs, Inc., 1301 Stylemaster Drive, Union, MO)	Wheat, nuts, corn, fruits and fruit juices, wine, oats, and coffee	DON, fumonisin, moniliformin, patulin, aflatoxins, (ergometrine, ergotamine, ergosine, etc.), zearalenone, citrinin, sterigmatocystin, ochratoxin, T2, and ergot alkaloids
Antibodies in a direct competition ELISA test	Agri-ScreenR series (Neogen Corporation, 620 Lesher Place, Lansing, MI)	Corn, whole cottonseed, oats, cornmeal, corn gluten meal, raw peanuts, corn/soy blend, wheat, rice, soy, barley, butter, cottonseed meal, peanut and mixed feeds	Deoxynivalenol (DON), T-2 toxin, ochratoxin, zearalenone, aflatoxin, and fumonisin

1.4.1.3 Gas Chromatography

Except for trichothecenes, gas chromatography (GC) is not commonly utilized for the identification of mycotoxins. Due to their lack of fluorescence and weak UV-Vis absorption, GC techniques were developed for the determination of trichothecenes (Krska *et al.*, 2007). It is common practice to use capillary column GC coupled with "trifluoroacetyl, heptafluorobutyry, or trimethylsilyl" derivatization for the simultaneous detection of multiple trichothecenes such as DON, T2, and HT-2. Most commonly, MS is used to verify peaks found by GC (Ismail *et al.*, 2010). To confirm patulin in apple juice, gas chromatography–mass spectrometry (GC-MS) can be utilized. The official methods published by the American Organization for Analysis of Compounds (AOAC) and the American Society of Brewing Chemists (ASBC) both report validated and recognized procedures for the assessment of trichothecenes using GC (Krska, 2016).

1.4.1.4 Capillary Electrophoresis

It is possible to use electrical potential also with the chromatographic technique of capillary electrophoresis to separate mycotoxins from matrix components (Turner *et al.*, 2009). There are cyclodextrin-based methods for detecting patulin, ochratoxins (A and B), aflatoxins, and zearalenone in apple juice (to boost the natural fluorescence) (Bueno *et al.*, 2015).

1.5 ANALYSIS OF ANTIBIOTIC RESIDUES

Therapeutic and subtherapeutic doses of medications like antibiotics, antifungals, tranquillizers, and anti-inflammatory drugs are sometimes given to animals raised for human consumption (Botsoglou and Fletouris, 2000). These low-level subtherapeutic pharmacological therapies can lower the prevalence of infectious illnesses brought on by bacteria and protozoa, accelerate weight gain, and reduce the amount of feed required to achieve weight gain (Kirtane *et al.*, 2021). The FDA's Center for Veterinary Medicine (CVM), a division, controls the production and distribution of medications used on animals, including those used to provide food for humans (Viola and DeVincent, 2006).

1.5.1 Identification and Evaluation

Sample preparation is often required for residue analysis in order to concentrate and purify the desired analytes. Antibiotic residues are typically extracted after a series of pre-treatment steps, including defatting, protein hydrolysis (meat or egg samples), protein precipitation (dairy samples), and hydrophilic washing (to remove excess sugar from honey) (Raza *et al.*, 2018). Liquid-liquid extraction and SPE are frequently employed for the extraction of many antibiotics. Next, the antibiotics undergo a preliminary purification process, during which ion-exchange cleanup systems, capitalizing on their acid/base makeup, are commonly used (Sanyal and Mathur, 2022). Tolerance thresholds have been established for some antibiotics, while others (like nitrofurans and chloramphenicol) have a strict no-tolerance policy (Hanekamp and Bast, 2015). For instance, the antibiotic chloramphenicol is currently the subject of significant concern in the United States, the European Union, and other nations. It has been discovered in imported seafood products and is utilized in various regions of the world to produce shrimp (e.g., shrimp, crayfish, and crab) (Chammem *et al.*, 2018). Because of its harmful effects on human health, the FDA has established a zero-tolerance policy for chloramphenicol in human food and banned its use in animals bred for food production (Baynes *et al.*, 2016). As a result, the analytical techniques must be as sensitive and focused as possible. For the examination of antibiotics, numerous analytical techniques in the categories of screening, determinative, and confirmatory have been created and improved.

1.5.2 Screening Procedures

The most common types of quantitative fast screening tests include microbial growth inhibition assays, receptor assays, enzyme-substrate assays, and immunoassays (Ahmed *et al.*, 2017). A class of antibiotics, a single individual antibiotic, or no specificity are all possible with some screening techniques. Screening tests for antibiotic residues in test materials initially relied heavily on microbial growth inhibition, but many now use alternative detection methods (Pikkemaat, 2009). Examples of screening assays are included in Table 1.3, along with the specific antibiotics found. Turbidity, the zone of inhibition, or acid production are frequently measured in microbial growth inhibition experiments (Shawkey *et al.*, 2003). When an indicator organism grows in a liquid culture, the turbidity increases, the presence of antibiotics inhibits growth, reducing turbidity. In a zone of inhibition assay, the substance is allowed to diffuse across a nutritional medium made of

Table 1.3: Commercial Antibiotic Residue Analysis Test Kits

Type of Test	Commercial Source	Matrix Examples	Antibiotics Tested
Competitive binding assay with enzyme-based color development	SNAP® test kit (IDEXX Laboratories, Inc., One IDEXX Drive, Westbrook, ME)	Raw cow milk from the individual cow and bulk tank testing.	chlortetracycline, tetracycline, β-lactams, oxytetracycline
Bacterial inhibition	Not applicable	Raw sheep, goat, and cow milk and other dairy products	Penicillin
Competitive radio-receptor binding assay	Charm II® (Charm Sciences, Inc., 659 Andover Street, Lawrence, MA)	Raw cow milk and liquid milk. Cream, condensed milk, and dairy powders. Meat and honey.	β-lactams (penicillin G, ampicillin, cephapirin, amoxicillin, ceftiofur, etc.), tetracycline, macrolides, cephalosporins, sulfa drugs, aminoglycosides
Competitive ELISA	Veratox® (Neogen Corporation, 620 Lesher Place, Lansing, MI)	Shrimp	Chloramphenicol

agar that has been uniformly inoculated with spores of a susceptible organism (Abedon, 2021). Any antibiotics found in the test material will prevent the organism from growing and germinating, leaving behind clear zones (Timmerer et al., 2020). The acid that bacteria make as they expand changes the medium's color in the assays for measuring acid production. If there was no color change, an inhibitory chemical was present in the test sample (Chaiharn et al., 2009).

Microbial growth inhibition assays take longer than many more recent screening procedures, but they are less expensive, can test a lot of samples, and have some sensitivity to various antibiotic classes. There are various microbiological methods for specific antibiotics as well as a non-specific microbiological method for antibiotics in the AOAC Official Methods. The Charm IIR test, which has various versions created to detect various classes of antibiotics, is an example of a receptor assay (Gustavsson et al., 2004). A limited number of specific binding sites on the surface of bacteria added to the test sample compete with labeled antibiotics and antibiotic residues in the milk sample, as determined by the Charm IIR test system (Conzuelo et al., 2013). Higher levels of antibiotic residue in the milk sample means less radiolabeled tracers will bind to the microbe. The binding receptor from a susceptible bacteria and the radiolabeled tracer antibiotics are added to the milk sample to begin the experiment (Wiesner and Vilcinskas, 2010). The microbial mass is subculture in scintillation liquid, the sample is incubated, and centrifuged; the fat is removed; and the radiolabeled tracer is quantified (Bulthaus, 2004). The technique works with meat in addition to milk and some other dairy products. Enzyme-substrate assays are designed to quantify the degree to which an antibiotic inhibits an enzyme's activity on a substrate. The PenzymeR III commercial kit is a good illustration of a test for raw milk because it is specific for lactam antibiotics, which equimolarly hinder D, D-carboxypeptidase. D-alanine is released when this enzyme reacts with a certain substrate, and this release can be detected in subsequent steps of the experiment by a change in color (Ollivaux et al., 2014). Some ELISA-type immunoassays and some lateral flow strips are used to detect antibiotic residues (Table 1.3). A lateral flow strip is used in the milk and cream testing assay Charm OSA (Rapid One Step Assay) MRL (Chiesa et al., 2020). Specially designed receptor-gold is utilized on the test strip to decrease false positives. The SNAPR milk testing kit is another illustration of an immunoassay that is sold for milk testing (IDEXX Laboratories, Inc., Westbrook, ME). The test kit's enzyme-labeled antibiotics and any remaining antibiotics in a milk sample are put in competition by the assay (Wang et al., 2021). Any antibiotics present in the milk will prevent color development by inhibiting the enzyme's ability to act on a substrate and change its color (Asif et al., 2020). An example of a competitive ELISA used for the detection of a particular antibody is the VeratoxR assay for the antibiotic chloramphenicol.

1.5.2.1 Methods for Determination and Confirmation

Antibiotic residues in food products can be quantitatively determined using the same general methods as other trace analytes (Wang and Leung, 2007). After sample preparation procedures, chromatographic separation, detection, and quantification are applied to the partly purified extract (including pre-treatment, extraction, and purification) (Wollgast and Anklam 2000). The most used chromatography technology is HPLC coupled with UV detection using variable wavelength or diode array detection (Karongo et al., 2020). HPLC has been used in conjunction with fluorescence, chemiluminescence, or post-column reaction detectors for antibiotic residue analysis. In the analysis of antibiotics, LC-MS and LC-MS/MS are increasingly employed for confirmatory and identification reasons at trace levels (Peres et al., 2010). For regulation, the FDA offers LC-MS/MS methods to identify fluoroquinolones in honey (FDA, 2006; FDA, 2009) as well as LC-MS techniques to identify chloramphenicol and related chemicals. The FDA has regulatory LC-MS/MS methods (Anonymous, 2009) for analyzing chloramphenicol and related chemicals in shrimp, crab, and crawfish, as well as LC-MS methods that aren't currently designed for regulatory purposes (FDA, 2009). LC-MS/MS has also been used to validate the presence of β-lactam residues in milk. Using LC-MS/MS, it was also possible to detect 14 different types of sulfonamides at concentrations below 10 ng/mL in milk as well as condensed milk and soft cheeses (Holstege et al., 2002; Cavaliere et al., 2003). For the measurement of macrolide antibiotic residues in a range of foods, "Ultra-high-performance liquid chromatography/quadrupole time-of-flight mass spectrometry (UHPLC/Q-Tof MS) has been compared to LC-MS/MS" (Clark et al., 2005). While LC-MS/MS had a lower limit of detection and improved precision, UPLC/Q-Tof MS offered superior confirmation of positive results (Wang and Leung, 2007).

1.6 ASSESSMENT OF GENETICALLY MODIFIED ORGANISM

Novel characteristics, such as herbicide tolerance or insect resistance, can be conferred on agriculturally significant plants by introducing DNA from another creature (a "transgene") into their genome (Wolt et al., 2016). A "genetically modified organism, or GMO", is the name given to the changed plant (EFSA, 2011). Four crops dominate current GMO production: "soybeans, corn, cotton, and canola" (a cultivar of rapeseed). The most typical GMO features are pest resistance and herbicide tolerance, but additional GMO crops include those with plants that have been altered to boost the nutritional value of the food or to improve postharvest quality (Bouis et al., 2003). PCR kits that target DNA sequences shared by numerous GMOs, such as the often-utilized promoter sequence, are readily available (Raitskin et al., 2019). References provide additional reading on the GMO and GMO detection issues (Heller, 2003; Ahmed, 2004; Jackson and Linskens, 2009).

1.6.1 Protein Methods

Different types of immunoassays are used in the protein techniques. Because they rely on the usage of antibodies for detection, immunoassays are very precise. Although conventional ELISA is utilized for GMO detection, lateral flow strips are a simpler and faster type of assay. Although unlike ELISA analyses, these single-step lateral flow immunochromatographic techniques are quick to run, they are not quantitative (Jawaid et al., 2015). Although immunoassays like Western blots were used for GMO protein detection, they have since been rendered obsolete by the widespread availability of more sensitive methods like enzyme-linked immunosorbent assays (ELISAs) and lateral flow assay (Bishop et al., 2019). Below, we'll only cover the ELISA and lateral flow assays. To screen for GMOs, the sandwich ELISA method is frequently employed.

In addition to the 96-well microwell plates that have been coated with antibodies, the ELISA plate kits also comprise the enzyme conjugate (antibodies that have an enzyme linked to them), the reagents (including enzyme substrate), and the standards (Hornbeck, 2015). These components are necessary for accurately quantifying the GMO proteins in a sample (including positive control). The kit's positive controls are utilized for both quantitative comparison with the samples and for determining when the studies should be stopped (Prange and Profrock, 2008). Various conjugates (in the same solution) and substrates (in different solutions) are included in kits designed for the detection of multiple proteins so that each protein can be identified independently in distinct wells within a single experiment. The plates are coated with antibodies to both proteins in order to make the first step universal (Doering et al., 2007). These assay kits have detection limits as low as 1–10 parts per billion (ppb); however, with the kit manufacturer's help, they can be diluted to reach higher detection thresholds. Numerous plates can be used to analyze hundreds of samples in less than two hours. Each plate is typically used to analyze 44 pairs of samples and four pairs of standards.

The self-contained immunochromatographic assay known as the lateral flow strip is simple to employ on the spot. Lateral flow strip testing can identify the GMO protein at or above a specific concentration level. The GMO lateral flow test kit assays are the assay of preference for qualitative field tests since they can be carried out without specialized tools or a high level of experience (Schüling *et al.*, 2018). Pipettes, a sample extraction jar, and the test kit are the only pieces of necessary equipment. Therefore, GMO material can be tested using lateral flow strips at every stage of processing or transportation, including the growing field, storage facilities, transit locations, and processing facilities. It only takes five minutes to read the findings of a positive test. (1) DNA extraction from the sample, (2) PCR amplification of the DNA, and (3) identification and quantification of the amplified DNA are the three independent procedures that must be completed in order to conduct the analysis (Brooks *et al.*, 2015). Positive (a) and negative (b) results from a lateral flow strip test for GMOs in Roundup Ready foods (b). All three processes—analysis, amplification, and detection/quantification—occur simultaneously in real-time quantitative polymerase chain reaction. (Reprinted with permission of Neogen Corporation, Lansing, MI.) The PCR amplification is crucial to the success and precision of the study, albeit the other two steps are also important. Table 1.4 displays a handful of the many PCR kits sold by various companies. Additionally, to the aforementioned gene-specific PCR kits, screening kits that amplify the widely used promoters and transcription terminators for the overall assay for GMO material are also readily accessible.

Table 1.4: Example Commercial Test Kits for GMO Analysis

Assay Type	Example	GMO, (Protein), and Trait	Commercial Source
Single-step lateral flow immunochromato-graphic assay	Soybeans and corn products	Roundup Ready® (CP4 EPSPS protein). Confers tolerance to RoundupR herbicides	Reveal® (Neogen Corporation, 620 Lesher Place, Lansing, MI)
Single-step lateral flow immunochromato-graphic assay	Cotton and seeds	BollGard® II (Bt-Cry 2 A and Bt-Cry 1Ab/1Ac proteins) Most leaf- and boll-feeding worm species are under control.	FlashKits® Cotton (Agdia Biofords, 5 Rue Henri Desbrueres, Evry, France)
Single-step lateral flow immunochromato-graphic assay	Corn and other matrices	StarLink® (Bt-Cry9C and PAT proteins). European corn borer resistance and phosphinothricin (PPT) herbicide tolerance	AgraStrip® GMO ST (Romer Labs, Inc., 1301 Stylemaster Drive, Union, MO)
PCR kit	Most of the raw materials and food samples that contain soybean	Roundup Ready® Soya. (cp4 epsps gene). Tolerance to Roundup herbicides	LightCycler® GMO Soya Quantification (Roche Diagnostics, 9115 Hague Road, Indianapolis, IN)
PCR kit 3	The majority of corn-containing food samples and raw materials	NaturGard® KnockOut® (cry 1ab and bar genes). Resistance to European corn borer and PPT pesticide tolerance	LightCycler® GMO Maize Quantification (Roche Diagnostics)
ELISA	Single seed and leaf tissue	YieldGard Plus, YieldGard Plus/RR2, YieldGard VT Triple, YieldGard Rootworm, and Yieldgard VT Rootworm/RR2 (Cry1Ab and Cry3Bb1 proteins). Corn borer and corn rootworm larval control	QualiPlate® Kit for Cry1Ab and Cry3Bb1 (Envirologix, Inc., 500 Riverside Industrial Parkway, Portland, ME)

1.6.2 PCR

1.6.2.1 DNA Extraction

Before the food matrix is subjected to any significant processing, the DNA should be extracted to speed up the analysis (Miraglia *et al.*, 2004). It is therefore suggested to test raw materials more frequently than tested highly processed goods. It is imperative to keep in mind that using extremely high heat and pressure can damage DNA, rendering the subsequent PCR and identification ineffective (Butler, 2011). The particular food matrix to be examined is a crucial factor as well. To liberate the DNA from the matrix, all extraction techniques disturb the matrix in some way. In a food matrix, this is typically done by grinding the sample into a fine powder (Capriotti *et al.*, 2012). The pulverized material is then spread out into an extraction solution and any undesirable components are eliminated. For instance, a detergent can remove lipids, and a protease can remove protein (Gibbs *et al.*, 1999). The DNA may then be precipitated using cold alcohol, such as ethanol or isopropanol, as a final step. The specific matrix will determine the exact methodology to be employed.

1.6.2.2 PCR Amplification

The PCR method is a way to make more copies of a certain DNA sequence. Through enzymatic replication, the PCR process cycles back and forth, gradually increasing the copy number. Millions of copies of the target sequence can be created without end (Santhanam *et al.*, 2020). Thermal cycling is used to alternately copy the target sequence, and the process is then repeated by melting the DNA into single strands. The technique is based on the application of two synthetic DNA fragments that are complementary to the target sequence's opposing ends. These are known as primers, and they can only be formed if the target sequence is known (Miura *et al.*, 2015). Primers are typically 18 to 35 bases in length. The promoter sequence, which is shared by all transgenic agricultural species frequently employed for commercial food production, would be complementary to the primers if a generic identification of any GMO material were required. The primers would have a sequence containing transgene DNA and plant DNA if a specific GMO product were to be identified. In order to prevent the discovery of bacterial DNA generated from bacteria that could be on or in the plants (both the Bt toxin and the Roundup® tolerance transgenes are originated from soil bacteria), it is imperative that this is done (Orford *et al.*, 2007). The PCR mixture also includes deoxynucleoside triphosphates (dNTPs), which are the nucleotide bases that make up DNA, a heat-stable DNA polymerase like Taq polymerase (from Thermus aquaticus), and a buffer solution to keep the reactions running smoothly (Ong *et al.*, 2006). Each of these components is included in the commercial kits. The reaction vials are then put into a thermal cycler after that. The mixed solution is normally stored in these vials (20–200 l). The mixture is first heated to melt the DNA and separate it into single strands. Next, it is cooled to allow the primers to anneal to the single-stranded target DNA. Finally, it is cooled to permit the DNA polymerase to replicate a new DNA strand complementary to the target strand by adding dNTPs beginning at the primers (Bruijns *et al.*, 2020). The process is then routinely repeated for 30 to 50 cycles, which is enough to create millions of copies of the DNA.

1.6.2.3 DNA Analysis

Once sufficient DNA has been generated by PCR, the sample can be evaluated using agar gel electrophoresis. The gel is stained after the run, and the presence and amount of DNA can be assessed by contrasting it with the location and intensity of staining of the standards (Duineveld *et al.*, 2001). The standards and sample pass through the gel. Some of the more modern kits contain chemicals that can be used to fluorescein or other tags to specifically tag double-stranded DNA. These kits are made to be used in conjunction with specialized PCR equipment that deposits the finished, labeled mixture together into capillary at which high-sensitivity fluorescence spectroscopy defines the intensity of the fluorescence, that is proportional to the abundance of DNA, eliminate the necessity of electrophoresis (Wilson, 2008). The specificity is determined by the primer sequence, which in this case is directly related to the type of "double-stranded DNA" present. These kits also include standards and any additional materials needed to accurately determine the amount of the target DNA in the sample (Czechowski *et al.*, 2004). After each round of amplification in "real-time PCR, the fluorescence" in the reaction tube is read, resulting in a more complex curve with many data points than would be possible with traditional PCR (Ismail *et al.*, 2010).

1.6.3 Comparison of Methods

Table 1.4 contains a list of assay type for GMO analysis such as Multiplex polymerase chain reaction (PCR) kits, single-step lateral flow immunochromatographic assays and enzyme-linked immuno-sorbent assays. The PCR kits are example of a complete assay kit designed to be used with specialized equipment, in this case, the "Roche LightCycler System®", and is titled "LightCycler® GMO Soya quantification kit from Roche Diagnostics" (Ismail *et al.*, 2010). After extraction and system setup, numerous samples can be evaluated using this equipment and the associated software in less than one hour, with the creation of a concluding report. This system's obvious drawback is the requirement for a fully stocked laboratory, specialized equipment, and a high degree of skill. The lateral flow immunochromatographic strips, on the other hand, may be utilized on-site without any large or expensive equipment and can be used by anyone with little training (Wang *et al.*, 2016). They also provide a comprehensive analysis without additional handling. The fact that the lateral flow strips are not quantitative is their main drawback (Andryukov, 2020). Although the ELISA tests are quantifiable immunoassays, access to lab space and equipment is constrained and some training is required. Therefore, each analysis should be tailored to the information needed and the requirements of the technique.

1.7 ANALYSES OF ALLERGENS

Allergens are proteins in food that trigger an immune system reaction. "Hives", "swelling of the face and tongue", "difficulty breathing", and possibly "lethal anaphylactic shock" are all indications of an allergic reaction (McNeil *et al.*, 2016). Food intolerance (such as lactose intolerance), pharmacologic reactions (primarily brought on by food additives like sulfites and benzoate), and toxin-mediated reactions are other types of negative reactions to food that should be noted as distinct from a food allergy, which brings on an immune system reaction (due to residues such as pesticides and mycotoxins). Commercially available test kits for food allergy analysis (based on DNA or proteins, among other things). In this article, food allergen analytical techniques are reviewed.

1.7.1 Protein Methods

1.7.1.1 General Considerations

Like the study of many other potentially harmful food elements present in trace levels, problems with sampling adequacy and detection limits affect allergen analysis (Crevel *et al.*, 2008). Another problem is how to effectively remove the various allergies. Because proteins are the analytes of interest, the extraction solution is typically a buffer with a range of pHs and salt concentrations. Some solutions can extract the same allergens at a wide concentration range, while others are unable or and this is mostly due to the differences in the composition of the extraction buffers used (Westphal *et al.*, 2004). For instance, adding salt significantly improves the effectiveness of extracting the main peanut allergen, which is not extracted by phosphate buffer. Additionally, it is crucial that the extraction solution is appropriate for the test being used and does not alter the chemical structure of the analyte (such as an immunoassay) (Buick *et al.*, 1990). The conditions used for food processing should be taken into account while choosing the extraction method. Protein recovery may be lowered as a result of processing-induced protein denaturation and aggregation, which can affect the solubility of the proteins (Westphal, 2008). Therefore, it is essential to choose the right extraction process for the target analyte in order to acquire trustworthy and accurate findings.

1.7.1.2 Analytical Methods Based on Proteins

Immunoassays, which are antibody-based tests, are often used in protein-based methods for the detection of food allergens due to their sensitivity and specificity. Immunoassays use monoclonal, polyclonal, or a combination of the two antibodies to specifically target the offending allergen (Hartmann, 2009). The bulk of commercially available test kits use polyclonal antibodies, which differ in their specificity and the quantity of proteins they target (Echan *et al.*, 2005). The presence of allergens in food can be detected using a number of immunoassay-based techniques. These include Western blots, dot immunoblotting, biosensor immunoassays (antibodies immobilised on a biosensor chip), and enzyme-linked immunosorbent tests (ELISAs) (similar to Western blotting); however, protein extract is spot effectively onto the nitrocellulose or PVDF membrane before enzyme-labeled protein specific antibody). Most screening and qualitative uses use Western blotting and dot immunoblotting. The most common ELISA sandwich- or competitive-based immunoassays for the statistical diagnosis of food allergies are available today. The competitive

ELISA is utilized for minuscule protein allergens with a molecular weight of less than 5 kDa. Numerous sandwiches and competitive ELISA approaches have been developed for a number of food allergies (Poms *et al.*, 2004). In order to identify food allergies, "lateral flow test strips (dipstick assays)" that are based on the "ELISA principle" are now commercially accessible. Because dipstick tests are quick, inexpensive, and instrument-free, they are used in screening processes (Chondrogiannis *et al.*, 2022).

1.7.2 DNA Techniques

There are advantages and disadvantages to using "DNA-based approaches" to discover food allergies. Since DNA-based methods do not specifically target the allergen in the sample, the recognition of the allergen-encoding DNA does not always correspond with the presence of the allergen, especially once the food has been supplemented with pure protein (Gabriel, 2016). During processing, such as when making soy protein isolate (purified proteins used as elements for fortification and functionality enhancement), protein and DNA may separate, which could result in inaccurate conclusions about the allergen's presence in the sample (Friedman and Brandon, 2001). However, the targeted DNA is less sensitive to different treatment and extraction parameters than proteins, making DNA-based approaches more sensitive and selective. For DNA-based techniques, the DNA must first be extracted before being amplified by PCR with a thermostable polymerase. Following agarose gel electrophoresis, the amplified material is subsequently viewed via fluorescence staining or Southern blotting (Notomi *et al.*, 2000). If internal standards were applied, this approach typically yields qualitative data or semiquantitative data. Real-time PCR or PCR-ELISA can be employed to quantify samples. The PCRELISA method links an amplified DNA fragment from an allergic meal to a specific protein-labeled DNA probe, which is being connected to a specific enzyme-labeled antibody (Thellin *et al.*, 2009). The reaction that produces color as an enzyme and substrate interact provides the basis for quantifying DNA.

1.8 PACKAGING MATERIAL RESIDUES

1.8.1 Bisphenol A

The FDA restricts the use of bisphenol A (BPA), an organic molecule containing two phenol groups, in food contact applications (Almeida *et al.*, 2018). Under certain conditions, bisphenol A (BPA), which is used extensively in the production of polycarbonate plastic, can be released. Since BPA can act similarly to the body's natural hormones, even at low dosages, there is concern about potential adverse health effects (Wang *et al.*, 2017). Researchers discovered a link between heightened urinary BPA levels and an increased risk of diabetes, cardiovascular disease, and extremely high levels of some liver enzymes in the first human study of BPA's effects (Rochester, 2013). The limit for all "Bisphenol A diglyceridyl ether (BADGE)" derivatives has been reduced by the European Union, and the Food and Drug Administration (FDA) is reevaluating BPA safety levels after presenting a draught evaluation on BPA in 2008 (Ismail *et al.*, 2010).

1.8.2 Methylbenzophenone

The molecules benzophenone and 4-methylbenzophenone, which is a metabolite of benzophenone, are present in the ink used to package food (Snedeker, 2014). Animal studies have shown that benzophenone causes liver and kidney enlargement, while similar information for 4-methylbenzophenone is lacking. Due to its volatility, these chemicals can migrate if there is no effective barrier between the printed surface of cardboard boxes and the food within (Aparicio and Elizalde, 2015). Following the discovery of 4-methylbenzophenone in cereal recently, the European Food Safety Authority looked into the toxicological data about benzophenone. Short-term exposure to 4-methylbenzophenone in contaminated breakfast cereals poses no health risk, according to EFSA's assessment. However, because a health risk for youngsters could not be ruled out, the agency plans to collect more information. In light of these worries, many labs now provide 4-methylbenzophenone testing using LC-MS/MS. As per the reports, the LOD for foods and packaging is 10 ppb and 1 ppm, respectively (Sultana, 2018).

1.8.3 Acrylamide

Acrylamide, like furan, is a relatively new food additive discovered using improved analytical methods and generated after specific forms of heat treatment (Wenzl *et al.*, 2007). Acrylamide is used in a variety of industries, such as water purification, gel electrophoresis, and paper production

(Bhunia *et al.*, 1987). Acrylamide has several negative health effects and is a known neurotoxin and carcinogen (Michalak *et al.*, 2020). The presence of acrylamide in food was first reported in 2002 by Swedish researchers, who found it in several fried and oven-baked items (Mucci *et al.*, 2003). Acrylic is formed when foods high in sugar and cooked at high temperatures combine with the amino acid asparagine and the carbonyl group of reducing sugars (Yaylayan *et al.*, 2003). The Food and Drug Administration (FDA), the Food and Agriculture Organization/World Health Organization (FAO/WHO), and many other international bodies researched acrylamide and made suggestions (Exon, 2006). NOAEL for acrylamide neuropathy was established by the FAO and WHO at 0.5 mg/kg body weight/day. Updating and verifying the LC-MS/MS method was part of the FDA's acrylamide action plan (Friedman, 2003). Both this LC-MS/MS method and a GC-MS method have been compared and both are accepted as the most valuable and authoritative techniques for acrylamide determination (Zhang *et al.*, 2005).

1.8.4 Benzene

Benzene is employed in a broad variety of industrial processes and is discharged into the environment via the exhaust of cars and the combustion of fossil fuels like oil and coal, both of which are recognized carcinogens (Lewtas, 2007). The FDA's MCL for bottled water and the EPA's MCL for benzene in drinking water are both 5 ppb. The soft drink industry informed the FDA in 1990 that benzene could form at low concentrations in some beverages that also include ascorbic acid and benzoate salts (Nyman *et al.*, 2008). More investigation revealed that heat and light facilitated benzene synthesis in the presence of these components. Many producers have reformulated their drinks to lessen or eliminate benzene generation. The FDA has maintained its vigilance over the benzene content in benzoate salts and ascorbic acid-containing drinks. Headspace sampling accompanied by gas chromatography-mass spectrometry is the FDA-approved method for benzene measurement in beverages (Steele *et al.*, 1994). A 0.2 ppb LOD can be achieved through cryogenic focusing of the headspace sample, while a 0.02 ppb LOD can be achieved through GC separations on a customized capillary column (Zoccali *et al.*, 2019).

1.8.5 Monochloropropane

1,2-Diol (3-MCPD) the most often found member of the chemical contamination class known as chloropropanols, for instance, is 3-MCPD, a known carcinogen and suspected genotoxin (Arris *et al.*, 2020). This substance is produced in the reaction of hydrochloric acid with any lingering fat in the protein source during the production of acid-hydrolyzed vegetable protein (acid-HVP) using heat and food-grade acids (Ismail *et al.*, 2010). Soups, savory snacks, and gravy mix frequently contain acid-HVP, a contaminant that is known to contain 3-MCPD (Vicente *et al.*, 2015). In the United States, the FDA considers acid-HVP with more than 1 ppm of 3-MCPD to be an unauthorized food additive because it is not GRAS (Dolan *et al.*, 2010). Many nations have established guidelines for the maximum allowable concentration of 3-MCPD in acid-HVP and Asian-style sauces, which range from 0.01 to 1 ppm (Ismail *et al.*, 2010). Currently, GC-MS is the method of choice, but only after extensive sample preparation that includes numerous extractions and derivatizations (Jiye *et al.*, 2005).

1.8.6 Furans

A colorless, flammable liquid called furan is used in some chemical manufacturing sectors; it also seems to be a known human carcinogen (Fiedler, 2003). More sensitive analytical methods have allowed for its detection; however, it is possible that it has been around for many years and only recently recognized (Stroebe *et al.*, 2012). It would appear that common methods of heat treatment (such as retorting canned food) are the source of furan. The FDA developed a plan to assess furan levels in various foods in 2005, but only with the aim of quantifying dietary exposure and determining the mechanisms responsible for furan production in foods (Javed *et al.*, 2021). To analyze furans, the FDA created a headspace (HS) GC-MS technique that relies on standard addition for quantification (Zoller *et al.*, 2007). Utilizing minor adjustments for various food categories and accelerating throughput (e.g., by employing headspace), many comparative studies have been based on this approach, with LODs of 0.02–0.12 ng/g being attained.

1.8.7 Perchlorate

Rocket fuel contains perchlorate, a chemical that is found naturally and can be produced in industrial settings under the right conditions (Trumpolt *et al.*, 2005). Perchlorate causes hypothyroidism when present in high enough concentrations to block iodide uptake by the thyroid gland (Wolff, 1998). Many items, including bottled water, milk, and lettuce, have been discovered to contain perchlorate. The FDA developed a quick, precise, sensitive, and focused assay for perchlorate using ion chromatography-tandem mass spectrometry (IC-MS/MS) (Ismail *et al.*, 2010). The procedure includes extraction, clean up by solid-phase extraction, and filtration before the IC-MS/MS determination. With a limit of detection (LOD) ranging from 0.5 ppb to 3 ppb per billion, depending on the food product, this approach has been updated to be able to detect perchlorate in a larger variety of foods (Sharma *et al.*, 2015).

1.9 CONCLUSION

Analysis of numerous pollutants, residues, and chemical elements in food is necessary due to consumer concerns and government laws centered on food safety. Pesticides, mycotoxins, antibiotics, GMOs, allergies, adulterants, packaging materials, hazardous chemicals, environmental contaminants, and other substances are all examples of such molecules. Meeting the objectives of industry and government for a safe and reliable food supply requires a combination of fast screening methods and more time-consuming quantitative procedures. If a screening test turns out positive for a substance of interest, more in-depth analyses are performed to confirm and quantify the finding. Because of the chemicals' low concentrations and the complexity of food matrices, sampling and sample preparation might present major difficulties. Homogenization, extraction, cleaning, and sometimes derivatization are typical steps in sample preparation. Immune-based techniques including ELISA, LFS, immunosensors, and immunoaffinity chromatography columns are becoming increasingly common in screening procedures. Some immunoassays go beyond simple screening and might be classified as quantitative. Enzyme tests, thin layer chromatography, and microbial growth inhibition are some other common screening techniques. The most widely used chromatographic approach for quantitative analysis of pesticides is gas chromatography (GC); however, for many of the substances of concern discussed in this chapter, high-performance liquid chromatography (HPLC) is the gold standard. Mass spectrometry detection is being added to GC and HPLC analyses, frequently with MS tandem systems. Protein-based methods (like immunoassays) or DNA-based methods (like PCR) are generally used to screen for GMOs and allergies, respectively. There is ongoing research and development into better and more precise methods of analysis for chemical residues and compounds of concern, with a primary emphasis on enhancing the speed, cost, and reliability of screening methods and lowering the detection limits of quantitative methods.

REFERENCES

Abedon, S. T. (2021). Detection of bacteriophages: Phage plaques. *Bacteriophages: Biology, Technology, Therapy*, 507–538.

Ahmed F. E. (2004). *Testing of genetically modified organisms in foods*. CRC, Boca Raton, FL

Ahmed, S., Ning, J., Cheng, G., Ahmad, I., Li, J., Mingyue, L., … & Yuan, Z. (2017). Receptor-based screening assays for the detection of antibiotics residues – A review. *Talanta*, 166, 176–186.

Alla, S. A. G., Loutfy, N. M., Shendy, A. H., & Ahmed, M. T. (2015). Hazard index, a tool for a long term risk assessment of pesticide residues in some commodities, a pilot study. *Regulatory Toxicology and Pharmacology*, 73(3), 985–991.

Almeida, S., Raposo, A., Almeida-González, M., & Carrascosa, C. (2018). Bisphenol A: Food exposure and impact on human health. *Comprehensive Reviews in Food Science and Food Safety*, 17(6), 1503–1517.

Alnassrallah, M. N., Alzoman, N. Z., & Almomen, A. (2022). Qualitative immunoassay for the determination of tetracycline antibiotic residues in milk samples followed by a quantitative improved HPLC-DAD method. *Scientific Reports*, 12(1), 1–14.

Alshannaq, A., & Yu, J. H. (2017). Occurrence, toxicity, and analysis of major mycotoxins in food. *International Journal of Environmental Research and Public Health*, 14(6), 632.

Andryukov, B. G. (2020). Six decades of lateral flow immunoassay: From determining metabolic markers to diagnosing COVID-19. *AIMS Microbiology*, 6(3), 280.

Anonymous (2009). *Code of federal regulations. Chloramphicol infection*. 21 CFR 522.390 (3). US Government Printing Office, Washington, DC.

Aparicio, J. L., & Elizalde, M. (2015). Migration of photoinitiators in food packaging: A review. *Packaging Technology and Science*, 28(3), 181–203.

Arris, F. A., Thai, V. T. S., Manan, W. N., & Sajab, M. S. (2020). A revisit to the formation and mitigation of 3-chloropropane-1, 2-diol in palm oil production. *Foods*, 9(12), 1769.

Asif, M., Awan, F. R., Khan, Q. M., Ngamsom, B., & Pamme, N. (2020). Paper-based analytical devices for colorimetric detection of *S. aureus* and *E. coli* and their antibiotic resistant strains in milk. *The Analyst*, 145(22), 7320–7329. https://doi.org/10.1039/d0an01075h

Baeyens, W. R. G., Schulman, S. G., Calokerinos, A. C., Zhao, Y., Campana, A. M. G., Nakashima, K., & De Keukeleire, D. (1998). Chemiluminescence-based detection: principles and analytical applications in flowing streams and in immunoassays. *Journal of Pharmaceutical and Biomedical Analysis*, 17(6-7), 941–953.

Bari, M. L., & Yeasmin, S. (2018). Foodborne diseases and responsible agents. In *Food safety and preservation* (pp. 195–229). London, United Kingdom: Academic Press.

Baynes, R. E., Dedonder, K., Kissell, L., Mzyk, D., Marmulak, T., Smith, G., ... & Riviere, J. E. (2016). Health concerns and management of select veterinary drug residues. *Food and Chemical Toxicology*, 88, 112–122.

Bhatnagar, D., Yu, J., & Ehrlich, K. C. (2002). Toxins of filamentous fungi. *Chemical immunology*, 81, 167–206.

Bhunia, A. K., Johnson, M. C., & Ray, B. (1987). Direct detection of an antimicrobial peptide of *Pediococcus acidilactici* in sodium dodecyl sulfate-polyacrylamide gel electrophoresis. *Journal of Industrial Microbiology and Biotechnology*, 2(5), 319–322.

Bishop, J. D., Hsieh, H. V., Gasperino, D. J., & Weigl, B. H. (2019). Sensitivity enhancement in lateral flow assays: A systems perspective. *Lab on a Chip*, 19(15), 2486–2499.

Botitsi, H. V., Garbis, S. D., Economou, A., & Tsipi, D. F. (2011). Current mass spectrometry strategies for the analysis of pesticides and their metabolites in food and water matrices. *Mass spectrometry reviews*, 30(5), 907–939.

Botsoglou, N. A., & Fletouris, D. J. (2000). *Drug residues in foods*. Marcel Dekker.

Bouis, H. E., Chassy, B. M., & Ochanda, J. O. (2003). 2. Genetically modified food crops and their contribution to human nutrition and food quality. *Trends in Food Science & Technology*, 14(5–8), 191–209.

Brooks, J. P., Edwards, D. J., Harwich, M. D., Rivera, M. C., Fettweis, J. M., Serrano, M. G., ... & Buck, G. A. (2015). The truth about metagenomics: Quantifying and counteracting bias in 16S rRNA studies. *BMC Microbiology*, 15(1), 1–14.

Bruijns, B., Tiggelaar, R., & Gardeniers, H. (2020). A microfluidic approach for biosensing DNA within forensics. *Applied Sciences*, 10(20), 7067.

Bruns, H. A. (2003). Controlling aflatoxin and fumonisin in maize by crop management. *Journal of Toxicology: Toxin Reviews*, 22(2-3), 153–173.

Bueno, D., Istamboulie, G., Muñoz, R., & Marty, J. L. (2015). Determination of mycotoxins in food: A review of bioanalytical to analytical methods. *Applied Spectroscopy Reviews*, 50(9), 728–774.

Buick, A. R., Doig, M. V., Jeal, S. C., Land, G. S., & McDowall, R. D. (1990). Method validation in the bioanalytical laboratory. *Journal of Pharmaceutical and Biomedical Analysis*, 8(8–12), 629–637.

Bulthaus, M. (2004). Detection of antibiotic/drug residues in milk and dairy products, ch. 12. In: Wehr H. M., & Frank J. F. (eds) *Standard methods for the examination of dairy products*, 17th edn. American Public Health Association, Washington, DC.

Butler, J. M. (2011). *Advanced topics in forensic DNA typing: Methodology*. MA, USA: Academic Press.

Candlish, A. A. G. (1991). *The determination of mycotoxins in animal feeds by biological methods* (Vol. 223). CRC Press, Boca Raton.

Capriotti, A. L., Caruso, G., Cavaliere, C., Foglia, P., Samperi, R., & Laganà, A. (2012). Multiclass mycotoxin analysis in food, environmental and biological matrices with chromatography/mass spectrometry. *Mass Spectrometry Reviews*, 31(4), 466–503.

Cavalier, C., Curini, R., Di Corcia, A., Nazzari, M., & Samperi, R. (2003). A simple and sensitive liquid chromatography-mass spectrometry confirmatory method for analyzing sulfonamide antibacterials in milk and egg. *Journal of Agricultural and Food Chemistry*, 2003 Jan 29; 51(3), 558–566. doi: 10.1021/jf020834w.

Chaiharn, M., Chunhaleuchanon, S., & Lumyong, S. (2009). Screening siderophore producing bacteria as potential biological control agent for fungal rice pathogens in Thailand. *World Journal of Microbiology and Biotechnology*, 25(11), 1919–1928.

Chammem, N., Issaoui, M., De Almeida, A. I. D., & Delgado, A. M. (2018). Food crises and food safety incidents in European Union, United States, and Maghreb Area: Current risk communication strategies and new approaches. *Journal of AOAC International*, 101(4), 923–938.

Chen, Y., & Schwack, W. (2014). High-performance thin-layer chromatography screening of multi class antibiotics in animal food by bioluminescent bioautography and electrospray ionization mass spectrometry. *Journal of Chromatography A*, 1356, 249–257.

Chiesa, L. M., DeCastelli, L., Nobile, M., Martucci, F., Mosconi, G., Fontana, M., … & Panseri, S. (2020). Analysis of antibiotic residues in raw bovine milk and their impact toward food safety and on milk starter cultures in cheese-making process. *Lwt*, 131, 109783.

Chondrogiannis, G., Réu, P., & Hamedi, M. M. (2022). Paper-based bacterial lysis enables sample-to-answer home-based DNA testing. *Advanced Materials Technologies*, 2201004.

Clark, S. B., Turnipseed, S. B., & Madson, M. R. (2005) Confirmation of sulfamethazine, sulfathiazole, and sulfadimethoxine residues in condensed milk and soft-cheese products by liquid chromatography/tandem mass spectrometry. *Journal of AOAC International*, 88, 736–743

Conzuelo, F., Campuzano, S., Gamella, M., Pinacho, D. G., Reviejo, A. J., Marco, M. P., & Pingarrón, J. M. (2013). Integrated disposable electrochemical immunosensors for the simultaneous determination of sulfonamide and tetracycline antibiotics residues in milk. *Biosensors and Bioelectronics*, 50, 100–105.

Crevel, R. R., Ballmer-Weber, B. K., Holzhauser, T., Hourihane, J. O. B., Knulst, A. C., Mackie, A. R., & Taylor, S. L. (2008). Thresholds for food allergens and their value to different stakeholders. *Allergy*, 63(5), 597–609.

Czechowski, T., Bari, R. P., Stitt, M., Scheible, W. R., & Udvardi, M. K. (2004). Real-time RT-PCR profiling of over 1400 Arabidopsis transcription factors: unprecedented sensitivity reveals novel root-and shoot-specific genes. *The Plant Journal*, 38(2), 366–379.

Dasriya, V., Joshi, R., Ranveer, S., Dhundale, V., Kumar, N., & Raghu, H. V. (2021). Rapid detection of pesticide in milk, cereal and cereal based food and fruit juices using paper strip-based sensor. *Scientific Reports*, 11(1), 1–9.

de Oliveira, D. N., Ferreira, M. S., & Catharino, R. R. (2014). Rapid and simultaneous in situ assessment of aflatoxins and stilbenes using silica plate imprinting mass spectrometry imaging. *PLoS One*, 9(3), e90901. https://doi.org/10.1371/journal.pone.0090901

Dey, D. K., Kang, J. I., Bajpai, V. K., Kim, K., Lee, H., Sonwal, S., … & Shukla, S. (2022). Mycotoxins in food and feed: toxicity, preventive challenges, and advanced detection techniques for associated diseases. *Critical Reviews in Food Science and Nutrition*, 1–22.

Doering, W. E., Piotti, M. E., Natan, M. J., & Freeman, R. G. (2007). SERS as a foundation for nanoscale, optically detected biological labels. *Advanced Materials*, 19(20), 3100–3108.

Dolan, L. C., Matulka, R. A., & Burdock, G. A. (2010). Naturally occurring food toxins. *Toxins*, 2(9), 2289–2332.

Duineveld, B. M., Kowalchuk, G. A., Keijzer, A., van Elsas, J. D., & van Veen, J. A. (2001). Analysis of bacterial communities in the rhizosphere of chrysanthemum via denaturing gradient gel electrophoresis of PCR-amplified 16S rRNA as well as DNA fragments coding for 16S rRNA. *Applied and Environmental Microbiology*, 67(1), 172–178.

Echan, L. A., Tang, H. Y., Ali-Khan, N., Lee, K., & Speicher, D. W. (2005). Depletion of multiple high-abundance proteins improves protein profiling capacities of human serum and plasma. *Proteomics*, 5(13), 3292–3303.

EFSA Panel on Genetically Modified Organisms (GMO). (2011). Guidance for risk assessment of food and feed from genetically modified plants. *EFSA Journal*, 9(5), 2150.

Exon, J. H. (2006). A review of the toxicology of acrylamide. *Journal of Toxicology and Environmental Health, Part B*, 9(5), 397–412.

FDA (2006) Preparation and LC/MS/MS analysis of honey for fluoroquinolone residues. 29 Sept 2006 (last updated 6/18/2009). http://www.fda.gov/Food/ScienceResearch/LaboratoryMethods/DrugChemicalResiduesMethodology/ucm071495.htm

FDA (2009) Analytical methods for residues of chlorampenicol and related compounds in foods. http://www.fda.gov/Food/ScienceResearch/LaboratoryMethods/DrugChemicalResiduesMethodology/ucm113126.htm; http://www.fda.gov/Food/ScienceResearch/LaboratoryMethods/DrugChemicalResiduesMethodology/ucm071463.htm

FDA (2009) Fluoroquinolones (last updated 5/14/2009). http://www.fda.gov/Food/ScienceResearch/vLaboratoryMethods/DrugChemicalResiduesMethodology/ucm071463.htm

Fiedler, H. (2003). *Dioxins and furans (PCDD/PCDF). In Persistent organic pollutants* (pp. 123–201). Springer, Berlin, Heidelberg.

Friedman, M. (2003). Chemistry, biochemistry, and safety of acrylamide. A review. *Journal of Agricultural and Food Chemistry*, 51(16), 4504–4526.

Friedman, M., & Brandon, D. L. (2001). Nutritional and health benefits of soy proteins. *Journal of Agricultural and Food Chemistry*, 49(3), 1069–1086.

Gabriel, M. S. D. F. (2016). Evaluation of Alt a 1 as specific marker of exposure to fungal allergenic sources and clinical relevance of a manganese-dependent superoxide dismutase and a serine protease as new *Alternaria alternata* allergens. PhD thesis. http://hdl.handle.net/10400.6/4195

Gavahian, M., Sarangapani, C., & Misra, N. N. (2021). Cold plasma for mitigating agrochemical and pesticide residue in food and water: Similarities with ozone and ultraviolet technologies. *Food Research International*, 141, 110138.

Gibbs F., Selim K., All I. Catherine N. Mulligan, B. (1999). Encapsulation in the food industry: A review. *International Journal of Food Sciences and Nutrition*, 50(3), 213–224.

Gustavsson, E., Degelaen, J., Bjurling, P., & Sternesjö, Å. (2004). Determination of β-lactams in milk using a surface plasmon resonance-based biosensor. *Journal of Agricultural and Food Chemistry*, 52(10), 2791–2796.

Hanekamp, J. C., & Bast, A. (2015). Antibiotics exposure and health risks: Chloramphenicol. *Environmental Toxicology and Pharmacology*, 39(1), 213–220.

Hartmann, M., Roeraade, J., Stoll, D., Templin, M. F., & Joos, T. O. (2009). Protein microarrays for diagnostic assays. *Analytical and Bioanalytical Chemistry*, 393(5), 1407–1416.

Heller K. J. (2003) *Genetically engineered food: Methods and detection*. Wiley-VCH, Weinheim, Germany.

Higueras, P., Oyarzun, R., Iraizoz, J. M., Lorenzo, S., Esbrí, J. M., & Martínez-Coronado, A. (2012). Low-cost geochemical surveys for environmental studies in developing countries: Testing a field portable XRF instrument under quasi-realistic conditions. *Journal of Geochemical Exploration*, 113, 3–12.

Holstege, D. M., Puschner, B., Whitehead, G., Galey, F. D. (2002). Screening and mass spectral confirmation of B-lactam antibiotic residues in milk using LC-MS/MS. *Journal of Agricultural and Food Chemistry*, 50(2), 406–411.

Hornbeck, P. V. (2015). Enzyme-linked immunosorbent assays. *Current Protocols in Immunology*, 110(1), 2–1.

Ingelse, B. A., Van Dam, R. C., Vreeken, R. J., Mol, H. G., & Steijger, O. M. (2001). Determination of polar organophosphorus pesticides in aqueous samples by direct injection using liquid chromatography–tandem mass spectrometry. *Journal of Chromatography A*, 918(1), 67–78.

Ismail, B., Reuhs, B. L., & Nielsen, S. S. (2010). Analysis of food contaminants, residues, and chemical constituents of concern. In *Food Analysis* (pp. 317–349). Springer, Boston, MA.

Jackson, J. F. & Linskens, H. F. (2009). *Testing for genetic manipulation in plants (molecular methods of plant analysis)*. Springer, New York.

Javed, F., Shahbaz, H. M., Nawaz, A., Olaimat, A. N., Stratakos, A. C., Wahyono, A., ... & Park, J. (2021). Formation of furan in baby food products: Identification and technical challenges. *Comprehensive Reviews in Food Science and Food Safety*, 20(3), 2699–2715.

Jawaid, W., Campbell, K., Melville, K., Holmes, S. J., Rice, J., & Elliott, C. T. (2015). Development and validation of a novel lateral flow immunoassay (LFIA) for the rapid screening of paralytic shellfish toxins (PSTs) from shellfish extracts. *Analytical Chemistry*, 87(10), 5324–5332.

Jiye, A., Trygg, J., Gullberg, J., Johansson, A. I., Jonsson, P., Antti, H., ... & Moritz, T. (2005). Extraction and GC/MS analysis of the human blood plasma metabolome. *Analytical Chemistry*, 77(24), 8086–8094.

Ju, Q., Noor, M. O., & Krull, U. J. (2016). based biodetection using luminescent nanoparticles. *Analyst*, 141(10), 2838–2860.

Kadokami, K., Miyawaki, T., Takagi, S., Iwabuchi, K., Towatari, H., Yoshino, T., ... & Li, X. (2022). Novel automated identification and quantification database using liquid chromatography quadrupole time-of-flight mass spectrometry for quick, comprehensive, cheap and extendable organic micro-pollutant analysis in environmental systems. *Analytica Chimica Acta*, 340656.

Kang, J. S. (2012). Principles and applications of LC-MS/MS for the quantitative bioanalysis of analytes in various biological samples. *Tandem Mass Spectrometry–Applications and Principles*, 441–492.

Karongo, R., Ikegami, T., Stoll, D. R., & Lämmerhofer, M. (2020). A selective comprehensive reversed-phase× reversed-phase 2D-liquid chromatography approach with multiple complementary detectors as advanced generic method for the quality control of synthetic and therapeutic peptides. *Journal of Chromatography A*, 1627, 461430.

Keding, G. B., Schneider, K., & Jordan, I. (2013). Production and processing of foods as core aspects of nutrition-sensitive agriculture and sustainable diets. *Food Security*, 5(6), 825–846

Kirtane, A. R., Verma, M., Karandikar, P., Furin, J., Langer, R., & Traverso, G. (2021). Nanotechnology approaches for global infectious diseases. *Nature Nanotechnology*, 16(4), 369–384.

Krska, R. (2016). MyToolBox: Safe food and feed through an integrated ToolBox for mycotoxin management. FINAL PROGRAM &, 5.

Krska, R., Welzig, E., & Boudra, H. (2007). Analysis of Fusarium toxins in feed. *Animal Feed Science and Technology*, 137(3–4), 241–264.

Lei, H., Wang, Z., Eremin, S. A., & Liu, Z. (2022). Application of antibody and immunoassay for food safety. *Foods*, 11(6), 826.

Lentz, T. J., Dotson, G. S., Williams, P. R. D., Maier, A., Gadagbui, B., Pandalai, S. P., ... & Mumtaz, M. (2015). Aggregate exposure and cumulative risk assessment—Integrating occupational and non-occupational risk factors. *Journal of Occupational and Environmental Hygiene*, 12(sup1), S112–S126.

Lewtas, J. (2007). Air pollution combustion emissions: Characterization of causative agents and mechanisms associated with cancer, reproductive, and cardiovascular effects. *Mutation Research/ Reviews in Mutation Research*, 636(1–3), 95–133.

Liu, X., Liu, Z., Bian, L., Ping, Y., Li, S., Zhang, J., ... & Wang, X. (2022). Determination of pesticide residues in chilli and Sichuan pepper by high performance liquid chromatography quadrupole time-of-flight mass spectrometry. *Food Chemistry*, 387, 132915.

Loos, G., Van Schepdael, A., & Cabooter, D. (2016). Quantitative mass spectrometry methods for pharmaceutical analysis. *Philosophical Transactions of the Royal Society A: Mathematical, Physical and Engineering Sciences*, 374(2079), 20150366.

McCalley, D. V. (2002). Analysis of the Cinchona alkaloids by high-performance liquid chromatography and other separation techniques. *Journal of Chromatography A*, 967(1), 1–19.

McNeil, M. M., Weintraub, E. S., Duffy, J., Sukumaran, L., Jacobsen, S. J., Klein, N. P., ... & DeStefano, F. (2016). Risk of anaphylaxis after vaccination in children and adults. *Journal of Allergy and Clinical Immunology*, 137(3), 868–878.

Michalak, J., Czarnowska-Kujawska, M., Klepacka, J., & Gujska, E. (2020). Effect of microwave heating on the acrylamide formation in foods. *Molecules*, 25(18), 4140.

Miraglia, M., Berdal, K. G., Brera, C., Corbisier, P., Holst-Jensen, A., Kok, E. J., ... & Zagon, J. (2004). Detection and traceability of genetically modified organisms in the food production chain. *Food and Chemical Toxicology*, 42(7), 1157–1180.

Miura, H., Gurumurthy, C. B., Sato, T., Sato, M., & Ohtsuka, M. (2015). CRISPR/Cas9-based generation of knockdown mice by intronic insertion of artificial microRNA using longer single-stranded DNA. *Scientific reports*, 5(1), 1–11.

Mucci, L. A., Dickman, P. W., Steineck, G., Adami, H. O., & Augustsson, K. (2003). Dietary acrylamide and cancer of the large bowel, kidney, and bladder: absence of an association in a population-based study in Sweden. *British Journal of Cancer*, 88(1), 84–89.

Nestola, M., & Schmidt, T. C. (2016). Fully automated determination of the sterol composition and total content in edible oils and fats by online liquid chromatography–gas chromatography–flame ionization detection. *Journal of Chromatography A*, 1463, 136–143.

Notomi, T., Okayama, H., Masubuchi, H., Yonekawa, T., Watanabe, K., Amino, N., & Hase, T. (2000). Loop-mediated isothermal amplification of DNA. *Nucleic Acids Research*, 28(12), e63–e63.

Nyman, P. J., Diachenko, G. W., Perfetti, G. A., McNeal, T. P., Hiatt, M. H., & Morehouse, K. M. (2008). Survey results of benzene in soft drinks and other beverages by headspace gas chromatography/mass spectrometry. *Journal of Agricultural and Food Chemistry*, 56(2), 571–576.

Ollivaux, C., Soyez, D., & Toullec, J. Y. (2014). Biogenesis of d-amino acid containing peptides/proteins: Where, when and how?. *Journal of Peptide Science*, 20(8), 595–612.

Ong, J. L., Loakes, D., Jaroslawski, S., Too, K., & Holliger, P. (2006). Directed evolution of DNA polymerase, RNA polymerase and reverse transcriptase activity in a single polypeptide. *Journal of Molecular Biology*, 361(3), 537–550.

Orford, S., Delaney, S., & Timmis, J. (2007). The genetic modification of cotton. *Cotton: Science and Technology*, 1, 103–129.

Peres, G. T., Rath, S., & Reyes, F. G. R. (2010). A HPLC with fluorescence detection method for the determination of tetracyclines residues and evaluation of their stability in honey. *Food Control*, 21(5), 620–625.

Pikkemaat, M. G. (2009). Microbial screening methods for detection of antibiotic residues in slaughter animals. *Analytical and Bioanalytical Chemistry*, 395(4), 893–905.

Poms, R. E., Klein, C. L., & Anklam, E. (2004). Methods for allergen analysis in food: A review. *Food Additives and Contaminants*, 21(1), 1–31.

Pou, K. J., Raghavan, V., & Packirisamy, M. (2022). Microfluidics in smart packaging of foods. *Food Research International*, 111873.

Prange, A., & Pröfrock, D. (2008). Chemical labels and natural element tags for the quantitative analysis of bio-molecules. *Journal of Analytical Atomic Spectrometry*, 23(4), 432–459.

Prieto, A., Basauri, O., Rodil, R., Usobiaga, A., Fernández, L. A., Etxebarria, N., & Zuloaga, O. (2010). Stir-bar sorptive extraction: A view on method optimisation, novel applications, limitations and potential solutions. *Journal of Chromatography A*, 1217(16), 2642–2666.

Raitskin, O., Schudoma, C., West, A., & Patron, N. J. (2019). Comparison of efficiency and specificity of CRISPR-associated (Cas) nucleases in plants: An expanded toolkit for precision genome engineering. *PLoS One*, 14(2), e0211598.

Raza, N., Kim, K. H., Abdullah, M., Raza, W., & Brown, R. J. (2018). Recent developments in analytical quantitation approaches for parabens in human-associated samples. *TrAC Trends in Analytical Chemistry*, 98, 161–173.

Rejczak, T., & Tuzimski, T. (2015). Recent trends in sample preparation and liquid chromatography/mass spectrometry for pesticide residue analysis in food and related matrixes. *Journal of AOAC International*, 98(5), 1143–1162.

Rochester, J. R. (2013). Bisphenol A and human health: A review of the literature. *Reproductive Toxicology*, 42, 132–155.

Sabui, P., Mallick, S., Singh, K. R., Natarajan, A., Verma, R., Singh, J., & Singh, R. P. (2022). Potentialities of fluorescent carbon nanomaterials as sensor for food analysis. *Luminescence*. 2023 Jul; 38(7), 1047–1063. doi: 10.1002/bio.4406.

Salvador, A., Carrière, R., Ayciriex, S., Margoum, C., Leblanc, Y., & Lemoine, J. (2020). Scout-multiple reaction monitoring: A liquid chromatography tandem mass spectrometry approach for multi-residue pesticide analysis without time scheduling. *Journal of Chromatography A*, 1621, 461046.

Sannino, A. (2008). Pesticide residues. *Comprehensive Analytical Chemistry*, 51, 257–305.

Santhanam, M., Algov, I., & Alfonta, L. (2020). DNA/RNA electrochemical biosensing devices a future replacement of PCR methods for a fast epidemic containment. *Sensors*, 20(16), 4648.

Sanyal, D., & Mathur, P. (2022). Advanced adsorbent mediated extraction techniques for the separation of antibiotics from food, biological, and environmental matrices. *Separation & Purification Reviews*, 51(3), 373–407.

Schüling, T., Eilers, A., Scheper, T., & Walter, J. (2018). Aptamer-based lateral flow assays. *AIMS Bioengineering 5 (2018), Nr. 2*, 5(2), 78–102.

Seiber, J. N., Woodrow, J. E., & David, M. D. (2021). Organophosphorus esters. In *Chromatographic Analysis of Environmental and Food Toxicants* (pp. 229–257). Boca Raton: CRC Press.

Shah, R., Kolanos, R., DiNovi, M. J., Mattia, A., & Kaneko, K. J. (2017). Dietary exposures for the safety assessment of seven emulsifiers commonly added to foods in the United States and implications for safety. *Food Additives & Contaminants: Part A*, 34(6), 905–917.

Sharma, R., Ragavan, K. V., Thakur, M. S., & Raghavarao, K. S. M. S. (2015). Recent advances in nanoparticle based aptasensors for food contaminants. *Biosensors and Bioelectronics*, 74, 612–627.

Shawkey, M. D., Pillai, S. R., & Hill, G. E. (2003). Chemical warfare? Effects of uropygial oil on feather-degrading bacteria. *Journal of Avian Biology*, 34(4), 345–349.

Singh, J., & Mehta, A. (2020). Rapid and sensitive detection of mycotoxins by advanced and emerging analytical methods: A review. *Food Science & Nutrition*, 8(5), 2183–2204.

Snedeker, S. M. (2014). Benzophenone UV-photoinitiators used in food packaging: potential for human exposure and health risk considerations. In *Toxicants in Food Packaging and Household Plastics* (pp. 151–176). Springer, London.

Soest, E. V., & Fritschi, L. (2004). Occupational health risks in veterinary nursing: An exploratory study. *Australian Veterinary Journal*, 82(6), 346–350.

Songa, E. A., & Okonkwo, J. O. (2016). Recent approaches to improving selectivity and sensitivity of enzyme-based biosensors for organophosphorus pesticides: A review. *Talanta*, 155, 289–304.

Sorbo, A., Pucci, E., Nobili, C., Taglieri, I., Passeri, D., & Zoani, C. (2022). Food safety assessment: Overview of metrological issues and regulatory aspects in the European Union. *Separations*, 9(2), 53.

Steele, D. H., Thornburg, M. J., Stanley, J. S., Miller, R. R., Brooke, R., Cushman, J. R., & Cruzan, G. (1994). Determination of styrene in selected foods. *Journal of Agricultural and Food Chemistry*, 42(8), 1661–1665.

Stroebe, W., Postmes, T., & Spears, R. (2012). Scientific misconduct and the myth of self-correction in science. *Perspectives on Psychological Science*, 7(6), 670–688.

Stroka, J., & Anklam, E. (2002). New strategies for the screening and determination of aflatoxins and the detection of aflatoxin-producing moulds in food and feed. *TrAC Trends in Analytical Chemistry*, 21(2), 90–95.

Sultana, A. (2018). *Chemical contaminants in rice, spice and vegetable samples* (Doctoral dissertation, University of Dhaka).

Tekel, J., & Hatrík, Š. (1996). Pesticide residue analyses in plant material by chromatographic methods: clean-up procedures and selective detectors. *Journal of Chromatography A*, 754(1–2), 397–410.

Thellin, O., ElMoualij, B., Heinen, E., & Zorzi, W. (2009). A decade of improvements in quantification of gene expression and internal standard selection. *Biotechnology Advances*, 27(4), 323–333.

Tian, W., Wang, L., Lei, H., Sun, Y., & Xiao, Z. (2018). Antibody production and application for immunoassay development of environmental hormones: A review. *Chemical and Biological Technologies in Agriculture*, 5(1), 1–12.

Timmerer, U., Lehmann, L., Schnug, E., & Bloem, E. (2020). Toxic Effects of Single Antibiotics and Antibiotics in Combination on Germination and Growth of Sinapis alba L. *Plants*, 9(1), 107.

Trichilo, C. L., & Schmitt, R. D. (1989). Tolerance setting process in the US Environmental Protection Agency. *Journal of the Association of Official Analytical Chemists*, 72(3), 536–538.

Trumpolt, C. W., Crain, M., Cullison, G. D., Flanagan, S. J., Siegel, L., & Lathrop, S. (2005). Perchlorate: sources, uses, and occurrences in the environment. *Remediation Journal: The Journal of Environmental Cleanup Costs, Technologies & Techniques*, 16(1), 65–89.

Turner, N. W., Subrahmanyam, S., & Piletsky, S. A. (2009). Analytical methods for determination of mycotoxins: a review. *Analytica chimica acta*, 632(2), 168–180.

Tyczewska, A., Woźniak, E., Gracz, J., Kuczyński, J., & Twardowski, T. (2018). Towards food security: Current state and future prospects of agrobiotechnology. *Trends in Biotechnology*, 36(12), 1219–1229.

Upadhayay, J., Rana, M., Juyal, V., Bisht, S. S., & Joshi, R. (2020). Impact of pesticide exposure and associated health effects. *Pesticides in crop production: physiological and biochemical action*, 69–88.

Van Aken, K., Strekowski, L., & Patiny, L. (2006). EcoScale, a semi-quantitative tool to select an organic preparation based on economic and ecological parameters. *Beilstein Journal of Organic Chemistry*, 2(1), 3.

Vicente, E., Arisseto, A. P., Furlani, R. P. Z., Monteiro, V. , Gonçalves, L. M. , Pereira, A. L. D. , &Toledo, M. C. F. (2015). Levels of 3-monochloropropane-1,2-diol (3-MCPD) in selected processed foods from the Brazilian market. *Food Research International*, 77, 310–314. https://doi.org/10.1016/j.foodres.2015.03.035

Viola, C., & DeVincent, S. J. (2006). Overview of issues pertaining to the manufacture, distribution, and use of antimicrobials in animals and other information relevant to animal antimicrobial use data collection in the United States. *Preventive Veterinary Medicine*, 73(2–3), 111–131.

Wang, J., & Leung, D. (2007). Analyses of macrolide antibiotic residues in eggs, raw milk, and honey using both ultra-performance liquid chromatography/quadrupole time-of-flight mass spectrometry and high-performance liquid chromatography/tandem mass spectrometry. *Rapid Communication in Mass Spectrometry,*, 21(19), 3213–3222.

Wang, K., Qin, W., Hou, Y., Xiao, K., & Yan, W. (2016). The application of lateral flow immunoassay in point of care testing: A review. *Nano Biomedicine and Engineering*, 8(3), 172–183.

Wang, X., Li, J., Jian, D., Zhang, Y., Shan, Y., Wang, S., & Liu, F. (2021). Based antibiotic sensor (PAS) relying on colorimetric indirect competitive enzyme-linked immunosorbent assay for quantitative tetracycline and chloramphenicol detection. *Sensors and Actuators B: Chemical*, 329, 129173.

Wang, Z., Liu, H., & Liu, S. (2017). Low-dose bisphenol A exposure: A seemingly instigating carcinogenic effect on breast cancer. *Advanced Science*, 4(2), 1600248.

Wenzl, T., Lachenmeier, D. W., & Gökmen, V. (2007). Analysis of heat-induced contaminants (acrylamide, chloropropanols and furan) in carbohydrate-rich food. *Analytical and Bioanalytical Chemistry*, 389(1), 119–137.

Westphal, C. D. (2008). Improvement of immunoassays for the detection of food allergens, ch. 29. In: Siantar, D. P., Trucksess, M. W., Scott, P. M., & Herman, E. M. (eds) *Food contaminants: mycotoxins and food allergens. ACS symposium series 1001*. American Chemical Society, Washington, DC.

Westphal, C. D., Pereira, M. R., Raybourne, R. B., & Williams, K. M. (2004). Evaluation of extraction buffers using the current approach of detecting multiple allergenic and nonallergenic proteins in food. *Journal of AOAC International*, 87(6), 1458–1465.

Wiesner, J., & Vilcinskas, A. (2010). Antimicrobial peptides: The ancient arm of the human immune system. *Virulence*, 1(5), 440–464.

Wilson, I. D., & Brinkman, U. T. (2003). Hyphenation and hypernation: The practice and prospects of multiple hyphenation. *Journal of Chromatography A*, 1000(1–2), 325–356.

Wilson, R. (2008). The use of gold nanoparticles in diagnostics and detection. *Chemical Society Reviews*, 37(9), 2028–2045.

Wolff, J. (1998). Perchlorate and the thyroid gland. *Pharmacological Reviews*, 50(1), 89–106.

Wollgast, J., & Anklam, E. (2000). Review on polyphenols in *Theobroma cacao*: Changes in composition during the manufacture of chocolate and methodology for identification and quantification. *Food Research International*, 33(6), 423–447.

Wolt, J. D., Wang, K., & Yang, B. (2016). The regulatory status of genome-edited crops. *Plant Biotechnology Journal*, 14(2), 510–518.

Yadav, I. C., & Devi, N. L. (2017). Pesticides classification and its impact on human and environment. *Environmental Science Engineering*, 6, 140–158.

Yang, Y. T., & Chen, B. (2016). Governing GMOs in the USA: Science, law and public health. *Journal of the Science of Food and Agriculture*, 96(6), 1851–1855.

Yaylayan, V. A., Wnorowski, A., & Perez Locas, C. (2003). Why asparagine needs carbohydrates to generate acrylamide. *Journal of Agricultural and Food Chemistry*, 51(6), 1753–1757.

Zhang, Y., Zhang, G., & Zhang, Y. (2005). Occurrence and analytical methods of acrylamide in heat-treated foods: Review and recent developments. *Journal of Chromatography A*, 1075(1–2), 1–21.

Zoccali, M., Tranchida, P. Q., & Mondello, L. (2019). Fast gas chromatography-mass spectrometry: A review of the last decade. *TrAC Trends in Analytical Chemistry*, 118, 444–452.

Zoller, O., Sager, F., & Reinhard, H. (2007). Furan in food: Headspace method and product survey. *Food Additives and Contaminants*, 24(sup1), 91–107.

2 Quality Control and Risk Assessment of Food Storage and Packaging

Mariya Divanshi AS, Ayantika Das, and Raghu HV

2.1 INTRODUCTION

Food is an organic substance that become the ideal source of nutrition for living beings. Food can be of plant origin or animal origin which are rich in carbohydrates, proteins, vitamins and minerals. Hence they undergo physical, chemical or microbiological spoilage quickly affecting the texture, color, nutritive value and its acceptability. Packaging and storage of food at required and optimum condition is of great importance. Packaging has been used since the BC time period. Hollowed fruit husk and animal skins were first used to carry water and woven grass baskets were used to keep other food materials. Archaeological proofs from 8000 BC shows the use of clay pots and jars for storing salt, oil, olives, grains etc. As cities were developed, casks, wooden barrels and woven grass panniers were used to pack things to bring in to markets. Thereby, the use of packaging for convenience to transport, protection and display in bulk level was developed. The drastic developments in human life have made a big leap from bulk-level packaging to consumer packs. This gave producers the freedom to own the style of packaging for their product promotion. Packaging nowadays plays a significant role in differentiating between the same products of different companies. The product with prompt packing will be free from any damage and extraneous matter; and will be safe and wholesome. About 70% of the packaging is used for food and drink packing while the remaining percent goes for health care, pharmaceutical, beauty products, chemicals, clothing and electrical items. The different level of packaging includes primary, secondary and tertiary. Out of this, the packaging that has a direct contact with the product is called primary packaging. Utmost care should be taken for deciding the type of packing material chosen for the primary packaging so as to be non-toxic, non-reacting and O_2 non-permeating. Environmental-friendly packaging is an ideal concept and not easy to achieve. The manufacturers should be responsible for the best way to reduce the usage of packaging material by recovering it from waste and recycling them. Maintaining optimal storage condition is also important in choosing appropriate packaging material. The storage condition will depend up on the type food to be stored, the nature of packaging material, process done etc. The producer can choose from different types of storage conditions accordingly. Perishable foods with higher moisture content are usually frozen while semi-perishable and non-perishable foods are stored in dry storage. Both packaging and storage can determine the quality and shelf life of the product. The packaging and storage have very important roles in maintaining the quality of food materials. The quality can be controlled and different risks upon tampering with the package and storage condition should be predictable for the product to be safe and wholesome for end use.

2.2 CLASSIFICATION OF FOOD

Food can be classified on different bases, such as its shelf life, nutritional value and function and different processing, as given in Table 2.1. On the basis of shelf life, the food can be perishable, semi-perishable and non-perishable (Doyle, 2009). Perishable foods have a number of days or three weeks of shelf life. Special preservation, packaging and storage conditions are inevitable to avoid food spoilage (Steele, 2004). Many foods can have a shelf life for months while stored under optimum conditions. This includes cheese, fruits and vegetables and are called semi-perishable foods. The foods that are processed or not processed but naturally can be stored for several years are called non-perishable foods, such a canned fruits, flours, dried beans and nuts. The food can have different functions, such as body building and repairing action, energy giving, regulatory action and protective action. This depends on the food that is carbohydrate rich, fat rich, protein rich or vitamin and mineral rich (Chopra, 2005). Food on the basis of processing can be divided into minimally processed or unprocessed food, processed foods and ultra-processed foods (Monteiro *et al.*, 2010). The different techniques of processing can be physical, chemical or biological methods.

2.3 FOOD STORAGE AND PACKAGING

Generally, there are two types of packaging materials for food: direct and indirect food contact materials. Direct food contact materials include bottles, cans and plastics, whereas boards, varnish and inks come under indirect materials. Since ancient times, paper and paperboard are considered to be the most essential means for packaging different food items. These materials include vegetable

DOI: 10.1201/9781003334859-2

Table 2.1: The Classification of Food on Different Bases

Food Classified Based on Shelf Life

1. Perishable   Eggs, meats, milk and milk products, poultry, seafood
2. Semi-perishable   Cheeses, fruits, vegetables
3. Non-perishable   Canned fruits, flour, dry beans, mayonnaise, peanut butter, nuts, sugar

Food Classified Based on Nutrition Value and Function

Nutritional value

1. Carbohydrate rich   Rice, starch rich vegetables, wheat
2. Fat rich   Butter, egg yolk, oils
3. Protein rich   Fish, egg, meat, milk, nuts
4. Vitamin and mineral rich   Fruits and vegetables

Function

1. Body building   Fish, meat, milk, nuts, pulses, vegetables
2. Energy giving   Butter, cereals, dry fruits, oil, starch foods, sugar
3. Regulatory   Beverages citrus fruits, raw vegetables, water
4. Protective   Cereals, fruits meat, milk, vegetables, whole grain

Food Classified on the Basis of Processes Done

1. Unprocessed   Cereals, coffee, dried beans, dried fruits, eggs, fish, fresh fruits, meats, poultry, pulses, nuts, seeds, tea yogurt, vegetables
2. Processed   butter, cosmetic additives food industry ingredients, milk cream, processed milk, noodles, sweeteners, raw pastas, vegetables
3. Ultra-processed Breads, biscuits, burgers, bottled vegetables, cakes, canned fish, cereal bars, chips, chicken nuggets, chocolates, cured meat, fish, hot dogs, ice cream, milk drinks, pastries, pre-prepared meat, pickled meat, poultry, processed meat, sausages, salted meat, smoked fish, sugared fruits, soft drinks, vegetable canned in brine

parchment paper such as butter paper, kraft paper, glassine paper, plastic-coated paper and wax-coated paper. The paperboards include solid fiber boards, box-boards and liner boards (De, 1980). Glass is another type of packaging material that is used in the form of jars, tumblers and bottles. It may be opaque or transparent. Aluminium foil is also used in the form of wrappers, cartons and boxes as it has good barrier properties and is opaque, hygienic, odourless, non-toxic, non-sorptive, shrink and grease-proof. This may be coated with plastic or lacquer to increase corrosion resistance in foil. Tinplate is used in the form of cans as it has excellent barrier properties and is corrosion resistant when lacquered internally for different variety of products. Plastics constitute a wide variety of packaging materials in the modern-day world. It can be used as injection-molded, blow-molded and thermoformed containers in cups, bottles, cartons and others. Due to its low cost and ease of fabrication, the use of plastics in packaging of food items has made a rapid advancement in today's period throughout the world. Although it is highly in use, it has a lot of disadvantages that include low barrier properties, non-heat resistant, fragile at low temperatures and hence, the deterioration of plastics causes great harm to the environment.

A unique variety of "food-grade" plastics is generally used for packaging milk and milk products as this prevents the transfer of any toxic material from package material to the product. Flexible plastic packaging films, which include wrappers, pouches and sachets, can be low polymer plastics like cellophane or high polymer plastics that include polypropylene, polyester and PVC (Table 2.2). Laminates, used as sachets, bags and cartons are made by combining complete surfaces of two or more webs of different films has improved barrier properties that provide a surface that will further strengthen the film material and heat-seal, too. Cellophane-polythene, paper-polythene, polyester-polythene and paper-aluminium foil-polythene are some examples of typical laminates used in food packaging. In food packaging, TiO_2 nanoparticles serve as effective photo-catalysts and antimicrobials. They work against the germs that cause food spoilage by causing lipid peroxidation, which is brought on by the creation of reactive oxygen species (ROS) molecules in the presence of

Table 2.2: Packaging Materials and Forms Used in the Food Industry

Packaging Materials	Forms Used
Paper and Paperboards	Wrappers, Cartons, Boxes
Glass	Bottles, Jars, Tumblers
Aluminium Foil	Wrappers, Cartons, Boxes
Tinplate	Cans
Plastics	Bottles, Cartons, Cups
Flexible Plastic Films	Wrappers, Sachets, Pouches
Laminates	Cartons, Sachets, Pouches

Table 2.3: Shelf Life of Food Products in Cold Storage

SL No.	Food Product	Shelf Life in Cold Storage
1.	Poultry products	3 days
2.	Crustaceans	2 days
3.	Meat products	3–5 days
4.	Cured meat products	2–3 days
5.	Minced meat and offal	2–3 weeks
6.	Seafood	3 days
7.	Fruit juices	7–14 days
8.	Milk	5–6 days
9.	Cream	20 days
10.	Cheese	variable (1–3 months)
11.	Soft cheeses (camembert, brie)	2–3 weeks
12.	Cottage, ricotta, cream cheeses	10 days
13.	Eggs	3–6 weeks
14.	Butter	8 weeks
15.	Oil and fat	variable (6 months)
16.	Margarine	variable (6 months)

light and ultraviolet radiation, thus causing cell death. TiO_2-based bio-nano-composites have been employed as effective packaging materials for a variety of foods that are oxygen-sensitive.

The process of keeping cooked or raw food commodities at appropriate conditions without the entry of microorganism or its multiplication is termed food storage. Basically, storage is of two types: storage at dry and low temperature. On the basis of different temperature, they are further divided into refrigerated, cold and frozen storage when stored at 3 to $10^{\circ}C$, 0 to $3^{\circ}C$ and -20 to $0^{\circ}C$, respectively. Food can be perishable, semi-perishable or non-perishable on the basis of the pace at which they spoil. Usually perishable foods are frozen while semi-perishable and non-perishable foods are stored in dry storage. The expected shelf life of the food will also vary with the type of food and its composition: moisture content, fat, protein, sugar and other nutrient content and storage temperature (Table 2.3). The storage area should be of proper temperature, humidity for prolonged storage, protect stored commodities from pest or microbial manifestation, be kept out of direct sunlight and heat, easily accessible by transport etc. The rotation of stored commodities can obey certain rules, such as first in first out, first manufactured first used or last in last out.

2.4 CHANGES IN STORAGE

The storage in food can be from days to months. This depends on certain parameters and characteristics of the food subjected to storage. The changes that occur in storage can be physical, chemical and microbiological (Steele, 2004). Visible changes occur in color, flavor, odor and texture. The nutritive value subsequently reduces with changes due to enzymatic deterioration of protein, fat, sugar and vitamins in food. The spoilage in storage will reduce the edibility of the food. The moisture content, pH, temperature of storage presence of air, microbes, pests and other chemicals also have a significant role in changes that occur during storage.

2.4.1 Physical Changes

The instability on storage to the physical characteristics can cause physical spoilage. The changes visible can be shrinkage due to loss of moisture, migration of moisture in and out, changes in different components, change in color due to some chemical reactions, physical separation of components, settling down of components etc. It is mainly affected by the moisture content of food, relative humidity and temperature of storage; crystallization of water or other components in food. The moisture transfer will depend on the water activity, water content, relative humidity and temperature of storage. The moisture is lost when the temperature is high and relative humidity is low (Balasubramanian and Viswanathan, 2010). An optimum movement of air and relative humidity along with optimum temperature is adequate for long life of the food package. The fruit and vegetable storage is mainly affected by the storage temperature, since it affects the metabolism of the respiring food, which slows down the ripening and increases the post-harvest life (Kadar, 1989). When respiring foods are subjected to freezing, the cells can be damaged and can cause mechanical injury. It usually happens when the food is frozen to a temperature of 5 to 15°C (Steele, 2004). The glass transition temperature is a value dependent on the physical stability of the food product. At this temperature, the solids present in food as an amorphous glassy matrix to a rubbery structure (Kumar et al., 2020). The water inside the food freezes and forms ice on frozen storage and this is a case of concern since the produced ice crystals can grow in storage, producing damage to the cells of the food and leakage of the contents (Sahagian and Goff, 2019). The growth of ice crystals into a large one can be reduced by the addition of water-binding agents or emulsifiers (Maity et al., 2018). Sugar-rich products also undergo crystallization of the sweetening agent that causes staling of products such as candies, cookies and ice cream.

2.4.2 Chemical Changes

The flavor, appearance and taste can be varied by different chemical and biochemical reactions in food in storage. Microbial growth and their metabolism will vary different parameters in food, such as pH, nutrient content, metabolite production, toxic compound production, oxidation reaction, bitter amino acid production by proteolysis, lipolysis, acid production etc. (Gram et al., 2002). Hence, chemical changes are connected with microbial spoilage. Other than this, discoloration and enzymatic browning can also occur due to various chemical reactions. The oxidation reaction than occurs in stored food is purely chemical in nature and may depend on the temperature. In fresh meat and fish, the basic unit of protein that is an amino acid will oxidize to ammonia in the presence of oxygen. Lipids, along with oxygen, will undergo rancidification, resulting in oiling off, discoloration, off flavor and harmful compound production (Enfors, 2008). The protein-rich food products are susceptible to proteolysis. The hydrolysis of peptide bonds in proteins resulting in the release of peptides and amino acids from them may cause off flavor in the product is termed proteolysis. It occurs in the presence of protease enzymes present in the product or will be of microbial origin (Rogers et al., 2013). The protein is a polymer compound of amino acids that will produce amino acids or upon the action of proteases, which can be bitter peptides. An anaerobic condition, if it persists after proteolysis in the food pack, will result in a mixture of amines, organic acids, indole, ammonia, phenols and sulfur compounds such as H_2S and mercaptans from amino acid. The process is called putrefaction (Butsbach and Danielle, 2010). Putrefaction can occur in protein-rich products such as meat stored at higher temperatures than 15°C that is favorable to the growth of microbes. A non-enzymatic browning called maillard browning occurs at high-temperature processed foods such as breakfast cereals, dry milk and dry egg (Phosanam et al., 2021). The browning is a reaction with the amino acid group of proteins, mainly lysine and the carbonyl group of reducing sugar present. The action can proceed to different pathways according to the pH of food products and will have end products such as methyl ketones, furfural, hydroxyl methyl furfural, carbonyls, aldehydes, ketones, etc. The process gets initiated at a high temperature, but the concentration of these compounds can increase in the time of storage, increasing the intensity of brown color of the food.

The main chemical change that can occur in fruits and vegetables in storage is the hydrolysis of the cell wall containing polysaccharides, called pectin. Pectin is found in dicotyledonous and some monocotyledonous plants. The natural softening of the cell wall on ripening is due to the pectin hydrolysis by innate pectinase enzymes present. The activation of pectinase can also occur while causing damage to fruits and vegetables. The damaged tissue can also activate another enzyme called pectin methyl esterase that de-esterifies the pectin substances in food. It affects mainly jam and jellies, which are rich in pectin (Walter and Taylor, 1991). The presence of metal ions elevates the rate of hydrolysis, since they are stored in glass jars instead of metal containers.

2.4.3 Microbial Changes

The change in microbial flora from the initial amount is of importance since it can cause spoilage in the food in storage. The microbial change can be increased in the amount of spoilage organisms and presence of foodborne pathogens. They may gain entry in food as post-processing contamination and will increase in number within the storage period. The microbial growth will also adversely affect the chemical changes also. The microbes present in them can produce certain enzymes such as protease, lipase and glycolytic enzyme that will hydrolyse protein, fat and sugar, respectively. The factors that are intrinsic to food that can be altered to prevent the growth of microbes include storage temperature, pH and water activity (Tianli et al., 2014). The nutrient present in each food also have a role on the type of organism that grows in it. Proper packaging can also limit the proliferation of organism by providing a stressful condition inside by giving a moisture and air barrier (Veld and Huis, 1996). The microbial changes can be affected by both intrinsic and extrinsic factors. The intrinsic factor, as mentioned earlier, includes nutrient content, oxidation reduction potential, pH and water activity while the extrinsic factors affecting food are microbial activity, air present, temperature and relative humidity of storage, as given in Figure 2.1). The intrinsic factors also include the endogenous enzyme and its substrates and oxygen and light present. Microbial spoilage can occur in food due to three agents, namely bacteria, yeast and mold. They all grow in a different pH, temperature, water activity and they are sensitive to heat. This affects different varieties of food; mold and yeast affect low-pH food and low water activity (0.6) foods like bottled, fermented food. Bacteria can attack food with a higher moisture content like fish, fresh meat, eggs, poultry and mild heated foods; spore formers can be present in pasteurized milk. Effective processing can destroy the spoilage microorganisms and proper packaging and storage can prevent post-processing contamination. The factors can also be maintained to an optimum condition to avoid microbial growth in storage.

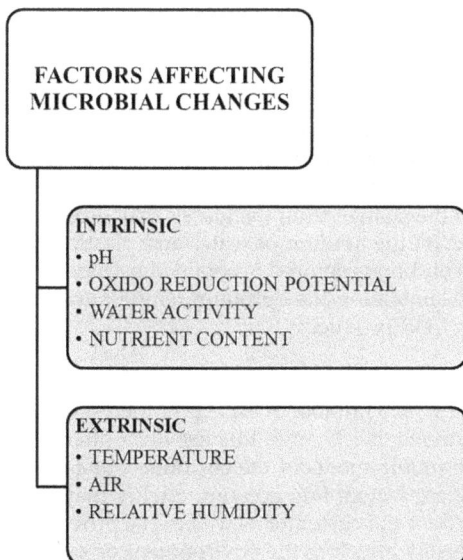

Figure 2.1 Different extrinsic and intrinsic factors affecting the microbial changes in storage of food.

2.5 FACTORS AFFECTING CHANGES IN STORAGE

There are certain factors that will affect the chemical, physical or microbiological changes that happen to the food in storage. They can be the factors inside the food or the factors from the environment, namely intrinsic and extrinsic factors, respectively. The factors are nutrient content, moisture content, temperature, pH, water activity and oxidation reduction potential.

2.5.1 Nutrient Content

Food is an organic substance that will provide certain functions such as body building and repairing, energy giving and protective or regulatory actions that will be rich in carbohydrate, protein, fat, vitamins and minerals. They all provide energy for the body metabolism in the consumers. According to the different nutrients present in food, they are divided into carbohydrate rich, protein rich, fat rich, vitamins, and minerals rich in nature (Chopra, 2005). Their different function also becomes a basis of classification. Many of the nutrients will contain essential amino acids, fatty acids and vitamins that cannot be synthesized by the human body. Hence, the only source for this is food consumption. The high amount of protein in food can have a function of body building and repairing while a high amount of carbohydrate and fat can act as high energy rich high-energy-rich food. The higher amount of nutrients in food can also become a factor for the growth of microbes in them. The bacteria that utilizes fat, protein and carbohydrate can live and multiply in the food on storage. Hence, the exclusion of these bacteria is important from the food package.

2.5.2 Moisture Content

The moisture content in food on storage plays a significant role in deterioration of food. On the basis of moisture content of food, the relative humidity of the storage the water gain or water loss will be factors. The transfer of water from food is directly related to the water activity of the food. Water activity is the ratio of vapor pressure of water in the system to vapor pressure of pure water. At normal temperature, the water activity of food is 1.0; at $-20^{\circ}C$ and $-40^{\circ}C$, the water activity is 0.82 and 0.68, respectively. According to the amount of moisture in each food, it can be divided into high moisture food, intermediate moisture food and low moisture food; accordingly, it can also be called perishable, semi-perishable and non-perishable food. The increase in the amount of moisture makes it favorable for the growth of bacteria and other microorganism. Therefore, it will increase the spoilage due to microbial growth causing off-flavor, moldy appearance and owing to moisture migration defect in its texture and body.

2.5.3 Temperature

The temperature affects significantly the respiring foods, mainly fruits and vegetables (Gao *et al.*, 2020). The temperature affects the ripening in storage and its post-harvesting life. If the storage temperature is lower, it will give slow ripening and increase the post-harvest life. The metabolic rate of respiring food is affected by temperature since different enzymes activate and inactivate at different temperatures. Low-temperature storage is preferable for the storage of respiring foods. Very low temperature, such as freezing temperature, can have undesirable effects like freeze damage. Pectin containing food commodities are more affected by freeze damage. The pectin hydrolysis is initiated on freeze-damaged cell wall structure. The formation of ice crystals in low-temperature freezing will damage the cell wall protection coat and increases saturation of different sugar components in food, thereby increasing the formation of crystals of the same. The glass transition temperature (G_t) is the temperature at which the change from the glassy structure of the solids in food into a rubbery stage. The G_t will depend on the amount of water and plasticizers present in the food (Mahato *et al.*, 2019). The optimum condition required is obtained by modifying the condition and it is called a modified atmosphere. Maintaining the optimum temperature throughout the storage will extend the shelf life of the food product.

2.6 QUALITY CONTROL AND RISK ASSESSMENT

The phrase "quality of a food product" can be correlated with a number of aspects of food products. Crosby (1979) defined quality as 'conformance to requirements'. Though this has been often put under question as products like heroin may satisfy the requirements of the customer but cannot (and should not) be called a quality product. Taking these factors into account, quality may be defined as "A state in which value entitlement is realized for the customer as well as the provider in every aspect of the business relationship without adversely affecting the environment or society"

(Rao, 2010). The best judge of quality is believed to be the consumers; however, the perceptions are usually subjective. They may judge a food product as poor, bad, good or excellent depending upon a number of attributes which often does not communicate the true impression of the quality. Here, the food producer and manufacturer have to understand quality in objective terms, which leads to the identification and quantification of quality parameters on the basis of variables and attributes in order to give an objective sense to the "quality" of the particular food. Food safety is the first requirement of the quality of a food product as an unsafe food, despite having other good qualities, as it is the poorest quality as it is a threat to human health. Thus, food safety may be defined as a subset of food quality (Jongen, 2002).

Food processing is seasonal in nature, both in terms of demand for products and availability of raw materials. Most crops have a well-established harvest time, though demands are often continuous throughout the year. Even in cases of food such as milk that ais available throughout the year, there are peaks and troughs in volume in lean and flush seasons in addition to their chemical and microbial properties (Food Processing Handbook, 2012). Quality programs are often in force in various food industries. A quality program may be defined as a set of activities performed to ensure the compliance to the quality and safety needs of the food product. The most basic and fundamental quality needs are those laid down by regulation and by consumers and customers. A quality system on the other hand is an integrated set of documented activities in relation to food safety and quality and inter-relationships and synergies that exist between them. The objective of a quality system is to provide a basis to act upon to maintain required levels of food safety and quality while taking into consideration costs involved and the well-being of all interested parties. Both quality programs and quality systems are often used in the food industry (Alli, 2003). The term 'total quality management (TQM)' was introduced in North America in the mid-1980s (Dahlgaard et al., 2008). The term is associated with a management approach to quality improvement in Japan to achieve long-term success. The TQM approach embodies both management principles and quality concepts such as customer focus, employee empowerment, leadership, strategic planning, improvement and process management. These principles and concepts were developed in the second half of the 20th century with significant contributions from several recognized experts in the field of quality control. The most widely recognized of these contributions is W. Edward Deming. During the 1980s and his 1990s, many North American companies adopted his TQM approach and developed a framework for using it in their quality management systems to gain competitive advantage in global markets (Alli, 2003 and Tari, 2005).

The probability of adverse effects on health as a result of food hazards and the severity of those effects is a risk. Therefore, risk assessment has two main purposes. If sufficient data are available, the risk is determined from the degree of contamination and frequency at the time of ingestion, and the amount of ingestion (Lammerding et al., 2000). After the initial stage, strategies and action need to be identified to be used to reduce health risks. This typically requires modeling food production, processing and handling, as well as changes in the farm-to-table chain. The steps in processing that are very important to the food safety view, where control measures or intervention can be made to provide the greatest reduction in the risk of foodborne illness, can be identified and acted upon. Therefore, it may be used for CCP identification for HACCP implementations (Forsythe, 2008).

Hazard analysis critical control point (HACCP) is widely recognized as the best method to ensure product safety and is internationally recognized as a tool to control food-related safety risks. There are many activities related to HACCP, but these mainly focus on creating an HACCP plan (Principles 1–5). Although creating the HACCP plan is an essential part of the HACCP process, it is only the beginning and needs to be implemented in the factory to become a working system. This implementation can be a difficult and time-consuming part of the HACCP process and is an area that many food companies struggle with. The transfer of ownership of the plan is important, as it is unlikely that the HACCP study team will be responsible for the daily execution of the HACCP plan in the factory. This is usually done by line operators or supervisors and it is important that they understand and are committed to the role of HACCP. Management control and staff training is essential to ensure supervisors, managers and operators can efficiently implement the HACCP plan. Training should cover topics such as sources of risk, the role of CCPs, the controls and monitoring procedures that individuals are responsible for and documentation requirements. The final element is the maintenance of the HACCP plan. Verification and periodic review should be in place for proper functioning of the system (Khandke et al., 1998).

HACCP can be used by everyone and is an excellent tool for reducing food safety risk. Many companies have not taken full advantage of this. The HACCP process itself is fairly logical and it is the hazard analysis step that can be difficult to get right without the proper expertise, i.e., knowledge of hazards and control measures. Determining critical limits also cause problems. PRPs are essential across HACCP for prevention of cross contamination from the environment or people (Wallace *et al.*, 2012). Just how essential needs to be determined through a hazard analysis and risk evaluation but typically PRPs after any pathogen-reducing step or in any high-risk ready-to-consume product environment will be critical for food safety assurance. Food safety programs (HACCP and PRPs) require ongoing management commitment if they are to be sustainable and authentic. This includes provision of resources and application of all the normal management practices that will provide an essential operating framework. There are many external pressures for using HACCP, but none more important than the real desire to keep consumers safe. Regulatory requirements, media interest, brand protection, and customer requirements are all external drivers for its use. There are many examples of failure to learn from some that may have been prevented (Bertolini *et al.*, 2007; Sun and Ockerman, 2005).

The quality management systems with different quality management tools are most effective for companies to increase their competitiveness. Dr. Edward Deming and Dr. Joseph Juran had studied and implemented the quality management system in companies 60 years ago (Priede, J., 2012). The qualiy management sysem developed by them was a long time ago and they are still an effective approach and the researchers are still developing it into the next sage. Many companies are aiming to become world-class organizations and to achieve business excellence through the strategic implementation of QMSs. Business knowledge has been accumulated that can help to reach the goal. However, a lot more is required, especially top-level support and ground-level leadership in order to plan, develop and deploy the programs based on customer need and expectations. Periodic review and evaluation of the QMSs are also necessary, as a static program can often be translated to a non-performing program (Lona *et al.*, 2013). A flexible and adaptable leadership is critical to any group environment, and it exists at all levels throughout an organization. Research studies (Thite, 2000) have highlighted that essential leadership traits and abilities, such as the ability to manage people, stress, emotions, bureaucracy and communication, are required to ensure success. Charismatic leadership behaviors are identified as among the most critical leadership behaviors in terms of satisfaction. Charismatic leaders attempt to fuse each member's personal goals with the organizational mission that promotes team commitment and cohesiveness leading to improved performance. The world has seen many charismatic leaders in the last century who have made a big impact on the success map (Lona *et al.*, 2013).

2.7 INNOVATIONS IN PACKAGING

Traditionally, the food packages were used as a passive barrier that will protect the food from environment and delay the deterioration of product inside. The modern packaging system not only has the function of containing food, protection and marketing, they will have a significant role in retaining the quality and safety of food in storage and in the supply chain (Ahvenainen, 2003). The innovative packaging is synonymously called active, clever, indicator, intelligent, interactive or smart packaging. But the concepts between them can be entirely different. The active or intelligent packaging should be in agreement with food contact material legislation that includes the evaluation of specific migration limits, overall migration limit and toxicological properties.

2.7.1 Active Packaging

Active packaging interacts with the food and surroundings, thereby increasing the shelf life. Incorporation of components that will absorb moisture, CO_2, CO_2 emitter, O_2 scavenger, antimicrobial substance, preservatives and scavenge ethylene will reassure the reduction of changes in packaging and storage. It is also intended to be a passive barrier to prevent the contamination or changes in the environment, along with the active packaging will have an active action due to the presence of active agents that will absorb, emit or release certain compounds for food preservation (Table 2.4). Moisture absorbers are also termed 'relative humidity controllers', since they reduce the moisture in the head space of packaging, mainly in fresh fruits and vegetable packages and meat products (Yildirim *et al.*, 2018). The package must be designed so as to maintain the properties of active agents, even after processing of the package and should not interfere with the properties of plastic. Usually as desiccants, inorganic substances such as bentonite, calcium sulfate, silica gel and molecular sieves, organic substances like cellulose, fructose, modified starch and sorbitol and also graphene oxide is used (Bovi *et al.*, 2018). Pure unsaturated hydrocarbons are used as ethylene removing systems. Ethylene is a hormone with an important role in ripening, degradation of

Table 2.4: Different Types of Agents Used in Active Packaging

Type of Active Agent	Food Applied	Packaging Materials Applied
Moisture absorber	Fruit and vegetables, meat products	LDPE, PET tray
Ethylene scavenger	Fruits and vegetables, fresh products	Film, sachet
Carbon dioxide absorbers	Fermented foods, fresh products, fruit, coffee	Film, sachet
Carbon dioxide emitter	Fish, fruit products, meat products, processed and precooked products	Cellulose based pads, food-grade packets and films
Oxygen scavenger	Beverages, fresh products	In can sealents and closure coating
Antimicrobial agent	All food products, chilled and frozen food products	Cool bags, paper products
Antioxidant	Cereal products	Film
Phase change material	Frozen, cold stored food, perishable products	Vegetable oil-based packaging material

chlorophyll, embryogenesis, growth cycle, respiration rate, root growth and development. Many of these functions may adversely affect the food quality in storage. Therefore, absorbing the ethylene inside the package can reduce the deterioration rate. The different ethylene-adsorbing agents include activated carbon, cloisite, Japanese Oya clay, montmorillonite, silica and zeolite. Higher amount of CO_2 inside the package also can cause adverse effects to the food. The products with higher microbial activity inside the package, such as cheese, kimchi, soy paste and yogurt, will produce higher amounts of CO_2 inside the package. This can be reduced by using scavengers like activated carbons, calcium hydroxide, potassium hydroxide, silica and zeolites (Han *et al.*, 2018). The microbial activity inside the package can be reduced by the addition of antimicrobial substances such as CO_2, ethanol, salts of acetate, benzoate, propionate, sorbate, metal oxides (CuO, MgO, TiO, ZnO), nanoparticles of metal (Au, Ag, Cu, Pt, Se), antioxidants, O_2 scavengers, bacteriocins and enzymes (Wyrwa *et al.*, 2017).

2.7.2 Intelligent Packaging

Intelligent packaging is defined as the materials or articles that will monitor the food inside the package and the environment surrounding the package. Intelligent packages use the technology as a communication function to enable the decision so as to provide information, warn of any problems, improve safety, guarantee quality and extend shelf life by continuously monitoring the internal and external environmental changes of the packages (Yam, 2012). It uses the metabolites produced on storage inside the package or the external temperature as the parameters to be monitored to find the status of quality, safety and shelf life of the food inside and also tracking and automatic identification (Clodoveo *et al.*, 2021). The package is the source to provide the information of the food since it moves along with the food. The invention of intelligent packaging has led to a more efficient and safer supply chain. The traceability of the food product is important in the HACCP (hazard analysis and critical control point) point of view. The three main technologies that are used in the intelligent packaging system include data carriers, indicators and sensors. The data carriers will efficiently manage the supply chain logistics while indicators and sensors will provide the information related to the quality of food inside the pack. The system can be placed in primary, secondary or tertiary packaging structure (Ghaani *et al.*, 2016). The indicators usually give information about the presence or absence of certain compounds that have resulted in the reaction between two or more compounds inside the food. This is immediately indicated as visual changes such as irreversible color change, giving qualitative or quantitative information (Brizio *et al.*, 2016). The indicators mainly used are freshness indicator, gas indicator and time temperature indicator. The storage temperature and time of storage carries an important role in the shelf life of the product. Hence, the time temperature sensor will provide a valuable support to rectify it. They can provide partial history or full history of the product; hence, they are categorized broadly as two. A full history indicator will provide the information about all the temperatures that can be related to the entire life of the food (Robertson, 2016). It is majorly applied in chilled and frozen food packaging. The principles mainly used for time temperature indicators include physical cage, chemical change, polymerization reaction, enzymatic reaction on the basis of biological activity of bacteria, enzyme ad spore, melting reaction or reaction in the presence of acid, indicated by appearance or disappearance of color and deformation of the pack (Taoukis *et al.*, 2003).

The freshness indicator indicates the spoilage or decay of the product inside the package. A change in the amount of biogenic amines, CO_2, ethanol, glucose, organic acid and volatile nitrogen compounds will indicate the growth of microorganism (Poyatos-Racionero *et al.*, 2018). The increase in amount of CO_2 and decrease in amount of O_2 can also indicate the growth of microbes. The change in gas composition inside the food package can depend on respiration of food product, permeability of the packaging material and the storage condition. Also, there are other gases reduced or increased with the growth of microorganism; these all can be detected using the gas indicators that have the gas scavengers in them. In modified atmosphere packaging, it is an essential requirement to find the package integrity and leaks across the package. Sensors are devices that respond to a physical, chemical or biological property of the food inside the package and provide a quantifiable measurement of mainly pH, humidity, light exposure and temperature. Chemical sensors are the devices that detect the presence of certain biological or chemical activity, substances, gas etc. with the help of a detector that is of biological or chemical in nature. The signal from the detector is either physical or chemical and is converted into an electric signal with the help of a transducer. The analyzed signal is processed and presented by a signal processor. It can be chemical-, biosensor-, electrochemical-based biosensor and edible sensors. The electrochemical sensor will have an electrode as the transduction element. Some of the established systems for chemical detection includes amperometric oxygen sensors, metal oxide semiconductor field-effect transistors (MOSFETs), organic conducting polymers, piezoelectric crystal sensors and potentio-metric carbon dioxide sensors. They are subjected to cross-sensitivity to carbon dioxide and hydrogen sulfide, contamination of sensor membranes and consumption of the analyte and, in most cases, these systems involve destructive analyses of packages (Ghaani *et al.*, 2016). Data carriers do not provide the information of quality of the food inside the package but help in traceability and protection against forging or theft prevention (Muller and Schmid 2019). The main data carries used include barcode labels, QR code labels and radiofrequency identification tags (RFID tags) on tertiary packaging such as containers and pallets.

2.8 RISK ASSESSMENT OF FOOD STORAGE AND PACKAGING

As discussed earlier, the term 'risk' defines the probability of causing adverse health effects on consumers and severity. Risk assessment is an integral component of an HACCP plan and has a pivotal role in the functioning of any food safety or quality system (Forsythe, 2008).

Risk assessment can be divided upon the basis of the type of hazards in food, such as:

a. Physical Risk Assessment

b. Chemical Risk Assessment

c. Microbiological/Biological Risk Assessment

d. Allergen Risk Assessment

Physical risk assessment is involved with the identification of the probability and severity of occurrence of a physical hazard and is often considered to be simpler and studies usually involve those at the farm/production unit level. Chemical risk assessment involves the risk of pesticide residues, antibiotics, heavy metals and toxins in food and usually requires deeper study to assess the risks involved in an objective way. The risk assessment on chemical hazard and biological hazards is still developing and a developing science. The assessments vary in different ways, such as:

a. Microorganisms may multiply or die in food, whereas concentrations of chemical and physical hazards don't change very much.

b. Microbial risks are primarily the result of single exposure, whereas chemical risks are often due to cumulative effects.

c. Microorganisms are rarely distributed in the food homogeneously.

d. Microorganisms can be transmitted by transmission through other agents without the direct ingestion of the food.

e. Immunity varies from person to person and from population to population so the effects are difficult to ascertain (Forsythe, 2008). Among the main chemical contaminants in food are carcinogens and risk assessment methods for carcinogens have evolved with the advancement of scientific knowledge. While former methods allowed only hazard identification and potencies, new techniques facilitate the modes of action and division into genotoxic and epigenetic (non-genotoxic, non- DNA reactive) categories. These provide new opportunities for detailed risk assessment and provides quantitative estimates of risk. The qualitative advice does not provide risk managers with information for prioritizing a "margin of exposure" approach for substances that are both genotoxic and carcinogenic but it has been used now by the World Health Organization and the European Food Safety Authority (Barlow and Schlatter, 2010).

The risk assessment of pesticide residues in food is currently performed on a compound-by-compound basis. If potential exposure of consumers is below the relevant health-based guidance value, the use of the particular pesticide is deemed to be acceptable. However, this often leads to the exposure of consumers to a number of pesticide residues at once time or in a short span of time. In the European Union, in 53–64% of food samples, pesticide residues were not detectable, 32–42% contained detectable residues, which were below the maximum residue levels (MRL) and 3.0–5.5% contained levels above the MRL, respectively. Of note, 14–23% of the samples with detectable residues contained more than one active ingredient. The consequence of such combined exposure has raised concerns amongst both consumers and regulators. This was recognized in the U.S. Food Quality Protection Act (FQPA) of 1996 and more recently in Europe in Regulation (EC) No. 396/2005 on MRLs. This emphasizes the importance "to carry out further work to develop a methodology to take into account cumulative and synergistic effects of pesticides" (Boobis et al., 2008).

Aflatoxins are various toxic carcinogens and mutagens produced by certain molds, particularly *Aspergillus* species. The fungi grow in soil, decaying vegetation, and various staple foods and commodities such as hay, sweetcorn, wheat, millet, sorghum, cassava, rice, chili peppers, cotton-seed, peanuts, tree nuts, sesame seeds, sunflower seeds and various spices. In short, the relevant fungi grow on almost any crop or food. When such contaminated food is processed or consumed, the aflatoxins enter the general food supply. They have been found in both pet and human food, as well as livestock feed. Animals fed contaminated feed can shed aflatoxin metabolites into eggs, dairy and meat. Most of the available data are on AFB1 and information on the other aflatoxins is scarce. AFB1 is readily absorbed and distributed to the liver. In humans, a mutational signature for aflatoxin exposure has been identified in HCC. AFB1 affects reproductive and developmental parameters (i.e., brain development, shortened time to delivery, low birth weight and adverse effects on spermatogenesis and folliculogenesis) at low doses (≥ 4 lg/kg bw per day) and these effects may occur following a short-term exposure. Aflatoxins reduce immunity, particularly at the cellular level (Schrenk et al., 2020).

The term 'microbial risk assessment' is relatively newly applied to microbial food safety issues. The publicaions about microbial risk assessment from 1994 to 1999 had only came from the USA and half of them had already been reviewed. Risk assessments that had been successfully completed were seven that targeted a specific pathogen (Klapwijk et al., 2000).

a. *B. cereus* risk in pasteurized milk.

b. Salmonella in chicken products.

c. *E. coli* O0157:H7 in ground (minced) beef

d. *S. Enteridis* in shell eggs and egg products

e. *L. monocytogenes* data survey

f. Rotavirus in drinking water

g. *Cryptosporidium* in drinking water.

The growth in the research of microbial risk assessment has given rise to the necessity of predictive food microbiology. The field of study that combines mathematics and statistics with microbiology to develop a model that will describe and predict the growth or decline of microorganisms in a certain environment is termed 'predicitive food microbiology' (Whiting, 1995). The factors are mainly temperature of processing, distribution and storage if subjected to any fluctuations and the microbial growth in food is affected. The model that shows the rate of microbial growth is termed a 'kinetic model' and the likelihood of specific pathogens in food is predicted by 'probability' models (Ross and McMeekin, 1994).

The main objective of predictive microbiology is to describe mathematically the growth of micro-organisms in food under prescribed growth conditions. The major factors affecting growth are pH, aw, atmosphere, temperature and pressure. A risk analysis is carried out to identify the factors, components, reason for the risk and establish a clear picture of intensity of risk, thereby finding a perfect solution to the risk. Risk analysis is a matter of concern if regulatory requirements are considered. It contributes to decision making. Risk assessment comes under risk management. Risk analysis and risk evaluation together become a risk assessment. The risk assessments are done using models rather than only monitoring the available data. There are about 35 models currently in use recognized by the USA-EPA for the assessment of risk on exposure. They all are different from one another in their level, scope and purpose of analysis. They can be used for assessing risk on exposure of one pollutant or multiple pollutants. They can be applicable to either human or ecological receptors and also applicable to both of them. There can be models aimed at universal population, subpopulation or individuals concerning chronic, sub-chronic and acute exposures. They are used by government authorities, private companies and researchers in an international level. The recent paper will discuss different models in risk assessment for exposure assessment. A human exposure model is used for assessing the risk of emitted toxins in the air. An interaction occurring between a cause and its target at a particular place, at a particular time is termed 'exposure', while dose is the amount of that cause or an agent crossing the contact boundary.

The model can be selected transport models, exposure models and integrated transport and exposure models. The transport model examines the movement and potential dose in particular units of these so-called risk agents at different environment. The dose referred to here can be a representation of exposure potential of the risk agents in an actual exposure. The exposure model observes the exposure factors, its activity patterns and yield at a general environment or in a microenvironment. The integrated transport and exposure model predict both and doses. The models used by the U.S. EPA can be either screening level or higher-tiered level (Figure 2.2). Most of the models are later types, since the screening level has limited spatial and temporal resolution. A model should be thoroughly reviewed by external expertise and those used for regulatory and research purposes are reviewed by the U.S. EPA's scientific advisory boards, panels or committees. Webb and Davey (1992) called for a terminology for models to give express meaning to the model description and development. Whiting, 1995 proposed a new classification system for PFM according to specific criteria under three levels of primary, secondary and tertiary models.

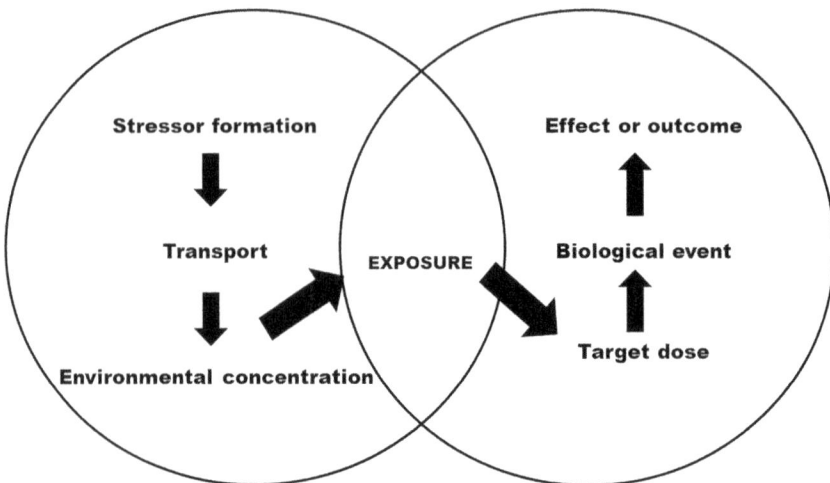

Figure 2.2 A view of response model.

Primary models describe the change in microbial number with time under particular environmental and climatic conditions. The response can be measured directly by total viable count, toxin formation, substrate level or metabolic products or indirectly through absorbance, optical density or impedance. If a bacterial growth curve is monitored by recording how its TVC changes with time, the data collected can be plotted using a primary model. This can then generate information about the microorganism, such as generation time, lag phase duration, exponential growth rate and maximum population density. Secondary models describe the response of one or more parameters of a primary model changing to one or more changes in cultural and environmental conditions (pH, aw, Eh, temperature). These data are then collated using a secondary model, so that the effect of temperature is described by a mathematical equation. This allows the end user to determine what generation time will be observed at a temperature T (Dickson *et al.*, 1992). Tertiary models basically take modeling to its final form. They are applications of one or more primary or secondary models, incorporated into a software. Tertiary models make predictive microbiology easily accessible and powerful to all areas of the food industry and research (Whiting, 1995).

In India, microbiological risk analysis has been conducted by a number of researchers. The study by Kundu *et al.* (2018) estimates illness (diarrhea) risks from fecal pathogens that can be transmitted via fecal-contaminated fresh produce. To do this, a quantitative microbial risk assessment (QMRA) framework was developed in National Capital Region, India, based on bacterial indicators and pathogen data from fresh produce wash samples collected at local markets. Produce wash samples were analyzed for fecal indicator bacteria (*Escherichia coli*, total *Bacteroidales*) and pathogens (*Salmonella*, *Shiga*-toxin producing *E. coli* (STEC), enterohemorrhagic *E. coli* (EHEC)). Based on the *E. coli* data and on literature values for *Cryptosporidium* and norovirus, the annual mean diarrhea risk posed by ingestion of fresh produce ranged from 18% in cucumbers to 59% in cilantro for *E. coli* O157:H7, and was <0.0001% for *Cryptosporidium*; for norovirus the risk was 11% for cucumbers and up to 46% for cilantro. The risks were drastically reduced, from 59% to 4% for *E. coli* O157:H7, and from 46% to 2% for norovirus for cilantro in post-harvest washing and disinfection scenario. The present QMRA study revealed the potential hazards of eating raw produce and how post-harvest practices can reduce the risk of illness. The results may lead to better food safety surveillance systems and use of hygienic practices pre- and post-harvest. The most prevalent pathogenic bacteria isolated were *S. aureus* (3.4 $\log_{10}$ CFU/g) and *B. cereus* (3.4 $\log_{10}$ CFU/g). *Salmonella* spp. was present in salads (3.2 $\log_{10}$ CFU/g) and hand washings of the food handler (3.5 $\log_{10}$ CFU/g). *Salmonella* contamination was found in salads served along with chicken fried rice and chicken noodles than in the food.

2.9 CONCLUSION

The packaging and storage conditions of food products have a very significant important role in its quality. The quality of food relies up on storage conditions such as temperature, relative humidity and atmospheric conditions. An optimum condition is required for obtaining the required shelf life of the food product. The changes occurring in a food inside the package can be physical, chemical or biological. The changes can adversely affect the food quality and hence should be dealt with utmost care. Scientific studies have opened the way for the development of different inventions in packaging and the effectiveness of active and intelligent packaging will dominate in a few years in the food industry. 'Risk' is the term of the probability of an adverse health effect produced by a hazard in the food product. Therefore, risk assessment is an integral component and plays a pivotal role in the functioning of any food safety or quality system. The quality control and risk assessment of food storage and package count on the whole organization personnel participation. An effective association of food industries with research institutes in research development, legislative and commercial functions is required to overcome the challenges of risk.

REFERENCES

Ahvenainen, R. 2003. Active and intelligent packaging: an introduction. In *Novel food packaging techniques*, 5–21. Woodhead Publishing.

Alli, I. 2003. *Food quality assurance: Principles and practices*. CRC Press.

Balasubramanian, S., and Viswanathan, R. 2010. Influence of moisture content on physical properties of minor millets. *Journal of Food Science and Technology* 47: 279–284.

Barlow, S., and Schlatter, J. 2010. Risk assessment of carcinogens in food. *Toxicology and Applied Pharmacology* 243(2): 180–190.

Bertolini, M., Rizzi, A., and Bevilacqua, M. 2007. An alternative approach to HACCP system implementation. *Journal of Food Engineering* 79(4): 1322–1328.

Boobis, A.R., Ossendorp, B.C., Banasiak, U., Hamey, P.Y., Sebestyen, I., and Moretto, A. 2008. Cumulative risk assessment of pesticide residues in food. *Toxicology Letters* 180(2):137–150.

Bovi, G.G., Caleb, O.J., Klaus, E., Tintchev, F., Rauh, C., and Mahajan, P.V. 2018. Moisture absorption kinetics of FruitPad for packaging of fresh strawberry. *Journal of Food Engineering* 223: 248–254.

Brizio, A.P.D.R. 2016 Use of indicators in intelligent food packaging. In *IEEE conference on intelligent transaction system*: 1256–2010.

Butzbach, D.M. 2010. The influence of putrefaction and sample storage on post-mortem toxicology results. *Forensic Science, Medicine, and Pathology* 6: 35–45.

Chopra, P. 2005. *Food & nutrition education*. APH Publishing.

Clodoveo, M.L., Muraglia, M., Fino, V., Curci, F., Fracchiolla, G., and Corbo, F.F.R. 2021. Overview on innovative packaging methods aimed to increase the shelf-life of cook-chill foods. *Foods* 10(9): 2086.

Crosby, P. B.1979.*Quality Is Free: The Art of Making Quality Certain*, New York: McGraw-Hill.

Dahlgaard, J.J., Khanji, G.K., and Kristensen, K. 2008. *Fundamentals of total quality management*. Routledge.

De, Sukumar 1980. Outlines of dairy technology. Oxford University Press, New Delhi, India.

Dickson, J.S., Siragusa, G.R., and Wray Jr, J.E. 1992. Predicting the growth of *Salmonella typhimurium* on beef by using the temperature function integration technique. *Applied and Environmental Microbiology* 58(11): 3482–3487.

Doyle, M.P. 2009. *Compendium of the microbiological spoilage of foods and beverages*. Springer Science & Business Media.

EFSA Panel on Contaminants in the Food Chain (CONTAM), Schrenk, D., Bignami, M., Bodin, L., et al. 2020. Risk assessment of aflatoxins in food. *Efsa Journal* 18(3): 06040.

Enfors S-O. 2008 *Food microbiology*. KTH-Biotechnology.

Forsythe, S.J. 2008. *The microbiological risk assessment of food*. John Wiley & Sons.

Gao, T., Tian, Y., Zhu, Z., and Sun, D.W. 2020. Modelling, responses and applications of time-temperature indicators (TTIs) in monitoring fresh food quality. *Trends in Food Science & Technology* 99: 311–322.

Ghaani, M., Cozzolino, C.A., Castelli, G., and Farris, S. 2016. An overview of the intelligent packaging technologies in the food sector. *Trends in Food Science & Technology* 51: 1–11.

Gram, L., Ravn, L., Rasch, M., Bruhn, J.B., Christensen, A.B., and Givskov, M. 2002. Food spoilage—Interactions between food spoilage bacteria. *International Journal of Food Microbiology* 78(1–2): 79–97.

Han, J.W., Ruiz-Garcia, L., Qian, J.P., and Yang, X.T. 2018. Food packaging: A comprehensive review and future trends. *Comprehensive Reviews in Food Science and Food Safety* 17(4): 860–877.

Huis in 't, and Veld, J.H. 1996. Microbial and biochemical spoilage of foods: An overview. *International Journal of Food Microbiology* 33(1): 1–18.

Jongen, W. ed. 2002. *Fruit and vegetable processing: Improving quality.* Elsevier.

Kader, A.A., Zagory, D., Kerbel, E.L., and Wang, C.Y. 1989. Modified atmosphere packaging of fruits and vegetables. *Critical Reviews in Food Science & Nutrition* 28(1): 1–30.

Khandke, S.S., and Mayes, T. 1998. HACCP implementation: A practical guide to the implementation of the HACCP plan. *Food Control* 9(2-3): 103–109.

Klapwijk, P.M., Jouve, J.L., and Stringer, M.F. 2000. Microbiological risk assessment in Europe: The next decade. *International Journal of Food Microbiology* 58(3): 223–230.

Kumar, P.K., Rasco, B.A., Tang, J., and Sablani, S.S. 2020. State/phase transitions, ice recrystallization, and quality changes in frozen foods subjected to temperature fluctuations. *Food Engineering Reviews* 12: 421–451.

Kundu, A., Wuertz, S., and Smith, W.A. 2018. Quantitative microbial risk assessment to estimate the risk of diarrheal diseases from fresh produce consumption in India. *Food Microbiology* 75: 95–102.

Lammerding, A.M., and Fazil, A. 2000. Hazard identification and exposure assessment for microbial food safety risk assessment. *International Journal of Food Microbiology* 58(3): 47–157.

Mahato, S., Zhu, Z., and Sun, D.W. 2019. Glass transitions as affected by food compositions and by conventional and novel freezing technologies: A review. *Trends in Food Science & Technology* 94: 1–11.

Maity, T., Saxena, A., and Raju, P.S. 2018. Use of hydrocolloids as cryoprotectant for frozen foods. *Critical reviews in food science and nutrition* 58(3): 420–435.

Monteiro, C.A., Levy, R.B., Claro, R.M., Castro, I.R.R.D., and Cannon, G. 2010. A new classification of foods based on the extent and purpose of their processing. *Cadernos de saude publica* 26: 2039–2049.

Müller, Patricia and Schmid, Markus. 2019. Intelligent Packaging in the Food Sector: A Brief Overview. *Foods*, 8, 16. 10.3390/foods8010016.

Phosanam, A., Chandrapala, J., Zisu, B., and Adhikari, B. 2021. Storage stability of powdered dairy ingredients: A review. *Drying Technology* 39(11): 1529–1553.

Priede, J. 2012. Implementation of quality management system ISO 9001 in the world and its strategic necessity. *Procedia-Social and Behavioral Sciences* 58: 1466–1475.

Poyatos-Racionero, Elisa, Ros-Lis, Jose Vicente, Vivancos, José-Luis, and Martínez-Máñez, Ramón. 2018. Recent advances on intelligent packaging as tools to reduce food waste. *Journal of Cleaner Production*, 172, 3398–3409. 10.1016/j.jclepro.2017.11.075.

Rao, Ursula. 2010. Making the Global City: Urban Citizenship at the Margins of Delhi. *Ethnos*, 75, 402–424. 10.1080/00141844.2010.532227.

Robertson, G.L. 2016. *Food packaging: Principles and practice.* CRC Press.

Rocha-Lona, L., Garza-Reyes, J.A., and Kumar, V. 2013. *Building quality management systems: selecting the right methods and tools.* CRC Press.

Rogers, L.D., and Overall, C.M. 2013. Proteolytic post-translational modification of proteins: Proteomic tools and methodology. *Molecular & Cellular Proteomics*, 12(12): 3532–3542.

Ross, T., and McMeekin, T.A .1994. Predictive microbiology. *International Journal of Food Microbiology*, 23, 241–264. 10.1016/0168-1605(94)90155-4.

Sahagian, M.E., and Goff, H.D. 2019. Fundamental aspects of the freezing process. *Freezing effects on food quality*: 2–50.

Steele, R. ed. 2004. *Understanding and measuring the shelf-life of food*. Woodhead Publishing.

Sun, Y.M., and Ockerman, H.W. 2005. A review of the needs and current applications of hazard analysis and critical control point (HACCP) system in foodservice areas. *Food Control* 16(4): 325–332.

Taoukis, P.S., and Labuza, T.P. 2003. Time-temperature indicators (TTIs). *Novel Food Packaging Techniques*: 103–126.

Tarí, J. 2005. Components of successful total quality management. *The TQM Magazine* 17(2): 182–194.

Thite, M. 2000. Leadership styles in information technology projects. *International Journal of Project Management* 18(4): 235–241.

Tianli, Y., Jiangbo, Z., and Yahong, Y. 2014. Spoilage by *Alicyclobacillus* bacteria in juice and beverage products: Chemical, physical, and combined control methods. *Comprehensive Reviews in Food Science and Food Safety* 13(5): 771–797.

Wallace, C.A., Holyoak, L., Powell, S.C. and Dykes, F.C. 2012. Re-thinking the HACCP team: An investigation into HACCP team knowledge and decision-making for successful HACCP development. *Food Research International* 47(2): 236–245.

Walter, R.H. 1991. *Analytical and graphical methods for pectin*, 189–225. Academic Press.

Webb, Katie and Davey, Graham C.L.1992. Disgust sensitivity and fear of animals: Effect of exposure to violent or revulsive material. *Anxiety, Stress & Coping* 5, 329–335. 10.1080/106158092 08248369.

Whiting, R.C. 1995. Microbial modeling in foods. *Critical Reviews in Food Science & Nutrition* 35(6): 467–494.

Wyrwa, J., and Barska, A. 2017. Innovations in the food packaging market: Active packaging. *European Food Research and Technology* 243: 681–1692.

Yam, K.L. 2012. Intelligent packaging to enhance food safety and quality. In *Emerging food packaging technologies*, 137–152. Woodhead Publishing.

Yildirim, S., Röcker, B., Pettersen, M.K., et al. 2018. Active packaging applications for food. *Comprehensive Reviews in Food Science and Food Safety* 17(1): 165–199.

3 Chromatography Coupled with Tandem Mass Spectrometry Application in the Assessment of Food Contaminants and Safety

Meng-Lei Xu and Yu Gao

3.1 INTRODUCTION

Today, mass spectrometry (MS) is considered the gold standard and the preferred approach for the analysis of contaminants in food. MS is coupled with various separation techniques, such as gas chromatography (GC) and liquid chromatography (LC), to detect physical, chemical, and biological contaminants, even at low concentrations (Xu et al., 2021). Since the 1970s, GC has been coupled with MS for the detection of food contaminants. Over the years, the technology behind the GC–MS technique has matured through optimization. Among the various contaminants in food, the GC-MS technique can detect volatile organic compounds (VOCs) and semi-volatile organic compounds (SVOCs). Alkanes, phthalate esters, polycyclic aromatic hydrocarbons, and polychlorinated biphenyls are some of the VOCs in food that can be detected by GC-MS. SVOCs, on the other hand, include pesticides, endocrine disrupters (ERs), and carcinogens. GC-MS method can also be used to target emerging contaminants such as brominated flame retardants and perfluorochemicals. LC-MS can also be used for the detection of thermally unstable chemical contaminants and pollutants with high polarity or high molecular weight. Proteins can also be analyzed using the LC-MS technique (Picó et al., 2015). There are some excellent reviews covering the LC-MS detection of food contaminants including agricultural chemical residues, veterinary drug residues, transformation and degradation products of these residues, and packaging migration substances (Santos Pereira, C. Cunha, and Fernandes, 2019; Chang et al., 2022; Malachová et al., 2014).

Contaminants detection follows three main steps: contaminants isolation and concentration, separation, and identification. Generally, no pretreatment is considered the best sample pretreatment, especially for detecting trace pollutants. With the development of MS technology, it is great to see two emerging trends. First, high-resolution MS can infer the composition of elements and accurate material mass, becoming a crucial tool for identifying unknown substances. Second, MS/MS with high separation can obtain the secondary mass spectrum, which provides high sensitivity (Morales-Gutiérrez et al., 2014). In addition, multi-residue analysis methods and adsorption extraction are characterized by being fast, simple, inexpensive, effective, robust, and safe. At the same time, the number of trace organic pollutants worldwide is increasing, many of which are considered emerging pollutants. Future multi residue methods should take more account of emerging pollutants extraction and cleanup methods (Xu et al., 2021). Mass spectrometers with MS/MS separation capabilities include triple quadrupole, ion trap, linear ion trap, and quadrupole time-of-flight (TOF). The need for sophisticated sample cleanup and tedious analytical techniques can be eliminated with MS/MS-based multi-contaminant methods. MS/MS can be operated in multiple reaction monitoring (MRM) mode when the number of contaminants to be analyzed in samples is high. Multi-contaminant analytical approaches can provide high sensitivity and specificity. They are becoming the new trend in food contaminant detection (Xu et al., 2021).

Current food contaminants and residue analysis trend involves applying the TOF-MS system. This trend arises from the user-friendly and cheaper nature of the TOF-MS system compared to other high-resolution and high-accuracy MS techniques. This aspect is especially crucial for analyzing complex food matrices. LC-TOF-MS system offers several advantages. First, the LC-TOF-MS technique can detect several compounds at a time without having to reduce the sensitivity. Second, the LC-TOF-MS approach can identify unknown peaks through accurate mass number and isotopic abundance evaluation. Third, data can be obtained for additional compounds not involved in the previous detection via post-processing (Saito et al., 2012). The GC-TOF-MS system can also be applied in food pollutants and residue analysis and can further be used to determine the composition of elements (Elbashir and Aboul-Enein, 2018). In GC-TOF-MS techniques, the specificity of identifying unknown objects has also been enhanced. High-quality resolution and accurate quality monitoring of the substance to be measured can effectively reduce the noise of chemical substances from various sources, thus improving the limit of quantitation (LOQ) (Wang et al., 2013). Food with complex matrix and different processing technologies raise the quality and safety concerns. This chapter aimed to summarize the application of chromatography coupled with MS in food contaminants assessment and food safety (Xu et al., 2022).

DOI: 10.1201/9781003334859-3

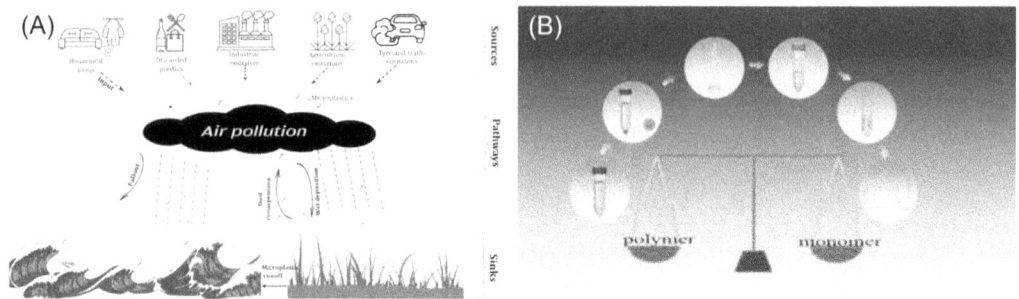

Figure 3.1 Sources (A) and determination by liquid chromatography tandem mass spectrometry (B) of microplastics.

Source: Mbachu *et al.*, 2020; Wang *et al.*, 2022.

3.2 PHYSICAL CONTAMINANTS DETECTION

Physical contamination of food is the inadvertent incorporation of impurities during production and processing, or radionuclides exceeding their limits. Plastics are widely used man-made polymers. Approximately 79% of the plastic that is used ends up in landfills or in the environment, leading to the widespread occurrence of microplastics (MPs). Since the COVID-19 outbreak, the large-scale use of face masks has been part of a complex infection prevention strategies. While, plastic polymers and additives such as phthalates are found in face masks (Zuri *et al.*, 2022). Large amounts of plastic can end up in the environment if mask disposal is inappropriate (Saberian *et al.*, 2021). Biodegradable plastics account for more than 64% of the global bioplastic production capacity. Polylactic acid (PLA) takes several years to a degree *in situ*, which may result in periodic accumulation of MPs in the environment. Since biodegradable plastics are more susceptible to degradation forces, more MPs could be generated from biodegradable plastics than from non-degradable feedstocks over the same period (Sintim *et al.*, 2020). PLA can be hydrolyzed under certain conditions. A quantification method for PLA MPs was developed by efficient depolymerization of PLA by LC-MS/MS with a LOQ of 18.7 ng/g (Wang *et al.*, 2022) (Figure 3.1B). Polyamide (PA), also namely nylon, a widely used plastic. MPs were developed by efficient depolymerization of PA6 and PA66 and determination of the emerging building block compounds, namely 6-aminocaproic acid (6-ACA) and adipic acid (AA), by LC-MS/MS. The results showed that with concentrations of 0.725–321 mg/kg, PA MPs were widely detected in fish guts and gills (Peng *et al.*, 2020). MS/MS techniques possess higher detection sensitivity than other methods, such as fluorescence microscopy, polarized light microscopy, and Fourier-transform infrared spectroscopy. However, information on the particle size distribution of MPs cannot be obtained by MS/MS approaches (Mbachu *et al.*, 2020) (Figure 3.1A).

3.3 CHEMICAL CONTAMINANTS DETECTION

Chemical contaminants in food always come from agrochemicals, and that arises from food processing and packaging (Xu *et al.*, 2014). These contaminants represent a major food safety concern, and several papers have reported synthesis strategies for their detection using chroma-tography coupled with MS/MS (Zhang *et al.*, 2023). Chemical contaminants can pollute food through direct pollution, brief pollution, food chain enrichment, cross-contamination, and accident (Xu *et al.*, 2017). Most food matrices are affected by chemical pollutants. The chemical contaminant residues detected by chromatography coupled with MS/MS methods in different food categories may vary, especially for the pre-treatment method. For pesticide residues, for example, several reports focused on vegetables and fruits than on other food items due to the use of plat protection product for pests and diseases control. Herein, detection method and pre-processing techniques are summarized according to the volatility of chemical pollutants, which can be divided into VOCs, SVOCs, and non-VOCs.

3.3.1 VOCs Chemical Contaminants Analysis

According to food matrices and chemical properties of pesticide residues, chromatography coupled with MS/MS methods in conjunction with a suitable sample preparation procedure can realize multiple residue detection. Several preparation methods, including solid-phase micro extraction (SPME), have been developed for VOCs detection in food. In general, SPME can be carried out in three different modes: direct immersion (DI-SPME), headspace (HS), and membrane-protected SPME. DI-SPME, as the name suggests, involves direct immersion of the SPME fiber in the sample for the extraction and concentration of target analytes (Hu *et al.*, 2022). With the potential to reduce matrix interference for food substrates, HS extraction is a promising technique frequently used for volatile and semi-volatile compounds in complex substrates (Yue *et al.*, 2020). The most commonly used techniques for separating a portion of volatiles matter from the non-volatiles are SPME and stir bar sorptive extractions (SBSE). VOCs are separated on a GC system. Mass spectrometry (MS, MS/MS, TOF-MS) is used to identify the chemical compounds (Starowicz, 2021). According to the available information, fast and non-solvent extractive techniques, such as SPME and SBSE, are the most popular techniques (Xu *et al.*, 2021).

3.3.2 SVOCs or Non-VOCs Compounds Analysis

Pesticides, ERs, carcinogens such as mycotoxins, and process contaminants are common SVOCs or non-VOC compounds. These compounds can cause cancer or impaired neurodevelopment in humans. As a general practice, the ideal techniques for sample preparation should be clean, selective, time-saving, inexpensive, easy to use, and environmentally friendly. For accurately detection of contaminants in multi-residue analysis, MS/MS is preferred over MS technology. The importance of preparation selectivity can be replaced by high separation in MS/MS. Therefore, a new trend is to extract multiple analytes at the same (Xu *et al.*, 2021). For this reason, multi-residue preparation techniques, e.g., QuEChERs, adsorption extraction using a combination of SBSE and single-drop micro extraction, are the most widely used, rapid, and solvent-free extraction approaches.

A new approach has been developed using smartphone-based imaging and image processing to estimate the acrylamide content in homemade toast made from white bread in a simple and fast way. In addition, in order to study the relationship between the concentration of acrylamide and color parameters of certain food products, a colorimetric analysis technique has been implemented (Sáez-Hernández *et al.*, 2022). Smartphones can provide vital information to the user with just a picture, making them a user-friendly tool for food safety. These findings are promising for the future development of a smartphone-based application that can provide an orientation to the consumer in an in situ, green, and easy way (Figure 3.2). However, the efficiency of extracting from problematic food matrices needs to be further improved. At the same time, general improvements in laboratory equipment and the miniaturization of the analytical equipments have led to a wider range of applications.

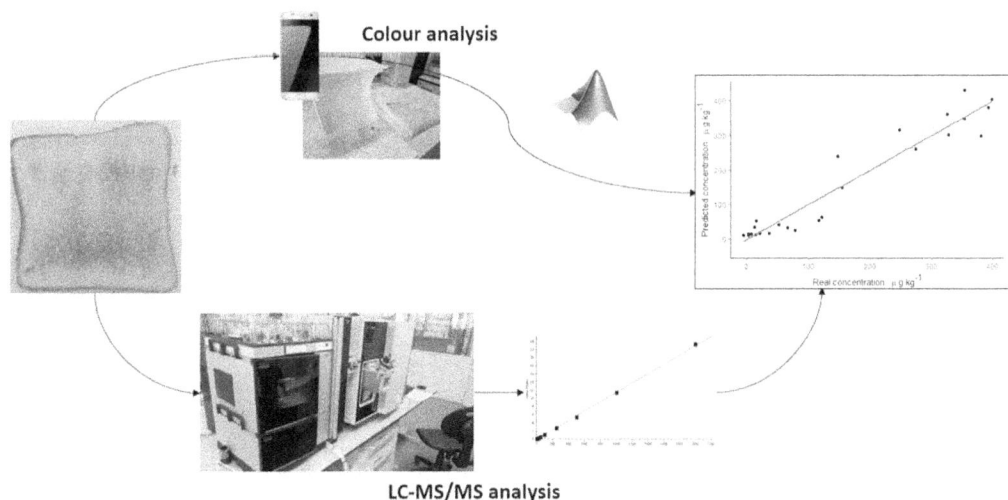

Figure 3.2 Smartphones as user-friendly tools in food safety compared to liquid chromatography tandem mass spectrometry.

Source: Sáez-Hernández *et al.*, 2022.

3.4 BIOLOGICAL CONTAMINANT DETECTION

Biological contaminants of food include microorganisms and their toxic metabolites, parasites and their eggs, viruses, and insect vectors. Microbial and toxic contaminants are the most common types of contaminants and are considered to contributors to food safety. Although we focus on the applications of MS in the analysis of biological contaminants in food, while, many target analytes are still chemical contaminants. For example, mycotoxins are indeed secondary metabolites of certain microorganisms, but the toxins themselves are of a chemical nature. Similar to mycotoxins, food allergens also fall under the category of chemical hazards; however, they are bioactive and with larger molecular weight than ordinary chemical pollutants, thus their mass spectrometry analysis is more complex.

3.4.1 Mycotoxins Analysis

Mycotoxins are secondary metabolites produced by certain fungi like *Aspergillus*, *Penicillium*, *Fusarium*, and *Alternaria* (Yang *et al.*, 2020). Mycotoxins detection strategies share a resemblance with the detection of chemical contaminants. Isolation and concentration are the first steps, followed by separation and identification. Increasing attention is also being paid to the simultaneous occurrence of several mycotoxins. MS techniques have become reliable tools for mycotoxin analysis in different food matrices because of their excellent sensitivity and selectivity. Efficient sample pre-treatment procedures are also needed, including extraction, purification, detection, and quantitation. Based on the best we know, LC techniques have been used for detecting mycotoxins with florescent properties such as aflatoxins and ochratoxins. Due to its excellent sensitivity, selectivity, accuracy, and precision, GC is often used for identification mycotoxins that have no chromophore and fluorescent groups, or that have weakly fluorescent and weakly absorbent groups (Xu *et al.*, 2012). However, the application of GC in the detection of mycotoxins is limited by the relatively slow analytical speed of GC. On the other hand, MS/MS has become the gold standard for the simultaneously separation and analyzing of multiple mycotoxins, in particular LC-MS/MS (Pallarés *et al.*, 2022). SPE and immunoaffinity columns (IACs) are the most commonly used techniques to purify or cleanup for mycotoxins. Moreover, IACs remain the most popular approach for mycotoxin analysis due to their high degree of selectivity and their ability to achieve the maximum possible removal of matrix interferences (Pallarés *et al.*, 2022). However, there are still some mycotoxins without a regulatory limit but with some evidence of toxicity. These mycotoxins are known as "emerging mycotoxins" and include some fusarium toxins such as enniatins and beauvericin (Gruber-Dorninger *et al.*, 2017). Due to their non-polar nature, the non-aqueous capillary electrophoresis method coupled to Q-TOF-MS has been proposed for their identification with LOQ from 4.0 to 8.3 µg/kg depending on the emerging mycotoxin in concern (Delgado-Povedano *et al.*, 2023) (Figure 3.3).

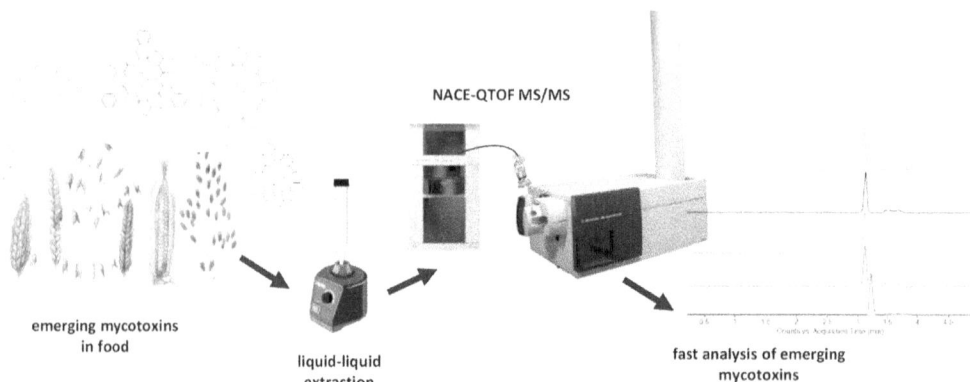

Figure 3.3 Non-aqueous capillary electrophoresis method coupled to quadrupole time-of-flight mass spectrometer applied to emerging mycotoxin detection.

Source: Delgado-Povedano *et al.*, 2023.

3.4.2 Food Allergen Analysis

Food allergy refers to a disorder of the immune system caused by antigens in food. It is mediated by antibodies of the immunoglobulin E type (Tuzimski and Petruczynik, 2020). Around 90% of allergic reactions occur to milk and dairy products, eggs and egg products, fish and crustaceans, peanut, leguminous plants, nuts, and wheat (Xi and Yu, 2020). Small changes in allergen levels in food can lead to variations in allergenicity, with potentially life-threatening consequences for allergy sufferers. It is particularly important to develop sensitive and effective tests for these food allergens. Enzyme-linked immunosorbent assay (ELISA) kits utilizing polyclonal antibodies are currently used for the detection of food allergens. However, accurate quantitation of food allergens cannot be ensured by ELISA alone. Recently, the use of LC-MS or LC-MS/MS techniques, rather than routinely performed ELISA, has become established for quantitating allergens in various matrices. A full MS/MS fragmentation spectrum for a given parent ion allows all the product ions to be monitored at the same time with a high degree of accuracy and resolution (Koeberl et al., 2014). Some signature peptides, e.g., GGLEPINFQTAADQAR have been confirmed and synthesized as quantitative peptides of proteins, e.g., ovalbumin. The relative isotope-labeled internal standards were used for quantitative analysis (Fan et al., 2023). More methods are being developed, including multi-allergen assays, for the co-determination of several major food allergens from complex processed food matrices (Korte and Brockmeyer, 2017). Accordingly, high-performance LC and nano-liquid chromatography-tandem mass spectrometry have been used for determining the intact protein or signature peptides after protein digestion (Fan et al., 2022).

3.4.3 Food Poisoning Analysis

Food poisoning includes non-infectious primary causes, such as consumption of foods containing biological and chemical toxins, as well as sub-acute illnesses. Poisoning can be caused by bacterial, fungal, chemical, or toxic animal and plant sources. Fungal and chemical food poisoning are caused by a variety of chemical or biological contaminants and were covered in previous chapters. Phytotoxins are a class of naturally organic substances. They have high biological activity and are toxic. Toxic plants like toadstools, cassava, green beans, sprouted potatoes, and fresh daylilies can be poisonous when improperly prepared. Legumes contribute significantly to phytotoxicity, containing major toxins such as plant erythrocyte lectin, trypsin inhibitor, saponin glycosides, and phytic acid. Grains from the nightshase (*Solanaceae*) and the lily family (*Liliaceae*) may be the cause of poisoning via their secondary metabolites. Specific animals' species or tissues are often responsible for animal food poisoning. Common examples include puffers, crustaceans, animal food poisoning, liver, tetrodotoxin, crustacean toxins, and ichthyotoxin. The sources of these phytotoxins or toxic food intoxications of animal origin are still chemicals; therefore, sharing similar detection approaches with chemical contaminants.

3.4.4 Food Authenticity Analysis

Another crucial research topic is food authenticity. Food adulteration can pose a serious health threat to consumers by using cheaper materials for economic gain. With the discovery of several major food adulteration incidents over the past decade, a key issue is the authentication of agro-food products. A significant incident was the contamination of chilli powder with dye in 2005 and the identification of the adulterated chilli powder exported from India to Britain in a Worcestershire sauce (Lohumi et al., 2015). The unrivaled specificity, accuracy, and sensitivity of MS techniques has led to their increasing use in food authentication and traceability. These characteristics are critical for establishing analytical strategies to detect food fraud and adulteration by monitoring selected constituents in food matrices. MS approaches are increasingly consolidating protein and peptide profiling (Valletta et al., 2021). Fish gelatins in seven commercial cyprinids, namely, black carps, grass carps, silver carps, bighead carps, common carps, crucian carps, and Wuchang breams, were analyzed using high-performance LC-MS/MS (HPLC-MS/MS) (Sha et al., 2023).

3.5 PROCESSED FOODS ANALYSIS

Specifically, food-processing contaminants of animal-based foods are mainly chemical pollutants formed when the components undergo chemical changes. Major food processing contaminants include N-nitrosos, polycyclic aromatic hydrocarbons, acrylamide, and heterocyclic amines. The mechanisms by which they form are known, but relevant reports remain sparse. Several regulatory analytical methods are used to determine nitrosamines (Gopireddy et al., 2022). A fully automated

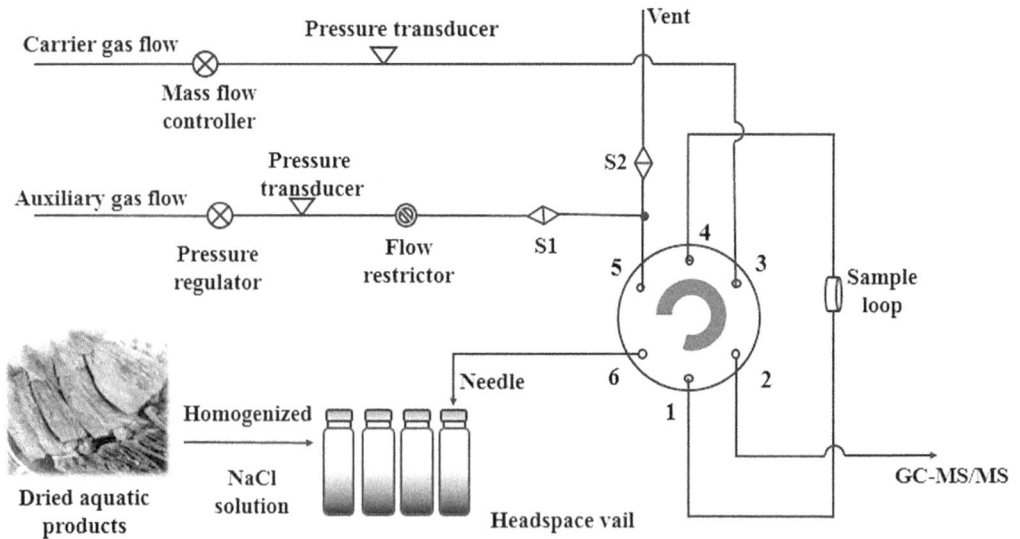

Figure 3.4 Fully automated analytical platform based on static headspace-gas chromatography-tandem mass spectrometry for the analysis of five N-nitrosamines in dried aquatic products of animal origin.

Source: Huang *et al.*, 2022.

one-step static HS sampling and GC-MS/MS can be used to analyze N-nitrosamines (LODs: 0.08–0.29 µg/kg) in dried aquatic products of animal origin (Huang *et al.*, 2022) (Figure 3.4).

Edible oil can pose a risk to human health if consumed or stored improperly, and is classified as either vegetable or animal oil depending on its source. Common oil quality and safety issues rancidity, trans fatty acids and the presence of glucosinolate, erucic acid and gossypol residues. These quality and safety problems can be analyzed through the detection of associated chemical substances. Malondialdehyde, for example, is a biomarker of lipid peroxidation. It has traditionally been associated with rancidity. Nevertheless, waste cooking oil from restaurants and street vendors-known as 'recycled cooking oil' contains considerable amounts of endogenous pollutants. Magnetic solid phase extraction coupled to ultra HPLC-MS/MS offers benefits of low LOD, broad linear range and fast throughput. In addition, the sorbent is easy to synthesize; the adsorption process is convenient, fast, and efficient; and the sorbent can be reused at least ten times without significant loss of recovery (Lu *et al.*, 2020). Traditional condiments that may contain harmful substances produced during their manufacture include sauces, vinegar, monosodium glutamate, and sugar. Chloropropanol is formed when hydrochloric acid is used to hydrolysis vegetable protein and is found in raw soy sauce, aged soy sauce, and oyster sauce. Four chloropropanols in paper straws were detected using matrix solid-phase dispersion coupled with GC-MS/MS with a LOD of 0.200 µg/L (Yuan *et al.*, 2022).

3.6 PROTEIN IN GENETICALLY MODIFIED FOODS

Advances in biotechnology have made it possible to grow genetically modified crops containing multiple beneficial traits, known as stacked trait products, to control plant diseases, pests, and weeds (Hill *et al.*, 2017). Genetically modified (GM) foods are a big concern globally due to the lack of information concerning their safety and health effects (Vidal *et al.*, 2015). The introduction of GM technology into the agricultural system requires due diligence and a thorough analysis of the associated risks and/or benefits associated with it for a variety of stakeholders on a case-by-case basis prior to its commercialization (Sendhil et al., 2022). Evaluation of the genetically engineered crop to determine whether the transgene insertion has resulted in unintended changes to the endogenous allergen profile, particularly in allergenic foods such as soy (Organisms, 2010).

Historically, the quantitative measurement of transgenic proteins in plant matrices has relied on immunochemistry techniques such as ELISA (Gu *et al.*, 2017). Although the methodology is highly accurate, it still lacks control tissue for confidently assessing antibody quality and needs high-quality protein standards. Protein analysis can take label-free or labeled approaches. Quantification of the herbicide resistance gene-related protein 5-enolpyruvylshikimate-3-phosphate by MRM in LC-MS/MS as a label-free method (Devi *et al.*, 2018). Tagged LC-MS/MS was developed and validated for the determination of endogenous Zea mays 14 in corn kernels surrogate peptide method. It is useful to document the natural variability of endogenous allergens such as Zea m 14 to understand the variability within maize grain, although the value of measuring endogenous allergen levels in GM crops has been questioned (Hill *et al.*, 2017).

3.7 CONCLUSION

To the best of our knowledge, most available studies on food contaminants detection focused on chemical and biological contaminants in raw materials. In addition, there is little data on the impact of the infusion process on pollutants and the migration of mycotoxins from raw materials into the surrounding food and beverage products. In the absence of GC-MS, LC-MS is the preferred instrument for food contaminants detection. Their relationship, however, is complementary rather than competitive because they have their specific scope of work. The development of mass spectrometry is not as significant for GC-MS, while triple quadrupole techniques and their coupling with GC systems have brought several advantages in multi-residue analysis. For multi-residue analysis, the application of liquid chromatography has a significantly increasing trend. This development of LC-MS is closely linked to this trend. Therefore, accurate quantitative results for a group of selected compounds can be obtained by MS/MS analysis of target contaminants. Moreover, TOF-MS can identify non-target and unknown compounds. MS/MS application plays a crucial role in the effective assessment of food contaminants and ensuring food safety, thereby serving as a vital tool in food quality and safety control.

REFERENCES

Chang, Jia, Jianhua Zhou, Mingyang Gao, Hongyan Zhang, and Tian Wang. 2022. "Research advances in the analysis of estrogenic endocrine disrupting compounds in milk and dairy products." *Foods* 11 (19). 10.3390/foods11193057.

Delgado-Povedano, María del Mar, Francisco J. Lara, Laura Gámiz-Gracia, and Ana M. García-Campaña. 2023. "Non-aqueous capillary electrophoresis–time of flight mass spectrometry method to determine emerging mycotoxins." *Talanta* 253:123946. 10.1016/j.talanta.2022.123946.

Devi, Shobha, Yi-Cheng Lin, and Yen-Peng Ho. 2018. "Quantitative analysis of genetically modified soya using multiple reaction monitoring mass spectrometry with endogenous peptides as internal standards." *European Journal of Mass Spectrometry* 25 (1):50–57. 10.1177/1469066718802548.

Elbashir, Abdalla Ahmed, and Hassan Y. Aboul-Enein. 2018. "Application of gas and liquid chromatography coupled to time-of-flight mass spectrometry in pesticides: Multiresidue analysis." *Biomedical Chromatography* 32 (2):e4038. 10.1002/bmc.4038.

Fan, Sufang, Junmei Ma, Chunsheng Li, Yanbo Wang, Wen Zeng, Qiang Li, Jinru Zhou, Liming Wang, Yi Wang, and Yan Zhang. 2022. "Determination of tropomyosin in shrimp and crab by liquid chromatography–tandem mass spectrometry based on immunoaffinity purification." *Frontiers in Nutrition* 9. 10.3389/fnut.2022.848294.

Fan, Sufang, Junmei Ma, Zhuo Liu, Yawei Ning, Meicong Cao, Qiang Li, and Yan Zhang. 2023. "Determination of egg and milk allergen in food products by liquid chromatography-tandem mass spectrometry based on signature peptides and isotope-labeled internal standard." *Food Science and Human Wellness* 12 (3):728–736. 10.1016/j.fshw.2022.09.006.

Gopireddy, Ramana Reddy, Arthanareeswari Maruthapillai, and Sudarshan Mahapatra. 2022. "A multi-analyte LC–MS/MS method for determination and quantification of six nitrosamine impurities in sartans like Azilsartan, Valsartan, Telmisartan, Olmesartan, Losartan and Irbesartan." *Journal of Chromatographic Science*: bmac059. 10.1093/chromsci/bmac059.

Gruber-Dorninger, Christiane, Barbara Novak, Veronika Nagl, and Franz Berthiller. 2017. "Emerging mycotoxins: Beyond traditionally determined food contaminants." *Journal of Agricultural and Food Chemistry* 65 (33):7052–7070. 10.1021/acs.jafc.6b03413.

Gu, Xin, Thomas Lee, Tao Geng, Kang Liu, Richard Thoma, Kathleen Crowley, Thomas Edrington, Jason M. Ward, Yongcheng Wang, Sherry Flint-Garcia, Erin Bell, and Kevin C. Glenn. 2017. "Assessment of natural variability of maize lipid transfer protein using a validated sandwich ELISA." *Journal of Agricultural and Food Chemistry* 65 (8):1740–1749. 10.1021/acs.jafc.6b03583.

Hill, Ryan C., Xiujuan Wang, Barry W. Schafer, Satyalinga Srinivas Gampala, and Rod A. Herman. 2017. "Measurement of lipid transfer proteins in genetically engineered maize using liquid chromatography with tandem mass spectrometry (LC-MS/MS)." *GM Crops & Food* 8 (4):229–242. 10.1080/21645698.2017.1349602.

Hu, Cong, Yuan Zhang, Yu Zhou, Zhi-fei Liu, and Xue-song Feng. 2022. "Unsymmetrical dimethylhydrazine and related compounds in the environment: Recent updates on pretreatment, analysis, and removal techniques." *Journal of Hazardous Materials* 432:128708. 10.1016/j.jhazmat.2022. 128708.

Huang, Minxing, Qiuxia Zeng, Zhipeng Liu, Xiaochu Chen, Yufeng Gao, Guihua Wang, and Goubin Yu. 2022. "Development of a fully automated analytical platform based on static headspace-gas chromatography-tandem mass spectrometry for the analysis of five N-nitrosamines in dried aquatic products of animal origin." *Journal of the Science of Food and Agriculture* 102 (15):7107–7114. 10.1002/jsfa.12072.

Koeberl, Martina, Dean Clarke, and Andreas L. Lopata. 2014. "Next Generation of Food Allergen Quantification Using Mass Spectrometric Systems." *Journal of Proteome Research* 13 (8):3499–3509. 10.1021/pr500247r.

Korte, Robin, and Jens Brockmeyer. 2017. "Novel mass spectrometry approaches in food proteomics." *TrAC Trends in Analytical Chemistry* 96:99–106. 10.1016/j.trac.2017.07.010.

Lohumi, Santosh, Sangdae Lee, Hoonsoo Lee, and Byoung-Kwan Cho. 2015. "A review of vibrational spectroscopic techniques for the detection of food authenticity and adulteration." *Trends in Food Science & Technology* 46 (1):85–98. 10.1016/j.tifs.2015.08.003.

Lu, Qing, Hao Guo, Dezeng Li, and Qingbiao Zhao. 2020. "Determination of capsaicinoids by magnetic solid phase extraction coupled with UPLC-MS/MS for screening of gutter oil." *Journal of Chromatography B* 1158:122344. 10.1016/j.jchromb.2020.122344.

Malachová, Alexandra, Michael Sulyok, Eduardo Beltrán, Franz Berthiller, and Rudolf Krska. 2014. "Optimization and validation of a quantitative liquid chromatography–tandem mass spectrometric method covering 295 bacterial and fungal metabolites including all regulated mycotoxins in four model food matrices." *Journal of Chromatography A* 1362:145–156. 10.1016/j.chroma.2014.08.037.

Mbachu, Oluchi, Graham Jenkins, Chris Pratt, and Prasad Kaparaju. 2020. "A new contaminant superhighway? A review of sources, measurement techniques and fate of atmospheric microplastics." *Water, Air, & Soil Pollution* 231 (2):85. 10.1007/s11270-020-4459-4.

Morales-Gutiérrez, F. J., M. P. Hermo, J. Barbosa, and D. Barrón. 2014. "High-resolution mass spectrometry applied to the identification of transformation products of quinolones from stability studies and new metabolites of enrofloxacin in chicken muscle tissues." *Journal of Pharmaceutical and Biomedical Analysis* 92:165–176. 10.1016/j.jpba.2014.01.014.

Organisms, Efsa Panel on Genetically Modified. 2010. "Scientific Opinion on the assessment of allergenicity of GM plants and microorganisms and derived food and feed." *EFSA Journal* 8 (7):1700. 10.2903/j.efsa.2010.1700.

Pallarés, N., J. Tolosa, E. Ferrer, and H. Berrada. 2022. "Mycotoxins in raw materials, beverages and supplements of botanicals: A review of occurrence, risk assessment and analytical methodologies." *Food and Chemical Toxicology* 165:113013. 10.1016/j.fct.2022.113013.

Peng, Chu, Xuejiao Tang, Xinying Gong, Yuanyuan Dai, Hongwen Sun, and Lei Wang. 2020. "Development and application of a mass spectrometry method for quantifying nylon microplastics in environment." *Analytical Chemistry* 92 (20):13930–13935. 10.1021/acs.analchem.0c02801.

Picó, Yolanda, Marinella Farré, and Damià Barceló. 2015. "Quantitative profiling of perfluoroalkyl substances by ultrahigh-performance liquid chromatography and hybrid quadrupole time-of-flight mass spectrometry." *Analytical and Bioanalytical Chemistry* 407 (15):4247–4259. 10.1007/s00216-015-8459-y.

Sendhil, R, Joan Nyika, Sheel Yadav, Joby Mackolil, Rama Prashat, G. Endashaw Workie, Raja Ragupathy, and P. Ramasundaram. 2022. "Genetically modified foods: bibliometric analysis on consumer perception and preference." *GM Crops & Food* 13 (1):65–85. 10.1080/21645698.2022.2038525.

Saberian, Mohammad, Jie Li, Shannon Kilmartin-Lynch, and Mahdi Boroujeni. 2021. "Repurposing of COVID-19 single-use face masks for pavements base/subbase." *Science of the Total Environment* 769:145527. 10.1016/j.scitotenv.2021.145527.

Sáez-Hernández, Roberto, Pablo Ruiz, Adela R. Mauri-Aucejo, Vicent Yusa, and M. L. Cervera. 2022. "Determination of acrylamide in toasts using digital image colorimetry by smartphone." *Food Control* 141:109163. 10.1016/j.foodcont.2022.109163.

Saito, Shizuka, Satoru Nemoto, and Rieko Matsuda. 2012. "Multi-residue analysis of pesticides in agricultural products by liquid chromatography time-of-flight mass spectrometry." *Food Hygiene and Safety Science (Shokuhin Eiseigaku Zasshi)* 53 (6):255–263. 10.3358/shokueishi.53.255.

Santos Pereira, Carolina, Sara C. Cunha, and José O. Fernandes. 2019. "Prevalent mycotoxins in animal feed: Occurrence and analytical methods." *Toxins* 11 (5). 10.3390/toxins11050290.

Sha, Xiao-Mei, Wen-Li Jiang, Zi-Zi Hu, Li-Jun Zhang, Zuo-Hua Xie, Ling Lu, Tao Yuan, and Zong-Cai Tu. 2023. "Traceability and identification of fish gelatin from seven cyprinid fishes by high performance liquid chromatography and high-resolution mass spectrometry." *Food Chemistry* 400:133961. 10.1016/j.foodchem.2022.133961.

Sintim, Henry Y., Andy I. Bary, Douglas G. Hayes, Larry C. Wadsworth, Marife B. Anunciado, Marie E. English, Sreejata Bandopadhyay, Sean M. Schaeffer, Jennifer M. DeBruyn, Carol A. Miles, John P. Reganold, and Markus Flury. 2020. "In situ degradation of biodegradable plastic mulch films in compost and agricultural soils." *Science of the Total Environment* 727:138668. 10.1016/j.scitotenv.2020.138668.

Starowicz, Małgorzata. 2021. "Analysis of volatiles in food products." *Separations* 8 (9). 10.3390/separations8090157.

Tuzimski, Tomasz, and Anna Petruczynik. 2020. "Review of new trends in the analysis of allergenic residues in foods and cosmetic products." *Journal of AOAC INTERNATIONAL* 103 (4):997–1028. 10.1093/jaoacint/qsaa015.

Valletta, Mariangela, Sara Ragucci, Nicola Landi, Antimo Di Maro, Paolo Vincenzo Pedone, Rosita Russo, and Angela Chambery. 2021. "Mass spectrometry-based protein and peptide profiling for food frauds, traceability and authenticity assessment." *Food Chemistry* 365:130456. 10.1016/j.foodchem.2021.130456.

Vidal, Nádia, Herbert Barbosa, Silvana Jacob, and Marco Arruda. 2015. "Comparative study of transgenic and non-transgenic maize (Zea mays) flours commercialized in Brazil, focussing on proteomic analyses." *Food Chemistry* 180:288–294. 10.1016/j.foodchem.2015.02.051.

Wang, Lei, Yawen Peng, Yali Xu, Junjie Zhang, Tao Zhang, Mengqi Yan, and Hongwen Sun. 2022. "An In Situ Depolymerization and Liquid Chromatography–Tandem Mass Spectrometry Method for Quantifying Polylactic Acid Microplastics in Environmental Samples." *Environmental Science & Technology* 56 (18):13029–13035. 10.1021/acs.est.2c02221.

Wang, Xian, Shujuan Wang, and Zongwei Cai. 2013. "The latest developments and applications of mass spectrometry in food-safety and quality analysis." *TrAC Trends in Analytical Chemistry* 52:170–185. 10.1016/j.trac.2013.08.005.

Xi, Jun, and Qiurong Yu. 2020. "The development of lateral flow immunoassay strip tests based on surface enhanced Raman spectroscopy coupled with gold nanoparticles for the rapid detection of soybean allergen β-conglycinin." *Spectrochimica Acta Part A: Molecular and Biomolecular Spectroscopy* 241:118640. 10.1016/j.saa.2020.118640.

Xu, Feng, Kong Weijun, Yang Meihua, and Ouyang Zhen. 2012. "Latest advancement for detection methods of mycotoxins in traditional Chinese medicine." *World Science and Technology* 14 (5):1944–1952. 10.1016/S1876-3553(13)60011-3.

Xu, Meng-Lei, Yu Gao, Xiao-Xia Han, and Bing Zhao. 2022. "Innovative application of SERS in food quality and safety: A brief review of recent trends." *Foods* 11 (14). 10.3390/foods11142097.

Xu, Meng-Lei, Yu Gao, Xiao Xia Han, and Bing Zhao. 2017. "Detection of pesticide residues in food using surface-enhanced Raman spectroscopy: A review." *Journal of Agricultural and Food Chemistry* 65 (32):6719–6726. 10.1021/acs.jafc.7b02504.

Xu, Meng-Lei, Yu Gao, Xiao Wang, Xiao X. Han, and Bing Zhao. 2021. "Comprehensive strategy for sample preparation for the analysis of food contaminants and residues by GC–MS/MS: A review of recent research trends." *Foods* 10 (10). 10.3390/foods10102473.

Xu, Meng-Lei, Jing-Bo Liu, and Jing Lu. 2014. "Determination and control of pesticide residues in beverages: A review of extraction techniques, chromatography, and rapid detection methods." *Applied Spectroscopy Reviews* 49 (2):97–120. 10.1080/05704928.2013.803978.

Yang, Yan, Guoliang Li, Di Wu, Jianghua Liu, Xiuting Li, Pengjie Luo, Na Hu, Honglun Wang, and Yongning Wu. 2020. "Recent advances on toxicity and determination methods of mycotoxins in foodstuffs." *Trends in Food Science & Technology* 96:233–252. 10.1016/j.tifs.2019.12.021.

Yuan, Rui, Wenbo Ding, Haixia Sui, and Wei Liu. 2022. "Migration studies of chloropropanols from paper straws: An improved method using GC-MS/MS." *International Journal of Environmental Analytical Chemistry*:1–14. 10.1080/03067319.2021.2019722.

Yue, Qi, Yu-Ying Huang, Xiao-Fang Shen, Cheng Yang, and Yue-Hong Pang. 2020. "In situ growth of covalent organic framework on titanium fiber for headspace solid-phase microextraction of 11 phthalate esters in vegetables." *Food Chemistry* 318:126507. 10.1016/j.foodchem.2020.126507.

Zhang, Can, Shuo Li, Jing Wu, Tengteng Ping, Ling Ma, Ke Wang, and Kaoqi Lian. 2023. "Developing a hydroxyl-functionalized magnetic porous organic polymer combined with HPLC-MS/MS for determining 31 amide herbicides in fruit wine." *Food Chemistry* 403:134442. 10.1016/j.foodchem.2022.134442.

Zuri, Giuseppina, Bernat Oró-Nolla, Ana Torres-Agulló, Angeliki Karanasiau, and Silvia Lacorte. 2022. "Migration of microplastics and phthalates from face masks to water." *Molecules* 27 (20). 10.3390/molecules27206859.

ABBREVIATIONS

6-ACA	6-aminocaproic acid
AA	adipic acid
DI-SPME	direct immersion solid-phase microextraction
ELISA	enzyme-linked immunosorbent assays
ERs	endocrine disrupters
GC	gas chromatography
GC-MS/MS	gas chromatography-tandem mass spectrometry
GM	genetically modified
HPLC-MS/MS	high performance liquid chromatography tandem mass spectrometry
HS	headspace
IACs	immunoaffinity columns
LC	liquid chromatography
LOQ	limit of quantitation
MPs	microplastics
MRM	multiple reaction monitoring
MS	mass spectrometry
MS/MS	tandem mass spectrometry
PA	polyamide
PLA	polylactic acid
Q-TOF-MS	quadrupole time-of-flight mass spectrometer
SVOCs	semi-volatile organic compounds
SPME	solid-phase microextraction
SBSE	stir bar sorptive extractions
VOCs	volatile organic compounds

4 FT-IR Analyses in Food Authentication
Food Safety and Quality Assurance

Rumana A. Jahan and Md. Nazim Uddin

4.1 INTRODUCTION

Food and agricultural products comprise complex and diverse chemical mixtures that historically made it challenging to evaluate food safety, nutrient content, stability, and sensory attributes. With the increasing food demand worldwide, their quality and safety are becoming more of a concern for consumers, producers, and governments as the global food supply system expands, incorporating an increasing number of involvements from "farm to fork" as well as the inherent vulnerability associated with such a process (US FDA, 2009; EC, 2018). These concerns are due to the health threat (including ethical grounds), business, and total health budget. The adverse effects of food fraud, including adulteration and authenticity, are illustrated by scandals like the use of carcinogenic industrial chemical dye in chili powder in Sudan in 2005 (Meikle, 2005), the melamine addition to infant formula milk in 2008 in China (Pei *et al.*, 2011), the horsemeat scandal in Europe in 2013 (Premanandh, 2013), and the replacement of expensive almonds with peanuts in UK restaurants in 2022 resulted in a fatal illness of customer. It has been difficult for policymakers and professionals to address the global public health threat posed by contaminated food and water since there is a lack of reliable data on the scope and kind of threats added to the incidence of related ailments.

Food adulteration can be defined as the process in which food quality is purposefully degraded either by adding low-grade quality material or by extracting valuable ingredients (Spink and Moyer, 2011). Although there is no harmonized definition of food fraud, it is generally accepted that it is committed intentionally for financial gain through consumer deception (EC, 2018). Intentional adulteration in food has become more widespread gradually due to the financial benefits, and several food products are commonly found fraudulent, such as honey and maple syrup, which is sometimes diluted with cheaper sweeteners, such as corn syrup or cane sugar, and then sold as pure products (Galvin-King *et al.*, 2018; Amiry *et al.*, 2017). Olive oil faces a similar problem by being diluted with cheaper vegetable oil (Rohman and Man, 2010; Georgouli et al., 2017). In addition, country-of-origin fraud concerning dried fish, legumes, other agricultural products, and spices is widespread (Lohumi *et al.*, 2017; Wielogorska *et al.*, 2018). Valuable foodstuffs are more susceptible to economically motivated adulteration (EMA) because of the high profits gained from these so-called premium food products, such as meat (Rohman *et al.*, 2011; Nunes *et al.*, 2016) and coffee (Reis *et al.*, 2017), are particularly susceptible to adulteration, primarily when they are produced and supplied through complex supply chains (Black *et al.*, 2016; Galvin-King *et al.*, 2018).

One of the major problems harming interactions between distributors and customers is the increase in food fraud (GMA and Kearney, 2010; PWC, 2016). Although the prevalence of food adulteration is unknown, it is believed that the damage to the world's food business could reach $40 billion yearly. The Consumer Brands Association (CBA) estimated that food fraud in the U.S. market impacts around $10–$15 billion annually. As an example, the expanding milk utilization due to its nutritious content has rendered it more vulnerable to fraud (Handford *et al.*, 2016). Conversely, inexpensive foods like corn and wheat are more likely to be susceptible to adulteration because of their poor profit margins (Smith *et al.*, 2017). Additionally, 21st-century agriculture faces numerous difficulties due to the growing demands for feedstock for the bioenergy industry as well as increased food production for people and livestock. Demands in the agricultural sector are also a result of changes in agriculture-dependent emerging nations, adopting more sustainable and efficient production techniques and climate change (FAO, 2009). Therefore, it is crucial to research the functional molecules and food's components, including their profile and speciation, authentication, traceability, and quality, as well as the detection and measurement of additives, adulterants, allergies, and chemical and microbiological pollutants. All of these reasons are dominant for the unwanted fact of food fraud. Therefore, evaluating these criteria along the entire chain, from the field/harvesting to storage, processing, packaging, and shelf life, is necessary, and these factors significantly impact the economy, agriculture, industry, and consumer health.

Food fraud may cause serious health problems, including cramping, nausea, diarrhea, vomiting, nerve damage, allergic responses, and paralysis, depending on the adulterant used (Sicherer *et al.*, 1998). For instance, there were allegations of cumin and paprika powder being mixed with peanuts

DOI: 10.1201/9781003334859-4

and almonds in the USA and Europe in 2015 (Agres, 2015). These two allergens might cause severe or fatal allergic reactions if accidentally consumed (Sicherer *et al.*, 1998). Fishes, soybeans, fruits, fish, milk, eggs, nuts, cereals, and shellfish are other major allergens in contaminated food that might have serious health repercussions (Añíbarro *et al.*, 2007). The controversies involving food fraud have raised the demand for food testing facilities to create quick as well as accurate analytical techniques for spotting fraudulent foods. FTIR, specifically mid-infrared (MIR) and near-infrared (NIR) vibrational spectroscopy, which offers a quick and accurate detection approach for elucidating organic molecules and the identification of pure substances, is the most frequently utilized analytical method for identifying food fraud nowadays. The use of FTIR and multi-dimensional instrumental approaches alone or as a component of stand-alone methodologies to analyze various food matrices will be critically discussed in this review.

4.2 DETECTION OF FOOD ADULTERATION BY INSTRUMENTAL ANALYSIS

4.2.1 Adulteration in Food and Authentication: Regulations and Quality Standards

Food replacement, addition, tampering, misrepresenting food along with its ingredients or packaging, as well as making false or deceptive claims about a product to profit financially, are all included in the category of "food fraud" (Spink, 2011; Lakshmi, 2012). Although there is no unified definition for food fraud currently, a few functioning criteria can be used to identify it (EC, 2018). These include willful disregard for existing food laws, malice, financial gain, and customer fraud (EC, 2018). The General Food Law Control EC 178/2002 outlines the major principles for authentication and regulation (EC, 2002) of food. For example, the General Food Law proclaims its goal is to avoid "fraudulent or misleading actions" and "the adulteration of food" under Article 8, "Protection of Consumers' Interests" (EC, 2002). Taking into account several known food scandals in Europe over the past 20 years, the European Food Safety Authority (EFSA) has been established to offer science and technology-based guidance (EC, 2002). The U.S. Department of Agriculture (USDA) and the U.S. Food and Drug Administration (USFDA) are the two central regulatory bodies in the United States for securing safe food (Johnson, 2014). The Federal Food, Drug, and Cosmetic Act (FFDCA), passed in 1938, is the main piece of legislation covering food, drugs, and consumer protection (US Congress, 1938). The Food Safety Modernization Act (FSMA), which became law in 2011, allowed the FDA to require recalls (US Congress, 2011). Given the known cases of foodborne diseases in the early 2000s, this was thought to be required.

A threshold for extraneous matter is now used by the majority of national, international, and industrial organizations to distinguish between accidental and intentional contamination. It is important to emphasize that not all contamination levels qualify as intentional adulteration. For instance, in the United Kingdom, the presence of 1% (w/w) or more foreign matter in food items qualifies as gross adulteration (Food Standards Agency, 2015). The European Spice Association (ESA) is an example of an industry organization with its own criteria, which is equivalent to the maximum 2.0% w/w for foreign materials set for herbs and 1.0% w/w for spices (ESA, 2015). The advancement of instrumental technology has made it possible to detect pollutants at very small levels using techniques like DNA tests and analytical chemical procedures, which are used to detect food fraud (Galvin-King et al., 2018). These restrictions are placed to distinguish between contamination and deliberate adulteration by the detection limit using the appropriate analytical techniques for identifying adulteration (Downey, 2016).

Authenticity testing is essential to verify that the food being sold is of the actual nature, composition, and quality that the buyer has grown to expect (Defra *et al.*, 2014). Food fraud occurs when manufacturers defraud customers by not correctly and truthfully disclosing a food product's ingredients (Defra *et al.*, 2014). In order to protect consumers and facilitate trade between Europe and other countries, the EU has standardized laws for food labeling, presentation, and advertising (EC, 2000). Three different categories of quality marks have been introduced by the European Council Regulation in 2006 to guarantee the authenticity of all kinds of agricultural products and foodstuffs (The Council of the European Union, 2006). These logos—the Traditional Specialty Guaranteed (TSG), Protected Designation of Origin (PDO), and Protected Geographical Indication (PGI)—have a particular connection to the geographic area from which the product originates. This rule dictates that food or beverages must meet stringent requirements (The Council of the European Union, 2006). The EU has also had explicit laws governing wine authenticity since 2011. In accordance with PDO and PGI, the wine that complies with the framework's standards is now classified (The European Commission, 2014b, 2014a).

4.2.2 History of Food Analysis

The evaluation of food's chemical makeup and physical attributes is referred to as food analysis. Laboratory analysis methods increasingly replaced the ancient approaches with modern instrumental techniques throughout the previous century, and food analysis has dramatically changed in the last 100 years. Innovative advancements in spectrometry, chromatography/separations, pH devices, and spectrophotometry frequently had direct applications to examining food. The feasible spectrum of food applications has increased due to significant advances in analytical accuracy, precision, detection limits, and sample throughput brought about by ongoing methodology improvements over this time. Food analysis, which goes beyond simple characterization, is a critical component of the construction and architecture of the contemporary global food distribution system. It is used for new product development, quality control, regulatory compliance, and problem solving. In addition to examining the quality of the finished food product, the food industry is increasingly focusing on analyzing, authenticating, and characterizing the raw materials and ingredients utilized in foods. Regulatory bodies, consumer advocacy organizations, and the public demand for safer foods contribute to this, especially in light of the effects of globalization on sources. Analysis of foods is continuously requesting the development of more robust, efficient, sensitive, and cost-effective analytical methodologies to guarantee the safety, quality, and traceability of foods in compliance with legislation and consumers' demands. The manufacturer's objective is to guarantee that substandard, adulterated, or incorrectly labeled ingredients never enter the production process. The early 20th-century procedures based on "wet chemistry" have evolved into the potent instrumental approaches currently employed in the food laboratory. Many of these crucial instrumental capabilities were put to use by food chemists through technological adaptation to create new analytical techniques and procedures for measuring food components. The practical range of food applications has increased as a result of these improvements in analytical accuracy, precision, detection limits, and sample throughput. Any forecast regarding novel advancements in food analysis for the 21st century must consider the convergence of several forces. The first thing to think about is how quickly knowledge is being shared via the Internet and other electronic media, which has accelerated the pace at which scientific discoveries naturally progress at their own rate. Innovation and experimentation from the past will also lead to new technological advancements. However, the present trend toward more portable analytical measurement uses in food production or field crops outside of traditional laboratory settings as well as faster results, lower detection thresholds, and smaller analytical devices, will continue. Analytical methods have historically been categorized based on how they operate. Each technique has its own benefits and limitations when used for food analysis, but they all provide specific information on the sample or components under examination based on a particular physical-chemical interaction. Additionally, widely used conventional analysis and detection techniques, including thin-layer chromatography (TLC) and high-performance liquid chromatography (HPLC), are typically destructive, labor-intensive, time-consuming, and dangerous. New analytical tools are therefore required that offer comparable analytical capabilities to lab instruments while also being sufficiently reliable and user-friendly to be employed closer to the sample's source by less experienced users.

4.3 EVOLUTION OF FOURIER-TRANSFORMED INFRARED (FT-IR) SPECTROSCOPY

Lord Rayleigh was the first scientist to discover in 1892 that a spectrum and its interferogram are connected via a Fourier transform. But after more than 50 years, Fellgett became the first scientist to translate an interferogram into its spectrum correctly. In 1965, Cooley and Turkey presented the quick Fourier transform technique based on the modern FTIR spectrometer. The Fourier transform is a mathematical technique for changing one function into another, and it is named after the French scientist and physicist Jean Baptiste Joseph Fourier. The first Fourier transform infrared (FTIR) spectrometer was commercially available in the 1970s, and JASCO developed its first FTIR spectrometer in 1982. As computers became widespread, FTIR spectroscopy became mainstream by the 1990s. (Daniels *et al.*, 1970; Skoog and Leary 1992). The fundamental idea behind FTIR spectrometers is that various gases absorb IR radiation at frequencies that are unique to each species. But since FTIR spectroscopy is a dispersed technique, measurements are made over a broad spectrum instead of a restricted range of frequencies. The most popular type of infrared spectroscopy relies on molecular vibrations brought on by specific infrared frequencies and energy absorption. Different chemicals have displayed distinctive infrared spectra, enabling FT-IR to recognize and categorize them. Because each IR active molecule has unique infrared spectra, the molecules could be identified and categorized using FT-IR (Cozzolino et al., 2011). FT-IR spectrometers have a few more advantages over other analytical techniques. The most important criteria is the drastic reduction of time required for data acquisition, specificity for particular compound, and also sensitivity. In addition, wavelength calibration in FT-IR ensures the precision of the analysis. The most common use is identifying unknown materials and confirming production materials (incoming or outgoing). The information content is particular in most cases, permitting acceptable discrimination between like materials. Chemical bonds in a molecule can be identified by FT-IR, which produces an infrared absorption spectrum, and the spectrum produces complete information about the sample. Early-day FT-IR instruments were expensive and large. The technological advances in the coming years made FT-IR spectrometers affordable and improved their performance features several-fold. FT-IR spectroscopy is a hugely popular technique due to its unique combination of sensitivity, flexibility, specificity, and robustness. It has become one of science's most commonly used analytical instrumental techniques, since it can handle solid, liquid, and gaseous analytes. The resultant spectrum is to be formed a few microns of sample penetration and indicate a small amount of homogeny because of how strongly organics absorb mid-IR. However, FTIR has some known limitations, including its relative intolerance of water (which quenches the IR signal even when present at just a few percent) and its sensitivity to the physical properties of the analysis matrix. It is nevertheless hugely popular and commonly used right across industries as diverse as food and beverage (Rohman et al., 2020), chemical, engineering, environmental (Freitag et al., 2009), pharmaceutical (Lawson et al., 2018), and biomass (Allison et al., 2009) and in clinical settings (Balan et al., 2019). Suitable instrumentation forms now include benchtops, hand-held, and online real-time devices.

Routine food assays are moving toward the automation of tools and analytical measurements based on smaller instrumentation, advancement in computer and data processing software, and above all, cost reduction. As diagnostic assay firms improve performance, the proportion of quick tests has risen considerably in recent years. As a result, advancements in rapid analytical methods for checking raw materials, product quality, and process monitoring have been proceeding. Food processing enterprises readily adopt these techniques to reduce the risk of food safety events and guarantee compliance with legal requirements. In this regard, infra-red (IR) spectroscopic technique has been used for its widespread application in different disciplines covering the analysis of both fresh and processed foods. Initiating from the IR spectrometers in the 1940s and 1950s, with the basic function to identify and clarify the structures of IR-active organic molecules, they were initially widely used in food research laboratories (Bureau et al., 2009; 2013; Clark, 2016; Cozzolino et al., 2011). However, with the constant advancements of computer systems and data processing capabilities, Fourier-transformed Infrared (FT-IR) spectrometry has become a more valuable and accurate technique based on the absorption of infrared radiation by the IR-active molecules of compounds. Mainly, when IR radiation passes through a solid, liquid, or gaseous compound, a portion of it is absorbed by the component molecules resulting in specific vibrational frequencies in the inherent molecules. The rest of the unabsorbed IR radiation is transmitted and picked up by a detector of the FT-IR spectrometer. Measuring the intensity of the absorbed or transmitted IR radiation processed by the in-built processing software using complex mathematical operations known as Fourier transformations, a unique spectrum of the compound is produced finally. Since no two molecular compounds generate identical IR spectra, the FT-IR spectrum is unique and considers each compound's fingerprint within the region 600–1,400 cm^{-1} (Larkin, 2018; Smith, 2011). This allows the utilization of FT-IR spectra for qualitative measurement and for identifying unknown compounds (Bureau et.al., 2019). Moreover, since the intensity of the absorbance in IR spectra is directly proportional to the concentration of a pure compound or that of a specific component when analyzing mixtures (Gauglitz and Vo-Dihn, 2003), this made the FT-IR technique highly applicable for quantitative analysis. The infrared (IR) region of the electromagnetic spectrum, which is located between 714 and 1×10^6 nm has three subregions in the infrared spectrum; the near-infrared (NIR) region (714–2,500 nm or 14,000–4,000 cm^{-1}), the mid-infrared (MIR) region (2,500–25,000 nm or 4,000–400 cm^{-1}), and the far-infrared region (25,000–1×10^6 nm or 400–10 cm^{-1}) (Larkin, 2018). In the regions of NIR and MIR, most uses concern analyses of fruits, vegetables, crops, and their processed products.

4.4 APPLICATION OF FT-IR SPECTROSCOPY IN FOOD ANALYSIS
4.4.1 Principle and Instrumentation

FT-IR is the most common form of infrared spectroscopy based on molecular vibrations resulting from the absorption of particular infrared frequencies and energy. Various compounds will have different infrared spectra, allowing FT-IR to identify and classify them. Because each IR active molecule has unique infrared spectra, the molecules could be identified and categorized using FT-IR. Thus, the assessment of produced food quality (Bureau, 2019; Clark, 2016; Cozzolino, 2011), quantification of bioactive materials and plant metabolites (Lu, 2012; Schulz, 2007), identification of adulteration and authenticating products for the type or specific geographic location (Cozzolino, 2017; Giusti *et al.*, 2011; Huang, 2009; Karoui, 2010), and biological contamination (Huang *et al.*, 2009; He and Sun, 2015) can be done. The FT-IR spectrometer uses an interferometer consisting of a source, a beam splitter, two mirrors, laser, and a detector. The energy goes to the beam splitter from the source, splitting the beam into two parts. One part is transmitted to a moving mirror, and the other is reflected in a fixed mirror. The moving mirror is able to move at a fixed velocity, controlled by the response of laser. The reflected beams from two mirrors are recombined again at the beam splitter, generating an interference pattern which then transmitted through the sample to the detector. This signal is then Fourier transformed to generate a spectrum. The working principle of an FT-IR spectrometer for food analysis is presented in a schematic diagram in Figure 4.1.

The primary tool used for infrared spectroscopy analysis is the FT-IR spectrometer, as depicted in Figure 4.1, which shows a schematic of its optical system, which is mainly made up of the fixed mirror, moving mirror, beam splitter, appendix, light source, etc., along with the detector to receive the signal and the signal processing and acquisition unit assembled to a computer. FT-IR detectors can detect infrared radiation in four ways: transmission, attenuated total reflection (ATR), mirror reflection, and diffuse reflection mode. The resulting infrared spectrum represents a graph with the frequency (wavelength) on the horizontal axis and the material's ability to absorb infrared light on the vertical axis. Depending on the various needs for detection and the physical conditions of the sample, users can select the best detection methods. Transmission spectroscopy, attenuation complete reflection spectroscopy, and other methods can be used to identify solid samples. It is possible to identify liquid samples using the transmission spectrum of an infrared liquid pool or the reflection spectrum of attenuated full-reflection accessories.

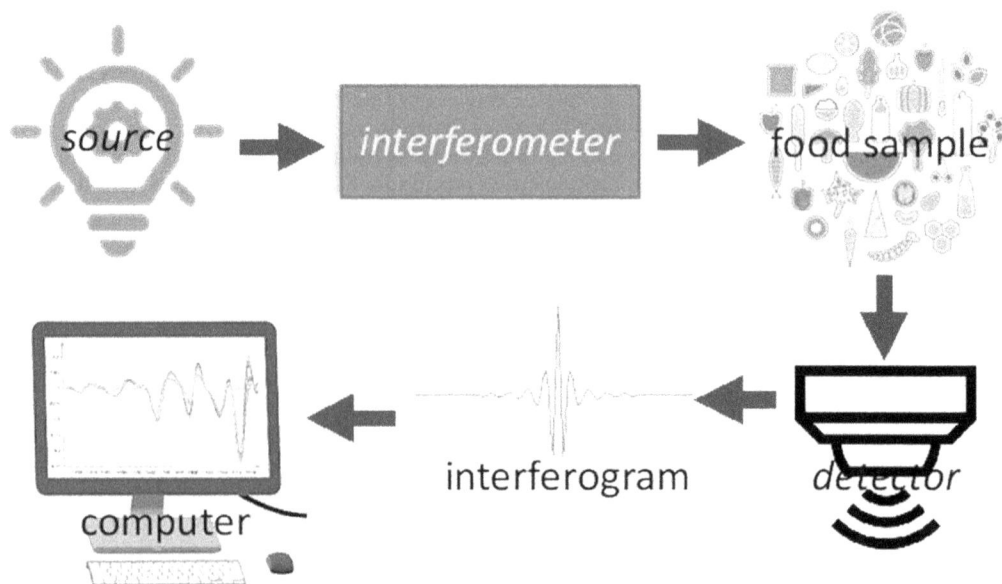

Figure 4.1 A schematic diagram of FT-IR working principle for food analysis.

4.4.2 Sampling Process for Food Analysis in FT-IR

Food samples as solids as well as in the form of liquids, pulps, and gels (freeze-dried to powders for removing interference from water) require little preparation generally for analysis using FT-IR. Solid foods can be analyzed using potassium bromide (KBr) discs, as this inorganic salt does not produce any vibrations in the IR region (Nyquist, 2012). In this process, a small portion of the solid analyte is mixed with KBr, followed by pressing the mix to a fused IR-transparent disc, which is then placed into the beam of light of an FT-IR spectrometer (Simmons, 1960; Gauglitz and Vo-Dihn, 2003). KBr discs are the preferred method in some investigations, such as for detecting fraud in saffron samples (Anastasaki, 2010; Ordoudi *et al.*, 2017). Routine solid powder analysis can be analyzed both in the classical transmission mode and by diffuse reflectance IR spectroscopy. Another mode of detection in Ft-IR spectroscopy known as the attenuated total reflectance (ATR) seems to be favored nowadays for investigating food authenticity or adulteration since it allows samples in all forms, including liquids, gels, composites, and also samples containing water. However, since water absorbs strongly in the IR region (Lu, 2012), giving rise to its absorption peaks in the IR spectrum. In the ATR unit, a crystal of small internal reflection element usually made from germanium (Ge), diamond, or zinc selenide (ZnSe) having a high refractive index and excellent IR transmitting properties is used (Gauglitz and Vo-Dihn, 2003) upon which the sample is placed. The technique operates depending on the difference between the refractive index of the sample and the reflective element at the interface as the IR beam directed to the sample surface of the sample undergoes total internal reflection, and the spectral information about the sample can be obtained from this reflected beam. On the other hand, in the ATR technique, sample preparation is minimal. For liquid samples, it only requires centrifugation to remove solid particulates or cellular debris. For analysis, volatile samples can be evaporated to a film directly on the ATR cell. Polymer or gel samples can be pressed against the reflecting element to be analyzed similarly.

4.5 RECENT ADVANCEMENTS IN FT-IR USED IN FOOD ANALYSIS

4.5.1 FT-MIR Spectroscopy

Recent studies have shown that FTIR-based methods, combined with chemometric techniques, can be successfully applied in the food industry to detect substances that affect the quality of food products or are employed for adulteration. Nowadays, the most frequently used technique for food analysis is FT-MIR spectroscopy, which uses a beam of MIR light through the sample and measures the transmission and absorption of the light (Su and Sun, 2019). FT-MIR spectroscopy is a non-destructive, label-free, susceptible, and specific technique that provides complete information on the chemical composition of biological samples. The technique can offer basic structural information and serve as a quantitative analysis tool. Several studies showed that FT-MIR contains the most powerful applications for fresh crops like fruits and vegetables as well as the processed products (Larkin, 2018; Clark, 2016). For example, as illustrated in Figure 4.2(A) (S. Bureau, 2019), the spectra of fruit juice interprets its applicability for fruits and vegetables analysis. MIR absorbance spectra of naval orange juices are shown in Figure 4.2(A). Orange juice contains simple sugars and citric acid as their major component and the fact is that the dominant components of such fruits don't have triple bonds; thus, the absorbance spectra with associated bond types appear in specific regions.

Accurate band assignment is necessary in matrices with many components for analyzing significant wave numbers, plots with variable loading, or plotting regression coefficient. When solid dissolves and reacts with solvent molecules or cations, the band assignments in polysaccharide compounds may expand or shift (Kanou *et al.*, 2017). The peak height at 1,723 cm^{-1} indicates the change in titratable acidity, which decreased significantly from 5.26 to 0.83% as the different juices were collected at various stages of maturation of that fruit. In contrast, an increase in soluble solids was reflected in the rise of other peaks (1,150–800 cm^{-1}), which were associated to an increment of free sugars. In contrast, the spectra of pure solutions of citric, malic, quinic, and tartaric acids are illustrated in Figure 4.2(B) (S. Bureau. 2019) to help comprehend the peak height for a particular wavenumber. In addition, the ATR-MIR spectra of major pure sugar, including sorbitol, fructose, glucose, and sucrose reference standards in 10% (w/w) dissolved water, are shown in Figure 4.2(C) (S. Bureau, 2019). The following papers by various authors contain some of the more thorough insights that apply to biological systems: (Castillo, 2017; Talari, 2017 and Wiercigroch, 2017). Most samples used in food products analysis are usually available in the form of liquids, pulps, gels, films, or solids which need to be powder by freeze-drying to reduce the interference of water. These

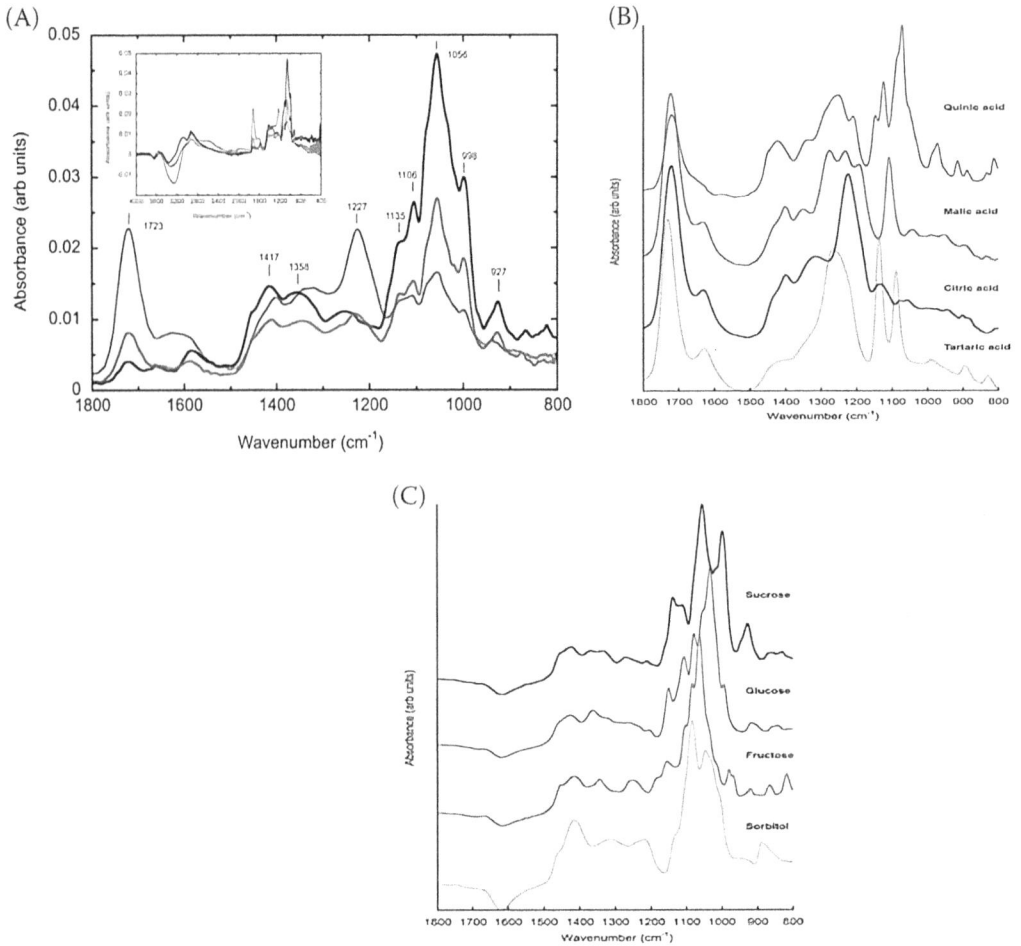

Figure 4.2 ATR-MIR ATR-MIR spectra (after subtraction of the water contribution) (1,800–800 cm^{-1}, 128 acquisitions, 4 cm^{-1} resolution) of (A) of navel orange juice at three stages during fruit development; early season (spectrum that has highest peak at 1723 cm^{-1}) (soluble solids 8.1%, titratable acidity (TA) 5.26% citric acid equivalents), mid-season (spectrum with middle peak at 1723 cm^{-1}) (9.1%, TA 1.83%) and harvest (spectrum with lowest peak at 1723 cm^{-1}) (14.0%, TA 0.83%), (B) pure acid and (C) pure sugar reference standards 10% (w/w) dissolved in water. Adapted from S. Bureau *et al.* (2019), with permission.

samples often don't need to be prepared in any particular ways. As mentioned, ATR is the most common and popular technique for food sample analysis as it requires almost no sample preparation, is rapid, and is easy to operate. Since most of the food samples contain water molecules to some extent which absorb significantly in the IR region, ATR can be applied practically to examine food adulteration and authenticity.

4.5.2 Rapid Food Analysis Using Handheld FT-IR Techniques

Compact, robust, and lightweight portable spectrometers are being developed due to the miniaturization of vibrational spectrometers. They can be configured to measure solids, liquids, or gases. The instrument's performance should ideally be similar to that of a laboratory FTIR counterpart, and it should also have the necessary resilience for the surrounding conditions, be a shock- and vibration-resistant, and have an interferometer with steady performance in all directions. These miniature, portable spectrometers are commonly configured for qualitative investigation, which entails the rapid recognition and classification of unknowns using spectrum databases. In hostile work environments, the first responders have employed such system as their main applications to identify various hazards, such as nerve agents, explosives, and poisonous industrial chemicals. In such circumstances, a fast, real-time response, minimal preparation for sample in a non-destructive manner require a speedy real-time response, minimal sample preparation using a non-destructive method, and also small sample quantities is required. Sorak *et al.* (2012) demonstrated a handheld technique to quantitatively analyze a pharmaceutical product, an alcohol mix, and additives in bitumen modified by polymers. The Agilent 4300 is an illustration of a handheld FT-IR instrument with all the components of a standard unit. It has a flexible application with five replaceable probes to enable users to analyze in different modes. If necessary, these can also be decontaminated later. Ayvaz *et al.* (2015, 2016a,b) screened potato breeding lines for specific nutritional characteristics and used a portable FT-IR to evaluate the quantity of sugar and amino acid in the tubers of raw potato. The small device demonstrated in every experiment that it could offer quick and reasonably priced prediction models to help with potato breeding and crop management. In a recent study, it has been demonstrated that the portable system has the ability to measure all of the parameters related to quality as textural aspects, in processed tomato juice when FTIR analysis of tomato liquids was performed using four different instruments. The handheld devices even offered superior quantitative determination compared to other reference techniques or FT-IR benchtop units (Sorak *et al.*, 2012)

4.6 FTIR SPECTROSCOPY USED FOR FOOD ADULTERATION AND AUTHENTICATION

Food adulteration is intentionally debasing food quality by either adding or replacing the food substances with undeclared alternative components or replacing some valuable components with substandard ones. Contrarily, food authentication is the process that makes sure the product matches the information on the label. Adulteration has an adverse effect on market growth because it undermines customer confidence and has a negative impact on human health. Therefore, food authentication is vital for food processors, retailers and consumers, and regulatory authorities. Food authenticity, however, frequently presents difficulties for the relevant authority due to the complexity of food and the expansion of adulterant types that make their identification challenging. Because expensive components may be fraudulently and/or mistakenly mislabeled, it is crucial for regulatory bodies, food processors, merchants, and consumers to ensure the authenticity of products by commodity, variety, and geographic origin. There is a need for development of a rapid technique to validate these parameters, and FTIR has been applied potentially in such cases in recent years. Combination of FT-NIR and FT-MIR with various statistical methods has been applied to authenticate herbal products, juices, crops, oils, dairy products, meat and other numerous food products. This method has a high degree of confidence in identifying food adulteration, monitoring biochemical and microbiological decomposition and shelf life, and determining changes in chemical components such as proteins and lipids. It is easily suitable for routine quality control or industrial applications (Kezban, 2020).

Recent fraud scandals have raised serious concerns about the authenticity of beef, milk, and other products. In order to identify and measure the amount of pork in some recipes for halal verification, Rohman *et al.* (2011) used FT-IR spectroscopy. Table 4.1 and Table 4.2 display several adulterants and levels of authenticity utilizing the FTIR analysis technique. According to the article, the level of pork adulterant may be satisfactorily quantified at the range between 1,200 and 1,000 cm^{-1} in the fingerprint region using the FT-IR/ATR spectroscopy technique. The ability to discriminate between halal and non-halal sausages in Chinese ham that contained pig was achieved by Xu *et al.* (2012) using FTIR spectroscopy in conjunction with chemometric data processing techniques. Rahmania *et al.* (2015) investigated the potential for FTIR spectroscopy with additional data analyses to be used to classify and quantify rat meat adulterants in beef product formulations, similar to how certain types of meat are forbidden in Muslim and Jewish communities, such as rat and dog meat. Nunes *et al.* (2016) discovered that injecting aqueous solutions of non-meat ingredients (NaCl, phosphates, carrageenan, maltodextrin, and collagen) into beef flesh constituted fraud utilizing FT-IR/ATR spectroscopy. They found that this technique offers a reliable tool to quantify rat meat in beef products. Other well-known sources of adulteration in meat and meat products include the addition of non-meat proteins and carbohydrates like hydrocolloids, cellulose, and starch, exogenous salts like phosphates, and components made from vegetable fats, all of which are used to enhance the ability of meat to hold water and, as a result, its weight increases (Nunes, 2016; Cavin et al., 2018).

Table 4.1: Summary of Studies on Different Adulterants/Authenticity Using FTIR as an Analysis Technique

Food	Quality Matter	Sampling Techniques	Wavelengths (cm^{-1})	References
Honey	Anatolian honey	ATR	1,800–700	Gok et al., 2015
Oil	Olive	ATR	700–740, 950–1,050, 1,100–1,250, 1,350–1,500, 1,700–1,800, 2,750–3,000	Gouvinhas et al., 2015
Beef	Non-beef NaCl, phosphates, carrageenan, maltodextrin	ATR	4,000–525	Nunes et al., 2016
Bovine meat/injected non-meat ingredients	(NaCl, phosphates, carrageenan, maltodextrin)/direct meat samples	ATR	4,000–525	Kezban et al., 2020
Tilapia fish	Use of sodium alginate in restructured tilapia fish product	ATR	2,000–800	Huang et al., 2017
Saffron	C. sativus stamens, calendula, safflower, turmeric	KBr discs	4,000–600	Petrakis and Polissiou, 2017
Milk	Hydrogen peroxide, synthetic urine, urea and synthetic milk	ATR	1,400–1800	Santos et al., 2013
	Melamine	ATR	4,000–650	Jawaid et al., 2013
Potato Chips	Acrylamide adulteration	ATR	4,000–700	Pedreschi et al., 2010

Table 4.2: Research Studies Showing the Results of Food Authenticity Using the FTIR Analysis Technique

Sample	Sampling Technique	Results	Source
Extra virgin olive oil	ATR	R2 = 0.99; detection limit 6%	Vlachos *et al.*, 2006
Adulteration with vegetable oils	ATR	Detection limit of 5% for binary mixture; error limit 1.04	Gurdeniz & Ozen, 2009
Adulterated with palm oil	ATR	R2 = 0.999; SECV = 0.285 (first derivation)	Rohman & Che Man, 2010
Evaluating origin	KBr discs	As able to correctly classify 80% (mean centered and first and second derivation)	Hennessy *et al.*, 2009
Authentication of fruits	KBr discs	Extraction improved SIMCA; 100% correct classification at commodity level	He *et al.*, 2007
Classifying wines as organic versus nonorganic	Bacchus flow cell	DPLS correctly classified 85%; LDA correctly classified 75%	Cozzolino *et al.*, 2009

4.7 CONCLUSIONS

The globalization of the food manufacturing chain has increased the risk of food adulteration. Authorities and food manufacturers are under increasing pressure to develop efficient systems for thorough food monitoring. Acrylamide, organic pollutants, Sudan colors, and melamine in milk and dairy products are just a few of the recent contamination problems that the food producers and consumers have had to solve. It is necessary to develop and validate appropriate analytical methods and put them into use by food producers and authorities as quality control parameters and risk management systems to analyze chemical food contaminants. One of the emerging and robust methods for determining the authenticity and safety of food is vibrational spectroscopic technologies, such as FT-IR spectroscopy. The ability of IR combined with other data analyzing program can be potentially utilized to distinguish between specific spectral signatures of food contamination (unexpected substances) and quantify such signals (recognized agents). As an established analytical method for quick, high-throughput, non-destructive investigation of various sample types, FT-IR spectroscopy produces a fingerprint of that sample's chemical or biological constituents. The advancements in FT-IR equipment and multivariate techniques have established their potential for complex multispectral data processing for biological system. FT-IR spectroscopy identifies the unknowns for detecting pollutants and adulterants in food, providing qualitative and quantitative information about the nature of substances, their structure, interactions, and molecular surroundings. As a result, the food industry can benefit from quick and specialized procedures like FTIR spectroscopy to evaluate quality and safety while also monitoring chemical contaminants such as micro/nano plastics. It would enable the food producer to evaluate the quality of their product immediately, enabling prompt corrective measures throughout manufacturing. Less operational expenses, small size, compactness, resilience, ease of use, and little prior knowledge are advantages of vibrational spectroscopy-based approaches.

REFERENCES

Allison, G. G., Thain, S. C., Morris, P., *et al.* (2009). Quantification of hydroxycinnamic acids and lignin in perennial forage and energy grasses by Fourier-transform infrared spectroscopy and partial least squares regression. *Bioresource Technology*, 100(3), 1252–1261. 10.1016/j.biortech.2008.07.043

Amiry, S., Esmaiili, M., Alizadeh, M. (2017). Classification of adulterated honeys by multivariate analysis. *Food Chemistry*, 224, 390–397. 10.1016/j.foodchem.2016.12.025

Anastasaki, E., et al. (2010). Differentiation of saffron from four countries by mid-infrared spectroscopy and multivariate analysis. *European Food Research and Technology*, 230(4), 571–577. 10.1007/s00217-009-1197-7

Añíbarro, B., Seoane, F. J., Múgica, M. V. (2007). Involvement of hidden allergens in food allergic reactions. *Journal of Investigational Allergology and Clinical Immunology*, 17(3), 168–172.

Ayvaz, H., Bozdogan, A., Giusti, M. M., Mortas, M., Gomez, R., Rodriguez-Saona, L. E. (2016a). Improving the screening of potato breeding lines for specific nutritional traits using portable mid-infrared spectroscopy and multivariate analysis. *Food Chemistry*, 211, 374–382.

Ayvaz, H., Santos, A. M., Moyseenko, J., Kleinhenz, M., Rodriguez-Saona, L. E. (2015). Application of a portable infrared instrument for simultaneous analysis of sugars, asparagine, and glutamine levels in raw potato tubers. *Plant Foods for Human Nutrition*, 70(2), 215–220.

Ayvaz, H., Sierra-Cadavid, A., Aykas, D. P., Mulqueeney, B., Sullivan, S., Rodriguez-Saona, L. E. (2016b). Monitoring multicomponent quality traits in tomato juice using portable mid-infrared (MIR) spectroscopy and multivariate analysis. *Food Control*, 66, 79–86.

Balan, V., Mihai, C. T., Cojocaru, F. D., *et al.* (2019). Vibrational spectroscopy fingerprinting in medicine: from molecular to clinical practice. *Materials*, 12(18), E2884. 10.3390/ma12182884

Black, C., *et al.* (2016). A comprehensive strategy to detect the fraudulent adulteration of herbs: the oregano approach. *Food Chemistry*, 210, 551–557. 10.1016/j.foodchem.2016.05.004

Bureau, S., Cozzolino, D., Clark, C. J. (2019). Contributions of Fourier-transform mid infrared (FT-MIR) spectroscopy to the study of fruit and vegetables: A review *Postharvest Biology and Technology*, 148, 1–14.

Bureau, S., Ruiz, D., Reich, M., Gouble, B., Bertrand, D., Audergon, J., Renard, C. (2009). Application of ATR-FTIR for a rapid and simultaneous determination of sugars and organic acids in apricot fruit. *Food Chemistry*, 115, 1133–1140. 10.1016/j.foodchem.2008.12.100

Bureau, S., Turion, B. Q., Signoret, V., Renaud, C., Maucourt, M., Bancel, D., Renard, C. (2013). Determination of the composition in sugars and organic acids inpeach using mid infrared spectroscopy: comparison of prediction results according to data sets and different reference methods. *Analytical Chemistry*, 85(23), 11312–11318. 10.1021/ac402428s

Castillo, R. D. P., Pena-Farfal, C., Neira, Y., Freer, J. (2017). Advances in analytical methodologies based on infrared spectroscopy for analysis of lignocellulosic materials: from classic characterisation of functional groups to FTIR imaging and micro-quantification. In Moore, E. (Ed.), *Fourier-transform Infrared Spectroscopy*. Nova Science Publishers, New York, pp. 33–65.

Cavin, C., Cottenet, G., Cooper, K. M., Zbinden, P. (2018). Meat vulnerabilities to economic food adulteration require new analytical solutions. *Chimia (Arau)*, 72(10), 697–703

Clark, C. C. (2016). Fast determination by Fourier-transform infrared spectroscopy of sugar acid composition of citrus juices for determination of industry maturity standards. *New Zealand Journal of Crop and Horticultural Science*, 44, 69–82.

Cooley, J. W., Turkey, J. W. (1965). An Algorithm for the Machine Calculation of Complex Fourier Series. *Mathematics of Computation*, 19, 297–301. 10.1090/S0025-5718-1965-0178586-1

Cozzolino, D., Holdstock, M., Dambergs, R. G., Cynkar, W. U., Smith, P. A. (2009). Mid-infrared spectroscopy and multivariate analysis: a tool to discriminate between organic and nonorganic wines grown in Australia. *Food Chemistry*, 116, 761–765.

Cozzolino, D. (2017). Vibrational and fluorescence spectroscopy. *Food Authentication*. John Wiley and Sons Ltd., New York, pp. 277–298.

Cozzolino, D., Cynkar, W. U., Shah, N., Smith, P. (2011). Multivariate data analysis applied to spectroscopy: potential application to juice and fruit quality. *Food Research International*, 44, 1888–1896.

Daniels, F., Williams, J. W., Bender, P., Alberty, R. A., Cornwell, C. D., Harriman, J. E. (1970). *Experimental Physical Chemistry*, 7th Ed. McGraw-Hill, New York, NY.

Defra, *et al.* (2014). Elliott review into the integrity and assurance of food supply networks–Final 558 report. *British Medical Journal*, 157(March), 146. 10.1136/bmj.1.4348.621-a

Downey, G. (2016). *Advances in Food Authenticity Testing*. Cambridge, UK.

ESA (2015). European Spice Association Quality Minima Document. Available at: https://www.esa-572spices.org/download/esa-qmd-rev-5-update-as-per-esa-tc-26-03-18.pdf

European Commission (2000). Directive 2000/13/E.C. on the approximation of the laws of the Member States relating to the labeling, presentation, and advertising of foodstuffs. *Official Journal of the European Union*, 109(16), 29–42. 2004R0726-v.7of 05.06.2013

European Commission (2002). Regulation (E.C.) 178/2002 of the European Parliament and of the Council of 28 January 2002 laying down the general principles and requirements of food law, establishing the European Food Safety Authority and laying down procedures in matters of food safety. *Official Journal of the European Union*, L 31, 1–40. 2004R0726-v.7of05.06.2013

European Commission (2018) Food Fraud. Available at: https://ec.europa.eu/food/safety/food-fraud_en (Accessed: 12 June 2018).

FAO (2009). Global agriculture towards 2050. *High Level Expert Forum-How to feed the world 2050*. http://www.fao.org/fileadmin/templates/wsfs/docs/Issues_papers/HLEF2050_Global_Agriculture.pdf

Food Standards Agency (2015). Adulteration of Food – Thresholds for Action and for, 590 (November), 1–7.

Freitag, S., Thain, S. C., Squier, A. H., Hogan, E. J., Crittenden, P. D. (2009). Assessing metabolic changes of the reindeer lichen C. portentosa to increasing environmental N inputs using metabolomic fingerprinting and profiling techniques. *Comparative Biochemistry and Physiology*, 153(2, Supplement), S57. 10.1016/j.cbpa.2009.04.518

Galvin-King, P., Haughey, S. A., Elliott, C. T. (2018). Herb and spice fraud; the drivers, challenges and detection. *Food Control*, 88, 85–97. 10.1016/j.foodcont.2017.12.031

Gauglitz, G., Vo-Dihn, T. (2003). *Handbook of Spectroscopy, Journal of the American Chemical Society*. 10.1021/ja033666c

Gauglitz, G., Vo-Dinh, T. (2003). *Handbook of Spectroscopy*. Wiley-VCH Verlag GmbH & Co. KGaA. 10.1002/3527602305

Georgouli, K., Martinez Del Rincon, J., Koidis, A. (2017). Continuous statistical modelling for rapid detection of adulteration of extra virgin olive oil using mid infrared and Raman spectroscopic data. *Food Chemistry*, 217, 735–742.

Giusti, M. M., Atnip, A., Sweeney, C., Rodriguez-Saona, L. E. (2011). Rapid authentication of fruit juices by infrared spectroscopic techniques. *Progress in Authentication of Food and Wine*. American Chemical Society, pp. 275–299.

GMA (Grocery Manufacturers Association) and Kearney, A. T. (2010). Consumer product Fraud: Deterrence and detection.

Gok, S., Severcan, M., Goormaghtigh, E., Kandemir, I., Severcan, F. (2015). Differentiation of Anatolian honey samples from different botanical origins by ATR-FTIR spectroscopy using multivariate analysis. *Food Chemistry*, 170, 234–240. 10.1016/j.foodchem.2014.08.040

Gouvinhas, I., de Almeida, J. M. M. M., Carvalho, T., Machado, N., Barros, A. I. R. N. A. (2015). Discrimination and characterization of extra virgin olive oils from three cultivars in different maturation stages using Fourier transform infrared spectroscopy in tandem with chemometrics. *Food Chemistry*, 174, 226–232. 10.1016/j.foodchem.2014.11.037

Gurdeniz, G., Ozen, B. (2009). Detection of adulteration of extra virgin olive oil by chemometric analysis of mid-infrared spectral data. *Food Chemistry*, 116, 519–525.

Handford, C. E., Campbell, K., Elliott, C. T. (2016). Impacts of milk fraud on food safety and nutrition with special emphasis on developing countries. *Comprehensive Reviews in Food Science and Food Safety*, 15(1), 130–142. 10.1111/1541-4337.12181

He, J., Rodriguez-Saona, L. E., Giusti, M. M. (2007). Midinfrared spectroscopy for juice authentication-rapid differentiation of commercial juices. *Journal of Agricultural and Food Chemistry*, 55, 4443–4452.

He, H.-J., Sun, D.-W. (2015). Microbial evaluation of raw and processed food products by visible/ infrared, Raman and fluorescence spectroscopy. *Trends in Food Science and Technology*, 46, 199–210.

Hennessy, S., Downey, G., O'Donnell, C. P. (2009). Confirmation of food origin claims by Fourier transform infrared spectroscopy and chemometrics: extra virgin olive oil from Liguria. *Journal of Agricultural and Food Chemistry*, 57, 1735–1741.

Huang, H., Grün, I. U., Ellersieck, M., Clarke, A. D. (2017). Measurement of total sodium alginate in restructured fish products using Fourier transform infrared spectroscopy. *EC Nutrition*, 11(1), 33–45

Huang, Y., Rasco, B. A., Cavinato, A. G. (2009). Fruit juices. *Infrared Spectroscopy for Food Quality Analysis and Control.* Elsevier, Amsterdam, The Netherlands, pp. 355–375.

Jawaid, S., Talpur, F. N., Sherazi, S. T. H., Nizamani, S. M., Khaskheli, A. A. (2013). Rapid detection of melamine adulteration in dairy milk by SB-ATR–Fourier transform infrared spectroscopy. *Food Chemistry*, 141(3), 3066–3071. 10.1016/j.foodchem.2013.05.106

Johnson, R. (2014). Food fraud and "economically motivated adulteration" of food and food ingredients. *Congressional Research Service Report*, January(R43358), 1–40.

Kanou, M., Kameoka, T., Suehara, K., Hashimoto, A. (2017). Mid-infrared spectroscopic analysis of saccharides in aqueous solutions with sodium chloride. *Bioscience, Biotechnology, and Biochemistry*, 81, 735–742.

Karoui, R., Downey, G., Blecker, C. (2010). Mid-infrared spectroscopy coupled with chemometrics: a tool for the analysis of intact food systems and the exploration of their molecular structure–quality relationships: a review. *Chemical Reviews*, 110, 6144–6168.

Kezban, C., Altuntas, E. G., İğci, N. (2020). Authentication and quality assessment of meat products by Fourier-transform infrared (FTIR) spectroscopy. *Food Engineering Reviews*, 13, 66–91. 10.1007/ s12393-020-09251-y

Lakshmi, V. (2012). Food adulteration. *International Journal of Science Inventions Today*, 1(2), 106–113.

Larkin P. J. (2018). *Infrared and Raman Spectroscopy: Principles and Spectral Interpretation*, 2nd ed. Elsevier, Amsterdam, Netherlands.

Lawson, G., Ogwu, J., Tanna, S. (2018). Quantitative screening of the pharmaceutical ingredient for the rapid identification of substandard and falsified medicines using reflectance infrared spectroscopy. *PLoS One*, 13(8), e0202059. 10.1371/journal.pone.0202059

Lohumi, S., *et al.* (2017). Quantitative analysis of Sudan dye adulteration in paprika powder using FTIR spectroscopy. *Food Additives & Contaminants: Part A*, 34(5), 678–686. 10.1080/19440049.2017.1290828

Lu, X., Rasco, B. A. (2012). Determination of antioxidant content and antioxidant activity in foods using infrared spectroscopy and chemometrics: a review. *Critical Reviews in Food Science and Nutrition*, 52, 853–875.

Meikle, J. (19 February 2005). Carcinogenic dye in hundreds of food products. The Guardian (London) – Final Edition 1.

Nunes, K. M., Andrade, M. V. O., Santos Filho, A. M. P., Lasmar, M. C., Sena, M. M. (2016). Detection and characterisation of frauds in bovine meat in natura by non-meat ingredient additions using data fusion of chemical parameters and ATR-FTIR spectroscopy. *Food Chemistry*, 205, 14–22.

Nyquist, R. A., Kagel, R. O. (2012). *Handbook of Infrared and Raman Spectra of Inorganic Compounds and Organic Salts: Infrared Spectra of Inorganic Compounds*. Academic Press, London.

Ordoudi, S. A., Cagliani, L. R., Melidou, D., Tsimidou, M. Z., Consonni, R. (2017). Uncovering a challenging case of adulterated commercial saffron. *Food Control*, 81, 147–155. 10.1016/j.foodcont.2017.05.046

Pedreschi, F., Segtnan, V. H., Knutsen, S. H. (2010). Online monitoring of fat, dry matter and acrylamide contents in potato chips using near infrared interactance and visual reflectance. *Food Chemistry*, 121, 616–620.

Pei, X., Tandon, A., Alldrick, A., Giorgi, L., Huang, W., Yang, R. (2011). The China melamine milk scandal and its implications for food safety741 regulation. *Food Policy*, 36(3), 412–420.

Petrakis, E. A., Polissiou, M. G. (2017). Assessing saffron (Crocus sativus L.) adulteration with plant-derived adulterants by diffuse reflectance infrared Fourier transform spectroscopy coupled with chemometrics. *Talanta*, 162, 558–566. 10.1016/j.talanta.2016.10.072

Premanandh, J. (2013). Horse meat scandal: A wake-up call for regulatory authorities. *Food Control*, 34(2), 568–569. 10.1016/j.foodcont.2013.05.033

PwC (2016). Food Fraud Vulnerability Assessment and Mitigation. PricewaterhouseCoopers, pp. 1–20. https://www.pwc.com/vn/en/publications/2016/food_fraud_vulnerability_assessment.pdf

Rahmania, H., Sudjadi Rohman, A. (2015) The employment of FTIR spectroscopy in combination with chemometrics for analysis of rat meat in meatball formulation. *Meat Science*, 100, 301–305.

Reis, N., *et al.* (2017). Simultaneous detection of multiple adulterants in ground roasted coffee by ATR-FTIR spectroscopy and data fusion. *Food Analytical Methods*, 10(8), 2700–2709.

Rohman, A., Che Man, Y. B. (2010). FTIR spectroscopy combined with chemometrics for analysis of lard in the mixtures with body fats of lamb, cow, and chicken. *International Food Research Journal*, 17, 519–526.

Rohman, A., Erwanto, Y., Man, Y. B. C. (2011). Analysis of pork adulteration in beef meatball using Fourier transform infrared (FTIR) spectroscopy. *Meat Science*, 88(1), 91–95.

Rohman, A., Ghazali, M. A. B., Windarsih, A., *et al.* (2020). Comprehensive review on application of FTIR spectroscopy coupled with chemometrics for authentication analysis of fats and oils in the food products. *Molecules*, 25(22), 5485. 10.3390/molecules25225485

Rohman, A., Man, Y. B. C. (2010). Fourier transform infrared (FTIR) spectroscopy for analysis of extra virgin olive oil adulterated with palm oil. *Food Research International*, 43(3), 886–892. 10.1016/j.foodres.2009.12.006

Santos, P. M., Pereira-Filho, E. R., Rodriguez-Saona, L. E. (2013). Rapid detection and quantification of milk adulteration using infrared microspectroscopy and chemometrics analysis. *Food Chemistry*, 138, 19–24.

Schulz, H., Baranska, M. (2007). Identification and quantification of valuable plant substances by I.R. and Raman spectroscopy. *Vibrational Spectroscopy*, 43, 13–25.

Sicherer, S. H., Burks, A. W., Sampson, H. A. (1998). Clinical features of acute allergic reactions to peanut and tree nuts in children. *Pediatrics*, 102(1), e6. 10.1542/peds.102.1.e6

Simmons, I. L. (1960). 'The Kbr Technique', 3, 1–8.

Skoog, D. A., Leary, J. J. (1992). *Principles of Instrumental Analysis, 4th Ed.* Harcourt Brace Jovanovich. Philadelphia, PA. Chapter 12.

Smith, B. (2011). *Fundamentals of Fourier Transform Infrared Spectroscopy*, 2nd ed. CRC Press, Boca Raton.

Smith, R., Manning, L., McElwee, G. (2017). Critiquing the inter-disciplinary literature on food fraud. *International Journal of Rural Criminology*, 3(2), 250–270.

Sorak, D., Herberholz, L., Iwascek, S., Altinpinar, S., Pfeifer, F., Siesler, H. W. (2012). New developments and applications of hand-held Raman, mid-infrared, and near-infrared spectrometers. *Applied Spectroscopy Reviews*, 47, 83–115.

Spink, J., Moyer, D. C. (2011). Defining the public health threat of food fraud. *Journal of Food Science*, 76(9), R157–R163. 10.1111/j.1750-3841.2011.02417.x

Su, W. H., Sun, D. W. (2019). Mid-infrared (MIR) spectroscopy for quality analysis of liquid foods. *Food Engineering Reviews*, 11, 142–158. 10.1007/s12393-019-09191-2

Talari, A. C. S., Martinez, M. A. G., Movasaghi, Z., Rehman, S., Rehman, I. U. (2017). Advances in Fourier-transform infrared (FTIR) spectroscopy of biological tissues. *Applied Spectroscopy Reviews*, 52, 456–506.

The Associated Press (2008). China's top food safety official resigns, NBC News. Available at: http://www.nbcnews.com/id/26827110/#.W30iicJJlhE

The Council of the European Union (2006). COUNCIL REGULATION (E.C.) No 510/2006 of 20 March 2006 on the protection of geographical indications and designations of origin for agricultural products and foodstuffs, 2006(510), 12–25.

The European Commission (2014a). COMMISSION DELEGATED REGULATION (E.U.) No 664/2014 of 18 December 2013 supplementing Regulation (E.U.) No 1151/2012 of the European Parliament and of the Council with regard to the establishment of the Union symbols for protected designations of origin, prote, 2014(886).

The European Commission (2014b). COMMISSION IMPLEMENTING REGULATION (E.U.) No 668/2014 of 13 June 2014 laying down rules for the application of Regulation (E.U.) No 1151/2012 of the European Parliament and of the Council on quality schemes for agricultural products and foodstuffs, 59(May), 35–59.

U.S. Food and Drug Administration (2009). Economically motivated adulteration; public meeting; request for comment' fed register, 74(64), 15497–15499.

United States Congress (1938). *Federal Food, Drug, and Cosmetic Act (FFCDA)*. United States Food and Drug Administration.

United States Congress (2011). *FDA Food Safety Modernization Act*. US FDA.

Vlachos, N., Skopelitis, Y., Psaroudaki, M., Konstantinidou, V., Chatzilazarou, A., Tegou, E. (2006). Applications of Fourier transform–infrared spectroscopy to edible oils. *Analytica Chimica Acta*, 573, 549–565.

Wielogorska, E., *et al.* (2018). Development of a comprehensive analytical platform for the detection and quantitation of food fraud using a biomarker approach. The oregano adulteration case study. *Food Chemistry*, 239, 32–39. 10.1016/j.foodchem.2017.06.083

Wiercigroch, E., Szafraniec, E., Czamara, K., Pacia, M. Z., Majzner, K., Kochan, K., Kaczor, A., Baranska, M., Malek, K. (2017). Raman and infrared spectroscopy of carbohydrates: a review. *Spectrochimica Acta A*, 185, 317–335.

Xu, L., Cai, C. B., Cui, H. F., Ye, Z. H., Yu, X. P. (2012). Rapid discrimination of pork in Halal and non-Halal Chinese ham sausages by Fourier transform infrared (FTIR) spectroscopy and chemometrics. *Meat Science*, 92(4), 506–510.

5 Analysis of Food Additives Using Chromatographic Techniques

Marco Iammarino and Aurelia Di Taranto

5.1 INTRODUCTION

More and more substances are permitted in food as additives for extending their shelf life, improving texture, modifying appearance, and for other technological functions. In Europe, more than 300 chemical substances are permitted as food additives in the Regulation No. 1129/2011/EC, with specific restrictions, where necessary (European Commission, 2011). A chemical substance is permitted for use as food additives after several toxicological studies aimed at ascertaining its safety for human consumption. The toxicological studies are focused on short-term genetic toxicity, acute oral toxicity, sub-chronic feeding, reproductive and developmental toxicity, and chronic toxicity/carcinogenicity. These studies and the related evaluations/opinions are carried out by dedicated Committees (e.g., the Expert Committee on Food Additives (JECFA), the EFSA Panel on Food Additives and Flavourings (FAF), etc.), for each food additive, individually, when the market requires the authorization for its use in food.

There is no compulsory test to take into account within food safety evaluations, regarding the contemporary presence of different food additives in the same food. This is a complex aspect due to the large number of authorized food additives and the wide range of foodstuffs that may be added with different food additives. The situation becomes more complicated when food consumed by younger and adolescents (especially sugary drinks, snacks, sweets, ice cream, fast foods, etc.) are taken into account, since a large use of food additives (especially food colorings and sweeteners) characterizes these products. Indeed, it is plausible that the synergic effect of several chemicals and/or the reactions among them may cause health risks for humans, also if their concentrations in the product are below the legal limits. In the last years, the attention of the scientific community (particularly EFSA) has focused on the so-called "cocktail effect" due to the presence of different toxic compounds (predominantly contaminants) in the same product. Some authors reported examples of enhanced toxicity due to the contemporary presence of different pesticides and of the combination polycyclic aromatic hydrocarbons/heterocyclic amines as products resulting from meat cooking (Jamin *et al.*, 2013; Ilboudo *et al.*, 2014). Given the lack of data about this topic and about the occurrence of the most important contaminants in largely consumed food throughout Europe, several "Total Diet Studies" have been developed in the last few years. These surveys have laid the foundation for subsequent toxicological evaluations of possible "cocktail effects". However, these "Total Diet Studies" substantially overlooked the food additives, since they were focused on the most important contaminants of the food chains (heavy metals, pesticides, dioxins, mycotoxins, etc.). It is also important to underline that, apart from some well-known food additives (e.g., nitrites/nitrates), for which many investigations have been carried out due to their potential toxic effects on humans, there is a substantial gap in knowledge about the occurrence in food of many others food additives (such as food dyes, sweeteners, butylated hydroxyanisole (BHA)/butylated hydroxytoluene [BHT], phosphates/polyphosphates, sulfiting agents, glutamates, etc.). In 2013, this gap has been highlighted by an External Scientific Report of the European Commission entitled: "Analysis of needs in postmarket monitoring of food additives and preparatory work for future projects in this field". In many cases, this gap is due to the lack of reference analytical methods to apply (Corporate author(s), 2013). Moreover, the EFSA released several scientific opinions related to the re-evaluation of certain food additives (i.e., phosphoric acid/phosphates/di-tri-and poly-phosphates, Indigo Carmine, etc.) reporting that the exposure may exceed the proposed ADIs, soliciting new studies about these topics (EFSA, 2014; 2019).

Among food additives, the most significant concern in food safety is the monitoring of food preservatives (FPs) used and their global intake in the diet. FPs are used for improving the safety of foods, delaying the quality parameters loss and extending the shelf life. Other than practical, economical, and devoid of off-flavors, they should be not toxic. However, some FPs, especially antimicrobials, are characterized by some aspects of toxicity on humans, but they are widely used, since the regulations permit their addition within specific limits, defined after accurate risk assessment. Given the large number of food types usually treated with FPs, such as fruit juice and nectars, flavored drinks, beer, wine (also alcohol-free) and vinegars, other alcoholic drinks with less than 15% of alcohol, snacks, desserts, flavored fermented milk products, milk powder and cheese products, fresh fruit and vegetables (also dried or frozen), jam, jellies and marmalades, mustard,

DOI: 10.1201/9781003334859-5

Table 5.1: Main Food Additives

Additive	E-code	Additive Group	Possible Health Effects due to High and Prolonged Intake
Sorbic acid	E200	Preservatives: Sorbates	Development of mutagens and genotoxic
Potassium sorbate	E202		agents, especially at low pH, as in the gastric
Calcium sorbate	E203		conditions
Benzoic acid	E210	Preservatives: Benzoates	Allergic reactions in sensitive subjects such as
Sodium benzoate	E211		rhinitis, hives, and dermatitis. Possible
Potassium benzoate	E212		generation of benzene in acidic beverages
Calcio benzoate	E213		
Sulfur dioxide	E 220	Preservatives/	Headache, gastrointestinal disturbances and
Sodium sulfite	E 221	antioxidants:	immunity-mediated reactions,
Sodium hydrogen sulfite	E 222	Sulfiting Agents	bronchoconstriction and fatal anaphylaxis. Cholinergic mediated bronchoconstriction
Sodium metabisulfite	E 223		Increased formation of reactive oxygen
Potassium metabisulfite	E 224		species and the oxidative stress. Vitamin deficiency, repro-toxicity and teratogenesis,
Calcium sulfite	E 226		gastrointestinal lesions, and esophageal
Calcium hydrogen sulfite	E 227		cancer
Potassium nitrite	E 249	Preservatives: Nitrite/	Formation of nitrosamine compounds.
Sodium nitrite	E 250	nitrate	Oxidation of hemoglobin to methemoglobin
Sodium nitrite	E 251		(Methemoglobinemy). Other adverse
Potassium nitrate	E 252		reactions in susceptible people
Polyphosphates	E 452	Thickeners/ emulsifiers/ stabilizers: Phosphates	Bile duct formation and reduced absorption of calcium
Ponceau 4R	E 124	Colorings: Ponceau 4R	Exacerbated hyperactivity in certain susceptible children with Attention Deficit/ Hyperactivity Disorder and other problem behaviors
Tartrazine	E 102	Colorings: Tartrazine	Asthma and chronic hives in a sensitive subpopulation of consumers

processed potato products, chewing gum, liquid egg, soups and broths, sauces, fresh meat preparations, meat products, unprocessed and processed fish, molluscs and crustaceans, fine bakery wares, processed nuts, breakfast cereals, and the overall intake in the diet can be very high.

The Commission Regulation (EU) No. 1129/2011 of 11 November 2011 amending Annex II to Regulation (EC) No 1333/2008 of the European Parliament and of the Council by establishing a Union list of food additives identifies the FPs by assigning a specific code. Most of FPs are used as antimicrobials (i.e., organic acids, sulfiting agents, nitrite/nitrate, etc.) within the pH range of the food. Nowadays, the most used FPs are sorbic acid and sorbates (E200-E203), benzoic acid and benzoates (E210-E213), sulfiting agents (E230-E238), nitrite/nitrate (E249-E252), and ascorbic acid and ascorbates (E300-E302). Moreover, other types of additives, such as polyphosphates, food dyes, and others are widely used as well, although some health effects of these compounds have been proven (EFSA, 2014; 2015a; 2015b; 2017a; 2019) (Table 5.1).

The inhibition of microorganisms' growth and their death is usually reached within a few days or weeks, depending on the type and concentration of FP used. A specific equation (5.1) is available related to the timescale for the killing of microorganisms in food treated with FPs:

$$K = 1/t \ \ln Z_0/Z_t \text{ or } Z_t = Z_0 \cdot e^{-Kt} \tag{5.1}$$

where K is the "death rate" constant, t is the time period, Z_0 is the number of living cells when the FP begins its action, and Z_t is the number of living cells after time t. This equation is valid when the cell material is genetically uniform and the FP concentration is high. The mechanism of action of FPs consists in delaying and inhibit the lag phase. However, FPs are not effective when the microbial population is high, since their action is especially focused on the inhibitory action of enzymes. Another particular characteristic of some FPs (weakly lipophilic acid) is the capability of moving freely through the membrane. In the cell cytoplasm, where the pH is high, these FPs acidify the environment by breaking down the pH component of the proton motive force. The cell reacts by expelling the protons, diverting the energy from growth functions, leading to cell fall (Batt and Tortorello, 2014; Caballero et al., 2016; MacDonald and Reitmeier, 2017).

Although many food additives have been approved for use in food and absolutely safe for human consumption, there is scientific evidence of several negative effects, particularly FPs, in the literature. In this regard, several food regulations worldwide define specific limits of addition for each food in which a FP is authorized. However, consumers have raised concerns about the presence of FPs in foods. Among FPs, the scientific community has focused the studies on a shortlist of compounds that seem the worthiest of attention. These FPs are sorbates, benzoates, sulfites, and nitrites/nitrates.

Sorbic acid and benzoic acids are two major food preservatives, added into foods for inhibiting the microbial growth and are also effective against molds and yeasts. The antimicrobial activity of sorbic acid and sorbates consists of the reaction with the sulfhydryl group of bacteria, molds, and yeasts enzymes (i.e., catalase, peroxidases, fumarase, succinic dehydrogenase, etc.) (Surekha and Reddy, 2014). This type of FP is widely used in cakes, sugary drinks, cheese, wine, and dried meats. The interaction between sorbate and nitrite results in the development of mutagens and genotoxic agents, especially at low pH, as in the gastric conditions. Two mutagens agents were identified: ethylnitrolic acid and 1,4-dinitro-2-methylpyrrole (Hartman, 1983). Moreover, a recent study linked the high potassium sorbate intake to genotoxicity for the human peripheral blood lymphocytes in vitro (Mamur et al., 2010). Benzoic acid and benzoates are especially used in acid foods, such as sauces (ketchup), soft drinks, canned tuna, etc., since at low pH they are effective. They also compromise the microbial cell membrane permeability and play an important role in inhibiting the enzymes of the oxidative phosphorylation, acetic acid metabolism, tricarboxylic acid cycle, and the amino acid uptake (Surekha and Reddy, 2014). Benzoic acid is considered harmless at employment doses, since it is completely eliminated in the urine as hippuric acid. However, an acceptable daily intake (ADI) corresponding to 300–400 mg for subjects weighing 60–80 kg, has been established by FAO/WHO (FAO/WHO, 2006). In fact, these additives may be responsible of allergic reactions in sensitive subjects such as rhinitis, hives, and dermatitis (Iammarino et al., 2011). The major concern related to the presence of these additives in food is the possible generation of carcinogenic benzene. Indeed, hydroxyl radicals, generated by metal-catalyzed reduction of oxygen and hydrogen peroxide by ascorbic acid (present in high amount in many sugary drinks, both as natural compound or added acidity regulator), can react with benzoic acid to form benzene. Many foods and beverages are characterized by ideal conditions (such as low pH) for this reaction (Gardner and Lawrence, 1993), so, the replacement of benzoic acid and benzoates in such food and beverages is needed.

Sulfites are a class of chemical compounds, SO_2 releasers, widely used as additives in the food industry, thanks to their antimicrobial, color stabilizing, anti-browning, and antioxidant properties. Sulfur dioxide is a highly reactive molecule that interacts with many cell components. The main activity is exercised on the molecular configuration of enzymes. In particular, the sulfite ion acts as a powerful nucleophile, cleaving the disulfide bonds of proteins, modifying the active sites of enzymes. Sulfur dioxide may also react with important coenzymes (i.e., nicotinamide adenine dinucleotide), prosthetic groups (i.e., thiamin, flavin, folic acid, heme, pyridoxyl, etc.), and cofactors. Regarding its action against yeasts, SO_2 and sulfite addition results in a significant decrease in adenosine triphosphate content prior to cell death. Another example of action is against *Escherichia coli*. In this case, the NAD-dependent formation of oxaloacetate from malate is inhibited by sulfiting agents (Surekha and Reddy, 2014). Nowadays, sulfites are included in the list of allergens and all food products containing sulfites above 10 mg kg^{-1} or 10 mg L^{-1} must show this information on the label. The symptomatology and pathogenesis of sulfite hypersensitivity are different depending on individual, exposure and source. These additives may cause headaches, gastrointestinal disturbances, and immunity-mediated reactions (skin irritations, urticaria, dermatitis, etc.); also, bronchoconstriction and fatal anaphylaxis. After ingestion, sulfites are mainly absorbed in the gastrointestinal tract (70–90%) and only a small amount through the lungs. A molybdenum-dependent enzyme, linked to the mitochondrial respiratory chain, the sulfite oxidase (SO_X), detoxifies sulfites converting them to sulfates, especially in the liver, kidney, and heart, where the catalyzing activity is the highest. Moreover, it has been indicated that the SO_X enzyme has lower activity in a subgroup of asthmatic patients. As a result, the excessive accumulation of sulfites may induce cholinergic mediated bronchoconstriction in some individuals. The release of chemical and cellular mediators may be mediated by IgE or non-IgE mechanisms. The toxic mechanisms of sulfite have been investigated for many years. These compounds could increase the expression of asthma-related genes (EGF, EGFR, and COX-2), and induce oxidized glutathione and reduced glutathione depletion in rat hepatocytes, increasing the formation of reactive oxygen species and the oxidative stress. Moreover, the formation of glutathione sulfonate may increase the toxicity of xenobiotics. Other studies demonstrated that sulfite addition may induce the increase in erythrocyte lipid peroxidation in sulfite oxidase-competent animals and the alteration of oral and gut microflora. As indicated above, sulfite can destroy the thiamine and denature other vitamins. So, the chronic exposure to sulfite, and the consequent vitamins deficiency, may lead to organ and tissue atrophy, fibrosis, teeth depigmentation, focal myocardial necrosis, and, more rarely, polyneuritis. Finally, although both EFSA and IARC reported no evidence of carcinogenicity, repro-toxicity, and teratogenesis for sulfites, many studies reported possible gastrointestinal lesions and esophageal cancer caused by disulfites (D'Amore *et al.*, 2020).

Sodium and potassium salts of nitrite and nitrate are other food additives largely used, due to contemporary action as antimicrobials (especially against *Clostridium botulinum*), organoleptic characteristics improvers (mainly appearance and color) and stabilizers against peroxidation of lipid (D'Amore *et al.*, 2019). Other than vegetables and water, the intake of such compounds is also due to cured meats consumption. Indeed, the link between high consumption of cured meats and gastric cancer was demonstrated by several studies. The reason of such correlation was found in the reaction between nitrite and free amines with consequent formation of carcinogenic compounds (nitrosamines). The amount of nitrate in the products is also significant, since it can be converted to nitrite under gastric acidic condition or due to reducing action of bacteria enzymes. Both the IARC, in 2015, and the EFSA, in 2017, confirmed the significant role of nitrate and nitrite as food preservatives of meat products on the incidence of colorectal cancers (EFSA, 2017b; IARC, 2015). Other health effects due to the presence of these additives in food are the oxidation of hemoglobin to methemoglobin that cannot bind and transport oxygen to tissues (Methemoglobinemia) and several adverse reactions in susceptible people (Iammarino and Di Taranto, 2012a; Iammarino *et al.*, 2013a; 2017c). By virtue of the foregoing, many researches focused on the replacement of these additives have been developed during the last decades (i.e., nanoparticles-packaging, addition of natural additives, such as antimicrobial molecules from plants, probiotic bacteria, etc.). However, a definitive solution has not been achieved yet, and nitrite and nitrate are still used in large amounts in foods. The antimicrobial action of nitrite consists in the release of nitrous acid and nitrogen oxides. This release inhibits active transport of aldolase and proline in *E. coli*, *Pseudomonas aeruginosa*, and *Enterococcus faecalis*. Instead, the anticlostridial action is based on the reaction between nitric oxide and iron of a siderophore compound involved in clostridia electron transport (Batt and Tortorello, 2014).

Finally, the following chapters deal with other widely used food additives, such as polyphosphates, food dyes, and antioxidants, which were connected to food safety concerns in various ways.

The following sections describe, case by case, a series of confirmatory analytical methods validated for the quali-quantitative detection and quantification of most important food additives in several types of food and beverages. These analytical procedures, fully described and validated following the most updated international guidelines (Iammarino, 2019), can be considered as useful tools for laboratories in charge of food inspection for both routine food control and research activity. Moreover, these methods can be applied for risk/benefits assessments, taking into account that this topic has gained much in importance during the last years and is an ongoing debate, especially at the FDA and EFSA. Indeed, there is no food or food additive 100% safe for all population groups, so that EFSA recently proposed composite metrics such as DALYs or QALYs for this kind of study (EFSA, 2010). The analytical approaches described in these paragraphs are then at the basis of such studies, since they allow the obtainment of comprehensive and reliable data sets on food additive levels in food and beverages (Iammarino *et al.*, 2022b; 2022c). In Table 5.1, a list of the main food additives studied in this chapter, together with their possible health effects, is reported.

5.2 FOOD ADDITIVES DETERMINATION BY ION CHROMATOGRAPHY
5.2.1 Nitrites and Nitrates in Food

Three different procedures for sample preparation were optimized. The first is applicable to fresh meats, meat products, and shellfish analysis. A 5-g portion of a homogenized sample is mixed with 100 mL of ultrapure water and placed for five minutes at 70°C. After cooling, the mixture is centrifuged for 5 min at 1500 x g at room temperature and the supernatant (~2 mL) is microfiltered (0.2 μm) before IC injection. This final microfiltration step is the same for all three procedures. The second procedure of a sample preparation is applicable to dairy products. A 4-g portion of homogenized sample is placed in a 50 mL Falcon^{TM} plastic tube and mixed with 40 mL of ultrapure water. The mixture is then vortexed for 1 min and then centrifuged for 5 min at 1500 x g at room temperature. The third procedure is applicable to leafy vegetables. A 100-g portion of sample is properly homogenized using a mixer capable of obtaining a proper level of homogenization, and then a 4-g portion of homogenized sample is mixed with 200 mL of ultrapure water and placed at 70°C for five minutes. After cooling, the mixture is filtered as described above.

The analytical determinations were carried out using two ion chromatography systems. The first was a Dionex DX500, composed of a GP50 quaternary gradient pump, an electrochemical detector (model ED40, set to conductivity mode) equipped with a temperature compensated conductivity cell, a Dionex anion self-regenerating suppressor set to 50 mA and an injection valve with a 25 μL injection loop. The bottle containing the mobile phase (9 mM sodium carbonate) was pressurized with pure nitrogen to 0.8 MPa. The system was interfaced to a personal computer for instrumentation control, data acquisition, and processing via proprietary network chromatographic software (PeakNet^{TM}, Dionex Corporation, Sunnyvale, CA). The second system was a Thermo Scientific^{TM} Dionex^{TM} ICS-6000 HPIC^{TM} System Thermo Fisher Scientific Inc., Waltham, MA, USA) composed of a SP Single Pump (ICS-6000), a Dionex anion self-regenerating suppressor (ADRS 600, 4 mm set at recommended voltage), a DC detector set to conductivity mode, and an injection valve with a 25-μL loop. The chromatographic separations were accomplished using an IonPac AS9-HC anion exchange column (250 mm × 4 mm i.d., particle size: 13 μm) coupled to a precolumn AG9-HC (Dionex Corporation, Sunnyvale, CA) eluted in isocratic mode with a flow rate of 1.0 mL min^{-1}, resulting in a total run time of 20 minutes. The calibration was obtained by using nitrite and nitrate standard solution in ultrapure water in a concentration range equal to 0.2–6.5 and 0.6–12.5 mg L^{-1} for nitrite and nitrate, respectively. This procedure was validated according to the relevant Normative in force in Europe (European Commission, 2002; 2017) and with Regulation 882/2004/ EC (European Commission, 2004). The most important validation parameters evaluated (Iammarino and Di Taranto, 2012a) are reported in Table 5.2, while in Figure 5.1 some chromatogram examples are shown.

This method was used for the development of many research papers in food safety. Several surveys were carried out, focused on the quantification of nitrite and nitrate endogenous levels in foodstuffs. Very interesting findings were reported relating to fresh meats (nitrate concentration up to 40 mg kg^{-1} in equine meat and 30 mg kg^{-1} in pork and cow meat), to different cheese types (nitrate concentration up to 58.6 mg kg^{-1}) and to mussels (nitrate concentrations up to 205.3 mg kg^{-1}) (Iammarino et al., 2013a). This analytical method was also applied for a comprehensive monitoring for evaluating the Italian meat products with the highest levels of nitrite and nitrate. The survey concluded that Bresaola was the meat product characterized by the highest levels of nitrite and nitrate, followed by Speck. Comparable levels were detected in salami, bacon, wurstel, ham, and seasoned sausage; with slight lower values in Mortadella, cooked meat, and cooked ham; and the lowest concentrations in canned meat (Berardi et al., 2021). Regarding vegetable analysis, this approach was used for evaluating the levels of nitrite and nitrate in most consumed leafy vegetables (spinach, lettuce, chard, and wild rocket). These studies reported that these samples can be characterized by nitrate levels higher than the legal limits and that nitrite levels can reach high values, also exceeding 200 mg kg^{-1} in leafy vegetables, so that the introduction of specific legal limits for nitrites in leafy vegetables was suggested (Iammarino et al., 2014; 2022a). Moreover, the certification of a reference material for nitrates (spinach powder) was obtained by an international interlaboratory study, by using, among others, this analytical method (Pagliano et al., 2019).

5.2.2 Sulfites in Meat Products, Seafood, Processed Vegetables

The extraction of sulfites from different food matrices and the preparation of standard solutions are obtained by using a specific solution containing 50 mM NaOH and 10 mM fructose. This

Table 5.2: Main Validation Parameters of the Analytical Method for the Determination of Nitrite and Nitrate in Food by Traditional Ion Chromatography

Compound	Determination Coefficient (R^2)	LOD (mg kg^{-1})	LOQ (mg kg^{-1})	Recovery*(%)	RSD_r*(%)	Application Field (Selectivity and Ruggedness)	Measurement Uncertainty (%)
Nitrite	0.999	1.5	4.5	98.3	4.1	Meat, seafood, cheese, vegetables	7.8
Nitrate	0.999	3.2	9.6	98.7	4.2		9.7

RSDr = Repeatability relative standard deviation
* = Mean value calculated on 4 spiking levels (n = 6 for each level)

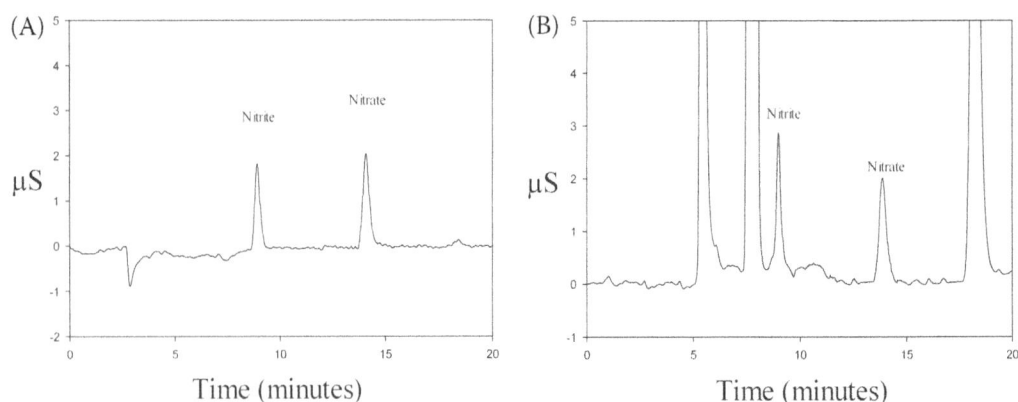

Figure 5.1 Nitrite and nitrate standard solution at a concentration of 3.0 mg L^{-1} (A); salami sample with nitrite and nitrate concentrations equal to 79.3 and 62.5 mg kg^{-1}, respectively (B).

composition allows the stabilization of sulfite ion against oxidation (at least 6 hours at 4°C). The sample is homogenized and then 4 g are weighed and placed in a 50-mL plastic tube together with 40 mL of this solution. The sample is placed on horizontal shaker for 30 min. After centrifugation for 5 min at 100 g at room temperature, about 2 mL of supernatant are filtered under a vacuum (if needed) and then microfiltered (0.2 μm) before IC injection. The same analytical equipment described in the previous paragraph (comprising the analytical column, without pre-column) can be used. The mobile phase, set at 1.0 mL min^{-1}, is composed of 8 mM Na_2CO_3 and 2.3 mM NaOH (A), and 24 mM Na_2CO_3 (B). The gradient program starts with an isocratic step at 100% A for 15 min, a gradient step to 50% A and 50% B in 1 min, and then 4 min at this eluent concentration. The system is then re-equilibrated for 20 min at 100% A (total run time: 40 min). The most important validation parameters evaluated (Iammarino *et al.*, 2010; 2013b) are reported in Table 5.3, while in Figure 5.2 some chromatogram examples are shown.

This analytical method was compared to the Monier-Williams standardized approach (where final titration is replaced by sulfate determination by ion chromatography) in terms of reliability and applicability for official food control activities. This study concluded that the accuracy and measurement uncertainty were comparable between two approaches, while method sensitivity is higher for ion chromatography method. However, the most significant finding of this study was the verification that some compounds containing sulfur, which can be found in meat (L-methionine, sulfides and 2-methyl-3-furanthiol) can cause false-positive responses when the Monier-Williams approach is used. So, the ion chromatography should be preferred as confirmatory technique (Carrabs *et al.*, 2017; Iammarino *et al.*, 2017a). This analytical procedure was applied for the development of several surveys useful for the evaluation of sulfite levels both in meat products (Iammarino *et al.*, 2017b) and shrimp (Iammarino *et al.*, 2013c). Moreover, through an in-depth monitoring on the market, developed by analyzing more than 2,000 meat products, it was possible to verify that particular ingredients of some meat preparations (mainly white wine) can lead to the detection of quantifiable levels of sulfite in such products, so that a maximum permitted level for sulfite in products with no added food additive, corresponding to 40 mg kg^{-1}, was proposed (Iammarino *et al.*, 2012). Regarding shrimp analysis, this method was applied for evaluating the effect of different cooking treatments (grilling, oven, frying, steaming, and stewed cooking) on the residual level of sulfites in shrimps. The study concluded that cooking leads to the decrease of sulfites levels in the products, with the highest percentage of reduction (55.3%) obtained by steaming and lowest using oven (13.9%) (Berardi *et al.*, 2022).

5.2.3 Polyphosphates Determination in Food of Animal Origin

This analytical method was developed for the simultaneous determination of all condensed phosphates listed in the European Regulation No. 1129/2011, namely (E450–diphosphate), E451 (triphosphate), E452 (tetrapolyphosphate), other than trimetaphosphate. These analytes are highly unstable and the degradation of more complex compounds (i.e., tetrapolyphosphate) leads to the

Table 5.3: Main Validation Parameters of the Analytical Method for the Determination of Sulfiting Agents in Food by Direct Ion Chromatography

Compound	Determination Coefficient (R^2)	LOD (mg kg^{-1})	LOQ (mg kg^{-1})	Recovery*(%)	RSD$_r$*(%)	Application Field (Selectivity and Ruggedness)	Measurement Uncertainty (%)
Sulfite	0.999	2.7	8.2	88.6	5.8	Fresh meat and meat products, processed vegetables, jam, seafood	9.9

RSDr = Repeatability relative standard deviation
* = Mean value calculated on 3 spiking levels (n = 6 for each level)

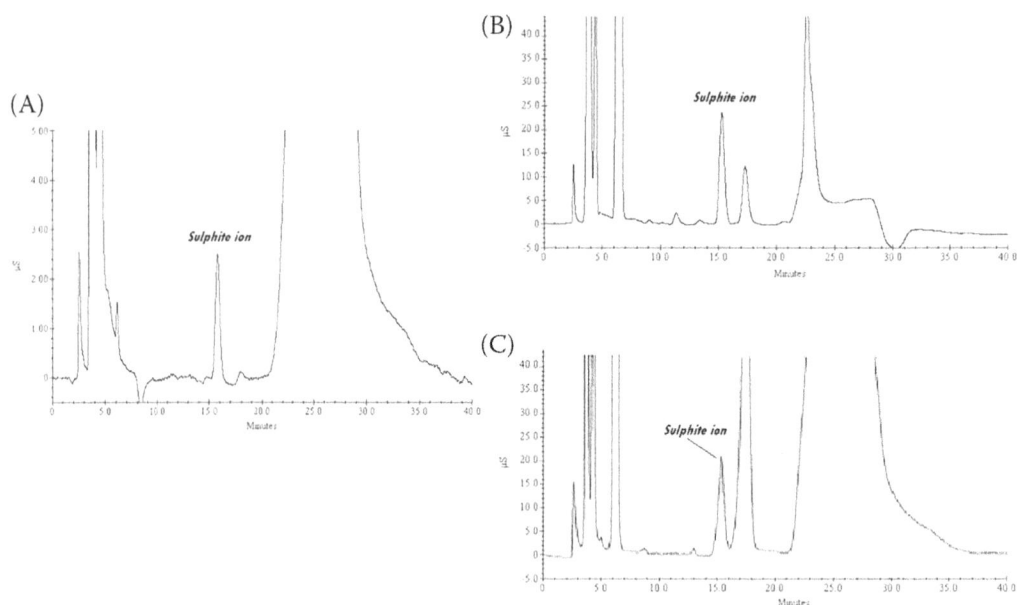

Figure 5.2 Sulfite standard solution at a concentration of 10.0 mg L^{-1} (A); fresh cattle/pork sausage with sulfite concentration of 835.5 mg kg^{-1} (as SO$_2$) (B); shrimp sample with sulfite concentration of 612.9 mg kg^{-1} (as SO$_2$) (C).

formation of simpler such as triphosphate and diphosphate. Thus, the standard solutions are prepared fresh daily and singularly for each analyte. The sample is homogenized, then 4 g are weighed and placed in a 50-mL plastic tube together with 40 mL of ultrapure water. The sample is vortexed for 1 min and centrifuged at 250 g for 5 min at room temperature. The supernatant (~2 mL) is then microfiltered (0.2 µm) prior to chromatographic analysis. No further cleanup step is required. The same chromatographic apparatus described above can be used for applying this technique. The chromatographic column is different, an IonPac AS11 (250 mm × 2 mm i.d., 9 µm) (Dionex Corporation, Sunnyvale, CA). The gradient elution is obtained by using two lines: 10 mM NaOH (A) and 80 mM NaOH (B) and a flow rate set at 0.5 mL min^{-1}. Starting from a Dionex Application Note (Dionex Corporation, 2016), the solvent gradient program was optimized for assuring method application to most important types of foodstuffs of animal origin. The gradient program is the following: 100% A (4 min), up to 20.5 mM in 1 min, isocratic (2 min), up to 45 mM in 1 min, isocratic (9 min), and a final re-equilibration step at 10 mM for 2 min. The calibration curves are obtained, injecting the following concentrations for diphosphate, triphosphate, and trimeta-phosphate: 1.25, 2.5, 5, 10, and 20 mg L^{-1}. Concentrations ten times higher are used for tetrapolyphosphate. The most important validation parameters evaluated for this method (Iammarino and Di Taranto 2012b) are reported in Table 5.4, while in Figure 5.3 some chromatogram examples are shown.

This analytical method was applied for discriminating fish samples (*Pangasius hypophthalmus*) treated with polyphosphate solutions, since this is a widespread food sophistication worldwide, with significant food safety implications. The application of this method was allowed to identify the samples submitted to this treatment while those not submitted showed no signal identifiable as polyphosphate so that no false-positive response was obtained. This method was also compared to the standardized approaches available for meat products and cheese (Ente Italiano di Normazione, 1997; International Organization for Standardization & International Dairy Federation, 2013), based on the indirect photometry and the total phosphorus determination after sample mineralization. This comparison verified that method trueness (recovery percentage) and the measurement uncertainty of two methods are comparable, while the precision of ion chromatography approach is higher. Another comparison was made by analyzing commercial samples, and the data were evaluated as "false positive" and "false negative" responses, based on the product label indications.

Table 5.4: Main Validation Parameters of the Analytical Method for the Determination of Condensed Phosphates in Food by Ion Chromatography

Compound	Determination Coefficient (R^2)	LOD (mg kg^{-1})	LOQ (mg kg^{-1})	Recovery* (%)	RSD$_r$* (%)	Application Field (Selectivity and Ruggedness)	Measurement Uncertainty (%)
Diphosphate	0.997	11.2	33.9	92.1	2.8	Meat products, cheese, seafood	5.3
Trimetaphosphate	0.999	8.1	24.6	94.2	2.5		7.5
Triphosphate	0.996	14.0	42.4	91.5	2.6		6.3
Tetrapolyphosphate	0.999	33.0	99.0	93.6	2.7		7.5

RSD$_r$ = Repeatability relative standard deviation
* = Mean value calculated on 4 spiking levels (n = 6 for each level)

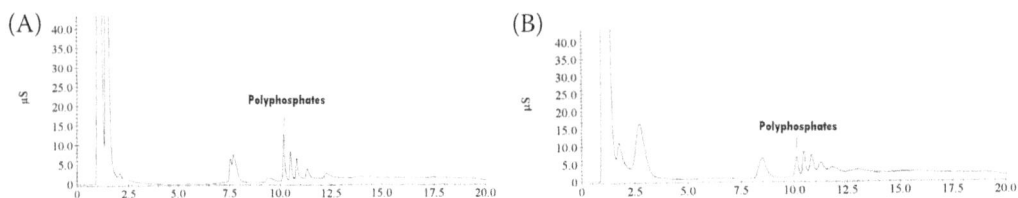

Figure 5.3 Polyphosphates standard solution at a concentration of 200.0 mg L^{-1} (A); processed cheese sample with polyphosphates concentration equal to 866.1 mg kg^{-1} (as P$_2$O$_5$) (B).

As the final result, the indirect photometry can be considered a reliable method as "screening technique", since the probability to obtain "false positive" results is significant. Ion chromatography may be used as a confirmatory technique, successfully (no "false-positive" responses), except for seafood analysis (Iammarino *et al.*, 2020a).

5.2.4 Organic Acids and Other Additives in Cheese

Different analytical methods were developed for the determination of several organic acids widely used in food as both food preservatives, stabilizers, and acidity regulators. One method is applicable for the simultaneous determination of lactic acid, acetic acid, sorbic acid, nitrite, benzoic acid, nitrate, and phosphate in all types of solid milk-based products. The same chromatographic system and analytical column reported above for sulfite analysis is used.

A 4-g portion of a sample, homogenized by blade homogenizer, is placed in a Falcon^{TM} tube and mixed with 40 mL of a NaOH 8.5 · 10^{-3}M solution. This particular solution allows an adequate extraction of all analytes from the matrix, increasing method sensitivity (by increasing acids dissociation) and stabilizing the nitrite ion. The analytes extraction is accomplished by placing the tubes in an ultrasonic bath (Ultrasound power: 80%; Heating: 40°C) for 10 minutes and then vortexing for 1 minute. The sample preparation is completed by a step of purification, carried out through a centrifugation (1,500 x g, 10 minutes at room temperature) and a microfiltration (0.2 μm) of supernatant. Prior to chromatographic analysis (injection volume: 25 μL), the excess of chloride is then removed by filtering ~ 1 mL of filtrate using OnGuard II Ag chromatography filters (Dionex Corporation, Sunnyvale, CA), previously activated with 1 mL of ultrapure water.

The analytes elution is obtained by using a flow rate of 1.0 mL min^{-1}. The mobile phase is composed by 0.9 mM Na$_2$CO$_3$ (A) and 28.5 mM Na$_2$CO$_3$ (B). The linear gradient is the following: from 0.9 mM to 3.7 mM in 5 minutes, from 3.7 mM to 9.2 mM in 1 min, an isocratic step for 19 minutes, then a linear gradient from 9.2 mM to 28.5 mM in 1 minute and 4 minutes at this eluent concentration. The system is then re-equilibrated for 10 min at the initial Na$_2$CO$_3$ concentration. The most important validation parameters evaluated for this method (Iammarino and Di Taranto 2013) are reported in Table 5.5, while in Figure 5.4 some chromatogram examples are shown.

This analytical method was applied for the development of a comprehensive monitoring on different types of fresh and ripened cheese, in order to evaluate the natural levels of benzoic acid, since this presence was reported in the literature (Jia *et al.*, 2023). This investigation allowed to suggest a maximum permitted level for this compound in cheese, equal to 40 mg kg^{-1} (Iammarino *et al.*, 2011). Another type of chromatographic separation was proposed for the analytical determination of citric acid in *mozzarella* cheese, by using the same chromatographic system described above, but a different analytical column, the IonPac AS11-HC column (250 mm × 4 mm i.d., particle size: 13 μm, Dionex Corporation, Sunnyvale, CA). This approach was optimized with an Application Note proposed by Thermo Fisher Scientific for the determination of organic acids in food (Thermo Fisher Scientific, 2012 Iammarino and Di Taranto, 2012b). The sample preparation consists in placing 4 g of homogenized sample in a 50-mL polypropylene tube together with 40 mL of deionized water. The analyte extraction is obtained by vortexing for 1 minute. Sample purification is carried out through a centrifugation (1,500 g, 10 minutes at room temperature) and subsequent microfiltration of supernatant (0.2 μm) prior to chromatographic analysis. The gradient elution, using a flow rate of 1.0 mL min^{-1}, is based on 38.25 mM NaOH (A) and 0.5 mM NaOH (B). The solvent program starts with a linear gradient from 0% to 10% A in 12 minutes, isocratic for 5 minutes, and then up to 100% A in 1 minute. The mobile phase composition remains constant for

Table 5.5: Main Validation Parameters of the Analytical Method for the Determination of 7 Food Additives in Cheese by Ion Chromatography

Compound	Determination Coefficient (R^2)	LOD (mg kg^{-1})	LOQ (mg kg^{-1})	Recovery*(%)	RSD$_r$*(%)	Application Field (Selectivity and Ruggedness)
Lactic Acid	0.992	0.6	1.9	92.3	9.4	Solid dairy products
Acetic Acid	0.998	3.6	11.9	79.6	3.0	
Sorbic Acid	0.990	59.4	196.0	72.8	10.2	
Nitrite	0.994	2.8	9.3	78.6	10.3	
Benzoic Acid	0.999	16.5	54.5	95.0	7.0	
Nitrate	0.997	2.1	6.8	88.1	10.9	
Phosphate	0.997	2.1	7.0	98.4	8.5	

RSD$_r$ = Repeatability relative standard deviation
* = Mean value calculated on 1 spiking level (n = 6)

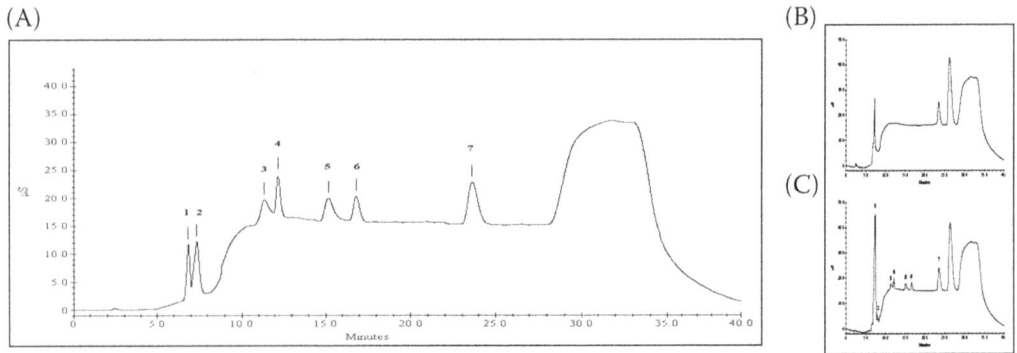

Figure 5.4 Seven additives standard solution: 1: Lactic acid, 2: Acetic acid, 3: Sorbic acid, 4: Nitrite, 5: Benzoic acid, 6: Nitrate, 7: Phosphate (A); "Blank" Cheese sample (B); cheese sample fortified with standard solution of 7 additives (C).

3 minutes, then the system was re-equilibrated for 4 minutes at 100% A. The instrumental calibration is obtained by analyzing four citric acid standard solutions in ultrapure water in a concentration range 6.25–50 mg L^{-1}. This method, coupled to the previous one, described for seven additives determination, was used for a study focused on the correct determination of lactic and citric acid in mozzarella cheese, since these two compounds are naturally present in such products. The study concluded that the addition of food additives can be ascertained if the concentration detected exceeds 0.65 and 4.0 g kg^{-1} for citric acid and lactic acid, respectively (Di Taranto et al., 2015).

5.3 FOOD PRESERVATIVES DETERMINATION BY CAPILLARY ION CHROMATOGRAPHY

5.3.1 Nitrites and Nitrates Determination in Food

Another approach useful for the detection of nitrate and nitrite in foodstuffs, alternative to the well-established traditional ion chromatography, was developed by using a novel chromatographic technique, the capillary ion chromatography (CIC). This technique is based on a novel technology, developed by using capillary column that, like conventional IC, utilize ion exchange, and eluent generation devices. When using these devices, reagents consumption is virtually zero and the operators simply add water to the system for obtaining the chromatographic separation (Bodsky and Hoefler, 2012).

In the case of nitrite and nitrate determination, this application can be obtained by using a capillary HPIC system (Dionex ICS-4000-Thermo Fisher Scientific, Waltham, USA), equipped with dual-stepper motor pump, 0.4 µL sample loop, eluent generator (EGC KOH), suppressor cartridge (ACES™ 300, set at 10 mA) and a conductivity detector. The chromatographic capillary column used is the Dionex IonPac® AS11-HC capillary column (250 mm x 0.4 mm i.d., 9 µm, Thermo Fischer Scientific), eluted using the following KOH gradient: 2 mM KOH (9 min), up to 15 mM in 1 min, up to 30 mM in 26 min and final re-equilibration step at 2 mM (2 min). The flow rate is 0.015 mL min^{-1} and the total run time is 38 min. The Chromeleon 7 software (Thermo Fisher Scientific, 2012) is used for instrument management, data acquisition, and elaboration. Regarding sample extraction and cleanup, 4 g of the sample, previously homogenized, weighed and placed in a 200-mL flask together with 80 mL of ultrapure water. The analyte extraction from the matrix is obtained by placing the samples in bain-marie at 80°C for 5 min. After cooling, the resulting suspension is microfiltered (0.22 µm) prior to chromatographic analysis. No further cleanup step is needed. The calibration curves can be obtained by injecting five standard solutions of nitrite and nitrate in the concentration range from 0.01 to 20 mg L^{-1}. The most important validation parameters, studied for this method (D'Amore et al., 2019), are reported in Table 5.6, while in Figure 5.5 some chromatogram examples are shown.

This analytical method was compared to the conventional approach by traditional ion chromatography, well established at an international level. The first comparison was focused on the selectivity parameter. Indeed, some authors identified some sugar phosphates present in animal

Table 5.6: Main Validation Parameters of the Analytical Method for the Determination of Nitrite and Nitrate in Food by Capillary Ion Chromatography

Compound	Determination Coefficient (R^2)	LOD (mg kg^{-1})	LOQ (mg kg^{-1})	Recovery* (%)	RSD$_r$* (%)	Application Field (Selectivity and Ruggedness)	Measurement Uncertainty (%)
Nitrite	0.999	1.2	3.6	100.8	2.7	Meat, seafood, cheese, vegetables	2.4
Nitrate	0.999	0.7	2.1	98.9	3.4		2.6

RSD$_r$ = Repeatability relative standard deviation
* = Mean value calculated on 3 spiking levels (n = 6 for each level)

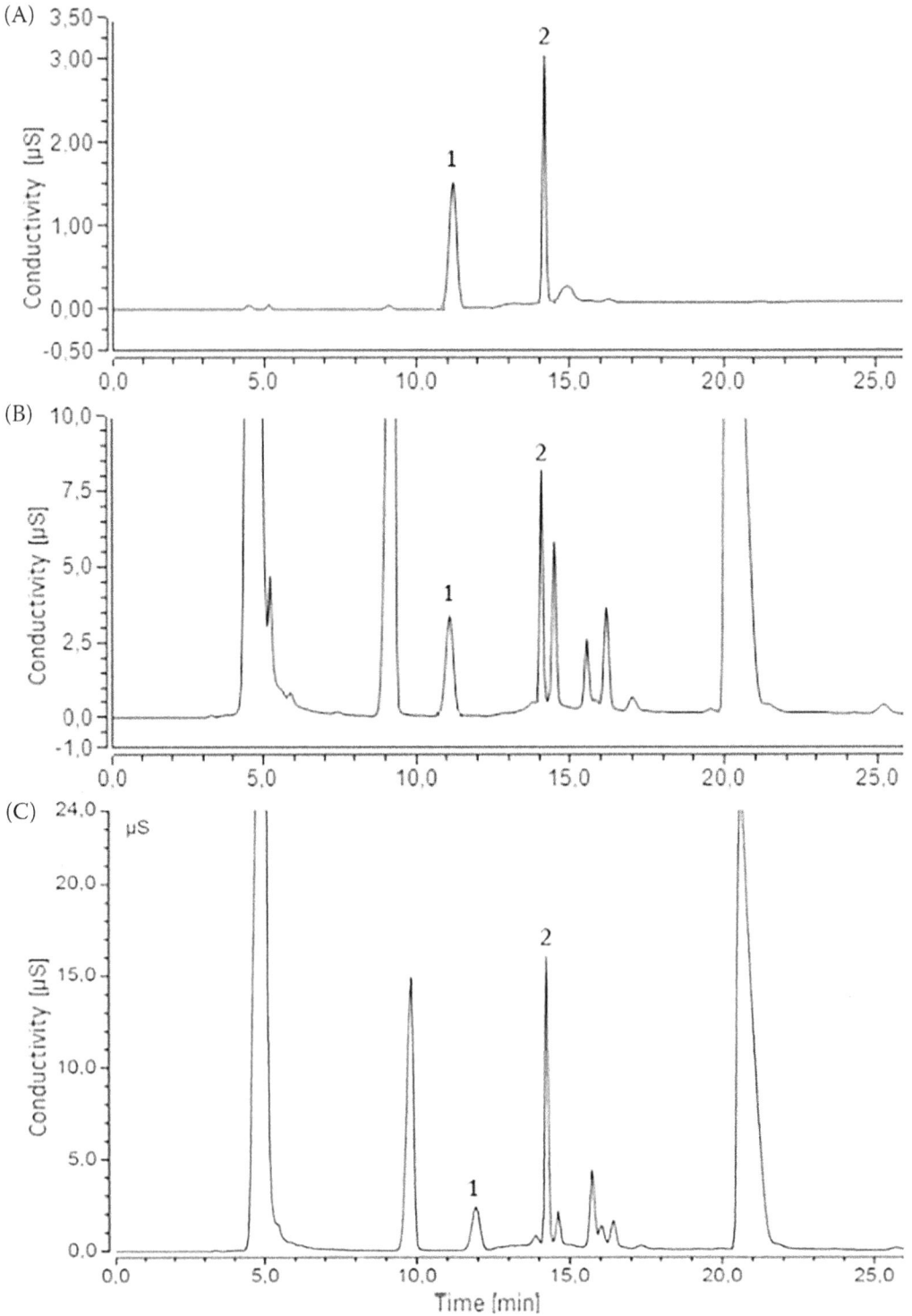

Figure 5.5 Nitrite (1) and nitrate (2) standard solution at a concentration of 2.0 mg L^{-1} (A); fresh cattle meat sample spiked with nitrite and nitrate (B); reference meat sample containing nitrite and nitrate (C).

muscles which can interfere with nitrate ion, since they are characterized by the same retention time. This drawback is avoided by using CIC. The second comparison was developed considering the poor chromatographic resolution between chloride and nitrite ions of traditional method. Also, in this case, CIC allows better resolution between two ions. Regarding validation parameters, method sensitivity of CIC is higher, method accuracy is comparable and measurement uncertainty is higher in conventional ion chromatography. Finally, the comparison highlighted that the reduced use of chemicals required for each analysis (15.2 μL of diluted KOH) can be considered another strength of this approach, in the view of the "Green chemistry" perspective (D'Amore et al., 2019).

5.3.2 Sulfite Determination in Solid Foods and Alcoholic Beverages

The most applied method for quantifying sulfiting agents in food is the Monier-Williams approach (Association of Official Analytical Chemists, 2000). This method is based on sample acidification with consequent distillation of SO_2, which is collected in H_2O_2 and converted to H_2SO_4. The final step of this method is the titration of such acid with NaOH (Montes et al., 2012). However, this last step has been called into question by several authors, since titrations can suffer from repeatability drawbacks in the determination of the end point (Hulanicki and Glab, 1975). The capillary ion chromatography has been tested for replacing the final titration of Monier-Williams method, successfully. The same chromatographic system and capillary analytical column described above for nitrate and nitrite determination in food by CIC is used. Regarding sample preparation, a distillation unit is used. Ten g of a homogenized sample (for solid foods) and 10 mL for alcoholic beverages are placed in a 300-mL sample tube with 50 mL of 4N HCl. Thirty mL of 3% (v/v) H_2O_2 are placed in the receiving vessel and then starting the distillation. The solution recovered by distillation is then concentrated on a hot plate at 200°C to a final volume of ~80 mL, transferred in a 100-mL volumetric flask and then filled to the mark with ultrapure water. About 2 mL of this solution are microfiltered (0.22 μm) and injected in the capillary chromatographic system. The elution is isocratic, the flow rate is 0.015 mL min^{-1} of 5 mM KOH and the run is completed in 15 min. The most important validation parameters, studied for this method (D'Amore et al., 2019), are reported in Table 5.7, while in Figure 5.6 some chromatogram examples are shown.

This analytical method was compared to the approach described at the previous paragraph "Sulphites in meat products, seafood, processed vegetables" for the analysis of fresh meats and meat products (direct ion chromatography). This comparison was carried out taking into account both the validation parameters and samples spiked with some sulfur-containing compounds. Indeed, the presence of such compounds in meats can lead to their final recovery as sulfuric acid and their wrong quantification as sulfite (false positive response). Regarding validation parameters, linearity, precision, recovery percentage, and measurement uncertainty resulted comparable between two methods, and method sensitivity of this optimized Monier-Williams method is slightly lower than the direct ion chromatography. The analysis made by spiking samples of meat preparations with some compounds containing sulfur, such as 2-methyl-3-furanthiol, sulfides, and L-methionine, demonstrated that these compounds can produce signals and then "false-positive" results, since the SO_2 detected can result higher than legal set for fresh meats, equal to 10.0 mg kg^{-1} (Iammarino et al., 2017a). Thus, this method is useful as "screening" technique, since it is robust and fast, but it should be coupled to the direct ion chromatography for confirmatory purpose.

5.3.3 Sorbic Acid and Benzoic Acid Determination if Food

These two food additives are used in a very wide range of foods. The most significant effort of this method development was made for optimizing a proper gradient elution suitable for analyzing all food types usually added with these 2 FPs. The capillary chromatography system described above is used. The following gradient elution was optimized, with a flow rate of 0.025 mL min^{-1}: 3mM KOH from 0 to 24 min, up to 40 mM in 1 min, isocratic for 5 min, back to 3 mM in 1 min and then a final re-equilibration step at this KOH concentration for 4 min (total run time: 35 min). Sample extraction is obtained by using the same solution described above for the determination of seven food additives by IC (NaOH $8,5 \cdot 10^{-3}$M, pH = 11.9) as extracting solvent. Two grams of homogenized sample are placed in a 250-mL volumetric flask together with 40 mL of extracting solution. The sample is placed in bain-marie at 70°C for 5 min. After cooling, about 2 mL of suspension are microfiltered (0.22 μm) before injection. The most important validation parameters, studied for this method (D'Amore et al., 2021), are reported in Table 5.8, while in Figure 5.7 some chromatogram examples are shown.

Table 5.7: Main Validation Parameters of the Analytical Method for the Determination of Sulfurous Anhydride in Food and Alcoholic Beverages by Distillation Coupled to Traditional Ion Chromatography

Compound	Determination Coefficient (R^2)	LOD (mg kg^{-1})	LOQ (mg kg^{-1})	Recovery* (%)	RSD$_r$* (%)	Application Field (Selectivity and Ruggedness)	Measurement Uncertainty (%)
Sulfite	0.999	3.3	9.9	81.0	4.6	Fresh meat and meat products, alcoholic beverages seafood	9.4

RSD$_r$ = Repeatability relative standard deviation
* = Mean value calculated on 3 spiking levels (n = 6 for each level)

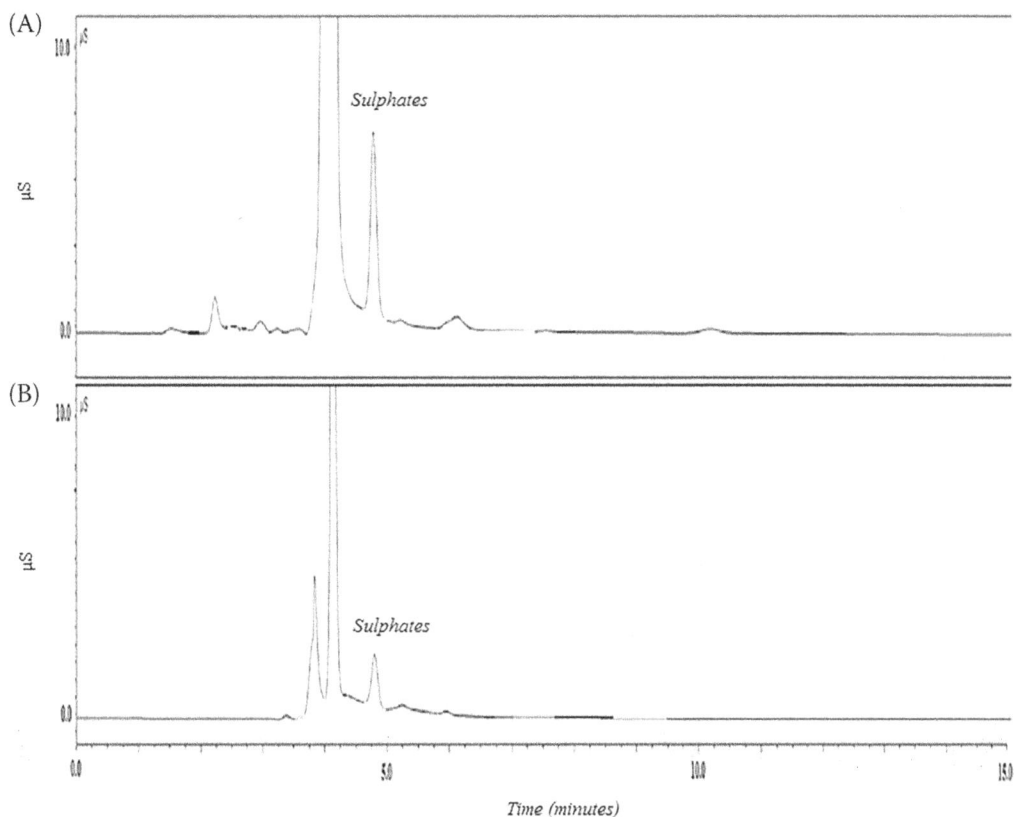

Figure 5.6 Chromatograms comparison. Cattle fresh raw meat sample spiked with sulfites (80 mg kg^{-1} as SO$_2$) (A); cattle fresh raw meat sample spiked with sulfides (30 mg kg^{-1}) (B). The ion detected by CIC is sulfate.

5.4 FOOD ADDITIVES DETERMINATION BY HPLC-UV-DIODE ARRAY DETECTION
5.4.1 Food Dyes Determination in Food

An optimized procedure of sample preparation and an effective chromatographic separation has been proposed for the determination of 12 permitted and not permitted food dyes in fresh meats and meat products (Iammarino *et al.*, 2019a; 2019b). The same procedure can also be applied for the simultaneous determination of eight food dyes in both solid and liquid foods. The following 12 dyes can be determined in meats: Carmine, Amaranth, Ponceau 4R, Allura Red AC, Carmoisine, Ponceau SX, Ponceau 3R, Erythrosine, Sudan I, Sudan II, Sudan III, Sudan IV; while the following widely used food coloring can be determined in solid foods (such as meats, cereal-based foods, sweets and food supplements) and beverages: Carmine, Ponceau 4R, Allura Red AC, Green S, Brilliant Black GN, Brilliant Blue, Tartrazine and Sunset Yellow. This analytical method was applied for fresh meats and meat products analysis by using a HPLC system: Waters^{TM} 2690 Separations Module (Milford, US) equipped with an autosampler, a column compartment, a micro-vacuum degasser and a Waters^{TM} 996 PDA Detector. The Waters® Millennium®32 software was used for data acquisition and elaboration. The same analytical method was also applied for other foodstuffs analysis, such as cereal-based products, sweets, beverages, and food supplements by using another HPLC system, a Shimadzu Nexera, composed of a CBM-40 system controller, a LC-40D XR solvent delivery pump, a SIL-40C XR autosampler, a DGU-405 degassing unit, a CTO-40S column oven, and a SPD-M40 PDA Detector (Shimadzu Corporation, Kyoto, Japan). The absorbance signal is detected at 520 nm for the determination of 12 dyes in meats, with other two wavelengths used for the determination of five food dyes in food: 600 nm and 455 nm for Green S/Brilliant Black/Brilliant Blue and Tartrazine/Sunset

Table 5.8: Main Validation Parameters of the Analytical Method for the Determination of Sorbic Acid and Benzoic Acid Food and Beverages by Capillary Ion Chromatography

Compound	Determination Coefficient (R^2)	LOD (mg kg^{-1})	LOQ (mg kg^{-1})	Recovery* (%)	RSD$_r$* (%)	Application Field (Selectivity and Ruggedness)	Measurement Uncertainty (%)
Sorbic Acid	0.999	1.6	4.9	97.6	2.1	Fresh and processed vegetables, jam, cheese, beverages, sauces, spices and soups	8.2
Benzoic Acid	0.999	4.1	12.6	92.2	2.5		7.1

RSD$_r$ = Repeatability relative standard deviation
* = Mean value calculated on 3 spiking levels (n = 6 for each level)

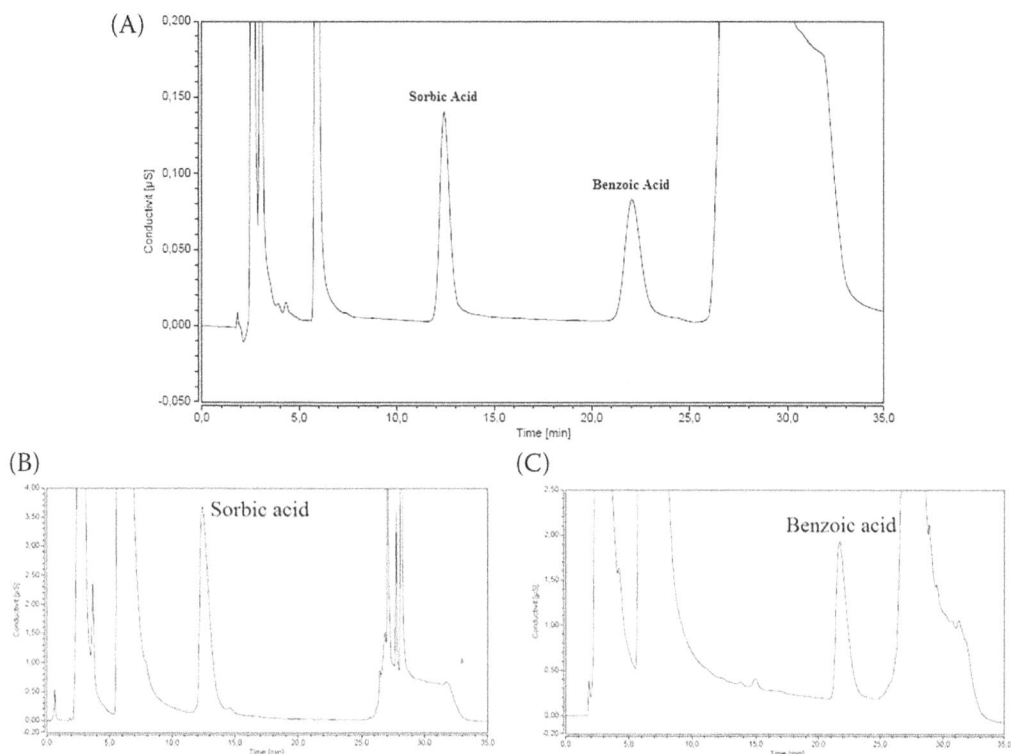

Figure 5.7 Chromatogram of sorbic and benzoic acids standard solution (2.0 mg L^{-1}). Chromatogram examples of commercial sample analysis: Mayonnaise sample added with 1420.9 mg kg^{-1} of sorbic acid (B); ketchup sample added with 779.3 mg kg^{-1} of benzoic acid (C).

Yellow, respectively (520 nm is confirmed for Carmine/Ponceau 4R and Allura Red AC). The diode array detector allows collecting the absorbance spectrum of each dye, in the range 190–700nm, which increase method selectivity, considerably. The analytical column used is C18 RP-Gold^{TM} (5-μm, 150 × 4.6 mm, Thermo Fisher, Waltham, USA), equipped with a drop-in guard cartridge (3-μm, 10 × 4 mm, Thermo Fisher). The optimized gradient elution is composed of 0.02M acetate (pH 7.0) (mobile phase A) and acetonitrile (mobile phase B). A flow rate of 1.2 mL min^{-1} and an injection loop of 10 μL are used. The elution gradient is the following: from 0% B to 15% B in 15 min, a gradient up to 34% B in 10 min, up to 80% B in 1 min, isocratic for 21 min, gradient to 0% B in 1 min and final re-equilibration step at this eluent composition for 4 min (total run time of 52 min). The optimized procedure of sample preparation consists of an extraction of 2 g (or 2 mL for beverage samples) of homogenized sample with 20 mL of acetonitrile:methanol:water:ammonia (50:40:9:1v/v/v/v) in a 50-mL polypropylene tube, using a vortex shaker for 2 min. The samples are then transferred to a 100-mL flask and placed in ultrasonic bath for 2 h (frequency: 100 Hz, T: 40°C). Finally, the samples are moved again in a 50-mL polypropylene tube for a re-extraction using vortex shaker for 1 min (this last step can be avoided for beverage samples). About 1.5 mL of supernatant are microfiltered (0.2 μm) directly in vial for HPLC analysis. The most important validation parameters, studied for this method (Iammarino et al., 2019a), are reported in Tables 5.9 and 5.10, while in Figures 5.8 and 5.9 some chromatogram examples are shown.

This analytical method was exploited to develop a comprehensive monitoring on the market for evaluating levels of dyes used in meat products. Indeed, these data are lacking, as highlighted in the first point of conclusion of the External Scientific Report of the European Commission entitled: "Analysis of needs in post-market monitoring of food additives and preparatory work for future projects in this field", which states "The groups of sweeteners and food colors were identified as priority substances to be addressed in post market monitoring" (Corporate author(s), 2013). This study concluded that only Carmine and Ponceau 4R are currently used in meats and, between the

Table 5.9: Main Validation Parameters of Analytical Method for Determination of 12 Food Dyes in Meat Products by HPLC/UV-DAD

Compound	Determination Coefficient (R^2)	LOD (mg kg^{-1})	LOQ (mg kg^{-1})	Recovery* (%)	RSD$_r$* (%)	Application Field (Selectivity and Ruggedness)	Measurement Uncertainty (%)
Carmine	0.999	4.3	13.2	100	6	Fresh meats, meat products and seafood	6.4
Amaranth	0.999	5.9	17.8	103	12		12.8
Ponceau 4R	0.999	4.9	14.9	95	14		13.6
Allura Red AC	0.998	3.7	11.3	93	15		12.4
Carmoisine	0.998	7.3	22.0	100	11		13.0
Ponceau SX	0.998	7.3	22.0	105	7		17.6
Ponceau 3R	0.997	4.1	12.3	99	10		14.4
Erythrosine	0.998	7.5	22.9	89	11		17.2
Sudan I	0.995	5.4	16.4	92	11		13.9
Sudan II	0.992	7.2	21.7	91	9		19.5
Sudan III	0.999	1.4	4.2	86	13		9.6
Sudan IV	0.999	6.1	18.6	90	12		10.1

RSD$_r$ = Repeatability relative standard deviation
* = Mean value calculated on 3 spiking levels (n = 6 for each level)

Table 5.10: Main Validation Parameters of the Analytical Method for the Determination of 8 Food Dyes in Food and Beverages by HPLC/UV-DAD

Compound	Determination Coefficient (R^2)	LOD (mg kg^{-1})	LOQ (mg kg^{-1})	Recovery* (%)	RSD$_r$* (%)	Application Field (Selectivity and Ruggedness)	Measurement Uncertainty (%)
Carmine	0.999	45.0	136.4	107.7	3.7	Cereal-based products, sweets, beverages, food supplements	10.2
Ponceau 4R	0.998	12.9	39.0	108.0	3.3		15.9
Allura Red AC	0.999	1.8	5.4	101.5	5.1		10.9
Green S	0.999	32.2	97.7	88.0	8.9		17.6
Brilliant Black	0.999	43.0	130.3	95.9	5.9		24.7
Brilliant Blue	0.999	39.2	118.7	103.5	4.9		5.6
Tartrazine	0.999	18.4	55.8	110.4	4.1		12.2
Sunset Yellow	0.999	12.0	36.4	102.1	4.9		7.0

RSD$_r$ = Repeatability relative standard deviation
* = Mean value calculated on 3 spiking levels (n = 6 for each level)

89

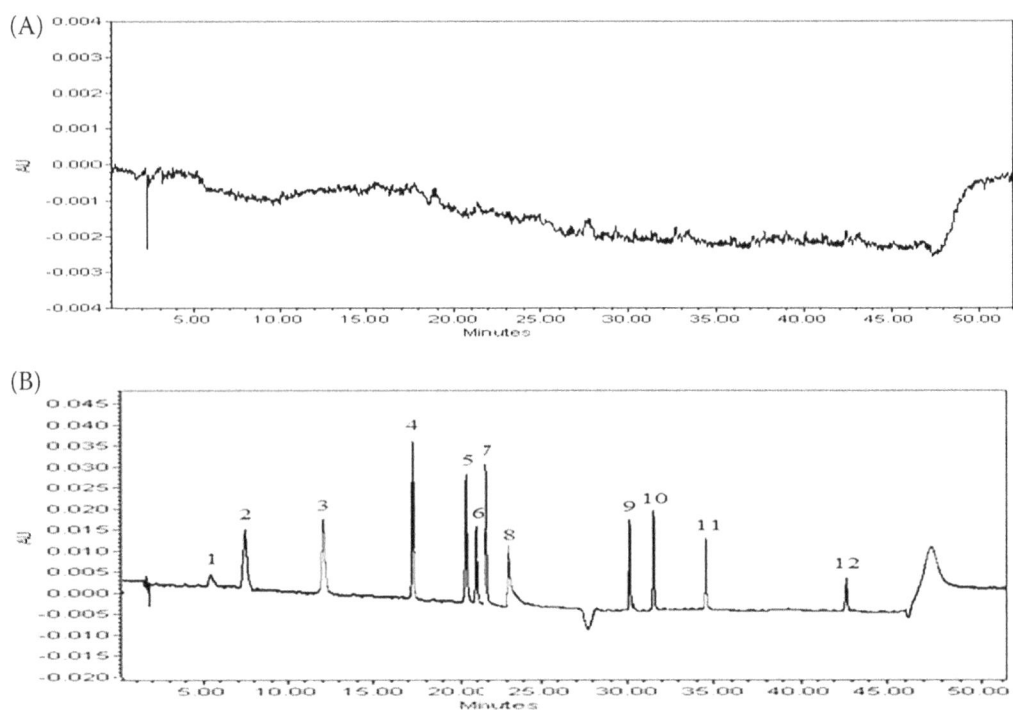

Figure 5.8 Chromatogram examples. "Blank" pork meat sample (A); pork meat sample spiked with standard solution of 12 dyes: 1-Carmine, 2-Amaranth, 3-Ponceau 4R, 4-Allura Red AC, 5-Carmoisine, 6-Ponceau SX, 7-Ponceau 3R, 8-Erythrosine, 9-Sudan I, 10-Sudan II, 11-Sudan III, 12-Sudan IV (B).

two, Carmine is used at mean levels 4–15 times higher than Ponceau 4R. None of analyzed samples resulted as "non-compliant" due to dye concentration exceeding than the legal limits defined in the Regulation No. 1129/2011/EC, and the highest concentrations detected resulted equal to 86.4 and 8.1 mg kg^{-1} for Carmine and Ponceau 4R, respectively. Another significant finding was the detection of Carmine in 4 samples of fresh meat preparations where the addition of food dyes is not permitted (Iammarino *et al.*, 2020b).

5.4.2 Ascorbic Acid and Nicotinic Acid in Meats

The same apparatus described above for food dyes determination in fresh meats and meat products was used for developing an analytical method for the determination of ascorbic acid and nicotinic acid in meats and meat products. Only the first additive is listed in the European Regulation No. 1333/2008 (E300-302, ascorbic acid and ascorbates), while nicotinic acid in not permitted, but it can be used for the same purpose in meats (antioxidant) in a fraudulent way.

The sample preparation consists of mixing 4 g of homogenized sample with 40 mL of phosphate buffer 10^{-2} M for extracting ascorbic acid and nicotinic acid at pH 3.5 and pH 9.0, respectively. The extraction of analytes is accomplished by using the vortex mixer for 1 min, then the mixture is centrifuged for 5 min at 250 x g at room temperature and the supernatant is microfiltered (0.2 μm) in vial for HPLC. The chromatographic separation is obtained through reversed phase liquid chromatography, using a Luna C18 column (250 × 4.6 mm i.d., particle size 5 μm, Phenomenex, Torrance, CA) chromatographic column, coupled to a HILIC Security Guard Cartridge (4 × 3.0 mm, Phenomenex), operating at a flow-rate of 1.5 mL min^{-1}, following a gradient elution of acetonitrile (A), water (B) and acetate buffer 100 mM, pH 5.8 (C). The gradient elution, with constant percentage of C (5%), starts at 5% A, isocratic for 2.5 min, then up to 45% A in 5 min, isocratic for 2.5 min and a final re-equilibration step at 5% A for 5 min (total run time: 15 min).

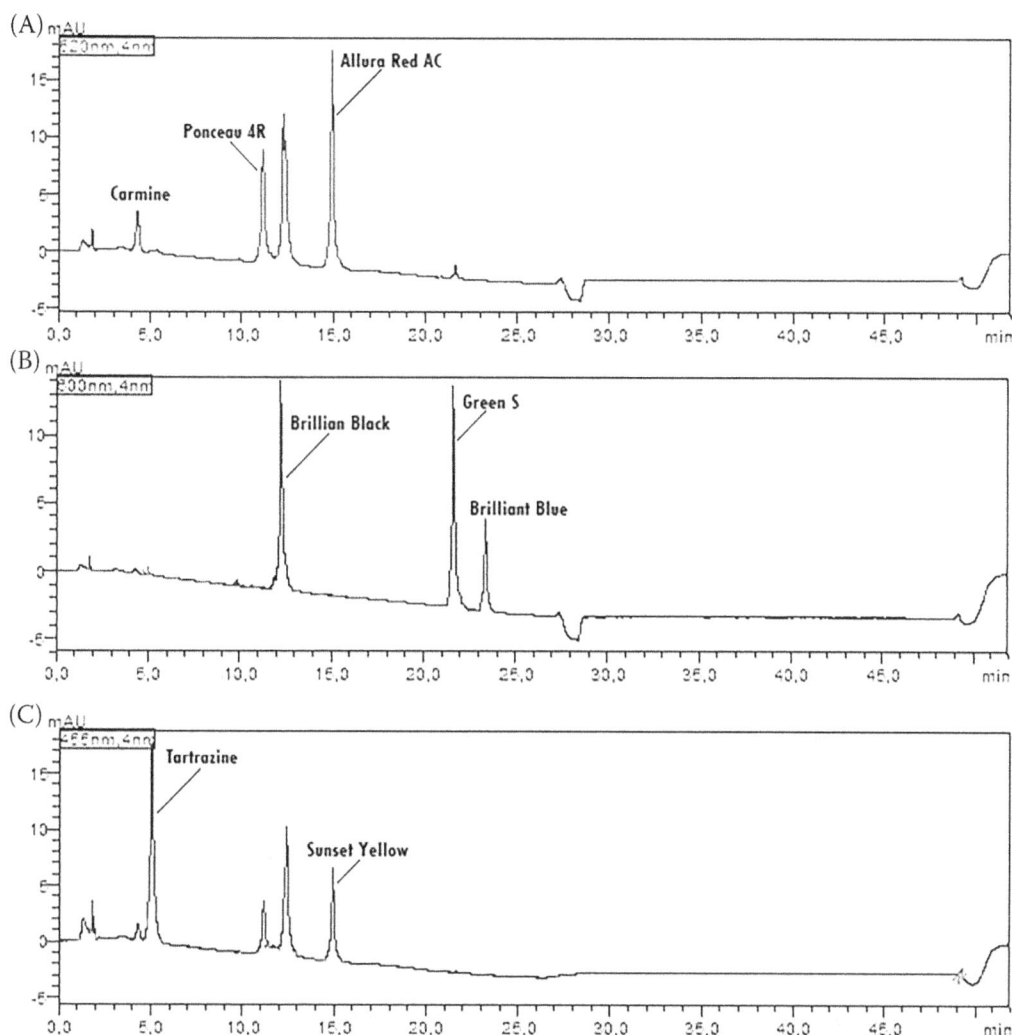

Figure 5.9 Chromatogram of a cereal-based sample spiked with 200 mg kg-1 of 8 dyes. Signals detected at 520 nm (A), 600 nm (B), and 455 nm (C).

The absorbance signal is detected at 260 and 215 nm for ascorbic acid and nicotinic acid, respectively. Setting the acquisition wavelength range from 190 to 320 nm, the diode array detector allows to obtain the absorbance spectrums of both compounds, improving method selectivity, substantially. The most important validation parameters, studied for this method (Iammarino and Di Taranto, 2015), are reported in Table 5.11, while in Figure 5.10 some chromatogram examples are shown.

This analytical method was applied for monitoring the ascorbic acid use in fresh meat preparations. A quantifiable concentration of ascorbic acid (> LOQ = 20.1 mg kg^{-1}) was detected in 33 samples out of 180 analzsed (18.3%). Nineteen of these samples were characterized by amounts greater than 160.0 mg kg^{-1}, so that the food additive addition was confirmed. Other 14 samples have shown an ascorbic acid concentration in the range LOQ – 50 mg kg^{-1}. In these cases, the addition of such low concentration of food additive seems unlikely, since the antioxidant effect would be very low. Thus, this presence was linked to the presence of tomato (particularly rich in ascorbic acid) in the products as food ingredient. Taking into account the distribution of the ascorbic acid concentrations detected in these samples, the measurement

Table 5.11: Main Validation Parameters of the Analytical Method for the Determination of Ascorbic Acid and Nicotinic Acid by HPLC/UV-DAD

Compound	Determination Coefficient (R^2)	LOD (mg kg^{-1})	LOQ (mg kg^{-1})	Recovery* (%)	RSD$_r$* (%)	Application Field (Selectivity and Ruggedness)	Measurement Uncertainty (%)
Ascorbic acid	0.999	6.6	20.1	104.2	4.4	Fresh meats and	5.6
Nicotinic acid	0.999	6.7	20.4	99.7	1.1	meat products	4.5

RSD$_r$ = Repeatability relative standard deviation
* = Mean value calculated on 3 spiking levels (n = 6 for each level)

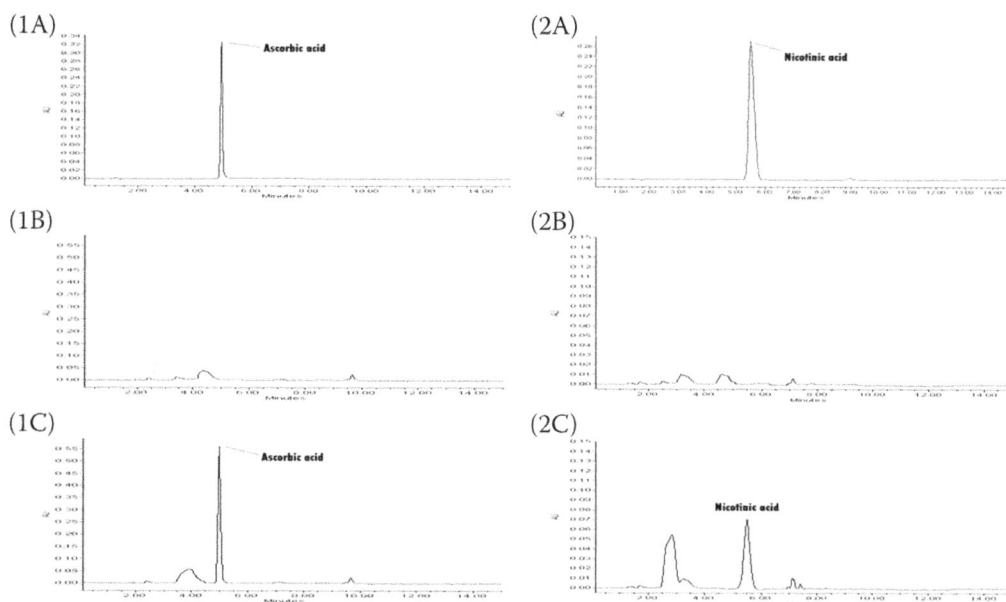

Figure 5.10 Chromatogram examples. Ascorbic acid standard solution 50 mg L^{-1} (1A); "Blank" cattle fresh sausage (1B); Cattle fresh sausage with measured ascorbic acid concentration of 693.3 mg kg^{-1} (1C); nicotinic acid standard solution 25 mg L^{-1} (2A), "blank" pork fresh sausage (2B); pork fresh sausage spiked with 50 mg kg^{-1} of nicotinic acid (2C).

uncertainty of the method (5.6%) and an appropriate tolerance, this study suggested a maximum permitted level of ascorbic acid in fresh meat preparations (in the presence of ascorbic acid sources) equal to 50.0 mg kg^{-1}. Below this value, the sample should be considered as "compliant" (Iammarino and Di Taranto, 2012c).

5.5 CONCLUSION

This chapter reports an accurate description of 10 chromatographic techniques applicable for determining the concentration of widely used food additives in most representative food types. All validation procedures applied to verify the full reliability of such approaches are described as well, together with all evaluated parameters and chromatogram examples. The chapter can be considered as a useful support for laboratories in charge of food control, relating to food additive determinations, and it supplies a list of analytical tools also useful for the development of monitoring and surveys to exploit during risk assessment studies.

ACKNOWLEDGMENTS

I would like to express my gratitude to the Istituto Zooprofilattico Sperimentale della Puglia e della Basilicata (Foggia, Italy), and all my colleagues and co-operators which contribute to the development, validation and publication of all the analytical methods described in this chapter. Many thanks to (in alphabetical order): Marzia Albenzio (University of Foggia, Italy), Giovanna Berardi (Istituto Zooprofilattico Sperimentale della Puglia e della Basilicata, Italy), Diego Centonze (University of Foggia, Italy), Antonio Eugenio Chiaravalle, Marianna Cristino, Teresa D'Amore (Istituto Zooprofilattico Sperimentale della Puglia e della Basilicata, Italy), Naceur M. Haouet (Istituto Zooprofilattico Sperimentale Umbria e Marche, Italy), Anna Rita Ientile, Mariateresa Ingegno, Michele Mangiacotti, Giuliana Marchesani (Istituto Zooprofilattico Sperimentale della Puglia e della Basilicata, Italy), Rosaria Marino, Annalisa Mentana (University of Foggia, Italy), Marilena Muscarella, Valeria Nardelli (Istituto Zooprofilattico Sperimentale della Puglia e della Basilicata, Italy), Donatella Nardiello (University of Foggia, Italy), Enea Pagliano (National Research Council Ottawa, Canada), Carmen Palermo (University of Foggia, Italy) and Domenico Palermo (Istituto Zooprofilattico Sperimentale della Puglia e della Basilicata, Italy).

REFERENCES

Association of Official Analytical Chemists. "Monier-Williams AOAC official method (optimized method) 990.28." *AOAC Official Methods of Analysis* (2000): 29–30.

Batt, C. A., and Tortorello, M. L. *Encyclopedia of Food Microbiology*, 2nd Ed. Cambridge: Academic Press, Elsevier Ltd, 2014.

Berardi G., Di Taranto, A., Vita, V., Marseglia, C., and Iammarino, M. "Effect of different cooking treatments on the residual level of sulphites in shrimps." *Italian Journal of Food Safety* 11(3), (2022): 10029. 10.4081/ijfs.2022.10029.

Berardi, G., Albenzio, M., Marino, R., D'Amore, T., Di Taranto, A., Vita, V., and Iammarino, M. "Different use of nitrite and nitrate in meats: A survey on typical and commercial Italian products as a contribution to risk assessment." *LWT* 150, (2021): 112004. 10.1016/j.lwt.2021.112004.

Bodsky, P., and Hoefler, F. "Fundamentals of Capillary Ion Chromatography." Accessed November 10, 2022 (2012). http://tools.thermofisher.com/content/sfs/posters/113852-PN70348_E-PN-IC-fundamentals-capillary.pdf

Caballero, B., Finglas, P. M., and Toldrá, F. *Encyclopedia of Food and Health*. Cambridge: Academic Press, Elsevier Ltd, 2016.

Carrabs, G., Smaldone, G., Carosielli, L., Girasole, M., Iammarino, M., and Chiaravalle, A. E. "Detection of sulfites in fresh meat preparation commercialised at retail in Lazio Region." *Italian Journal of Food Safety* 6, 6482, (2017): 93–95. 10.4081/ijfs.2017.6482.

Corporate author(s). "Analysis of needs in post-market monitoring of food additives and preparatory work for future projects in this field." *EFSA Supporting Publications*, 10(4), (2013): EN-419. 10.2903/sp.efsa.2013.EN-419.

D'Amore, T., Di Taranto, A., Berardi, G., and Iammarino, M. "Development and validation of an analytical method for nitrite and nitrate determination in meat products by capillary ion chromatography (CIC)". *Food Analytical Methods*, 12, (2019): 1813–1822. 10.1007/s12161-019-01529-0.

D'Amore, T., Di Taranto, A., Berardi, G., Vita, V., and Iammarino, M. "Going green in food analysis: a rapid and accurate method for the determination of sorbic acid and benzoic acid in foods by capillary ion chromatography with conductivity detection". *LWT – Food Science and Technology*, 141, (2021): 110841. 10.1016/j.lwt.2020.110841.

D'Amore, T., Di Taranto, A., Berardi, G., Vita, V., Chiaravalle, A. E., and Iammarino, M. "Sulfites in meat: occurrence, activity, toxicity, regulation and detection. A comprehensive review." *Comprehensive Review in Food Science and Food Safety*, 19(5), (2020): 2701–2720. 10.1111/1541-433 7.12607.

Di Taranto, A., Ingegno, M., Ientile, A. R., and Iammarino, M. "Food additives in mozzarella cheese: a contribution for a correct analytical determination." *Journal of Food and Nutrition Sciences. Special Issue: Emerging Issues in Food Safety, Food Additives: Risk Assessment, Analytical Methods and Replacement in Foodstuffs*, 3(1-1), (2015): 13–17. 10.11648/j.jfns.s.2015030101.13.

Dionex Corporation. "Determination of Polyphosphates Using Ion Chromatography." Accessed November 12, 2022 (2016). https://assets.thermofisher.com/TFS-Assets/CMD/Application-Notes/AU-172-IC-Polyphosphates-LPN2496-EN.pdf.

Ente Italiano di Normazione (UNI). "UNI Standard 10591:1997 Meat and meat products. Determination of total phosphorus content. Spectrometric method." Milan: Ente Italiano di Normazione, 1997.

European Commission. "Commission Decision No 2004/92/EC of 21 January 2004 on emergency measures regarding chilli and chilli products." *Official Journal of the European Union*, L27, (2004): 52–54.

European Commission. "Commission Regulation (EU) No 1129/2011 amending Annex II to Regulation (EC) No 1333/2008 of the European Parliament and of the Council by establishing a Union list of food additives." *Official Journal of the European Union*, L295, (2011): 1–177.

European Commission. "Decision (EC) No. 657/2002 of 12 August 2002 implementing Council Directive 96/23/EC concerning the performance of analytical methods and the interpretation of results." *Official Journal of the European Union*, L221, (2002): 8–36.

European Commission. "Regulation (EU) 2017/625 of the European Parliament and of the Council of 15 March 2017." *Official Journal of the European Union*, L95, (2017): 1–142.

European Food Safety Authority. "Re-evaluation of glutamic acid (E 620), sodium glutamate (E 621), potassium glutamate (E 622), calcium glutamate (E 623), ammonium glutamate (E 624) and magnesium glutamate (E 625) as food additives." *EFSA Journal*, 15(7), (2017a): 4910. 10.2903/j.efsa.2017.4910.

European Food Safety Authority. "Re-evaluation of phosphoric acid–phosphates–di-, tri- and polyphosphates (E 338–341, E 343, E 450–452) as food additives and the safety of proposed extension of use." *EFSA Journal*, 17(6), (2019): 5674. 10.2903/j.efsa.2019.5674.

European Food Safety Authority. "Re-evaluation of potassium nitrite (E 249) and sodium nitrite (E 250) as food additives." *EFSA Journal*, 15(6), (2017b): 4786. 10.2903/j.efsa.2017.4786.

European Food Safety Authority. "Refined exposure assessment for Ponceau 4R (E 124)." *EFSA Journal*, 13(4), (2015a): 4073. 10.2903/j.efsa.2015.4073.

European Food Safety Authority. "Scientific Opinion - Guidance on human health risk-benefit assessment of foods." *EFSA Journal*, 8(7), (2010): 1673. 10.2903/j.efsa.2010.1673.

European Food Safety Authority. "Scientific Opinion on the re-evaluation of Indigo Carmine (E 132) as a food additive." *EFSA Journal*, 12(7), (2014): 3768. 10.2903/j.efsa.2014.3768.

European Food Safety Authority. "Scientific Opinion on the re-evaluation of cochineal, carminic acid, carmines (E 120) as a food additive." *EFSA Journal*, 13(11), (2015b): 4288. 10.2903/j.efsa.2015.4288.

FAO/WHO – Food and Agriculture Organization of the United Nations/World Health Organization. "Toxicological evaluation of certain food additives. 67th meeting of the Joint FAO/WHO Expert Committee on Food Additives (JECFA), 20-29 June 2006." Rome (Italy), 2006.

Gardner, L. K., and Lawrence, G. D. "Benzene production from decarboxylation of benzoic acid in the presence of ascorbic acid and a transition-metal catalyst." *Journal of Agricultural and Food Chemistry*, 41(5), (1993): 693–695. 10.1021/jf00029a001.

Hartman, P. E. "Review: Putative mutagens and carcinogens in foods. II: Sorbate and sorbate-nitrite interactions." *Environmental Mutagenesis*, 5(2), (1983): 217–222. 10.1002/em.2860050209.

Hulanicki, A., and Głab, S. "Total systematic error in redox titrations with visual indicators—I: Basic principles." *Talanta* 22(4-5), (1975): 363–370. 10.1016/0039-9140(75)80082-8.

Iammarino M., Mentana, A., Centonze, D., Palermo, C., Mangiacotti, M., and Chiaravalle, A. E. "Simultaneous determination of twelve dyes in meat products: development and validation of an analytical method based on HPLC-UV-Diode array detection." *Food Chemistry*, 285, (2019a): 1–9. 10.1016/j.foodchem.2019.01.133.

Iammarino, M. *Simplified guidelines for chromatographic methods validation*. Beau Bassin, Mauritius: LAP Lambert Academic Publishing, 2019.

Iammarino, M., and Di Taranto, A. "Determination of polyphosphates in products of animal origin: application of a validated ion chromatography method for commercial samples analyses." *European Food Research and Technology*, 235, (2012b): 409–417. 10.1007/s00217-012-1766-z.

Iammarino, M., and Di Taranto, A. "Development and validation of an ion chromatography method for the simultaneous determination of seven food additives in cheeses." *Journal of Analytical Sciences, Methods and Instrumentation*, 3, (2013): 30–37. 10.4236/jasmi.2013.33A005.

Iammarino, M., and Di Taranto, A. "Monitoring on the presence of ascorbic acid in not prepacked fresh meat preparations by a validated HPLC method." *Journal of Food Research*, 1(2), (2012c): 22–31. 10.5539/jfr.v1n2p22.

Iammarino, M., and Di Taranto, A. "Nitrite and nitrate in fresh meats: a contribution to the estimation of admissible maximum limits to introduce in directive 95/2/EC." *International Journal of Food Science and Technology*, 47(9), (2012a): 1852–1858. 10.1111/j.1365-2621.2012.03041.x.

Iammarino, M., and Di Taranto, A. "Validation of an analytical method for the determination of ascorbic acid and nicotinic acid in fresh meat preparations by HPLC-UV-DAD." *Journal of Food and Nutrition Sciences – Special Issue: Emerging Issues in Food Safety, Food Additives: Risk Assessment, Analytical Methods and Replacement in Foodstuffs*, 3(1–1), (2015): 7–12. 10.11648/j.jfns.s.2015030101.12.

Iammarino, M., Berardi, G., Vita, V., Elia, A., Conversa, G., and Di Taranto, A. "Determination of nitrate and nitrite in Swiss chard (*Beta vulgaris* L. subsp. *vulgaris*) and wild rocket (*Diplotaxis tenuifolia* (L.) DC.) and food safety evaluations." *Foods*, 11(17), (2022a): 2571. 10.3390/foods11172571.

Iammarino, M., Di Taranto, A., and Centonze, D. "Determination of sulphiting agents in raw and processed meat: comparison between a modified Monier-Williams method and the direct analysis by ion chromatography with conductometric detection." *Food Analytical Methods*, 10, (2017a): 3956–3963. 10.1007/s12161-017-0960-9.

Iammarino, M., Di Taranto, A., and Cristino, M. "Endogenous levels of nitrites and nitrates in wide consumption foodstuffs: results of five years of official controls and monitoring." *Food Chemistry*, 140(4), (2013a): 763–771. 10.1016/j.foodchem.2012.10.094.

Iammarino, M., Di Taranto, A., and Cristino, M. "Monitoring of nitrites and nitrates levels in leafy vegetables (spinach and lettuce): a contribution to risk assessment." *Journal of the Science of Food and Agriculture*, 94, (2014): 773–778. 10.1002/jsfa.6439.

Iammarino, M., Di Taranto, A., and Ientile, A. R. "Monitoring of sulphites levels in shrimps collected in Puglia (Italy) by ion-exchange chromatography with conductivity detection." *Food Additives and Contaminants: Part B*, (2013b): 84–89. 10.1080/19393210.2013.848943.

Iammarino, M., Di Taranto, A., and Ientile, A. R. "Sulphur dioxide in meat products: 3-year control results of an accredited Italian laboratory." *Food Additives & Contaminants: Part B*, 10(2), (2017b): 99–104. 10.1080/19393210.2017.1280539.

Iammarino, M., Di Taranto, A., and Muscarella, M. "Investigation on the presence of sulphites in fresh meats preparations: estimation of an allowable maximum limit." *Meat Science*, 90, (2012): 304–308. 10.1016/j.meatsci.2011.07.015.

Iammarino, M., Di Taranto, A., Muscarella, M., Nardiello, D., Palermo, C., and Centonze, D. "Development of a new analytical method for the determination of sulfites in fresh meats and shrimps by ion-exchange chromatography with conductivity detection." *Analytica Chimica Acta*, 672(1–2), (2010): 61–65. 10.1016/j.aca.2010.04.002.

Iammarino, M., Di Taranto, A., Muscarella, M., Nardiello, D., Palermo, C., and Centonze, D. "Food additives determination in shrimps: development of innovative analytical methods by ion chromatography with conductivity detection." In *Shrimp: Evolutionary History, Ecological Significance and Effects on Dietary Consumption*, edited by Carmel A. Delaney, 89–110. Hauppauge NY: Nova Science Publisher, 2013c.

Iammarino, M., Di Taranto, A., Palermo, C., and Muscarella, M. "Survey of benzoic acid in cheeses: contribution to the estimation of an admissible maximum limit." *Food Additives and Contaminants: Part B*, 4(4), (2011): 231–237. 10.1080/19393210.2011.620355.

Iammarino, M., Haouet, N., Di Taranto, A., Berardi, G., Benedetti, F., Di Bella, S., and Chiaravalle, A. E. "The analytical determination of polyphosphates in food: A point-to-point comparison between direct ion chromatography and indirect photometry." *Food Chemistry*, 325, (2020a): 126937. 10.1016/j.foodchem.2020.126937.

Iammarino, M., Marino, R., and Albenzio, M. "How meaty? Detection and quantification of adulterants, foreign proteins and food additives in meat products." *International Journal of Food Science and Technology*, 52, (2017c): 851–863. 10.1111/ijfs.13350.

Iammarino, M., Mentana, A., Centonze, D., Palermo, C., Mangiacotti, M., and Chiaravalle, A. E. "Chromatographic determination of 12 dyes in meat products by HPLC-UV-Diode array detection." *MethodsX*, 6, (2019b): 856–861. 10.1016/j.mex.2019.04.018.

Iammarino, M., Mentana, A., Centonze, D., Palermo, C., Mangiacotti, M., and Chiaravalle, A. E. "Dyes use in fresh meat preparations and meat products: a survey by a validated method based on HPLC-UV-Diode array detection as a contribution to risk assessment." *International Journal of Food Science and Technology*, 55, (2020b): 1126–1135. 10.1111/ijfs.14275.

Iammarino, M., Palermo, C., and Tomasevic, I. "Advanced analysis techniques of food contaminants and risk assessment - editorial." *Applied Sciences*, 12(10), (2022b): 4863. 10.3390/app12104863.

Iammarino, M., Panseri, S., Unlu, G., Marchesani, G., and Bevilacqua, A. "Editorial: novel chemical, microbiological and physical approaches in food safety control." *Frontiers in Nutrition*, 9, (2022c): 1060480. 10.3389/fnut.2022.1060480.

IARC – International Agency for Research on Cancer. "IARC Monographs evaluate consumption of red meat and processed meat." Accessed 5 November 2022 (2015). https://www.iarc.who.int/wp-content/uploads/2018/07/pr240_E.pdf.

Ilboudo, S., Fouche, E., Rizzati, V., Toé, A. M., and Gamet-Payrastre, L. "In vitro impact of five pesticides alone or in combination on human intestinal cell line Caco-2." *Toxicology Reports*, 1, (2014): 474–489. 10.1016/j.toxrep.2014.07.008.

ISO (International Organization for Standardization), & IDF (International Dairy Federation). "ISO/TS 18083 – IDF/RM 51 (First Edition) – Processed cheese products – Calculation of content of added phosphate expressed as phosphorus." Geneva, 2013.

Jamin, E. L., Riu, A., Douki, T., Debrauwer, L., Cravedi, J.-P., Zalko, D., and Audebert, M. "Combined genotoxic effects of a polycyclic aromatic hydrocarbon (B(a)P) and an heterocyclic amine (PhIP) in relation to colorectal carcinogenesis." *PLoS ONE*, 8(3), (2013): e58591. 10.1371/journal.pone.0058591.

Jia, W., Wang, X., and Shi, L. "Interference of endogenous benzoic acid with the signatures of sulfonic acid derivatives and carbohydrates in fermented dairy products." *Fundamental Research, in press*, 2023. 10.1016/j.fmre.2022.09.033.

MacDonald, R., and Reitmeier, C. *Understanding Food Systems – Agriculture, Food Science, and Nutrition in the United States*. Cambridge: Academic Press, Elsevier Ltd, 2017.

Mamur, S., Yüzbaşıoğlu, D., Ünal, F., and Yılmaz, S. "Does potassium sorbate induce genotoxic or mutagenic effects in lymphocytes?" *Toxicology in Vitro*, 24, (2010): 790–794. 10.1016/j.tiv.2009.12.021.

Montes, C., Vèlez, J. H., Ramìrez, G., Isaacs, M., Arce, R., and Aguirre, M. J. "Critical comparison between modified Monier-Williams and electrochemical methods to determine sulfite in aqueous solutions." *The Scientific World Journal*, 2012, (2012): 168148. 10.1100/2012/168148.

Pagliano, E., Meija, J., Campanella, B., Onor, M., Iammarino, M., D'Amore, T., Berardi, G., D'Imperio, M., Parente, A., Mihai, O., and Mester, Z. "Certification of nitrate in spinach powder reference material SPIN-1 by high-precision isotope dilution GC-MS." *Analytical and Bioanalytical Chemistry*, 411(16), (2019): 3435–3445. 10.1007/s00216-019-01803-4.

Surekha, M., and Reddy, S. M. "Preservatives, classification and properties." In *Encyclopedia of Food Microbiology*, 2nd Ed, edited by Carl A. Batt, and Mary L. Tortorello, 69–75. Cambridge: Academic Press, Elsevier Ltd, 2014. 10.1016/B978-0-12-384730-0.00257-3.

Thermo Fisher Scientific. "Product Manual for Dionex IonPac™ AS11-HC and AG11-HC Columns." Accessed 5 November, 2022 (2012). https://assets.thermofisher.com/TFS-Assets/CMD/manuals/man-031333-ionpac-as11-hc-columns-man031333-en.pdf.

6 Aflatoxin Detection in Dairy Products

Luiz Torres Neto, Maria Lúcia Guerra Monteiro, Flavio Dias Ferreira, and Carlos Adam Conte Junior

6.1 AFLATOXINS

The mycotoxins are a group of secondary metabolites produced by fungi spp., including *Aspergillus*, *Penicillium*, *Fusarium*, and *Alternaria* (Negash, 2018; Rushing & Selim, 2019). Aflatoxins are one of many naturally occurring mycotoxins in soils, foods, humans, and animals. The genus *Aspergillus* produces 20 different aflatoxins, mainly six different types: B_1 (AFB_1), B_2 (AFB_2), G_1 (AFG_1), G_2 (AFG_2), M_1 (AFM_1), and M_2 (AFM_2) (Pal *et al.*, 2021; Quadri *et al.*, 2012). Furthermore, *Aspergillus flavus* and *Aspergillus parasiticus* produce the most toxigenic strains of aflatoxins AFB_1-AFB_2 and AFG_1-AFG_2, respectively (Kumar *et al.*, 2021). Aflatoxins cause reduced animal growth and productivity leading to morbidity and mortality and meat and milk contaminated rejection in the international market due to their harmful effects (Abebe *et al.*, 2018; Wagacha and Muthomi, 2008). This context results in a cost of $1 billion in the United States with loss by maize, groundnut, and wheat growers. In addition, achieving $160 million loss each year to maize producers in the United States and a total of $450 million in sub-Saharan Africa, accounting for 38% of global aflatoxin-related agricultural losses (Gbashi *et al.*, 2019; IARC, 2012b; Pal, 2017). Therefore, aflatoxin is a worldwide concern with economic and health impacts wherein 4.5 billion global population are at risk of excessive exposure to aflatoxins (Abrar *et al.*, 2013).

The *A. flavus* is ubiquitously found in soil and contaminates many of the world's crops, including cereals, oilseeds, spices, and nuts. The colonization of this fungus in plantations causes expressive economic losses by contaminating seeds with aflatoxin (Amaike and Keller, 2011; Klich, 2007). The aflatoxin produced by this fungus (AFB_1) is the most toxic aflatoxin to humans and animals since its association with hepatocellular carcinoma (HCC), leading to liver cancer (Qureshi *et al.*, 2015; Rajarajan *et al.*, 2013; Rushing & Selim, 2019). Furthermore, the IARC concluded that there was enough evidence in humans to classify this aflatoxin as a carcinogen (HCC) as well as growth suppression, immune system modulation, and malnutrition (IARC, 2012a, 2012b). From AFB_1, other metabolites are formed through the P450 system, including AFM_1, aflatoxin Q_1 (AFQ_1), aflatoxin P_1 (AFP_1), aflatoxicol (AFL), aflatoxicol H_1 (AFH_1), and aflatoxin $B_{2}a$ ($AFB_{2}a$) (Rushing and Selim, 2019). Briefly, AFB_1 reaches the intestine and is rapidly absorbed and transported via the portal bloodstream to the liver; thus, this aflatoxin is subjected to reduction, epoxidation, hydroxylation, and demethylation (Yiannikouris and Jouany, 2002). The AFM_1, a major metabolite produced by CYP1A2, is the most carcinogenic among the hydroxylated ones (Cullen *et al.*, 1987; Sinnhuber *et al.*, 1974), which is excreted in urine and secreted in milk in mammalian species (Saha Turna and Wu, 2021).

6.2 AFLATOXIN M_1 IN DAIRY PRODUCTS

The AFM_1 is the most crucial undesirable milk contaminant, being a consequence of the presence of AFB_1 in feed offered to dairy cows (Min *et al.*, 2021a; Prandini *et al.*, 2009), wherein approximately 0.3–6.2% of AFB_1 is converted into AFM_1 (Alahlah *et al.*, 2020). The AFM_1 results in carcinogenicity, mutagenicity, genotoxicity, teratogenicity, and immunosuppression, even at low concentrations (Nemati *et al.*, 2010). In this way, AFM_1 is classified as group 1 in terms of toxicity (IARC, 2012a). Studies have shown that the presence of AFM_1 in milk products is a health issue in many countries because all age groups daily consume these products. Furthermore, AFM_1 presents heat stability and thus cannot be degraded or destroyed by standard food processing procedures. Studies evaluated the presence of AFM_1 in raw, pasteurized, powder, organic, concentrated, and ultra-high temperature (UHT) milk samples and showed AMF_1 concentrations above the acceptable European limit (0.05 mg/kg) regardless of heat treatment such as pasteurization and sterilization. Although the literature is ambiguous about the effect of heat processing on the amount of AFM_1 in dairy products, most studies indicate that treatments using heat such as pasteurization and sterilization, do not cause changes in the AFM_1 levels in these products. Likewise, evaporation, concentration, or drying did not affect the AFM_1 content in milk (Flores-Flores *et al.*, 2015). In that regard, AFM_1 still may contaminate other dairy products, such as cheese and yogurt, generating even more health concerns for consumers (Campagnollo *et al.*, 2016; Iqbal *et al.*, 2015; Iqbal *et al.*, 2010).

DOI: 10.1201/9781003334859-6

Studies report that AFM_1 levels in milk can be reduced by increasing the quality control of animal feed. Flores-Flores *et al.* (2015) reported some factors that influence the concentration of AFM_1 in milk: I - Animal feeding: milk from animals fed by grazing presents lower levels of AFM_1 compared with milk from animals fed with compound feed and/or stored foodstuff; II - Animal feed storage: stored foodstuff has higher contamination of AFB_1 due to the humid conditions thereby facilitating the growth of fungi and accumulation of toxins. Despite being a well-known problem, the presence of AMF_1 in milk and dairy products is still a global concern. Furthermore, little is known about the influence of dairy processing steps on AFM_1 levels, a gap to be explored. Nonetheless, monitoring the incidence of AFM_1 in dairy products is necessary to reduce world public health risks (Li *et al.*, 2018; Škrbić *et al.*, 2014), mainly due to the high levels of AFM_1 still observed in these products in several countries in the world (Mollayusefian *et al.*, 2021; Saha Turna and Wu, 2021).

6.3 LIMITS OF AFM_1 IN DAIRY PRODUCTS

The limits of the presence of AFM_1 in milk vary in different countries worldwide. The maximum acceptable threshold established in Brazil and MERCOSUL is 500 ng/L for fluid and 5 ng/g for powder milk (Brasil, 2022). The same limits are observed in the regulations of Asian countries (500 ng/L) and United States (500 ng/kg) (Anukul *et al.*, 2013; FDA, 2007). Otherwise, stricter limits are found in other countries, such as Iran (100 ng/L), European Union (50 ng/L; the lowest limit of AFM_1 in liquid milk) (EU, 2010; ISIRI, 2010), and Morocco (50 ng/kg for raw milk, UHT milk, and milk intended for dairy products manufacturing) (Mannani *et al.*, 2021).

6.4 AFM_1 IN MILK

The AFM_1 concentrations in milk vary highly across most countries, as seen in Table 6.1. Studies from Asia show a high and variable occurrence of AFM1 in milk, attributed to differences in feeding systems, farm management practices, and analytical methods (Asi *et al.*, 2012). In addition, Pakistan and Bangladesh had the highest concentrations of AFM_1, including exceeded limits (500 ng/L; Section 1.3.). Regarding Africa, an occurrence similar to Asia wherein Rwanda, Malawi, and Ethiopia showed levels above 500 ng/L; however, Egypt was the only one demonstrating levels below 50 ng/mL (Table 6.1). The study by Iqbal *et al.* (2015) justified the high incidence of AFM_1 in milk from African countries to the lack of awareness and constraints in analytical facilities. In Europe and America, the levels and incidence of AFM_1 in milk were low compared to other continents. Ecuador was the only country where milk containing AFM_1 concentration was above 50 ng/mL (78.83 ng/L/Kg). The low level observed in these continents may be related to the strict regulations concerning these mycotoxins in feed and milk products and the adoption of good storage practices (Iqbal *et al.*, 2015).

The scenario found in Asian and African countries is attributed to the mismanagement of animal feed. For example, there is a fresh feed shortage during the cold season, in which concentrated feeds containing wheat, corn, and cotton seeds are used; however, they may have poor storage conditions (Ghiasian *et al.*, 2007; Abolfazl Kamkar, 2005; Tajkarimi *et al.*, 2008). Weather is another factor that affects AFM_1 levels, where high milk contamination is observed in areas with humid climates than in regions with arid/semi-arid climates. These conditions influence the AFB_1 formation in feed during the pre-harvest, harvest, and storage (Hashemi, 2016; Kos *et al.*, 2018; Tajkarimi *et al.*, 2007; Unusan, 2006). In Europe, fewer studies were observed, but the strict regulations may be responsible for the lower AFM_1 indices (Iqbal *et al.*, 2015). Furthermore, in America, good agricultural practices such as the use of pest-resistant crops, proper cultivation practices, adequate use of fertilizers, irrigation, and crop rotation can be responsible for preventing and controlling fungal growth and mycotoxin formation in dairy farming (Goncalves *et al.*, 2015; Gonçalves *et al.*, 2017; Pires *et al.*, 2022).

6.5 AFM_1 IN CHEESE

Among dairy products, cheese is the only one susceptible to the direct growth of fungi and mycotoxin production. Otherwise, the process used for cheese preparation can affect the AFM_1 levels in the final product (Rahmani *et al.*, 2018; Sengun *et al.*, 2008). Indeed, Kamkar *et al.* (2008) and López *et al.* (2001) related that the presence of aflatoxin in cheese may be possibly due AFM_1 residue in milk, the growth of the fungi on cheese followed by the consequent production of aflatoxins and the presence of AFM_1 in powdered milk enriching the milk used in cheese manufacture. Studies report that cheese could be the most potent source of aflatoxin among dairy products due to the highest concentration of AFM_1 (three–five times) than in corresponding milk (Prandini *et al.*, 2009; Scaglioni *et al.*, 2014). This higher concentration is related to the AFM1 association with the casein

Table 6.1: Occurrence of AFM$_1$ in Milk and Estimated Daily Intake (EDI)

Continent	Country	Milk Type	Quantification Method	Incidence	Range (ng/L and ng/kg)	Mean (ng/L and ng/kg)	EDI* (ng/kg bw/day)	Reference
Asia	China	Raw	ELISA	94/797	≤ 50–486	35.7	0.128	Min et al., 2021b
	Bangladesh	UHT	ELISA	05/25	25.07–48.95	35.46	0.127	Tarannum et al. (2020)
		Pasteurized	ELISA	13/25	18.11–672.18	99.77	0.356	
		Raw	ELISA	35/50	22.79 – 1,489.28	699.07	2.497	
	Iran	Pasteurized	ELISA	–	113.3–270.6	177.74	0.635	Mokhtari et al. (2022)
		Raw	ELISA	–	57–228	134.61	0.481	
	Japan	Whole milk*	HPLC-FLD; LC-MS/MS	19/37	< 10–70	45	0.161	Ono et al. (2020)
	Pakistan	Powder	ELISA	4/4	412.5 – 1,935.0	922.5	3.295	Yunus et al. (2020)
		UHT	ELISA	11/15	145.5–642.9	365.7	1.306	
		Pasteurized	HPLC-FLD	13/13	56.9 – 3,935.5	1,167.5	4.170	
	India	Raw	HPLC-FLD	19/46	ND-2,913	273	0.975	Hattimare et al. (2022)
		Pasteurized		6/15	ND-1,212	278	0.993	
		UHT		5/12	ND-1,523	416	1.486	
		Powder		2/10	ND-2,608	486	1.736	
Africa	Rwanda	Raw	ELISA	–	Max. 14,500	890	3.179	Nishimwe et al. (2022)
	Malawi	Raw	VICAM afla test fluorometry procedure	112/112	10–5,000	551	1.968	Njombwa et al. (2021)
	Ethiopia	Pasteurized	ELISA	21/64	11–1798	324	1.157	Zebib et al. (2022)
		Raw	ELISA	24/64	03–2,177	319	1.139	
		Raw	ELISA	52/52	550–1,410	690	2.464	Tadesse et al. (2020)
		Pasteurized	ELISA	56/56	29–2,159	970	3.464	
America	Egypt	Raw	HPLC-FLD	–	5.36–103.02	40.27	0.144	Esam et al. (2022)
	Brazil	Goat raw milk	HPLC-FLD	108/108	5.60–48.20	21.9	0.078	de Matos et al (2021)
	Ecuador	Raw	VICAM® Assay Tests	–	23–763	78.83	0.282	Torres et al. (2022)
Europe	Greece	Pasteurized	ELISA	32/22	2.04 17.84	7.72	0.028	Malissiova et al. (2022)

Note

* EDI = (AFM1 concentration) X (daily consumption) / (body weight = 70 kg; Saha Turna and Wu, 2021). Milk consumption: 81 g/day (Malissiova et al., 2022). ND – Not detectable.

Table 6.2: Occurrence of AFM₁ in Cheese and Estimated Daily Intake (EDI)

Continent	Country	Cheese Type	Quantification Method	Incidence	Range (ng/L and ng/Kg)	Mean (ng/L and ng/Kg)	EDI* (ng/kg bw/day)	Reference
Africa	Ghana	Wagashie cheese	HPLC-FLD	93/182	50–3,600	656	0.562	Kortei & Annan (2022)
	Ethiopia	Cheese (Industrial)	ELISA	72/72	18–5,580	2,210	0.002	Tadesse et al. (2020)
		Cheese (Local)	ELISA	10/10	8–3,860	770	0.660	
		Cottage cheese	ELISA	24/32	14–539	137	0.117	Zebib et al. (2022)
	Egypt	Ras cheese	HPLC-FLD	–	59.72–108.14	86.97	0.075	Esam et al. (2022)
		Processed cheese	HPLC-FLD	–	<5.0–27.75	10.77	0.009	
America	Brazil	Minas Frescal cheese	HPLC-FLD	20/28	–	113	0.097	Gonçalves et al. (2021)
	Nicaragua	White cheese	VERATOX kit	152/152	5–485	30	0.026	Peña-Rodas et al. (2020)
Europe	Greece	Feta cheese	ELISA	07/25	2.10–4.09	2.98	0.003	Malissiova et al. (2022)
	Italy	Fior di Latte cheese	UPLC - FLD	–	–	83.5	0.072	Pecorelli et al. (2020)
		Primosale cheese	UPLC – FLD	–	–	265.5	0.228	

Note
* EDI = (AFM1 concentration) X (daily consumption) / (body weight = 70 kg; Saha Turna and Wu, 2021). Cheese consumption: 60 g/day (Malissiova *et al.*, 2022). ND – Not detectable.

fraction in milk, which is concentrated in a significant proportion in the curd portion after draining of whey because of this binding (Kaan Tekinşen & Cenap Tekinşen, 2005; Scaglioni *et al.*, 2014). Furthermore, countries like Turkey introduced a legal limit for AFM_1 at 250 ng/kg for cheese (Turkish Food Codex, 2002).

Cheese is a product with diverse technological variations in its production, characteristic of the place of origin. However, few studies evaluate these different technologies' influence on AFM_1 indices. Table 6.2 shows the studies that assess the levels of AFM_1 in cheeses from other countries. Most parts of the studies have been identified in countries on the African continent, showing high incidence ranging of 51 at 100% of samples (Table 6.2), in the same way, with AFM_1 concentration between <5.0 at 5580 ng/L/Kg, showing AFM_1 levels higher than limits determined in several countries (Section 1.2.). Regarding to the average levels of AFM_1 only the processed cheese rated in Egypt showed acceptable aflatoxin levels within the strictest range ($\leq$ 50 ng/L; EU, 2010). Lower levels were observed in the Americas and Europe (2.10 at 485 ng/L/Kg) compared to Africa; however, most samples showed concentrations above the European limit (EU, 2010). Malissiova *et al.* (2022) related that the occurrence of AFM_1 in cheese is relatively high worldwide. However, the gradual decrease of AFM_1 in stored Karish cheese was reported by Marshaly *et al.* (1989). Indeed, some factors such as low pH, the presence of lactic acid bacteria, or the formation of by-products could contribute to the reduction of AFM_1 levels in dairy products; furthermore, other studies also suggest that during cheese production, a high percentage of AFM_1 is transferred in whey than curd (Pietri *et al.*, 2016). Despite the behaviors noted above, a high stability of AFM1 was also observed during the ripening and storage of some cheeses such as Brick, Limburger, Camembert, Tilsit, Cheddar, Gouda, Manchego, Parmesan, and Mozzarella (Galvano *et al.*, 1996). According to Mohammadi *et al.* (2022), the concentration and the occurrence of AFM_1 in fresh cheese types were higher than in the ripened samples. However, this behavior was not observed in current studies (Table 6.2), with Ras cheese (86.97 ng/L/Kg) showing AFM_1 levels similar to Fior di Latte cheese (83.5 ng/L/Kg) and lower than the Minas Frescal cheese (113 ng/L/Kg). In addition to factors such as cheese-making procedures and ripening conditions, the levels of AFM_1 in milk hygiene and storage conditions at dairies and the geographical region also influence the levels of this aflatoxin in cheese (Ghiasian *et al.*, 2007; Kaan Tekinşen and Cenap Tekinşen, 2005).

6.6 AFM_1 IN YOGURTS

Despite being broad, information about the influence of the processing of yogurts on AMF_1 is quite contradictory in relation to the data found as seen in the Table 6.3. However, some probiotic bacteria have been reported to be effective in binding and removing AFM_1 from contaminated milk. This effect can be affected by different factors such as the strain used, the concentration of AFM_1, pH, heat processes, ionic strength, fermentation temperature, protein content, titratable acidity, storage temperature, and time (Arab *et al.*, 2012; Sarlak *et al.*, 2017; Sevim *et al.*, 2019). Using bacteria for AFM_1 decontamination in dairy products is a promising area of study; nevertheless, this section will focus on detecting AFM_1 naturally contained in yogurt samples. Among dairy products, yogurt is one of the least studied products concerning the presence of AFM_1 (Tadesse *et al.*, 2020). Most of the current studies have been identified in countries on the Asian continent, showing high incidence between 53–893 ng/L/Kg (Table 6.3), but only one study shows samples with AFM_1 levels higher than limits determined in European limit ($\leq$ 50 ng/L; EU, 2010). However, yogurt samples from America and Africa showed high mean concentrations of AFM_1, with samples from Ethiopia showing levels above 1,600 ng/L/kg (Table 6.3).

Studies already reported the lower presence of AFM_1 in yogurt samples than in milk being attributed to lactic acid bacteria (El Khoury *et al.*, 2011; Guo *et al.*, 2019); this could be mainly attributed to fermentation factors such as low pH and the formation of organic acids or other by-products, or even to the presence of yogurt starter bacteria (Hassanin, 1994; Sarimehmetoğlu and Küplülü, 2004). The reduced pH during the process of fermentation causes denaturation of milk proteins (caseins), leading to the formation coagulum that leads to the adsorption of AFM_1 to proteins. Furthermore, the exposure of more hydrophobic sites of the complex casein fractions with denatured whey proteins can bind to a greater extent with aflatoxins (Brackett and Marth, 1982; Dosako *et al.*, 1980; Tamime and Robinson, 2007). However, the study of Tadesse *et al.* (2020) attributed the high presence of aflatoxins in their yogurt samples to previously contaminated milk. It concluded that the fermentation process during yogurt might not be presented as an impact on the level of AFM_1 in your samples. Contrary to other studies, varying concentrations of AFM_1 in yogurt were found concerning milk (Murshed, 2020). Egmond *et al.* (1977) and Munksgaard *et al.* (1987) reported a significant increase in free AFM_1 in

Table 6.3: Occurrence of AFM$_1$ in Yogurt and Estimated Daily Intake (EDI)

Continent	Country	Yogurt Sample	Quantification Method	Incidence	Range (ng/L and ng/Kg)	Mean (ng/L and ng/Kg)	EDI* (ng/kg bw/day)	Reference
Africa	Ethiopia	Yogurt industrial	ELISA	83/83	9–4010	1,631	1.398	Tadesse et al. (2020)
		Yogurt local (Ergo)	ELISA	10/10	7–4760	1,628	1.395	
America	Brazil	Yogurt	HPLC-FLD	7/72	17–130	130	0.111	Pires et al. (2022)
Asia	China	Yogurt	ELISA	194/319	10.0–66.7	20.7	0.018	Xiong et al. (2022)
		Yogurt	HPLC	–	–	37.34	0.032	Cai et al. (2021)
	Bangladesh	Fermented milk	ELISA	5/5	–	16.9	0.014	Sumon et al. (2021)
	Yemen	Yogurt	HPLC-FLD	54/62	53–893	399	0.342	Murshed (2020)

Note
* EDI = (AFM1 concentration) X (daily consumption) / (body weight = 70 kg; Saha Turna and Wu, 2021). Yogurt consumption: 21 g/day (Miller et al., 2022). ND – Not detectable.

fermented milk. This effect may be attributed to different final fermentation pHs of yogurts, various initial concentrations of starter bacteria and AFM_1 in the milk, other fermentation conditions, changes in some physicochemical properties of caseins, and/or application of unreliable analytical methods (Sarlak *et al.*, 2017).

6.7 ESTIMATED DAILY INTAKE (IDE) OF DAIRY PRODUCTS

Milk is a highly nutritious food with lactose; fat; vitamins A, D, and E; and several minerals, including calcium, magnesium, and potassium, together with caseins and whey proteins (Pereira, 2014). In addition, recent epidemiological evidence shows that aflatoxins are harmful to human health, including primary liver cancer, child growth impairment, and immune suppression (Hasninia *et al.*, 2022), and chronic dietary exposure constitutes the most significant risk of aflatoxins exposure in humans (Topi *et al.*, 2022), including the presence of AFM_1 in dairy products such as cheese, butter, yogurt, and others. The estimated daily intake (EDI) is a critical risk assessment stage, aiming to determine the concentration of aflatoxin to which an individual or a population is exposed (Saha Turna and Wu, 2021). In that regard, the EDI was determined for milk, cheese, and yogurt samples evaluated in the studies. The milk samples of Bangladesh, Pakistan, India, Rwanda, Malawi, and Ethiopia showed high levels of AFM_1 (above 1 ng/kg BW/day), regardless of the processing used (powder, UHT, and pasteurized) (Table 6.1). The cheese sample was the dairy product that showed the lowest EDI index with high values found in Ghana and Ethiopia, with 0.562 and 0.660 ng/kg BW/day, respectively (Table 6.2). Moreover, Ethiopia showed the country that shows the highest EDI in the yogurt samples (1.398 and 1.395 ng/kg BW/day), highlighting the need for greater preventive care regarding the presence of AFB_1 in milk samples.

The risk of AFM_1 exposure in milk consumers varies greatly among different countries, attributing to AFM_1 concentrations, dairy product consumption, and body weight considered in the EDI calculation. Xiong *et al.* (2022) report that the difference in milk AFM_1 concentrations is a critical factor regarding the risk of AFM_1 exposure in milk and dairy products. Currently, the Joint Expert Committee on Food Additives (JECFA) of the Food and Agriculture Organization and World Health Organization has not set a tolerable daily intake (TDI) for any of the aflatoxins, including AFM_1. Despite this, exposure to this mycotoxin should be kept as low as reasonably achievable (Saha Turna and Wu, 2021).

6.8 CONCLUSION

According to current literature data, high levels of AMF_1 are still found in milk samples in many countries. About 79% of the milk samples contain AFM_1 levels higher than the maximum acceptable level established in the EU for this toxin. This high AFM_1 contamination results from high aflatoxin contamination in dairy cattle feedstuffs. The influence of dairy product processing on AFM_1 levels still needs to be further explored because of the greater number of samples of cheese and yogurt contaminated with high levels of AFM_1. With respect to the toxic effects of AFM_1 on human health, strict regulatory monitoring and legislation must be applied to reduce exposure to aflatoxins in animal feed, which is the best strategy to reduce population exposure to AFM_1.

REFERENCES

Abebe, B. Abriham, K. Yobsan, T. 2018. Review on aflatoxin and its impacts on Livestock. *Journal of Dairy & Veterinary Sciences*, 6:555685.

Abrar, M., Anjum, F. M., Butt, M. S., Pasha, I., Randhawa, M. A., Saeed, F., and Waqas, K. 2013. Aflatoxins: biosynthesis, occurrence, toxicity, and remedies. *Critical Reviews in Food Science and Nutrition*, 53:862–874.

Alahlah, N., El Maadoudi, M., Bouchriti, N., Triqui, R., and Bougtaib, H. 2020. Aflatoxin M1 in UHT and powder milk marketed in the northern area of Morocco. *Food Control*, 114:107262.

Amaike, S., and Keller, N. P. 2011. Aspergillus flavus. *Annual Review of Phytopathology*, 49:107–133.

Anukul, N., Vangnai, K., and Mahakarnchanakul, W. 2013. Significance of regulation limits in mycotoxin contamination in Asia and risk management programs at the national level. *Journal of Food and Drug Analysis*, 21:227–241.

Arab, M., Sohrabvandi, S., Mortazavian, A. M., Mohammadi, R., and Tavirani, M. R. 2012. Reduction of aflatoxin in fermented milks during production and storage. *Toxin Reviews*, 31:44–53.

Asi, M. R., Iqbal, S. Z., Ariño, A., and Hussain, A. 2012. Effect of seasonal variations and lactation times on aflatoxin M1 contamination in milk of different species from Punjab, Pakistan. *Food Control*, 25:34–38.

Brackett, R. E., and Marth, E. H. 1982. Association of aflatoxin M1 with casein. *Zeitschrift Fr Lebensmittel-Untersuchung Und -Forschung*, 174:439–441.

Brasil. 2022. Resolução n° 274 de 15/10/2002 da ANVISA. Aprova o Regulamento Técnico sobre limites máximos de aflatoxinas no leite, amendoim e milho. https://bvsms.saude.gov.br/bvs/saudelegis/anvisa/2002/res0274_15_10_2002.html. (accessed October 01, 2022).

Cai, C., Zhang, Q., Nidiaye, S., Yan, H., Zhang, W., Tang, X., and Li, P. 2021. Development of a specific anti-idiotypic nanobody for monitoring aflatoxin M1 in milk and dairy products. *Microchemical Journal*, 167:106326.

Campagnollo, F. B., Ganev, K. C., Khaneghah, A. M., Portela, J. B., Cruz, A. G., Granato, D., Corassin, C. H., Oliveira, C. A. F., and Sant'Ana, A. S. 2016. The occurrence and effect of unit operations for dairy products processing on the fate of aflatoxin M1: A review. *Food Control*, 68:310–329.

Cullen, J. M., Ruebner, B. H., Hsieh, L. S., Hyde, D. M., and Hsieh, D. P.1987. Carcinogenicity of dietary aflatoxin M1 in male Fischer rats compared to aflatoxin B1. *Cancer Research*, 7.1913 7.

de Matos, C. J., Schabo, D. C., do Nascimento, Y. M., Tavares, J. F., Lima, E. de O., da Cruz, P. O., de Souza, E. L., Magnani, M., and Magalhães, H. I. F. 2021. Aflatoxin M1 in Brazilian goat milk and health risk assessment. *Journal of Environmental Science and Health, Part B*, 56:415–422.

Dosako, S., Kaminogawa, S., Taneya, S., and Yamauchi, K. 1980. Hydrophobic surface areas and net charges of α s1 -, κ-casein and α s1 -casein: κ-casein complex. *Journal of Dairy Research*, 47:123–129.

Egmond, H. P. Van Paulsch, W. E., Veringa, H. A., Schuller, P. E., Sizoo, E. A., and Wilhelmina, E. 1977. The effect of processing on the aflatoxin M1 content of milk and milk products. *Archives de l'Institut Pasteur de Tunis*, 4:381–390.

El Khoury, A., Atoui, A., and Yaghi, J. 2011. Analysis of aflatoxin M1 in milk and yogurt and AFM1 reduction by lactic acid bacteria used in Lebanese industry. *Food Control*, 22:1695–1699.

Esam, R. M., Hafez, R. S., Khafaga, N. I. M., Fahim, K. M., and Ahmed, L. I. 2022. Assessment of aflatoxin M1 and B1 in some dairy products with referring to the analytical performances of enzyme-linked immunosorbent assay in comparison to high-performance liquid chromatography. *Veterinary World*, 1:91–101.

EU. 2010. Commission regulation (EU) No 165/2010 of 26 February 2010. amending regulation (European Commission) No 1881/2006 setting maximum levels for certain contaminants in foodstuffs as regards aflatoxins.

FDA. 2007. CPG Sec. 527.400 Whole Milk, Lowfat Milk, Skim Milk in Aflatoxin M1.

Flores-Flores, M. E., Lizarraga, E., López de Cerain, A., and González-Peñas, E. 2015. Presence of mycotoxins in animal milk: A review. *Food Control*, 53:163–176.

Galvano, F., Galofaro, V., and Galvano, G. 1996. Occurrence and stability of Aflatoxin M1 in milk and milk products: a worldwide review. *Journal of Food Protection*, 59:1079–1090.

Gbashi, S., Edwin Madala, N., De Saeger, S., De Boevre, M., Adekoya, I., Ayodeji Adebo, O., and Berka Njobeh, P. 2019. The Socio-economic Impact of Mycotoxin Contamination in Africa. In *Mycotoxins – Impact and Management Strategies. IntechOpen.* https://www.intechopen.com/chapters/62483. (accessed October 01, 2022).

Ghiasian, S. A., Maghsood, A. H., Neyestani, T. R., and Mirhendi, S. H. 2007. Occurrence of aflatoxin M1 in raw milk during the summer and winter seasons in Hamedan, Iran. *Journal of Food Safety*, 27:188–198.

Goncalves, B. L., Corassin, C. H., and Oliveira, C. A. F. 2015. Mycotoxicoses in Dairy Cattle: A Review. *Asian Journal of Animal and Veterinary Advances*, 10:752–760.

Gonçalves, B. L., Gonçalves, J. L., Rosim, R. E., Cappato, L. P., Cruz, A. G., Oliveira, C. A. F., and Corassin, C. H. 2017. Effects of different sources of Saccharomyces cerevisiae biomass on milk production, composition, and aflatoxin M1 excretion in milk from dairy cows fed aflatoxin B1. *Journal of Dairy Science*, 100:5701–5708.

Gonçalves, B. L., Ulliana, R. D., Ramos, G. L. P. A., Cruz, A. G., Oliveira, C. A. F., Kamimura, E. S., and Corassin, C. H. 2021. Occurrence of aflatoxin M1 in milk and Minas Frescal cheese manufactured in Brazilian dairy plants. *International Journal of Dairy Technology*, 74:431–434.

Guo, L., Wang, Y., Fei, P., Liu, J., and Ren, D. 2019. A survey on the aflatoxin M1 occurrence in raw milk and dairy products from water buffalo in South China. *Food Control*, 105:159–163.

Hashemi, M. 2016. A survey of aflatoxin M1 in cow milk in Southern Iran. *Journal of Food and Drug Analysis*, 24:888–893.

Hasninia, D., Salimi, G., Bahrami, G., Sharafi, K., Omer, A. K., Rezaie, M., and Kiani, A. 2022. Human health risk assessment of aflatoxin M1 in raw and pasteurized milk from the Kermanshah province, Iran. *Journal of Food Composition and Analysis*, 110:104568.

Hassanin, N. I. 1994. Stability of aflatoxin M1 during manufacture and storage of yoghurt, yoghurt-cheese and acidified milk. *Journal of the Science of Food and Agriculture*, 65:31–34.

Hattimare, D., Shakya, S., Patyal, A., Chandrakar, C., and Kumar, A. 2022. Occurrence and exposure assessment of Aflatoxin M1 in milk and milk products in India. *Journal of Food Science and Technology*, 59:2460–2468.

IARC. 2012a. Chemical agents and related occupations. IARC Working Group on the Evaluation of Carcinogenic Risks to Humans, Lyon, 100F. https://www.ncbi.nlm.nih.gov/books/NBK304416/. (accessed October 12, 2022).

IARC. 2012b. *Economics of mycotoxins: evaluating costs to society and cost-effectiveness of interventions.* IARC Scientific Publications, Lyon, 119–129. http://www.ncbi.nlm.nih.gov/pubmed/23477200. (accessed October 12, 2022).

Iqbal, S. Z., Jinap, S., Pirouz, A. A., and Faizal, A. R. A. 2015. Aflatoxin M1 in milk and dairy products, occurrence and recent challenges: A review. *Trends in Food Science & Technology*, 46:110–119.

Iqbal, S. Z., Paterson, R. R. M., Bhatti, I. A., and Asi, M. R. 2010. Survey of aflatoxins in chillies from Pakistan produced in rural, semi-rural and urban environments. *Food Additives and Contaminants: Part B*, 3:268–274.

ISIRI. 2010. Food and feed mycotoxins maximum tolerated level. Institute of Standards and Industrial Research of Iran, 5925.

Kaan Tekinşen, K., and Cenap Tekinşen, O. 2005. Aflatoxin M1 in white pickle and Van otlu (herb) cheeses consumed in southeastern Turkey. *Food Control*, 16:565–568.

Kamkar, A. 2005. A study on the occurrence of aflatoxin M1 in raw milk produced in Sarab city of Iran. *Food Control*, 16:593–599.

Kamkar, A., Karim, G., Aliabadi, F. S., and Khaksar, R. 2008. Fate of aflatoxin M1 in Iranian white cheese processing. *Food and Chemical Toxicology*, 46:2236–2238.

Klich, M. A. 2007. Aspergillus flavus: the major producer of aflatoxin. *Molecular Plant Pathology*, 8:713–722.

Kortei, N. K., and Annan, T. 2022. Aflatoxin M1 contamination of Ghanaian traditional soft cottage cheese (Wagashie) and health risks associated with its consumption. *Journal of Food Quality*, 2022:1–12.

Kos, J., Janić Hajnal, E., Šarić, B., Jovanov, P., Mandić, A., Đuragić, O., and Kokić, B. 2018. Aflatoxins in maize harvested in the Republic of Serbia over the period 2012–2016. *Food Additives & Contaminants: Part B*, 11:246–255.

Kumar, A., Pathak, H., Bhadauria, S., and Sudan, J. 2021. Aflatoxin contamination in food crops: causes, detection, and management: a review. *Food Production, Processing and Nutrition*, 3:17.

Li, S., Min, L., Wang, G., Li, D., Zheng, N., and Wang, J. 2018. Occurrence of Aflatoxin M1 in raw milk from manufacturers of infant milk powder in China. *International Journal of Environmental Research and Public Health*, 15:879.

López, C., Ramos, L., Ramadán, S., Bulacio, L., and Perez, J. 2001. Distribution of aflatoxin M1 in cheese obtained from milk artificially contaminated. *International Journal of Food Microbiology*, 64:211–215.

Malissiova, E., Soultani, G., Tsokana, K., Alexandraki, M., and Manouras, A. 2022. Exposure assessment on aflatoxin M1 from milk and dairy products-relation to public health. *Clinical Nutrition ESPEN*, 47:189–193.

Mannani, N., Tabarani, A., El Adlouni, C., Abdennebi, E. H., and Zinedine, A. 2021. Aflatoxin M1 in pasteurized and UHT milk marked in Morocco. *Food Control*, 124:107893.

Marshaly, R. I., Deeb, S. A., and Safwat, N. M. 1989. Distribution and stability of aflatoxin M1 during processing and storage of Karish cheese. *Alexandria Journal of Agriculture Research*, 31:219–228.

Miller, V., Reedy, J., Cudhea, F., Zhang, J., Shi, P., Erndt-Marino, J., Coates, J., Micha, R., Webb, P., Mozaffarian, D., Abbott, P., Abdollahi, M., Abedi, P., Abumweis, S., Adair, L., Al Nsour, M., Al-Daghri, N., Al-Hamad, N., Al-Hooti, S., and Zohoori, F. V. 2022. Global, regional, and national consumption of animal-source foods between 1990 and 2018: findings from the Global Dietary Database. *The Lancet Planetary Health*, 6:e243–e256.

Min, L., Fink-Gremmels, J., Li, D., Tong, X., Tang, J., Nan, X., Yu, Z., Chen, W., and Wang, G. 2021a. An overview of aflatoxin B1 biotransformation and aflatoxin M1 secretion in lactating dairy cows. *Animal Nutrition*, 7:42–48.

Min, L., Tong, X., Sun, H., Ding, D., Xu, B., Chen, W., Wang, G., and Li, D. 2021b. Aflatoxin M1 contamination in raw milk and its association with herd types in the ten provinces of Southern China. *Italian Journal of Animal Science*, 20:1562–1567.

Mohammadi, S., Behmaram, K., Keshavarzi, M., Saboori, S., Jafari, A., and Ghaffarian-Bahraman, A. 2022. Aflatoxin M1 contamination in different Iranian cheese types: A systematic review and meta-analysis. *International Dairy Journal*, 133:105437.

Mokhtari, S. A., Nemati, A., Fazlzadeh, M., Moradi-Asl, E., Ardabili, V. T., and Seddigh, A. 2022. Aflatoxin M1 in distributed milks in northwestern Iran: occurrence, seasonal variation, and risk assessment. *Environmental Science and Pollution Research*, 29:41429–41438.

Mollayusefian, I., Ranaei, V., Pilevar, Z., Cabral-Pinto, M. M. S., Rostami, A., Nematolahi, A., Khedher, K. M., Thai, V. N., Fakhri, Y., and Mousavi Khaneghah, A. 2021. The concentration of aflatoxin M1 in raw and pasteurized milk: A worldwide systematic review and meta-analysis. *Trends in Food Science & Technology*, 115:22–30.

Munksgaard, L., Larsen, J., Werner, H., Andersen, P., and Viuf, B. 1987. Carry over of aflatoxin from cows' feed to milk and milk products. *Milchwissenschaft*, 3:165–167.

Murshed, S. 2020. Evaluation and assessment of Aflatoxin M1 in milk and milk products in Yemen using high-performance liquid chromatography. *Journal of Food Quality*, 2020:1–8.

Negash, D. 2018. A review of aflatoxin: occurrence, prevention, and gaps in both food and feed safety*Journal of Applied Microbiological Research*, 1:190–197.

Nemati, M., Mehran, M. A., Hamed, P. K., and Masoud, A. 2010. A survey on the occurrence of aflatoxin M1 in milk samples in Ardabil, Iran. *Food Control*, 21:1022–1024.

Nishimwe, K., Bowers, E. L., de Dieu Ayabagabo, J., Habimana, R., Mutiga, S., and Maier, D. E. 2022. Preliminary sampling of aflatoxin M1 contamination in raw milk from dairy farms using feed ingredients from Rwanda. *Mycotoxin Research*, 38:107–115.

Njombwa, C. A., Moreira, V., Williams, C., Aryana, K., and Matumba, L. 2021. Aflatoxin M1 in raw cow milk and associated hepatocellular carcinoma risk among dairy farming households in Malawi. *Mycotoxin Research*, 37:89–96.

Ono, T., Ueta, M., Tsuji, M., Kawamoto, C., and Yamane, N. 2020. Examination of aflatoxin M1 contamination in milk and milk drinks available in Fukuyama City. *Food Hygiene and Safety Science (Shokuhin Eiseigaku Zasshi)*, 61:148–153.

Pal, M. 2017. Are Mycotoxins silent killers of humans and animals? *Journal of Experimental Food Chemistry*, 03:1000e110.

Pal, M., Lema, A. G., Ejeta, D. I., and Gowda, L. 2021. Global public health and economic concern due to aflatoxins. *Global Journal of Research in Medical Sciences*, 1:5–8.

Pecorelli, I., Branciari, R., Roila, R., Ranucci, D., Bibi, R., Van Asselt, M., and Valiani, A. 2020. Evaluation of Aflatoxin M1 enrichment factor in different cow milk cheese hardness category. *Italian Journal of Food Safety*, 9:8419.

Peña-Rodas, O., Martinez-Lopez, R., Pineda-Rivas, M., and Hernandez-Rauda, R. 2020. Aflatoxin M1 in Nicaraguan and locally made hard white cheeses marketed in El Salvador. *Toxicology Reports*, 7:1157–1163.

Pereira, P. C. 2014. Milk nutritional composition and its role in human health. *Nutrition*, 30:619–627.

Pietri, A., Mulazzi, A., Piva, G., & Bertuzzi, T. 2016. Fate of aflatoxin M1 during production and storage of parmesan cheese. *Food Control*, 60:478–483.

Pires, R. C., Portinari, M. R. P., Moraes, G. Z., Khaneghah, A. M., Gonçalves, B. L., Rosim, R. E., Oliveira, C. A. F., and Corassin, C. H. 2022. Evaluation of Anti-Aflatoxin M1 effects of heat-killed cells of Saccharomyces cerevisiae in Brazilian commercial yogurts. *Quality Assurance and Safety of Crops & Foods*, 14:75–81.

Prandini, A., Tansini, G., Sigolo, S., Filippi, L., Laporta, M., and Piva, G. 2009. On the occurrence of aflatoxin M1 in milk and dairy products. *Food and Chemical Toxicology*, 47:984–991.

Quadri, S. H. M., Niranjan, M., Chaluvaraju, K., Shantaram, U., and Enamul, H. 2012. An overview of chemistry, toxicity, analysis, and control of aflatoxins. *International Journal of Chemical and Life Sciences*, 2:1071–1078.

Qureshi, H., Hamid, S. S., Ali, S. S., Anwar, J., Siddiqui, A. A., and Khan, N. A. 2015. Cytotoxic effects of aflatoxin B1 on human brain microvascular endothelial cells of the blood-brain barrier. *Medical Mycology*, 53:409–416.

Rahmani, J., Alipour, S., Miri, A., Fakhri, Y., Riahi, S.-M., Keramati, H., Moradi, M., Amanidaz, N., Pouya, R. H., Bahmani, Z., and Mousavi Khaneghah, A. 2018. The prevalence of aflatoxin M1 in milk of Middle East region: A systematic review, meta-analysis and probabilistic health risk assessment. *Food and Chemical Toxicology*, 118:653–666.

Rajarajan, P. N., Rajasekaran, K. M., and Asha Devi, N. K. 2013. Aflatoxin contamination in agricultural commodities. *Indian Journal of Pharmaceutical and Biological Research*, 1:148–151.

Rushing, B. R., and Selim, M. I. 2019. Aflatoxin B1: A review on metabolism, toxicity, occurrence in food, occupational exposure, and detoxification methods. *Food and Chemical Toxicology*, 124:81–100.

Saha Turna, N., and Wu, F. 2021. Aflatoxin M1 in milk: A global occurrence, intake, & exposure assessment. *Trends in Food Science & Technology*, 110:183–192.

Sarimehmetoğlu, B., and Küplülü, Ö. 2004. Binding ability of aflatoxin M1 to yoghurt bacteria. *Ankara Üniversitesi Veteriner Fakültesi Dergisi*, 15:195–198.

Sarlak, Z., Rouhi, M., Mohammadi, R., Khaksar, R., Mortazavian, A. M., Sohrabvandi, S., and Garavand, F. 2017. Probiotic biological strategies to decontaminate aflatoxin M1 in a traditional Iranian fermented milk drink (Doogh). *Food Control*, 71:152–159.

Scaglioni, P., Becker-Algeri, T., Drunkler, D., and Badiale-Furlong, E. 2014. Aflatoxin B1 and M1 in milk. *Analytica Chimica Acta*, 829:68–74.

Sengun, I., Yaman, D., and Gonul, S. 2008. Mycotoxins and mould contamination in cheese: a review. *World Mycotoxin Journal*, 1:291–298.

Sevim, S., Topal, G. G., Tengilimoglu-Metin, M. M., Sancak, B., and Kizil, M. 2019. Effects of inulin and lactic acid bacteria strains on aflatoxin M1 detoxification in yoghurt. *Food Control*, 100:235–239.

Sinnhuber, R. O., Lee, D. J., Wales, J. H., Landers, M. K., and Keyl, A. C. 1974. Hepatic carcinogenesis of aflatoxin M1 in rainbow trout (Salmo gairdneri) and its enhancement by cyclopropene fatty acids. *Journal of the National Cancer Institute*, 5:1285–1288.

Škrbić, B., Živančev, J., Antić, I., and Godula, M. 2014. Levels of aflatoxin M1 in different types of milk collected in Serbia: Assessment of human and animal exposure. *Food Control*, 40:113–119.

Sumon, A. H., Islam, F., Mohanto, N. C., Kathak, R. R., Molla, N. H., Rana, S., Degen, G. H., and Ali, N. 2021. The Presence of Aflatoxin M1 in milk and milk products in Bangladesh. *Toxins*, 13:440.

Tadesse, S., Berhanu, T., and Woldegiorgis, A. Z. 2020. Aflatoxin M1 in milk and milk products marketed by local and industrial producers in Bishoftu town of Ethiopia. *Food Control*, 118:107386.

Tajkarimi, M., Aliabadi-Sh, F., Salah Nejad, A., Poursoltani, H., Motallebi, A. A., and Mahdavi, H. 2008. Aflatoxin M1 contamination in winter and summer milk in 14 states in Iran. *Food Control*, 19:1033–1036.

Tajkarimi, M., Shojaee Aliabadi, F., Salah Nejad, M., Pursoltani, H., Motallebi, A. A., and Mahdavi, H. 2007. Seasonal study of aflatoxin M1 contamination in milk in five regions in Iran. *International Journal of Food Microbiology*, 116:346–349.

Tamime, A. Y., and Robinson, R. K. 2007. *Tamime and Robinson's Yoghurt*. Sawston: Woodhead Publishing Limited

Tarannum, N., Nipa, M. N., Das, S., and Parveen, S. 2020. Aflatoxin M1 detection by ELISA in raw and processed milk in Bangladesh. *Toxicology Reports*, 7:1339–1343.

Topi, D., Spahiu, J., Rexhepi, A., and Marku, N. 2022. Two-year survey of aflatoxin M1 in milk marketed in Albania, and human exposure assessment. *Food Control*, 136:108831.

Torres, B., Ron, L., and Gomez, C. 2022. Dietary Aflatoxin B1-related risk factors for the presence of Aflatoxin M1 in raw milk of cows from Ecuador. *Open Veterinary Journal*, 12:129.

Turkish Food Codex 2002. Gida Maddelerinde belirli bulasanlarin maksimum seviyelerinin belirlenmesi hakkinda teblig. Resmi Gazete, 23 Eylul 2002. Sayi 24885. Basbakanlik Basimevi, Ankara, Turkey.

Unusan, N. 2006. Occurrence of aflatoxin M1 in UHT milk in Turkey. *Food and Chemical Toxicology*, 44:1897–1900.

Wagacha, J. M., and Muthomi, J. W. 2008. Mycotoxin problem in Africa: Current status, implications to food safety and health and possible management strategies. *International Journal of Food Microbiology*, 124:1–12.

Xiong, J., Wen, D., Zhou, H., Chen, R., Wang, H., Wang, C., Wu, Z., Qiu, Y., and Wu, L. 2022. Occurrence of aflatoxin M1 in yogurt and milk in central-eastern China and the risk of exposure in milk consumers. *Food Control*, 137:108928.

Yiannikouris, A., and Jouany, J.-P. 2002. Mycotoxins in feeds and their fate in animals: a review. *Animal Research*, 51:81–99.

Yunus, A. W., Ullah, A., Lindahl, J. F., Anwar, Z., Ullah, A., Saif, S., Ali, M., Zahur, A. Bin, Irshad, H., Javaid, S., Imtiaz, N., Farooq, U., Ahsan, A., Fatima, Z., Hashmi, A. A., Abbasi, B. H. A., Bari, Z., Khan, I. U., and Ibrahim, M. N. M. 2020. Aflatoxin contamination of milk produced in peri-urban farms of Pakistan: Prevalence and contributory factors. *Frontiers in Microbiology*, 11:159.

Zebib, H., Abate, D., and Woldegiorgis, A. Z. 2022. Aflatoxin M1 in raw milk, pasteurized milk and cottage cheese collected along value chain actors from three regions of Ethiopia. *Toxins*, 14:276.

7 Analysis of Heavy Metals in Seafoods

Long Wu, Wei Zeng, Ting Wu, Xuemei Tang, Wenjing Kang, Yin Liu, and Miaomiao Yang

7.1 INTRODUCTION

Seafood has rich nutrients for human health, including proteins, lipids, and other bioactive elements (Larsen *et al.*, 2011). As the components in seafood are quite different from those in the on-land animals, seafood can greatly enrich people's diet (Jobling, 2016). Nowadays, seafood has been regarded as one of the most important sources of nutrients in the food supply (Farmery *et al.*, 2022). In addition to raw seafood, it can be processed into a wide variety of products, with many forms (e.g., smoked, canned, cured, dried, fresh, frozen etc.). The processed seafood can be endowed with unique flavor and taste, but the contaminants of seafood may arise. Cooking is a good means to deal with seafood to avoid microbial contamination. Although a thorough cooking of food can destroy most contaminants like pathogens, metals or the compounds in food can be rather stable (Nivetha *et al.*, 2022). In this regard, we should take care about the content of metals in specific foods, especially the heavy metals. Actually, the human body requires various trace heavy metals in biological systems. High levels of heavy metals in food can induce cytotoxic effects or even genotoxicity. On the other hand, some metals can go through different forms in cooking and storing processes. Metal ions can form complexes with organic compounds at a low concentration level of mg kg^{-1}, which may result in unpleasant color variations (Saleh *et al.*, 2001). Therefore, much attention should be paid on the contaminants of heavy metals in seafood.

Heavy metals refer to a group of metals and non-metals, whose density is generally greater than 4,000 kg m^{-3} (Vardhan *et al.*, 2019), including mercury (Hg), cadmium (Cd), arsenic (As), lead (Pb), zinc (Zn), copper (Cu), and so on. According to the trace elements in human body, they can be divided into non-essential and essential trace elements (Shah, 2021). For non-essential trace elements, their accumulation in the organism will cause high toxicity. Meanwhile, a low concentration of essential trace elements plays an important role in the normal physiological process of human body, but a high concentration will pose severe health risks (Jaishankar *et al.*, 2014). It is generally believed that heavy metals will accumulate in aquatic animals through the following ways: 1) absorb the dissolved heavy metals through the gills; 2) ingest bait or food containing heavy metals; 3) through the infiltration exchange between the body surface and water (Ali and US SA, 2014). Due to their non-degradation, heavy metals can be transmitted and accumulated along the food chain, thus producing biological amplification effect. From this it is clear that heavy metals not only do harm to marine organisms, but also pose potential risks on human health when people consume heavy metal–contaminated seafood. Moreover, many heavy metals that coexist in organisms may show antagonism and synergy effects, making it more difficult to determine their toxicity. In addition, some microorganisms in the ocean can convert certain heavy metals into more toxic compounds (Regnell and Watras, 2018). The above facts indicated that heavy metal contamination can cause the problems in seafood, and it directly affects human health with the consumption of such seafood. Intensive attention has been paid on the influences of heavy metals on seafood safety and human health. Still, it is of great importance to inspect and control the heavy metal–contaminated seafood products. In this chapter, we will introduce the common heavy metals in seafood products, and then summarize all kinds of analytical methods (traditional and rapid ones) for the detection of heavy metals in seafood, especially the newly developed biosensors (Figure 7.1). In addition, some suggestions are given on the control of heavy metals in seafood. At the end, the current challenges as well as prospects of analysis of heavy metals will be presented to provide innovative idea tactics for seafood safety.

7.2 HEAVY METALS IN FOOD TOXICITY

7.2.1 Mercury (Hg) Toxicity

Mercury (Hg) is a heavy metal with forms of Hg(OH)$_2$ and HgCl$_2$ in water. Until the 1950s, Hg was considered a contaminant due to widespread poisoning deaths. After that, Hg exposure was thought to have harm to human health. Most Hg ends up in the ocean due to natural processes like atmospheric deposition and discharge of sewage plants (Sundseth *et al.*, 2017). Methylmercury (CH$_3$Hg), another form of Hg, is easily accumulated in aquatic organisms and transmitted to humans through the food chain (Li *et al.*, 2017). Due to human activities and the polluted environment, the acidity of rivers and lakes reduces the loss of volatile Hg and increases its binding

DOI: 10.1201/9781003334859-7

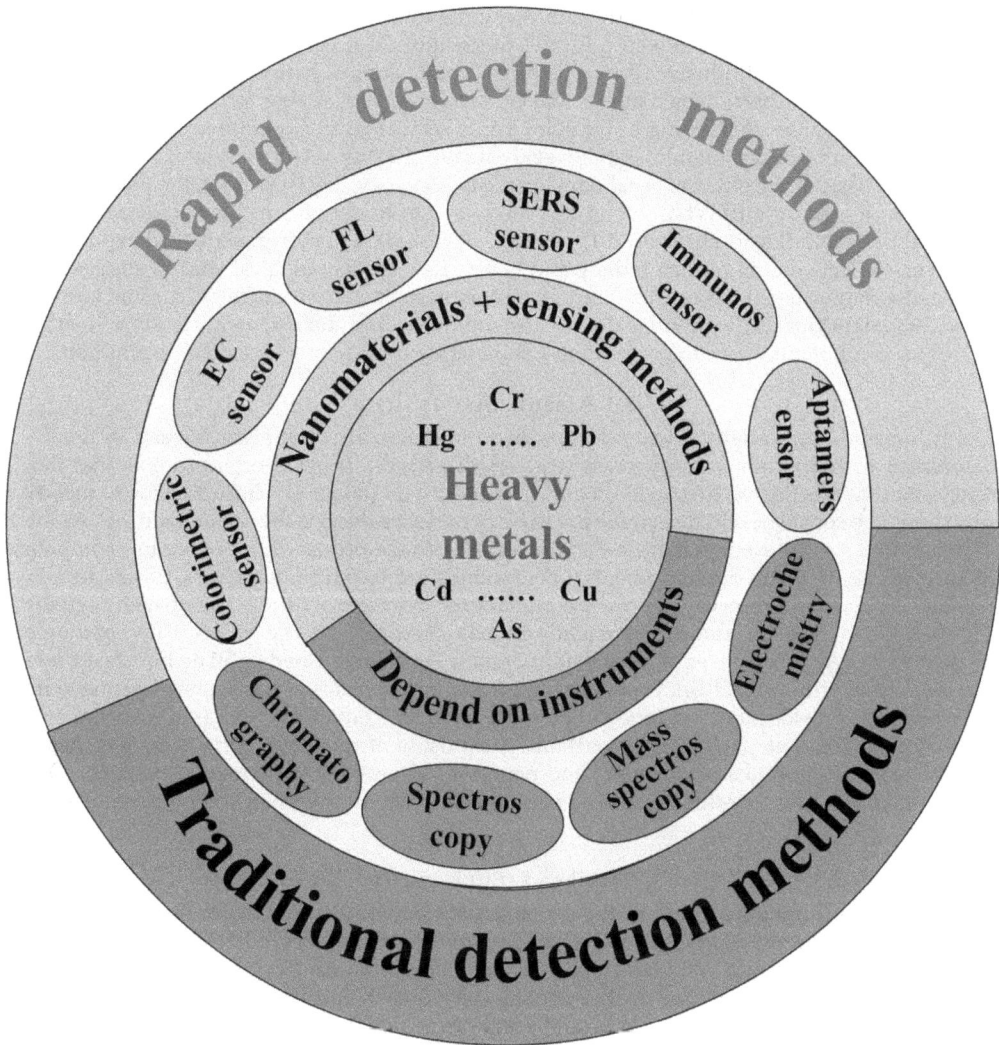

Figure 7.1 Schematic presentation of various detection methods for the detection of heavy metals in seafood.

to particulates in water, which facilitate the methylation of Hg at a low pH environment. In the case of consumption of seafoods, CH_3Hg is the most important toxic compound compared to other inorganic forms of Hg (Szefer, 2013).

The damage of Hg on health is largely determined by the metallic form and the exposure time and dosage (Boyd *et al.*, 2000). When suffering from Hg poisoning, common symptoms may occur, including skin rashes, impaired motor skills, fatigue, and anxiety. When the situation gets worse, memory loss or speech problems may happen. In the case of severe CH_3Hg poisoning, it has fatal effects on the brain development of fetuses (Hong *et al.*, 2012). CH_3Hg exposure in children may cause various serious health problems, such as renal disease, mental impairment, cerebellar ataxia, and physical growth disorder. The risk assessment of long-term and low-dose exposure to CH_3Hg is still not clear, but CH_3Hg as a hazardous substance in seafood should be particularly noticed.

7.2.2 Cadmium (Cd) Toxicity

Cadmium (Cd), a silver-white metal in monomers, is a non-essential element for the human body. It is widely distributed in nature with the characteristics of high toxicity, difficult to degrade, and easy to enrich. Man-made emission of Cd is mainly from industrial waste, which can be transported between the environment and the food chain (Chunhabundit, 2016). Excess Cd will have toxic effects on the living organisms in water. A trace amount of Cd is easily dissolved in the water and then absorbed by fish and other marine organisms. The biological half-life of Cd is more than ten years (Suwazono et al., 2009), which means that Cd can exist in organisms for a long time, especially in the liver and kidneys. According to the Joint FAO/WHO Expert Committee on Food Additives (JECFA) in 1992, the provisional tolerable weekly intake (PTWI) of Cd was set at 7 µg kg^{-1} body weight on the basis of the effect of renal damage (Galal-Gorchev, 1993). Besides, the European Food Safety Authority set Cd's PTWI as 2.5 µg kg^{-1} body weigh based on benchmark dose derived urinary Cd threshold (Pastorelli et al., 2012). Studies revealed that environmental Cd exposures can pose adverse effects on the human kidney and bone. Also, epidemiological studies suggested that Cd exposure may cause unfavorable clinical consequences, including diabetes, chronic kidney disease, hypertension, cardiovascular disease, osteoporosis, and diabetic nephropathy (Ju et al., 2012). So, it is a potential risk to go through Cd dietary exposure via seafood consumption.

7.2.3 Arsenic (As) Toxicity

Recently, arsenic (As) in natural water has received intensive attention, which exists in the environment with different valency states (e.g., As(III), As(V)). It is important to note that the valency state of As is related to toxicity in the aqueous system (Jain and Ali, 2000). Due to this, all its forms should be considered to determine the toxicity of As. Seafood is the major source of As intake in humans, and As is present in marine-derived foods with the primary form of organic compounds (Taylor et al., 2017). Some organic arsenical compounds are found in herbicides, pesticides, fungicides, and wood preservatives. So, the agricultural practices with the use of such ingredients could do harm to the environment and aquatic animals. According to the JECFA, PTWI of inorganic As is set as 15 µg kg^{-1} body weight, and that of organic As is established as 50 µg kg^{-1} body weight value (Saei-Dehkordi et al., 2010). Compared to inorganic arsenicals, organic arsenicals have much higher upward translocation and tends to be more easily in conversion and accumulation of As (Thirunavukkarasu et al., 2002). With constant As exposure, it will have various adverse health effects like skin diseases, carcinogenesis, and neurological diseases (Rahaman et al., 2021). On the basis of intake risk of As, health assessment should be carried out to set an acceptable consumption of seafood.

7.2.4 Lead (Pb) Toxicity

Lead (Pb) can be commonly found in the contaminated air, water, dust, food, or consumer products. Pb from the atmosphere or soil can end up in lakes and rivers with the wind and rain, and thus can be directly exposed to humans via drinking water (Bhateria and Jain, 2016). Exposure of Pb can occur via inhalation, ingestion, or occasionally skin contact. Specifically, Pb may be taken in via direct contact with the mouth, nose, eyes, and breaks in the skin. Most occurrences of Pb poisoning arise out of the ingestion and absorption of Pb through the digestive tract. The physical factors and physicochemical nature of consumed materials have an effect on the Pb absorption. Pb tends to aggregate in the soft tissue, such as the liver, spleen, kidneys, lungs, and brain. Exposure to Pb poisoning will lead to the increase of blood enzyme levels and blocks the protein synthesis. Further, the excess Pb can induce kidney toxicity by changing the kidney's excretory activity and damaging its structure (Rana et al., 2018). Pb poisoning also has various deleterious effects on neurological, cardiovascular, and reproductive systems. The FAO/WHO Expert Committee has approved the maximum daily Pb intake of 7 µg kg^{-1} body weight or 490 µg of Pb for adults (Abedi et al., 2020). Although infants and children are especially vulnerable to low Pb levels, no such guideline is provided for them.

7.2.5 Zinc (Zn) and Copper (Cu) Toxicity

Zinc (Zn) and copper (Cu) are essential trace metals with low toxicity in humans. However, excessive absorption of Zn can suppress copper (Cu) and iron absorption and may cause serious or irreversible adverse health issues in most individuals (Maret and Sandstead, 2006). High levels of intake of Zn may cause symptoms such as nausea, vomiting, pain, cramps, and diarrhea. It was reported that under long-term intake of high levels of Zn (100 mg elemental Zn/d), Cu deficiency may occur, and blood lipoprotein levels can be changed with increased levels of low-density lipoprotein, and decreased levels of high-density lipoprotein (Foster et al., 2010). A high level of Zn^{2+} in solution is highly toxic to bacteria, plants, invertebrates, and even vertebrate fish. Cu is essential in the human body as it is a component of many proteins. But excess Cu may lead to Cu poisoning with acute symptoms, such as vomiting, hematemesis, hypotension, melena, coma, jaundice, and gastrointestinal distress (Ishola *et al.*, 2017). Long-term Cu exposure can even damage the liver and kidneys. Actually, mammals have efficient regulation mechanisms (absorption and excretion) to adjust the body's Cu at an appropriate level, so that they can be protected from excess dietary Cu levels (Araya *et al.*, 2007). For this reason, it is difficult to set a limit of safe Cu consumption. However, the consumption of Cu-contaminated marine organisms are a potential threat to human health.

A study evidenced that the sediments in lakes had a relatively high concentration of heavy metals (Cr, Ni, Cu, Cd, and Pb) (Yin *et al.*, 2014). The observed Cr, Ni, Cu, and Pb in snail tissues were found to be significantly correlated with those metal concentrations in sediments of the lake. It can be deduced that benthic organisms such as shrimp, lobster, and crab are among the most affected by heavy metals in sediments due to their direct interaction. Of course, the route of heavy metal uptake is dependent on different ecological and feeding habits of marine organisms. Therefore, it is essential to appeal to analytical techniques to determine the content of heavy metals in seafood and estimate the risk of the consumption of different seafoods.

7.3 DETECTION TECHNIQUES FOR HEAVY METALS
7.3.1 Traditional Detection Techniques
7.3.1.1 Atomic Absorption Spectrometry (AAS)

As a traditional analytical technique for detection of heavy metals, atomic absorption spectroscopy (AAS) includes flame atomic absorption spectrometry (FAAS), graphite furnace atomic absorption spectrometry (GFAAS), and cold vapor atomic absorption spectroscopy (CVAAS). FAAS and GFAAS are widely used for the determination of heavy metals in seafood, while CVAAS is the most commonly used to measure total Hg content in seafood (Ferreira *et al.*, 2018). However, when the sample is complex or the concentration of analytes is at a low level, it is difficult to get accurate results by direct analysis of samples. Therefore, separation/enrichment steps before analysis and determination are necessary. Together with AAS, the common sample pretreatment technologies are acid digestion, microwave digestion, solid phase extraction (SPE), and dispersive liquid-liquid microextraction (DLLME). Arulkumar et al. reported the assay of the toxic heavy metal in ten species of fish from the Thondi fishery off the southeast coast of India by AAS after digestion with HNO_3-$HClO_4$-H_2SO_4 (5:2:1) mixture. Concentrations of Cd, Pb, Cu, and Zn were found to be within the recommended permissible levels for human consumption in national and international regulatory guidelines (Arulkumar *et al.*, 2017). Compared with acid digestion, microwave digestion technology is more widely used in sample pretreatment due to the higher safety and efficiency. Tuzen analyzed the toxic elements (Hg, As, Pb, Cd, and Ni) in ten different fish species from the Black Sea of Turkey by FAAS, GFAAS, and CVAAS after microwave digestion with a mixture of HNO_3 and H_2O_2 (Tuzen, 2009). The results showed that the contents of these elements were 25~84 µg kg^{-1}, 0.11~0.32 µg g^{-1}, 0.28 ~ 0.87 µg g^{-1}, 0.10 ~ 0.35 µg g^{-1}, and 1.14~3.60 µg g^{-1}, and the concentrations of toxic heavy metals in these fish did not cause much damage to humans. Likewise, after the treatment of microwave-assisted acid digestion, Cd and Pb existed in important fishes from the southern Kingdom of Morocco were determined using GFAAS. The results showed that the contents of Cd and Pb in fish muscle were 0.009~0.036 µg g^{-1} and 0.013~0.114 µg g^{-1}, respectively, which generally did not cause health problems to consumers (Chahid *et al.*, 2014). In the field of SPE, it is urgent to develop a variety of selective solid phase extraction adsorbents for heavy metal adsorption in seafood. Abolhasani et al. functionalized MCM-48 nanoporous silica with 1-(2-pyridinazo)-2-naphthol (PAN) and used it as an adsorbent for simultaneous separation of ultra-trace heavy metals ions and determination of these ions in seafood using FAAS. Under the optimized experimental conditions, the limits of detection (LODs) of Pb, Cd, Ni, and Cu ions were 0.9, 0.3, 0.6, and 0.4 ng mL^{-1}, respectively. Besides, the method showed excellent recoveries for the determination of Pb, Cd, Ni, and Cu ions in shrimp, crab, and fish (Abolhasani and Behbahani, 2015).

A novel Fe_3O_4@SiO_2@polypyrrole magnetic nanocomposite was developed for the efficient enrichment of heavy metal ions. The magnetic nanosorbent was applied for the selective extraction of Cd(II) and Ni(II) ions, and then the content of heavy metals was determined by FAAS. A result was obtained with LODs of 0.3 ng mL^{-1} and 1.2 ng mL^{-1} for Cd(II) and Ni(II), respectively. This nanocomposite was applied to the rapid extraction of trace quantities of heavy metal from fish and shrimp, and satisfactory recoveries were obtained (Abolhasani *et al.*, 2015). DLLME has become an ideal pretreatment technique owing to its simplicity, effectiveness, and low consumption. All the system parameters were elaborately optimized for DLLME-SQT (slotted quartz tube)-FAAS method and used for the detection of Pb in mussel samples. They have found the sensitivity of this method was increased by about 141 times over the conventional FAAS. The LOD was found to be 270 µg kg^{-1} for mussel (Erarpat et al., 2017).

7.3.1.2 Atomic Fluorescence Spectrometry (AFS)

AFS is a simple and excellent element analysis technique, which not only has high detection sensitivity, but also can detect multiple elements simultaneously. In recent years, the combination of AFS and other methods has become the most widely used technology for the determination of trace and ultrafine elements in seafood. Liang et al. have used AFS to directly measure the total Hg content in gastropod and bivalve species collected from eight coastal sites along the Chinese Bohai Sea, and the methylmercury content was determined by HPLC-AFS. Methylmercury levels ranged from 4.8 to 168.4 ng (Hg) g^{-1}, while total Hg contents ranged from 6.7 to 453.0 ng (Hg) g^{-1} (Liang et al., 2003). Chemical vapor generation (CVG) is considered one of most popular derivation procedures for mercury speciation (Shade and Hudson, 2005). Zu et al. developed an electrochemical cold vapor generation (ECVG) coupled with AFS for the determination of ultra-trace amount of methylmercury in six common seafood samples (i.e., tunny, sleeve-fish, yellow-fin, hairtail, sea shrimp, and kelp), and a very short detection time (60 s) was achieved using a homemade electrochemical flow cell. The methylmercury contents of seafood samples were unequally distributed from 3.7 to 45.8 ng g^{-1}, and the recoveries were from 87.6 to 103.6% (Zu and Wang, 2016).

Hydride generation (HG) is a derivatization and sample introduction technique for analytical atomic spectrometry, and the inclusion of online hydride generation leads to an improvement on sensitivity (Marschner et al., 2018). Mato-Fernandez et al. proposed a pressurized liquid extraction procedure for extracting arsenical species from marine biological material (mussel and fish), and the analysis of $AsC_5H_{11}O_2$ was achieved using high-performance liquid chromatography (HPLC) coupled to ultraviolet cracking and hydride generation-atomic fluorescence spectrometry (UV-HG-AFS). The $AsC_5H_{11}O_2$ concentration found in mussel and fish samples was around 2.8~12.8 mg kg^{-1} (Mato-Fernández et al., 2007). Table 7.1 listed the results of AFS for the detection of various heavy metals in seafood. The summarized samples demonstrate that AFS is an accurate and stable method for detection of heavy metals in highly complex samples, such as seafood.

7.3.1.3 X-Ray Fluorescence (XRF) Spectrometry

A great advantage of XRF techniques is that the multi elemental analysis can be directly carried out on solid samples. It avoids the tedious and laborious digestion steps and the possible analyte losses and/or sample contamination (Marguí et al., 2009). Wang et al. established a low-cost high-definition X-ray fluorescence (HDXRF) spectroscopy method for rapid and sensitive detection of multiple elements in scallop (Wang et al., 2022). In the work, low LODs were obtained for As, Cd, Ni, Pb, Sn, and Zn with 0.072, 0.070, 0.502, 0.063, 0.033, and 4.383 mg kg^{-1}. The results were further evaluated by other technique, demonstrating the good analytical performance of the HDXRF technique in scallops. A total reflection X-ray fluorescence (TXRF) method for determining Hg in several seafood samples (mussel, prawn, edible crab, hake, and sole) has been developed (Romero et al., 2014). The method was based on the trapping of Hg vapors using silver nanoparticles (Ag NPs) immobilized on quartz reflectors. The concentrations of Hg varied in the range of 0.1~0.7 mg g^{-1}. Besides, the method can be used for field sampling. Since it does not require a drying step before analysis, the preconcentrated analyte can be stabilized for at least 30 days without any losses by evaporation.

Alonso-Hernandez et al. adopted an energy dispersive X-ray fluorescence (EDXRF) for the detection of total As content in muscle tissues of species of fish, crustaceans, and molluscs (Alonso-Hernández et al., 2012). In this research, fish, crustaceans and molluscs give an average As value of 10.2, 26.5, and 22 µg g^{-1} dry wt, respectively. In addition, a novel method of dispersive micro-solid phase extraction (DMSPE) combined with EDXRF or TXRF was proposed for the determination of ultratrace Hg(II) ions in complex seafood samples (Musielak et al., 2022). Based on preconcentration and separation of Hg(II) ions with graphene oxide/thiosemicarbazide, DMSPE coupled with EDXRF and TXRF showed very high enrichment factors and low detection limits in both liquids (60, 2.1 pg mL^{-1}) and solid samples (73, 1.8 ng g^{-1}). DMSPE coupled with TXRF has lower LODs due to the better sensitivity of TXRF measurement.

Table 7.1: Detection Results of Various Heavy Metals in Seafood by AFS

Technique	Analytes	Pretreatment	LODs	Found	Recovery (%)	Ref.
RP-HPLC-UV-CV-	Hg^{2+}; CH_3Hg^+; $C_2H_5Hg^+$	Acid digestion Condensation and filtration.	7.6 ng g^{-1}; 10.8 ng g^{-1}; 18.2 ng g^{-1}	<0.14 µg g^{-1}; <0.26 µg g^{-1}; <0.21 µg g^{-1}	94~100; 95~99; 93~100	(Grijalba et al., 2018)
UV-AFS	Hg^{2+}; CH_3Hg^+	Ultrasound assisted acid leaching	0.015 mgL^{-1}; 0.081 mgL^{-1}	<1.05 µg kg^{-1}; <13.22 µg kg^{-1}	96~105; 86~108	(Hu et al., 2018)
EVG-AFS	Hg^{2+}; CH_3Hg^+	Extraction (HCl, double-frequency ultrasonic)	0.098 µgL^{-1}; 0.073 µgL^{-1}	<2.45 µg kg^{-1}; <34.62 µg kg^{-1}	87.3~109.6; 89.2~110.6	(Zhang et al., 2012)
LC-UV-HG-AFS	Hg^{2+}; CH_3Hg^+	Microwave digestion	1 ng g^{-1}; 0.3 ng g^{-1}	<2.33 mg kg^{-1}; <2.23 mg kg^{-1}	Sum of Hg: 88~117;	(Zmozinski et al., 2014)
LC-AFS	$AsC_5H_{11}O_2$	Methanol extraction	4~22 ng g^{-1}	<1947 µg kg^{-1}	/	(Simon et al., 2004)
CVG-HPLC-AFS	Hg^{2+}; CH_3Hg^+	Alkaline digestion	0.085 µg L^{-1}; 0.033 µg L^{-1}	<78.9 µg kg^{-1}	/	(Yin et al., 2008)
CVG/PVG-AFS	CH_3Hg^+	Alkaline digestion	2.77 µg kg^{-1}; 1.06 µg kg^{-1}	<5.09 mg kg^{-1}; <5.25 mg kg^{-1}	92.6~104; 93.2~105	(Lancaster et al., 2019)

7.3.1.4 Inductively Coupled Plasma Mass Spectrometry (ICP-MS)

Inductively coupled plasma mass spectrometry (ICP-MS), known as a standard detection method for heavy metals, has been one of the most popular techniques for detecting ultra-trace levels of metals and metalloids in a large variety of samples. ICP-MS has the advantages of high sensitivity, high precision, low LOD, and multi-element measurement capabilities. For example, Nam et al. used ICP-MS for the determination of total As in bluefin tuna, yellowfin tuna, bigeye tuna, and swordfish (Nam *et al.*, 2010). The results indicated that the concentrations of total As in the seafoods ranged from 0.74 to 6.87 mg kg^{-1}. Combining separation methods with detection techniques can effectively avoid the interferences in samples and enhance the detection sensitivity. For instance, Hight et al. developed a green method to detect methylmercury in seafood by HPLC-ICP-MS without hazardous solvents (Hight and Cheng, 2006). Hg compounds were extracted from seafood by 1% w/v L-cysteine·HCl·H$_2$O under 60 °C heating for 120 min, and L-cysteine (0.1%, w/v) plus L-cysteine·HCl·H$_2$O (0.1%, w/v) were adopted as a mobile phase to determine total Hg. The method showed low LOQs for CH$_3$Hg (7 µg kg^{-1}) and inorganic Hg (5 µg kg^{-1}) in edible seafood, indicating acceptable detection performance for seafoods in practical applications. With diluted HNO$_3$ solution for extraction and (NH$_4$)$_2$HPO$_4$ in 1% methanol as a mobile phase, Schmidt et al. developed LC-ICP-MS/MS for As speciation in shark, shrimp, squid, oyster, and scallop (Schmidt *et al.*, 2018). As a result, the recoveries of arsenite (As(III)), arsenate (As(V)), monomethylarsonic acid (MMA), acid dimethylarsinic (DMA), and arsenobetaine (AsB) in all samples ranged from 90 to 104%. The method exhibited good sensitivity with LOQs of 30, 26, 12, 6, and 6 ng g^{-1} for As(III), As(V), MMA, DMA, and AsB, as well as excellent accuracy and precision for As speciation analysis. After dissolution, derivatization, and extraction of the seafood, GC coupled to ICP triple quadrupole mass spectrometry (GC-ICP-MS/MS) was implemented in the analysis of mono methylmercury (MMHg) in several types of seafood, such as mussel, squid, crab, whale, cod, and dogfish (Valdersnes *et al.*, 2016). All the samples are determined with MM Hg concentrations of 0.035~3.58 mg kg^{-1} and repeatability relative standard deviations of 2.1 to 8.7%, revealing that it may serve as a potential method for MM Hg. Table 7.2 lists the applications for heavy metals detection in seafood by ICP-MS.

7.3.1.5 Inductively Coupled Plasma-Optical Emission Spectrometry (ICP-OES)

As a highly sensitive, accurate, and precise detection technique, ICP-OES owns the advantages of wide linear range, simultaneous multielement determination, and easy online determination. An online microcolumn separation/preconcentration combined with ICP-OES was proposed for the detection of trace amounts of MeHg$^+$ and Hg^{2+} in clam, oyster, scallops, fish, and shrimp (Xiong and Hu, 2007). Prior to the detection, the chelating resin was used as the microcolumn filler for the quantitative adsorption of MeHg$^+$ and Hg^{2+}. The method indicated that the recoveries of mercury species spiked in seafood samples with the range of 89.9~102.4% for MeHg$^+$ and 87.0~104.6% for Hg^{2+}. The accuracy of proposed method was further verified by using dogfish muscle as a certified reference material. In addition, ICP-OES can be applied to monitor the heavy metals in seafood and evaluate the risk of seafood consumption in humans. For example, Lehel et al. adopted ICP-OES for the determination of Hg, Cd, As, and Cr in shellfish, oysters, and squid were collected from a local fishery product market in Hungary (Lehel *et al.*, 2018). Comparing obtained results of heavy metals with provisional tolerable intake values, it is suggested that the consumption of investigated samples will not cause harm to the human body. Based on a novel ion imprinted polymer grafted on Fe$_3$O$_4$ nanoparticles, Najafi et al. achieved a simple extraction and preconcentration of Hg(II) ions in fish samples (Najafi *et al.*, 2013). The magnetic sorbent was successfully applied in detection of trace amounts of Hg(II) ions coupled with ICP-OES. The method showed a LOD of 0.03 ng mL^{-1} with RSD of 1.47%, and in fish sample tissue, Hg(II) was detected as low to 5.5 ng g^{-1}, which demonstrated the good performance of a proposed method.

In another work, the microwave digestion with HNO$_3$-H$_2$O$_2$ was used for smooth weakfish samples treatment (Silva *et al.*, 2017). After that, the detection of As, Pb, and Cd in smooth weakfish samples was performed using ICP-OES. It was found that the amount of Pb was below the LOD (1.9 ng g^{-1}), and As and Cd content in samples are ranged from 120.06~266.78 ng g^{-1} and 120.06~243.96 ng g^{-1}, respectively. Similarly, based on microwave-assisted digestion and ICP-OES, Milenkovic et al. analyzed the content of Cd, Hg, and Pb in packaged fish and seafood products (Milenkovic *et al.*, 2019). The concentrations of Cd, Hg, and Pb for sea fish ranged from 0.01 to 0.81 mg kg^{-1}, 0.01 to 1.47 mg kg^{-1}, and 0.10 to 6.56 mg kg^{-1}. The study concluded that the constant consumption of seafood may cause potential health risk, especially the Hg- and Pb-contaminated fishery products.

Table 7.2: Detection of Heavy Metals in Seafood with ICP-MS-Based Techniques

Technique	Analytes	Pretreatment	Found	LODs	Ref.
ICP-MS	Pb; Cd; Hg	Microwave-assisted acid digestion	<0.274 mg kg^{-1}; <0.364 mg kg^{-1}; <0.557 mg kg^{-1}	4, 1, 2 µg kg^{-1}	(Miedico et al., 2015)
ICP-MS/OES	Pb; Cd; As; Hg	Acid digestion (68% HNO3 + 32% H$_2$O$_2$)	<0.320 mg kg^{-1}; <0.256 mg kg^{-1}; <3.559 mg kg^{-1}; <0.052 mg kg^{-1}	2.4, 1.2, 2.8, 2.1 µg kg^{-1}	(Habte et al., 2015)
ICP-MS/OES	Pb; Cd; As; Hg	Microwave-assisted acid digestion	0.290, 2.51, 7.77, 0.036 mg kg^{-1}	2.43, 1.26, 2.82, 2.10 µg kg^{-1}	(Nho et al., 2016)
IC-ICP-MS	As(V), As(III), MMA, DMA, AsB	Microwave-assisted digestion	Total As: 55.57 mg kg^{-1} (seaweed); 10.01 mg kg^{-1} (fish)	8.0~12.0 µg (As) kg^{-1}	(Lin et al., 2020)
HPLC-ICP-MS	Hg^{2+}, CH$_3$Hg$^+$, C$_2$H$_5$Hg$^+$	Microwave-assisted acid digestion	Total Hg: 44.84 µg kg^{-1}	0.12, 0.08, 0.20 g L^{-1}	(Liu et al., 2018)
LC-CVG-ICP-MS	Hg^{2+}, CH$_3$Hg$^+$	Acid digestion and extraction with L-cysteine	<0.08 mg kg^{-1}; <1.05 mg kg^{-1};	1.7, 2.3 ng g^{-1}	(Schmidt et al., 2013)
ICP-MS	As, Cd, Cu, Cr, Ni, Pb, Zn	Microwave-assisted digestion	Dry mass: 12.8, 0.63, 10.9, 0.66, 1.83, 0.61, 46.2 mg kg^{-1}	0.07, 0.06, 0.3, 0.4, 0.08, 0.4, 0.6 µg g^{-1}	(Barbosa et al., 2019)
HPLC- ICP-MS	iAs = As(III) + As(V)	Microwave-assisted acid digestion	<0.663 mg kg^{-1}	4 µg kg^{-1}	(Pétursdóttir et al., 2012)

7.3.2 Rapid Detection Methods

Different from the conventional detection methods that require time-consuming steps and heavy instruments, rapid ones have been developed with flexible construction, fast response, and sensitive analysis. Based on the fundamental principles and setups of traditional instruments, rapid detection methods focus more on the specific functions of nanomaterials and strategies of method design. As they are derived from the traditional techniques combined with various nanomaterials and biosensing strategies, they are also known as biosensors. As a newly emerging analytical method, biosensors are built on the basis of various biological elements, such as protein, DNA, antibody, and so on. With the interaction force of physical adsorption, electrostatic attraction, biometric identification, and chemical coupling, the biological elements are connected with sensing platform. When analytes are captured by biological recognition elements, the physical variation (light, electricity, magnetism) is directly or indirectly converted into an output signal. Besides, the molecular properties (e.g., structures, types) are related to the characteristics of the output signals, with which analytes can be qualitatively detected. Further, the signal intensity reflects the amounts of analytes in a system, so it is often used for quantitative detection. Based on the above description, with different biological elements, biosensors with diverse functions can be constructed and applied in different analytical fields. In general, biosensors consist of immobilized sensitive biomaterials as identification components, appropriate physicochemical transducers and signal amplifiers (Figure 7.2). Biosensors can act as a receiver and a converter at the same time, so the signal can be effectively detected and output for rapid detection. Till now, many biosensors have been proposed in food safety, including colorimetric biosensor, optical biosensor (e.g., fluorescence, SERS), electrochemical sensor, magnetic relaxation switching sensors, and enzyme-linked immunosensors. Biosensors are flexible for the analysis of food samples, due to researchers can easily design the optimal detection scheme according to the properties of analytes. Therefore, biosensors are developed rapidly and greatly favored by scientists from all fields of analytical science. For the analysis of food contaminants, especially the heavy metal ions, the biosensor has the advantages of fast response speed, flexible design, and high sensitivity. In this section, we will introduce some examples of biosensors for the detection of heavy metals in seafood.

7.3.2.1 Colorimetric Sensor

A colorimetric sensor is a method that can quantitatively detect the concentration of heavy metal ions by the color changes of a solution or with a colorimeter. Based on different organic micromolecular, polymer dyes, quantum dots, and metal-organic frame nanomaterials, a simple and sensitive colorimetric sensor can be built for the detection of heavy metals. Due to the intrinsic oxidation property of Cu^{2+}, a simple colorimetric strategy was developed for detection of Cu^{2+} in seawater and shellfish (Yin $et\ al.$, 2015). In this system, L-cysteine can interact with 1-chloro-2, 4-dinitrophenylbenzene (CDNB) to form the yellow product of 2, 4-dinitrophenylcysteine (DNPC). The presence of Cu^{2+} can catalyze the oxidation of L-cysteine to L-cysteine, thus reducing the production of DNPC and causing the color of the solution changing from yellow to colorless. The results showed that the method had good sensitivity for detection of Cu^{2+} with a LOD of 0.5 nmol L^{-1}. In addition, based on Rhodamine conjugate polymer (P(RD-CZ)), a sensor platform was established for the detection of Hg^{2+} in anglerfish (Ayranci $et\ al.$, 2017). In another example, based on the single-atom nanozyme, a colorimetric method was proposed for the detection of Cr(VI) (Mao $et\ al.$, 2021). In the work, single-atom Fe acts as a peroxidase mimetic and 8-hydroxyquinoline (8-HQ) as an inhibitor to prevent TMB from oxidation. The chelation of Cr(VI) with 8-HQ is used to reproduce color change in TMB oxidation. This method was successfully applied to detect Cr(VI) in tuna samples with a LOD of 3 nM. Although colorimetric methods are not restricted by laboratory instruments, they may also suffer from some limitations, such as easy to be disturbed by the complex background of actual samples.

7.3.2.2 Electrochemical Sensors

Electrochemical analysis is a kind of analytical method that is established by using the relationship between the composition and content of the analyzed solution in the electrolytic cell and its electrochemical properties (e.g., resistance, potential, current). From the structure, the electrochemical detection system is usually composed of three electrodes, including a working electrode (WE), a reference electrode (RE), and a counter electrode (CE) (Figure 7.3). Heavy metal ions can be reduced or oxidized on a WE surface by applying a potential. The correlation between heavy metal

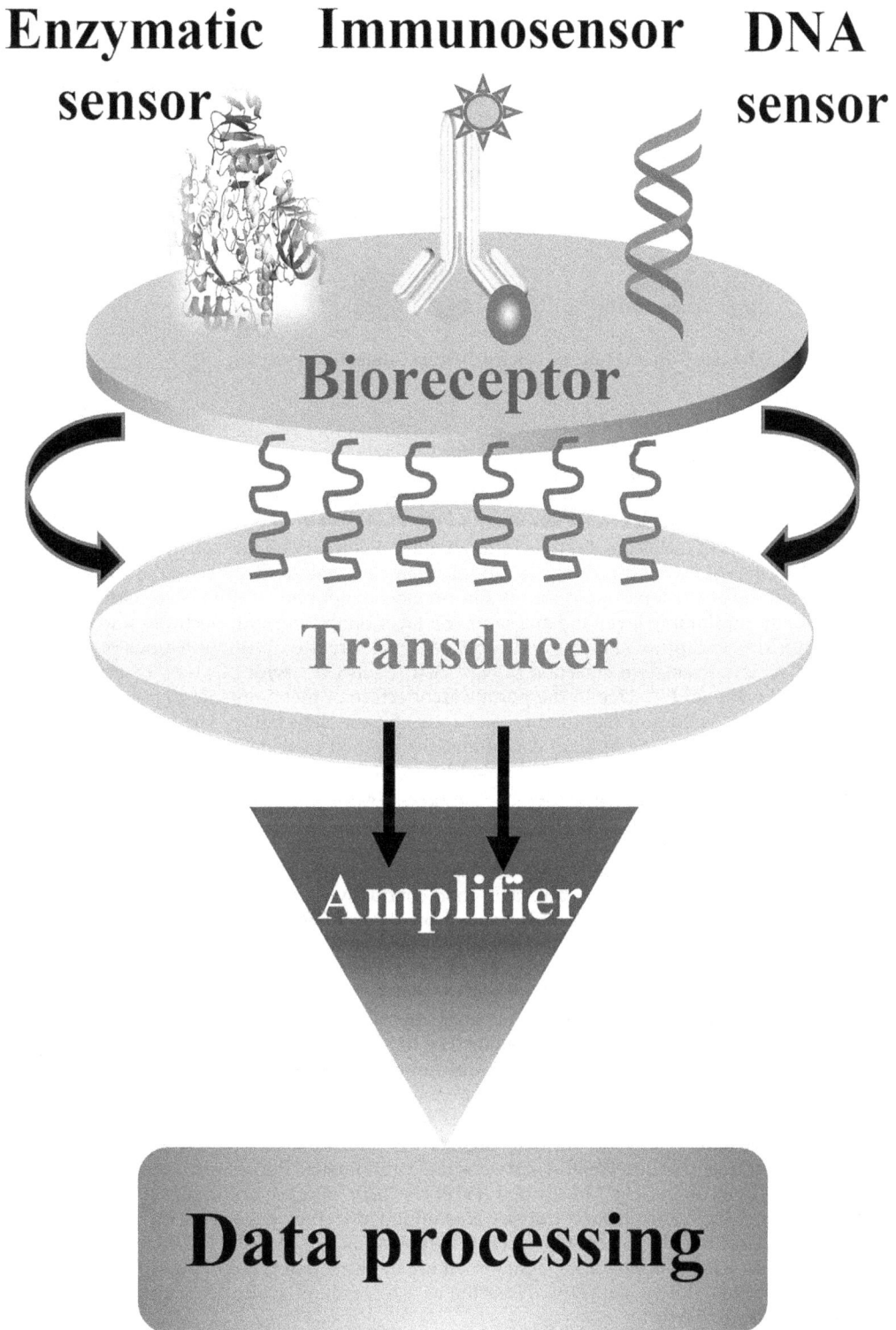

Figure 7.2 Schematic presentation of the structure of biosensors (enzymatic biosensor, immunosensor and DNA sensor, etc.).

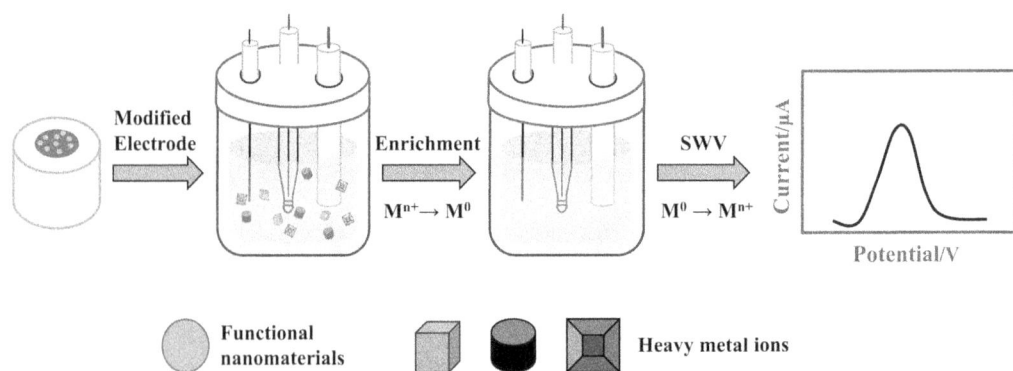

Figure 7.3 Schematic illustration of electrochemical detection system.

concentration and current can be obtained by electrochemical signals. The surface of WE play a crucial role in the measurement results, as it is sensitive to the electron transfer. The appropriate modification of electrode surface has great influence on the sensitivity and accuracy of detection. Therefore, the nanomaterials can be modified on the electrode to improve the conductivity and specific surface area of the electrode, thus improving the selectivity and sensitivity. For example, for sensitive detection of heavy metal ions, a modification step was conducted for the electrode (Lu *et al.*, 2018). Vertically ordered mesoporous silica-nanochannel film (VMSF) was firstly decorated and acted as the antifouling layer and anti-jamming layer on the working electrode, and GQD were confined in VMSF to capture analytes and enhance signal intensity. Without tedious pretreatment, the method achieved sensitive detection of Hg^{2+}, Cu^{2+}, and Cd^{2+} with LODs of 9.8 pmol L^{-1}, 8.3 pmol L^{-1}, and 4.3 nmol L^{-1}. Due to the porous architecture of metal-organic frameworks (MOFs), Cu-MOFs were prepared and modified on electrodes for the detection of Hg^{2+} in canned tuna (Singh *et al.*, 2020). Under the optimal conditions, the method provided good reproducibility and stability with an LOD of 0.0633 nmol L^{-1}. In addition, portable electrochemical devices are attractive for field monitoring due to their designability and disposability. For instance, an electrochemical device with stacked flat electrodes was designed for the detection of Pb^{2+} and Cd^{2+} in food samples (Pang *et al.*, 2023). Gold nanoparticles were firstly deposited on carbon paper electrodes (CPE) and then modified with Co-based MOFs; a functional disposable electrode was prepared with large specific surface area and good conductivity. With the development of various modified materials, electrochemical analysis has a broader prospect in the field of heavy metal ion detection.

7.3.2.3 Fluorescence Sensors

Due to their high selectivity, sensitivity, and instrument portability, fluorescent probes have been widely used in detection of food contaminants. For the fluorescence detection of heavy metals, different mechanisms are reported on the fluorescent probe interacting with the heavy metal ions, mainly including intramolecular charge transfer enhancement and fluorescence quenching induced by electron, charge, or energy transfer. The former mechanism will lead to a fluorescence "turn on" response, and the latter one gives "turn off" signals. For instance, based on terbium (III)-referenced N-doped carbon dots (N-CDs-Tb-DPA) composites, a ratiometric fluorescence sensor was constructed for detection of Hg^{2+} in seafood (He *et al.*, 2020). Because of the strong trapping ability of N-CDs toward Hg^{2+}, the electron transfer was inhibited in the presence of Hg^{2+} with the fluorescence quenching at 436 nm. However, the fluorescence of Tb-DPA (54 3nm) remained the same, which was used as a reference signal to eliminate the interference of background. The mercaptan-protected aldehyde of DAC-Hg acted as a recognition group for selectively sensing of Hg^{2+} with a LOD of 5.0 nM.

7.3.2.4 Enzymatic Biosensors

Using enzymes as bioreceptors, enzymatic biosensors are coupled with the physical transducer to generate optical, electronic, or magnetic signals that are proportional to the analyte concentration in the sample. In recent years, a variety of enzymes have been used to construct enzymatic biosensors for detection of heavy metal, such as urease, phosphodiesterase, peroxidase, xanthine oxidase, and glucose oxidase. Based on urease and magnetic nanoparticles (MNPs), a new enzymatic biosensing platform was developed for the detection of Hg^{2+}, Cd^{2+}, and Pb^{2+} in fish gill and muscle tissue (Swain et al., 2020). Wherein, urea and phenol red are the color-developing indicators, and when heavy metal ions exist, the urease activity is inhibited, which could be observed from the change in color of the phenol red indicator. The method showed a good stability for Hg^{2+}, Cd^{2+}, and Pb^{2+} with LODs of 0.5, 0.1, and 0.1 ng L^{-1}. With a portable microfluidic device, the method can be further developed for the rapid on-site testing of samples. In another study, a label-free DNAzyme double amplification biosensor was proposed for the detection of Pb^{2+} (Zhang et al., 2021). In this work, the recognition probe was constructed with two complementary oligonucleotides: the RNA base substrate and DNAzyme probe. In the presence of Pb^{2+}, it will trigger the cleavage of the RNA substrate, and two 3'-terminals were digested by Exonuclease I. As a result, the double-stranded DNA structure binds to Sybr Green I emits stronger fluorescence than that of digested DNA sequences.

7.3.2.5 Immunosensors

Owing to the high selectivity, rapid detection, good detection sensitivity, and stability, immunoassays such as ELISA have been regarded as a "golden standard" in detection. To achieve immunoassay for heavy metals, the preparation of monoclonal antibodies and complete antigens are some of the key issues. Commonly, non-toxic complexes are firstly prepared by chelating agents and metallothionein coordination with heavy metal ions. After that, the metal complexes are coupled to a carrier protein to form a complete immunogen, followed by a series of steps to prepare a heavy metal–specific monoclonal antibody. The detection follows the same principle of antigen-antibody recognition mode. For instance, based on monoclonal antibodies, an indirect competitive ELISA (ic-ELISA) and chemiluminescent enzyme immunoassay (CLEIA) were constructed for the detection of Pb^{2+} (Xu et al., 2020). The LODs of 0.7 ng mL^{-1} and 0.1 ng mL^{-1} were obtained for ic-ELISA and CLEIA), respectively. The prepared antibodies are demonstrated with high sensitivity, specificity, and accuracy for the detection of Pb^{2+} in water, food, and feed samples. Still, antibodies may suffer from some disadvantages, such as being difficult to prepare and high cost. Besides, they may behave with relatively low stability under drastic conditions including high temperature, oxygen, low or high pH, and so on.

7.3.2.6 Aptamer Sensors

For the detection of heavy metals, aptamers have the advantages such as low cost, easy preparation, high affinity and specificity, and good stability (Sarkar et al., 2022). As a kind of bioreceptor, aptamers are often combined with different transducers to construct different biosensors, including colorimetric, fluorescent, electrochemical, and surface-enhanced Raman scattering (SERS) aptasensors, and so on. Based on DNA-templated Ag-Au nanoparticles, two aptamers were designed for the specific recognition of CH_3Hg^+ and $C_2H_5Hg^+$ in fish muscle samples (Chen et al., 2018). In the strategy, CH_3Hg^+ and $C_2H_5Hg^+$ preferentially bind to the aptamer and induce the growth of Ag-Au with different sizes, thus giving rise to a color change from yellow to purple. With naked eye observation, the method showed LODs of 1.0 µg Hg g^{-1} for CH_3Hg^+ and the total concentration of CH_3Hg^+ and $C_2H_5Hg^+$. Besides, based on aptamer-modified Cu nanoclusters (CuNC) and Au nanoclusters (AuNC), a ratiometric fluorescent probe was proposed for detection of Hg^{2+} (Shi et al., 2021). In the presence of Hg^{2+}, the binding of the two aptamers with Hg^{2+} caused the aggregation of CuNC and AuNC, further resulting in a change in fluorescence intensity via fluorescence resonance energy transfer (FRET). The ratiometric signal output method showed good stability for evaluating Hg^{2+} levels in aquatic products. As SESR is a surface sensitive technique, the analytes should be close to the surface of a SERS substrate. For example, based on aptamers regulated graphene oxide (GO) catalysis towards $HAuCl_4$, a gold nanoparticle (AuNPs)–based SERS was constructed for detection of Hg^{2+} using Victoria blue 4R as a probe (Li et al., 2018). In the absence of Hg^{2+}, the aptamer will bind with GO to form complexes and inhibit its catalytic activity. When Hg^{2+} exists, they will compete with aptamer and GO catalysis facilitate the formation of AuNPs with higher signals. The method behaved a linear range of 0.25~10 nmol L^{-1} with a LOD of 0.08 nmol L^{-1}, indicating the high sensitivity and selectivity of SERS method.

7.4 CONCLUSION

In seafood samples, Pb, Cd, and Hg are the primary metals, and As is the primary metalloid to cause health concerns. The heavy metals in seafood are introduced from their sources to the toxic effects, aiming to emphasize their potential risks in seafood. Due to the widespread presence of heavy metals in the environment, existing analytical techniques are described to provide possible solutions for seafood safety, including the traditional instrumental techniques and rapid detection techniques. Thanks to their advantages like rapid response, high sensitivity, and flexible design, biosensors have become the most popular methods in analysis of heavy metals. When combined with nanomaterials, the detection performance (sensitivity, accuracy, and stability) can be greatly enhanced. However, most studies only focus on designing different biosensors, which make them difficult to apply in practical applications. Therefore, user-friendly biosensors are required to achieve more functions, such as high throughout detection, miniaturization, and intelligent integration. Moreover, to prevent superfluous testing, it is essential to make the evolving criteria for heavy metal levels in seafood clear about the choice of metals and stated permissible daily exposures.

REFERENCES

Abedi, Abdol-Samad, Esmat Nasseri, Fatemeh Esfarjani, Fatemeh Mohammadi-Nasrabadi, Motahareh Hashemi Moosavi, and Hedayat Hoseini. 2020. A systematic review and meta-analysis of lead and cadmium concentrations in cow milk in Iran and human health risk assessment. *Environmental Science and Pollution Research* 27 (10):10147–10159.

Abolhasani, Jafar, and Mohammad Behbahani. 2015. Application of 1-(2-pyridylazo)-2-naphthol-modified nanoporous silica as a technique in simultaneous trace monitoring and removal of toxic heavy metals in food and water samples. *Environmental Monitoring and Assessment* 187 (1):1–12.

Abolhasani, Jafar, Rahim Hosseinzadeh Khanmiri, Ebrahim Ghorbani-Kalhor, Akbar Hassanpour, Ali Akbar Asgharinezhad, Nafiseh Shekari, and Ahoura Fathi. 2015. An Fe_3O_4@SiO_2@polypyrrole magnetic nanocomposite for the extraction and preconcentration of Cd (II) and Ni (II). *Analytical Methods* 7 (1):313–320.

Ali, Afshan S, and R US SA Ahmad. 2014. Effect of different heavy metal pollution on fish. *Research Journal of Chemical and Environmental Sciences* 2 (1):74–79.

Alonso-Hernández, Carlos Manuel, Miguel Gómez-Batista, Misael Díaz-Asencio, Juan Estévez-Alvares, and Román Padilla-Alvares. 2012. Total arsenic in marine organisms from Cienfuegos bay (Cuba). *Food Chemistry* 130 (4):973–976.

Araya, Magdalena, Manuel Olivares, and Fernando Pizarro. 2007. Copper in human health. *International Journal of Environment and Health* 1 (4):608–620.

Arulkumar, Abimannan, Sadayan Paramasivam, and Rajendran Rajaram. 2017. Toxic heavy metals in commercially important food fishes collected from Palk Bay, Southeastern India. *Marine Pollution Bulletin* 119 (1):454–459.

Ayranci, Rukiye, Dilek Odaci Demirkol, Suna Timur, and Metin Ak. 2017. Rhodamine-based conjugated polymers: potentiometric, colorimetric and voltammetric sensing of mercury ions in aqueous medium. *Analyst* 142 (18):3407–3415.

Barbosa, Isa dos S, Geysa B Brito, Gabriel L Dos Santos, Luana N Santos, Leonardo SG Teixeira, Rennan GO Araujo, and Maria Graças A Korn. 2019. Multivariate data analysis of trace elements in bivalve molluscs: characterization and food safety evaluation. *Food Chemistry* 273:64–70.

Bhateria, Rachna, and Disha Jain. 2016. Water quality assessment of lake water: a review. *Sustainable Water Resources Management* 2:161–173.

Boyd, Alan S, Donna Seger, Stephen Vannucci, Melissa Langley, Jerrold L Abraham, and Lloyd E King Jr. 2000. Mercury exposure and cutaneous disease. *Journal of the American Academy of Dermatology* 43 (1):81–90.

Chahid, Adil, Mustapha Hilali, Abdeljalil Benlhachimi, and Taoufiq Bouzid. 2014. Contents of cadmium, mercury and lead in fish from the Atlantic sea (Morocco) determined by atomic absorption spectrometry. *Food Chemistry* 147:357–360.

Chen, Zhiqiang, Xusheng Wang, Xian Cheng, Weijuan Yang, Yongning Wu, and FengFu Fu. 2018. Specifically and visually detect methyl-mercury and ethyl-mercury in fish sample based on DNA-templated alloy Ag–Au nanoparticles. *Analytical Chemistry* 90 (8):5489–5495.

Chunhabundit, Rodjana. 2016. Cadmium exposure and potential health risk from foods in contaminated area, Thailand. *Toxicological Research* 32 (1):65–72.

Erarpat, Sezin, Gözde Özzeybek, Dotse Selali Chormey, and Sezgin Bakırdere. 2017. Determination of lead at trace levels in mussel and sea water samples using vortex assisted dispersive liquid-liquid microextraction-slotted quartz tube-flame atomic absorption spectrometry. *Chemosphere* 189:180–185.

Farmery, Anna K, K Alexander, Kelli Anderson, Julia L Blanchard, Chris G Carter, Karen Evans, Mibu Fischer, Aysha Fleming, Stewart Frusher, and Elizabeth A Fulton. 2022. Food for all: designing sustainable and secure future seafood systems. *Reviews in Fish Biology and Fisheries* 32 (1):101–121.

Ferreira, Sergio LC, Marcos A Bezerra, Adilson S Santos, Walter NL dos Santos, Cleber G Novaes, Olivia MC de Oliveira, Michael L Oliveira, and Rui L Garcia. 2018. Atomic absorption spectrometry–A multi element technique. *TrAC Trends in Analytical Chemistry* 100:1–6.

Foster, Meika, Peter Petocz, and Samir Samman. 2010. Effects of zinc on plasma lipoprotein cholesterol concentrations in humans: a meta-analysis of randomised controlled trials. *Atherosclerosis* 210 (2):344–352.

Galal-Gorchev, H. 1993. Dietary intake, levels in food and estimated intake of lead, cadmium, and mercury. *Food Additives & Contaminants* 10 (1):115–128.

Grijalba, Alexander Castro, Pamela Y Quintas, Emiliano F Fiorentini, and Rodolfo G Wuilloud. 2018. Usefulness of ionic liquids as mobile phase modifiers in HPLC-CV-AFS for mercury speciation analysis in food. *Journal of Analytical Atomic Spectrometry* 33 (5):822–834.

Habte, Girum, Ji Yeon Choi, Eun Yeong Nho, Sang Yeol Oh, Naeem Khan, Hoon Choi, Kyung Su Park, and Kyong Su Kim. 2015. Determination of toxic heavy metal levels in commonly consumed species of shrimp and shellfish using ICP-MS/OES. *Food Science and Biotechnology* 24 (1):373–378.

He, Xie, Yong Han, Xueli Luo, Weixia Yang, Chunhua Li, Wenzhi Tang, Tianli Yue, and Zhonghong Li. 2020. Terbium (III)-referenced N-doped carbon dots for ratiometric fluorescent sensing of mercury (II) in seafood. *Food chemistry* 320:126624.

Hight, Susan C, and John Cheng. 2006. Determination of methylmercury and estimation of total mercury in seafood using high performance liquid chromatography (HPLC) and inductively coupled plasma-mass spectrometry (ICP-MS): Method development and validation. *Analytica Chimica Acta* 567 (2):160–172.

Hong, Young-Seoub, Yu-Mi Kim, and Kyung-Eun Lee. 2012. Methylmercury exposure and health effects. *Journal of Preventive Medicine and Public Health* 45 (6):353–363.

Hu, Pingyue, Xiu Wang, Li Yang, Haiyan Yang, Yuyi Tang, Hong Luo, Xiaoli Xiong, Xue Jiang, and Ke Huang. 2018. Speciation of mercury by hydride generation ultraviolet atomization-atomic fluorescence spectrometry without chromatographic separation. *Microchemical Journal* 143:228–233.

Ishola, Adejumo Babatunde, IM Okechukwu, UG Ashimedua, D Uchechukwu, EA Michael, O Moses, IH Okwudili, HM Vaima, AU Itakure, and OK Ifeanyichukwu. 2017. Serum level of lead, zinc, cadmium, copper and chromium among occupationally exposed automotive workers in Benin city. *International Journal of Environment and Pollution Research* 5 (1):70–79.

Jain, CK, and I Ali. 2000. Arsenic: occurrence, toxicity and speciation techniques. *Water Research* 34 (17):4304–4312.

Jaishankar, Monisha, Tenzin Tseten, Naresh Anbalagan, Blessy B Mathew, and Krishnamurthy N Beeregowda. 2014. Toxicity, mechanism and health effects of some heavy metals. *Interdisciplinary Toxicology* 7 (2):60.

Jobling, Malcolm. 2016. Fish nutrition research: past, present and future. *Aquaculture International* 24 (3):767–786.

Ju, Yun-Ru, Wei-Yu Chen, and Chung-Min Liao. 2012. Assessing human exposure risk to cadmium through inhalation and seafood consumption. *Journal of Hazardous Materials* 227:353–361.

Lancaster, Shaun T, Christoph-Cornelius Brombach, Warren T Corns, Jörg Feldmann, and Eva M Krupp. 2019. Determination of methylmercury using liquid chromatography–photochemical vapour generation–atomic fluorescence spectroscopy (LC-PVG-AFS): a simple, green analytical method. *Journal of Analytical Atomic Spectrometry* 34(6):1166–1172.

Larsen, Rune, Karl-Erik Eilertsen, and Edel O Elvevoll. 2011. Health benefits of marine foods and ingredients. *Biotechnology Advances* 29 (5):508–518.

Lehel, József, András Bartha, Dávid Dankó, Katalin Lányi, and Péter Laczay. 2018. Heavy metals in seafood purchased from a fishery market in Hungary. *Food Additives & Contaminants: Part B* 11 (4):302–308.

Li, Chongning, Xiaoliang Wang, Aihui Liang, Yanghe Luo, Guiqing Wen, and Zhiliang Jiang. 2018. A simple gold nanoplasmonic SERS method for trace Hg^{2+} based on aptamer-regulating graphene oxide catalysis. *Luminescence* 33 (6):1113–1121.

Li, Rui, Han Wu, Jing Ding, Weimin Fu, Lijun Gan, and Yi Li. 2017. Mercury pollution in vegetables, grains and soils from areas surrounding coal-fired power plants. *Scientific Reports* 7 (1):46545.

Liang, Li-na, Jian-bo Shi, Bin He, Gui-bin Jiang, and Chun-gang Yuan. 2003. Investigation of methylmercury and total mercury contamination in mollusk samples collected from coastal sites along the Chinese Bohai Sea. *Journal of Agricultural and Food Chemistry* 51 (25):7373–7378.

Lin, Yaohui, Ying Sun, Xusheng Wang, Shilong Chen, Yongning Wu, and FengFu Fu. 2020. A universal method for the speciation analysis of arsenic in various seafood based on microwave-assisted extraction and ion chromatography-inductively coupled plasma mass spectrometry. *Microchemical Journal* 159:105592.

Liu, Hao, Jiaoyang Luo, Tong Ding, Shanyong Gu, Shihai Yang, and Meihua Yang. 2018. Speciation analysis of trace mercury in sea cucumber species of Apostichopus japonicus using high-performance liquid chromatography conjunction with inductively coupled plasma mass spectrometry. *Biological Trace Element Research* 186 (2):554–561.

Lu, Lili, Lin Zhou, Jie Chen, Fei Yan, Jiyang Liu, Xiaoping Dong, Fengna Xi, and Peng Chen. 2018. Nanochannel-confined graphene quantum dots for ultrasensitive electrochemical analysis of complex samples. *ACS Nano* 12 (12):12673–12681.

Mao, Yu, Shengjie Gao, Lili Yao, Lu Wang, Hao Qu, Yuen Wu, Ying Chen, and Lei Zheng. 2021. Single-atom nanozyme enabled fast and highly sensitive colorimetric detection of Cr (VI). *Journal of Hazardous Materials* 408:124898.

Maret, Wolfgang, and Harold H Sandstead. 2006. Zinc requirements and the risks and benefits of zinc supplementation. *Journal of Trace Elements in Medicine and Biology* 20 (1):3–18.

Marguí, Eva, I Queralt, and M Hidalgo. 2009. Application of X-ray fluorescence spectrometry to determination and quantitation of metals in vegetal material. *TrAC Trends in Analytical Chemistry* 28 (3):362–372.

Marschner, Karel, Stanislav Musil, Ivan Mikšík, and Jiří Dědina. 2018. Investigation of hydride generation from arsenosugars-Is it feasible for speciation analysis? *Analytica Chimica Acta* 1008:8–17.

Mato-Fernández, MJ, JR Otero-Rey, J Moreda-Pineiro, E Alonso-Rodríguez, P López-Mahía, S Muniategui-Lorenzo, and D Prada-Rodríguez. 2007. Arsenic extraction in marine biological materials using pressurised liquid extraction. *Talanta* 71 (2):515–520.

Miedico, Oto, Marco Iammarino, Ciro Pompa, Marina Tarallo, and Antonio Eugenio Chiaravalle. 2015. Assessment of lead, cadmium and mercury in seafood marketed in Puglia and Basilicata (Italy) by inductively coupled plasma mass spectrometry. *Food Additives & Contaminants: Part B* 8 (2):85–92.

Milenkovic, Biljana, Jelena M Stajic, Natasa Stojic, Mira Pucarevic, and Snezana Strbac. 2019. Evaluation of heavy metals and radionuclides in fish and seafood products. *Chemosphere* 229:324–331.

Musielak, Marcin, Maciej Serda, and Rafal Sitko. 2022. Ultrasensitive and selective determination of mercury in water, beverages and food samples by EDXRF and TXRF using graphene oxide modified with thiosemicarbazide. *Food Chemistry* 390:133136.

Najafi, Ezzatolla, Forouzan Aboufazeli, Hamid Reza Lotfi Zadeh Zhad, Omid Sadeghi, and Vahid Amani. 2013. A novel magnetic ion imprinted nano-polymer for selective separation and determination of low levels of mercury (II) ions in fish samples. *Food Chemistry* 141 (4):4040–4045.

Nam, Sang-Ho, Hae-Joon Oh, Hyung-Sik Min, and Joung-Hae Lee. 2010. A study on the extraction and quantitation of total arsenic and arsenic species in seafood by HPLC–ICP-MS. *Microchemical Journal* 95 (1):20–24.

Nho, Eun Yeong, Naeem Khan, Ji Yeon Choi, Jae Sung Kim, Kyung Su Park, and Kyong Su Kim. 2016. Determination of toxic metals in cephalopods from South Korea. *Analytical Letters* 49 (10):1578–1588.

Nivetha, N, B Srivarshine, B Sowmya, Mangaiyarkarasi Rajendiran, Panchamoorthy Saravanan, R Rajeshkannan, M Rajasimman, Thi Hong Trang Pham, Venkat Kumar Shanmugam, and Elena-Niculina Dragoi. 2022. A comprehensive review on bio-stimulation and bio-enhancement towards remediation of heavy metals degeneration. *Chemosphere* 312:137099.

Pang, Yue-Hong, Qiu-Yu Yang, Rui Jiang, Yi-Ying Wang, and Xiao-Fang Shen. 2023. A stack-up electrochemical device based on metal-organic framework modified carbon paper for ultra-trace lead and cadmium ions detection. *Food Chemistry* 398:133822.

Pastorelli, AA, M Baldini, P Stacchini, G Baldini, S Morelli, E Sagratella, S Zaza, and S Ciardullo. 2012. Human exposure to lead, cadmium and mercury through fish and seafood product consumption in Italy: a pilot evaluation. *Food Additives & Contaminants: Part A* 29 (12):1913–1921.

Pétursdóttir, Ásta H, Helga Gunnlaugsdóttir, Hrönn Jörundsdóttir, Adrien Mestrot, Eva M Krupp, and Jörg Feldmann. 2012. HPLC-HG-ICP-MS: a sensitive and selective method for inorganic arsenic in seafood. *Analytical and Bioanalytical Chemistry* 404 (8):2185–2191.

Rahaman, Md Shiblur, Md Mostafizur Rahman, Nathan Mise, Md Tajuddin Sikder, Gaku Ichihara, Md Khabir Uddin, Masaaki Kurasaki, and Sahoko Ichihara. 2021. Environmental arsenic exposure and its contribution to human diseases, toxicity mechanism and management. *Environmental Pollution* 289:117940.

Rana, Mohammad Nasiruddin, Jitbanjong Tangpong, and Md Masudur Rahman. 2018. Toxicodynamics of lead, cadmium, mercury and arsenic-induced kidney toxicity and treatment strategy: a mini review. *Toxicology Reports* 5:704–713.

Regnell, Olof, and Carl J Watras. 2018. Microbial mercury methylation in aquatic environments: a critical review of published field and laboratory studies. *Environmental Science & Technology* 53 (1):4–19.

Romero, Vanesa, Isabel Costas-Mora, Isela Lavilla, and Carlos Bendicho. 2014. Silver nanoparticle-assisted preconcentration of selenium and mercury on quartz reflectors for total reflection X-ray fluorescence analysis. *Journal of Analytical Atomic Spectrometry* 29 (4):696–706.

Saei-Dehkordi, S. Siavash, Aziz A. Fallah, and Amin Nematollahi. 2010. Arsenic and mercury in commercially valuable fish species from the Persian Gulf: influence of season and habitat. *Food and Chemical Toxicology* 48 (10):2945–2950.

Saleh, Mahmoud A, Emmanuel Ewane, Joseph Jones, and Bobby L Wilson. 2001. Chemical evaluation of commercial bottled drinking water from Egypt. *Journal of Food Composition and Analysis* 14 (2):127–152.

Sarkar, Dhruba Jyoti, Bijay Kumar Behera, Pranaya Parida, Vijay Kumar Aralappamavar, Shirsak Mondal, Jyotsna Dei, Basanta Kumar Das, Subhankar Mukherjee, Souvik Pal, and Pabudi Weerathunge. 2022. Aptamer-based NanoBioSensors for seafood safety. *Biosensors and Bioelectronics* 219:114771.

Schmidt, Lucas, Cezar A Bizzi, Fabio A Duarte, Valderi L Dressler, and Erico MM Flores. 2013. Evaluation of drying conditions of fish tissues for inorganic mercury and methylmercury speciation analysis. *Microchemical Journal* 108:53–59.

Schmidt, Lucas, Julio Alberto Landero, Diogo La Rosa Novo, Fabio Andrei Duarte, Marcia Foster Mesko, Joseph A Caruso, and Erico Marlon Moraes Flores. 2018. A feasible method for As speciation in several types of seafood by LC-ICP-MS/MS. *Food Chemistry* 255:340–347.

Shade, Christopher W, and Robert JM Hudson. 2005. Determination of MeHg in environmental sample matrices using Hg– Thiourea complex ion chromatography with on-line cold vapor generation and atomic fluorescence spectrometric detection. *Environmental Science & Technology* 39 (13):4974–4982.

Shah, Sofia B. 2021. Heavy metals in the marine environment—an overview. *Heavy Metals in Scleractinian Corals.* Springer, Cham, 1–26.

Shi, Yongqiang, Wenting Li, Xuping Feng, Lei Lin, Pengcheng Nie, Jiyong Shi, Xiaobo Zou, and Yong He. 2021. Sensing of mercury ions in Porphyra by Copper@ Gold nanoclusters based ratiometric fluorescent aptasensor. *Food Chemistry* 344:128694.

Silva, TS, C Conte, JO Santos, ES Simas, SC Freitas, RLS Raices, and SL Quitério. 2017. Spectrometric method for determination of inorganic contaminants (arsenic, cadmium, lead and mercury) in Smooth weakfish fish. *LWT-Food Science and Technology* 76:87–94.

Simon, Stéphane, Huong Tran, Florence Pannier, and Martine Potin-Gautier. 2004. Simultaneous determination of twelve inorganic and organic arsenic compounds by liquid chromatography–ultraviolet irradiation–hydride generation atomic fluorescence spectrometry. *Journal of Chromatography A* 1024 (1–2):105–113.

Singh, Sima, Arshid Numan, Yiqiang Zhan, Vijender Singh, Tran Van Hung, and Nguyen Dang Nam. 2020. A novel highly efficient and ultrasensitive electrochemical detection of toxic mercury (II) ions in canned tuna fish and tap water based on a copper metal-organic framework. *Journal of Hazardous Materials* 399:123042.

Sundseth, Kyrre, Jozef M Pacyna, Elisabeth G Pacyna, Nicola Pirrone, and Rebecca J Thorne. 2017. Global sources and pathways of mercury in the context of human health. *International Journal of Environmental Research and Public Health* 14 (1):105.

Suwazono, Yasushi, Teruhiko Kido, Hideaki Nakagawa, Muneko Nishijo, Ryumon Honda, Etsuko Kobayashi, Mirei Dochi, and Koji Nogawa. 2009. Biological half-life of cadmium in the urine of inhabitants after cessation of cadmium exposure. *Biomarkers* 14 (2):77–81.

Swain, Krishna Kumari, R Balasubramaniam, and Sunil Bhand. 2020. A portable microfluidic device-based Fe3O4–urease nanoprobe-enhanced colorimetric sensor for the detection of heavy metals in fish tissue. *Preparative Biochemistry & Biotechnology* 50 (10):1000–1013.

Szefer, Piotr. 2013. Safety assessment of seafood with respect to chemical pollutants in European Seas. *Oceanological and Hydrobiological Studies* 42 (1):110–118.

Taylor, Vivien, Britton Goodale, Andrea Raab, Tanja Schwerdtle, Ken Reimer, Sean Conklin, Margaret R Karagas, and Kevin A Francesconi. 2017. Human exposure to organic arsenic species from seafood. *Science of the Total Environment* 580:266–282.

Thirunavukkarasu, OS, T Viraraghavan, KS Subramanian, and S Tanjore. 2002. Organic arsenic removal from drinking water. *Urban Water* 4 (4):415–421.

Tuzen, Mustafa. 2009. Toxic and essential trace elemental contents in fish species from the Black Sea, Turkey. *Food and Chemical Toxicology* 47 (8):1785–1790.

Valdersnes, Stig, Peter Fecher, Amund Maage, and Kaare Julshamn. 2016. Collaborative study on determination of mono methylmercury in seafood. *Food Chemistry* 194:424–431.

Vardhan, Kilaru Harsha, Ponnusamy Senthil Kumar, and Rames C Panda. 2019. A review on heavy metal pollution, toxicity and remedial measures: current trends and future perspectives. *Journal of Molecular Liquids* 290:111197.

Wang, Yifan, Sizhe Dong, Jinrong Xiao, Qiuhui Hu, and Liyan Zhao. 2022. A rapid and multi-element method for the determination of As, Cd, Ni, Pb, Sn, and Zn in scallops using high definition X-ray fluorescence (HDXRF) spectrometry. *Food Analytical Methods* 15 (10):2712–2724.

Xiong, Chaomei, and Bin Hu. 2007. Online YPA4 resin microcolumn separation/preconcentration coupled with inductively coupled plasma optical emission spectrometry (ICP-OES) for the speciation analysis of mercury in seafood. *Journal of Agricultural and Food Chemistry* 55 (25):10129–10134.

Xu, Long, Xiao-yi Suo, Qi Zhang, Xin-ping Li, Chen Chen, and Xiao-ying Zhang. 2020. ELISA and chemiluminescent enzyme immunoassay for sensitive and specific determination of lead (II) in water, food and feed samples. *Foods* 9 (3):305.

Yin, Hongbin, Yongjiu Cai, Hongtao Duan, Junfeng Gao, and Chengxin Fan. 2014. Use of DGT and conventional methods to predict sediment metal bioavailability to a field inhabitant freshwater snail (Bellamya aeruginosa) from Chinese eutrophic lakes. *Journal of Hazardous Materials* 264:184–194.

Yin, Kun, Bowei Li, Xiaochun Wang, Weiwei Zhang, and Lingxin Chen. 2015. Ultrasensitive colorimetric detection of Cu2+ ion based on catalytic oxidation of l-cysteine. *Biosensors and Bioelectronics* 64:81–87.

Yin, Yong/guang, Jingfu Liu, Bin He, Jianbo Shi, and Guibin Jiang. 2008. Simple interface of high-performance liquid chromatography–atomic fluorescence spectrometry hyphenated system for speciation of mercury based on photo-induced chemical vapour generation with formic acid in mobile phase as reaction reagent. *Journal of Chromatography A* 1181 (1–2):77–82.

Zhang, Wang-Bing, Xin-An Yang, Yong-Ping Dong, and Jing-Jing Xue. 2012. Speciation of inorganic-and methyl-mercury in biological matrixes by electrochemical vapor generation from an L-cysteine modified graphite electrode with atomic fluorescence spectrometry detection. *Analytical Chemistry* 84 (21):9199–9207.

Zhang, Yong, Chengyong Wu, Hongxin Liu, Mohammad Rizwan Khan, Zhifeng Zhao, Guiping He, Aimin Luo, Jiaqi Zhang, Ruijie Deng, and Qiang He. 2021. Label-free DNAzyme assays for dually amplified and one-pot detection of lead pollution. *Journal of Hazardous Materials* 406:124790.

Zmozinski, Ariane V, Sergio Carneado, Carmen Ibáñez-Palomino, Àngels Sahuquillo, José Fermín López-Sánchez, and Márcia M Da Silva. 2014. Method development for the simultaneous determination of methylmercury and inorganic mercury in seafood. *Food Control* 46:351–359.

Zu, Wenchuan, and Zhenghao Wang. 2016. Ultra-trace determination of methylmercuy in seafood by atomic fluorescence spectrometry coupled with electrochemical cold vapor generation. *Journal of Hazardous Materials* 304:467–473.

8 Assessment of Biological Contaminants by Using ELISA/PCR Technique

Priyakshi Nath, Tejpal Dhewa, and Sanjeev Kumar

8.1 INTRODUCTION

Microorganisms are widely spread in nature, benefiting or harming any living thing. Human illness from microbial-contaminated food is brought by foodborne pathogens (Bintsis, 2017). Pathogenic organisms release toxins that have the power to alter metabolism, causing catatonia and death and exhibiting a negative impact on a nation's economic development. Before the World Health Organization (WHO) and its partner took the initiative in 2006 to quantify the worldwide burden of foodborne illness, the WHO created the Foodborne Disease Burden Epidemiology Reference Group (FERG) in 2007 to further the effort (WHO Report 2015). Approximately 7.69% of the world's population, or 7.8 billion people, suffer from foodborne illnesses, and 7.5% of deaths each year, or about 56 million deaths worldwide, are caused by foodborne illnesses (WHO Report, 2016). Outbreaks of foodborne illness usually occur in communities in which food is prepared for a sizeable population and served to a mass population, as in places such as old age homes, orphanages, boarding houses, ceremonial functions, restaurants/cafeterias, community kitchen practices, or any food pantry, etc. Currently, more than 200 distinct foodborne illnesses are recognized. According to reports, bacterial pathogens are more common and widespread in foodborne illnesses than viruses and other parasitic infections (Lee and Yoon, 2021).

According to a study on foodborne illness, there are over 48 million foodborne-related illnesses, 28,000 hospitalizations, and about 3,000 fatalities per year in the United States of America due to foodborne illness (Scharff *et al.*, 2016). The enterohemorrhagic *E. coli* O104:H4 strain, which might have been accidentally or purposefully introduced into the food chain, was responsible for one of the greatest epidemics of acute gastroenteritis/haemolytic uraemic syndrome (HUS) in Germany, which occurred in 2011 (Belojevic and Radosavljevic, 2014). The outbreak quickly spread to other nations, including Spain, France, the Netherlands, Denmark, Greece, Austria, Sweden, Canada, Poland, and the United Kingdom (UK), and it resulted in approximately 2,987 cases of the non-haemolytic uremic syndrome and 855 cases of the hemolytic uremic syndrome among people. On investigation, researchers found the presence of enteroaggregative features of *E. coli* and its ability to secrete *Shiga* toxin; consequences showed HUS, often preceded by bloody diarrhea (Frank *et al.*, 2011). *Staphylococcal* foodborne disease (SFD) is one of the common foodborne diseases that occur prominently in the USA. SFD occurs when food is contaminated by *Staphylococcus aureus* that secretes enterotoxins known as *Staphylococcal* enterotoxins (Kadariya *et al.*, 2014). *Staphylococcal* enterotoxins (SEs) are heat stable and belong to the pyrogenic toxin superantigens family. *Staphylococcal* food poisoning (SFP) incidents are caused by inappropriate handling and subsequent storage of foods such as meat, poultry products, dairy products, bread products, etc., under the circumstances more suitable for *Staphylococcus* contamination (Argudín *et al.*, 2010). Superantigens for pyrogenic toxins are present in *Streptococcus pyogenes* and *Staphylococcus aureus*, and they play roles in developing many human and animal diseases. These toxins have been linked to toxic shock syndrome (TSS) and *Staphylococcal* food poisoning (SFP). They can suppress the immune system and promote non-specific T-cell proliferation (Schlievert *et al.*, 2000). SFP occurs when food contaminated with *Staphylococcal* enterotoxins produced by *Staphylococcus*: coagulase positive strains namely *Staphylococcus aureus* and *Staphylococcus intermidus* are consumed (Hennekinne, Buyser, and Dragacci, 2012). *Staphylococcal* enterotoxins (SEs) promote transcytosis, permitting toxicogens to enter into the bloodstream and allowing SEs to communicate with APC (antigen-presenting cells) and T cells (Kadariya *et al.*, 2014). In a food poisoning epidemic that occurred in a military facility in India, 94 people who had eaten raita (a curd dish) for lunch were later revealed to have coagulase-positive *Staphylococcus aureus* infections. Following an inquiry, it was discovered that *staphylococci* infections were present in 1/3rd of the food handlers, and the same microbial species was also isolated from the kitchen's refrigerator, floor, and shelf. In this unintentional outbreak, the infected individuals mostly experienced nausea, vomiting, stomach discomfort, fever, and diarrhea (Mustafa *et al.*, 2009). European Union foodborne illness occurred because of *Campylobacter* in fresh poultry and broiler meat, where 45.2 infected cases/100,000 people appeared. *Campylobacter* is one of the major foodborne illness-causing agents in European countries, followed by *Salmonella*. *Campylobacter* species are present extensively throughout nature, and when *Campylobacter* enters the human body through food, they colonize the intestinal part of the body

DOI: 10.1201/9781003334859-8

(Bintsis, 2017). Among all the species of *Campylobacter*, *Campylobacter jejuni* is one of the prominent agents causing diarrhoeal illness in a human. According to a report, *C. jejuni* causes about eight lakh foodborne illness cases, hospitalization cases around 8,500, and nearly 80 cases of death each year in the USA (Scallan *et al.*, 2011). Cytolethal distending toxin (CDT) is a virulence tripartite toxin produced by *Campylobacter*, comprising three adjacent subunits encoded namely as *cdt*A, *cdt*B, and *cdt*C genes and these clusters of genes are responsible for full CDT activity (Méndez-Olvera et al. 2016). *cdt*A and *cdt*C genes are responsible for toxin binding and internalization to the host cell while *cdt*B gene plays an essential role in encoding the active component of the toxin (Igwaran and Ifeanyi, 2019). Gene *cdt*B produces protein CdtB, which has the capacity to initiate a cascade that leads to blocking of cell cycle. The protein produced by the gene cdtA and cdtC act as a dimeric subunit and helps to bind CdtB and eventually transports the complex structure into the cell interior. On entering the cell, CdtB then enters the nucleus and causes breakage of double-stranded DNA. As a consequence, cytolethal distending toxin blocks G2/M phase of the cell cycle prior to cell division and induces cytoplasmic swelling followed by the death of the cell (Carvalho *et al.*, 2013). *Clostridium perfringens* are nonmotile, encapsulated, rod-shaped bacterial foodborne illness-causing agents secreting proteins that encode toxins and release spores resistant to different environmental stresses such as heat, radiation, etc. (Bintsis, 2017). In 2018, the foodborne outbreak in West Midland, England, happened because of enterotoxigenic spores of *Clostridium perfringens* present in cheese sauce. About 34 people suffered from diarrhea and abdominal cramps because of consuming this contaminated food (Bhattacharya and Beaufoy, 2020). *Salmonella* spp. is another potential foodborne illness-causing agent. Salmonellosis occurs when a person ingests food or water contaminated with *Salmonella* spp. (Bintsis, 2017). In 2011, a food poisoning incidence occurred in a military establishment, in which 53 cases were reported of fever, headache, vomiting, abdominal cramp and diarrhea after eating a potato-bitter gourd vegetable that was contaminated with *Salmonella* spp., although other potential microbial organisms were also detected and suspected to cause the incidence (Kunwar *et al.*, 2013). Due to poor hygienic practices, a food poisoning outbreak in India resulted in 291 cases. It was discovered that the food was relatively contaminated with *Salmonella* spp. and enteropathogenic *E. coli* when left out in the open for an extended period (Bajaj *et al.*, 2019). However, WHO reported the highest annual rate of viral infections responsible for causing foodborne illness with 0.028 mortality rates (WHO Report 2015). Norovirus (NoV) is an enteric pathogen that causes acute gastroenteritis diseases mostly in humans. It can contaminate varieties of food sources such as berries, vegetables, meat, and seafood, even foodstuffs might get easily contaminated by Norovirus due to poor handling or during processing of the foodstuffs by the food-handlers. NoV infection generally takes place through direct ingestion of contaminated food or water, or coming in contact with NoV in the environment (Hardstaff *et al.*, 2018). In 2011, incidences of water samples infected with NoV that sickened people even up to two months after contamination were documented (Seitz *et al.*, 2011). According to a study, green bell peppers were found to be contaminated in the field or by employees who were infected and came in contact with the crops during harvesting, picking, and packaging the final product (Seitz *et al.*, 2011). In 2018, a nursing home experienced an outbreak of viral acute gastroenteritis because of the NoV infection found in turkey meat. This outbreak involved person-to-person virus transmission. There were around 11 afflicted individuals who experienced nausea, vomiting, fever, and watery and bloodless diarrhea (Parrón *et al.*, 2019). In a report by Canadian Food Inspection Agency (CFIA), 279 cases of foodborne illness appeared in Canada due to consumption of raw oysters that were contaminated with Norovirus as of by March 30, 2022. To avoid food poisoning caused by this virus, it is essential to cook the food at a temperature of at least 145°F. The aforementioned information conveys the idea that foodborne illnesses have emerged as global health concerns and are spreading at an alarming rate. Public health organizations are concerned about ensuring food safety due to the rising demand for and availability of street foods and ready-to-eat items (Law *et al.*, 2015). In order to prevent the spread of foodborne infections, other than providing assurance of good food quality, and ensuring a reliable supply of food, it is crucial to inspect food for the presence of any foodborne pathogens. No matter the population age or the region/country, foodborne pathogens have the potential to produce serious outbreaks of disease. The safety of food products is regularly monitored by numerous organizations around the world, including the WHO (World Health Organization), UNEP (United Nations Environment Program), FAO (Food and Agriculture Association), etc. It is crucial to quickly diagnose any pathogens present in contaminated food products, affected people after consuming contaminated food, or other sources that act as carriers of the pathogens.

In this chapter, ELISA (an antibody-based immunoassay) and PCR (polymerase chain reaction), along with a few other cutting-edge methods, are covered in detail. However, the conventional methods have some drawbacks, including a lengthy and labor-intensive pathogen detection process, the use of bulkier and larger instruments, higher costs, inadequate staff training, etc. Scientists have concentrated on developing novel techniques that may quickly diagnose diseases with high specificity, sensitivity, accuracy rate, and cost-reliability. Researchers are working hard to create better, more effective ways to quickly and easily identify infections.

8.2 ORIGIN OF CONTAMINATION

Humans can contract a foodborne illness by consuming tainted food or drink, or they can cause transmission of the pathogen from person to person. Regularly eaten foods like fresh fruits, vegetables, milk, seeds, herbs, dairy products, meat, eggs, bacon, etc., travel directly from the farmer to the customer or through a wider distributor to small-scale or large-scale companies (Baraketi and Lacroix, 2018). The risk for food contamination increases when the fresh raw products are directly consumed by consumers that have not undergone any heat-treatment process. The production of fresh raw materials offers consumers a variety of nutrient supplements. Contamination can happen at any level, including the pre- to post-harvest, transportation, preparation, and packaging stages, from the farm to the customer or directly to the industry. The leftover parts of the animal after slaughter are exposed to bacteria in the animal's intestinal tract throughout the meat processing process, eventually contaminating the remaining animal carcasses. As the meat is processed, it comes in touch with the air, water, food handlers, and distribution systems, all potentially contaminating the finished meat products. In the case of poultry products, de-feathering and evisceration are crucial processes that could result in food product contamination (Baraketi and Lacroix 2018). The foodborne pathogens that contaminate these meat products mostly include *Salmonella* spp., *Clostridium perfringes*, *Campylobacter* spp., *E. coli*, *Listeria monocytogenes*, *Staphylococcus aureus*, etc. (Bantawa *et al.*, 2018). The food code monologue outlines key points to safeguard consumer health, including knowledge demonstration, food handlers' health control, equipment contamination prevention, proper time management, and proper regulation of physical parameters like temperature conditions. It also outlines how to control risk factors (Food Code, 2017). The majority of supermarkets in both developed and developing nations sell ready-to-eat (RTE) foods, such as fruits, vegetables, dairy products, processed meats, etc., that are typically raw and readily available to customers. Before and after meal preparation, utensils and kitchen surfaces must be thoroughly cleaned to preserve hygiene and prevent any pathogenic contamination of the raw food ingredients. It is advised to divide big quantities of food into smaller portions for quick cooling in the refrigerator. It is also advised to reheat canned food products before consumption, simply to prevent cross-contamination (Baraketi and Lacroix, 2018). In addition, it has been noted that people tend to dine out more frequently than eating home-cooked meals in the majority of countries. This is one of the main causes of a higher risk of contracting foodborne illness. Inadequate hygienic conditions and improper cleansing of kitchen-ware, according to research, can lead to pathogenic diseases. A study reported that none of the kitchen participants typically washed their hands after breaking raw eggs. However, they did cleanse their hands properly every time they handled raw fish or meat in the kitchen (Onyeneho and Hedberg, 2013). Barker et al. demonstrated the decontamination of kitchenware using an improved cleaning method in which the use of hypochlorite at 5,000 ppm showed superior results to the detergent-based cleaning method with an addition of proper rinsing step. This study showed how crucial it is to undergo cleaning of kitchen surfaces and kitchen utensils to prevent cross-contamination of pathogens in the food items (Barker *et al.*, 2003).

8.3 METHODS

Conventional methods, which rely on cultivating the microscopic organisms on agar plates, are thought to be the most straightforward, frequently inexpensive, and sensitive approach. However, these procedures rely on the microorganisms' ability to proliferate in particular culture conditions employing fundamental procedures, which include pre-enrichment, selective enrichment, and selective plating, followed by a biochemical screening process and a serological confirmation phase (Mandal, 2011). Preliminary results from such a cumulative sequence of steps take two to three days, and the final confirmation of the identified pathogen takes longer than a week (Lin, Wang, and Oh 2013). Despite being the most straightforward and reliable pathogen identification strategy, conventional methods take much time and demand a lot of work in the steps, such as preparing culture media, inoculating plates and counting colonies to isolate and identify the pathogenic strain

(Mandal, 2011). Due to their low sensitivity rate, they sometimes even give false negative results, making it impossible to detect the original pathogenic strains (Law *et al.*, 2015). The next section discusses the conventional methods, including the culture-based technique, which identifies specific diseases based on microbial growth, the PCR detection method, and antibody-based immunoassay, which measures the quantitative reaction of an antigen with its antibody.

8.3.1 Culture-Based Approach

It is one of the earliest traditional methods for finding harmful microbial strains. The major benefit of this technology is that it is inexpensive and has a high success rate. It confirms the presence or absence of the pathogenic strain in the food supply (Priyanka *et al.*, 2016). Based on sorbitol fermentation, the investigation of *E. coli* O157:H7 culture on a SMAC (Sorbitol MacConkey Agar) medium revealed heavy growth culture with 100% sensitivity, 85% specificity, and 86% accuracy of *E. coli* in stool detection rate (March and Ratnam, 1986). The primary drawback of this method is the prolonged turnaround time. Additionally, due to emerging serotypes such as sorbitol-fermenting non-O157 and O157 STEC, false negative results were obtained occasionally (Hirvonen *et al.*, 2012). However, the limitations of SMAC medium can be controlled by using a chromogenic medium to isolate *Shiga* toxin *E.coli* (STEC) with enhanced specificity and sensitivity. CHROMagar STEC medium permits growth and speculative identification of mauve colonies around 75% of STEC isolate, comprising a vast collection of different types of serotypes of enterohemorrhagic *E. coli* (EHEC) (Gouali *et al.*, 2013). The major advantage of using the CHROMagar medium is that it provides easier discrimination of colonies based on color because of the presence of a chromogenic substance in the medium (Priyanka *et al.*, 2016). CHROMagar medium, thus, is more effective than a SMAC medium. A study by Tzschoppe et al. on the detection and isolation of EHEC strains and aggregative EHEC O104:H4 strains from ready-to-eat vegetables revealed that the growth of a STEC culture on a CHROMagar medium was closely linked to the essential STEC serotypes and the pre-existing terB (tellurite resistance) gene. While the presence of the terB gene was less common in the diarrheagenic eae-negative and stx-positive strains (13.5%), apathogenic *E. coli* strains (12.0%), and sorbitol-fermenting nonmotile O157 strains (0.0%), the distribution of terB was significantly higher (87.2%) among the diarrheagenic eae- and stx-positive strains. According to the study, tellurite was one of the CHROMagar STEC medium's selective components. By cause of regional variations in availability, there may be variances in the specificity and sensitivity of *E. coli* isolated stains from diarrheagenic and non-diarrheagenic strains (Tzschoppe *et al.*, 2012). Antibiotics are a common example of a selected component or drug that is employed in chromogenic media to stimulate microbial growth in the culture. These antibiotics can potentially impede the growth of other off-target bacteria in the medium (Perry, 2017). The use of multiple distinct chromogenic enzymes and selective components, such as the bacterial detection procedure, would improve the identification and specificity of the pathogen detected (Nehra *et al.*, 2022). One of the most often employed enzymes is a bacterial hydrolase, such as β-galactosidase (β-Gal), which is primarily targeted because of its chromo-fluorogenic substrate activity (Lozano-torres *et al.*, 2021). When included in the media, chromogenic and fluorogenic substrates allow for a change in colour intensity (mostly bright color), following an enzymatic reaction and enable differentiation of colonies based on the target enzyme. When pathogens from food products are discovered using fluorogenic media, a small amount of sample is required for detection and once fluorescence is seen in the culture or agar media, the pathogen already present in the sample is screened (Nehra *et al.*, 2022). In general, the enzyme's activity is not species-specific. When a particular enzyme hydrolyzes a fluorogenic substrate, fluorophores are produced. When the rate of fluorescence is measured using spectro-photometry under UV light, the presence of a pathogen can be determined by the discharge of detectable signals (Manafi, 1996). A microbial pathogen enters a state of dormancy when exposed to unfavourable environmental conditions, such as nutrient deprivation, changes in osmotic pressure, severe temperatures, exposure to food preservatives, etc. (Fakruddin *et al.*, 2013). Since these microorganisms are difficult to cultivate on standard nutritional media while in a dormant state, several fluorescent dyes have successfully identified these infections. Fluorescein isothiocyanate (FITC) has been used as a marker in a study to detect gram-negative bacteria based on its binding to the unlabeled antibody (Zeng *et al.*, 2018). The detection of pathogens using fluorescent dyes has shown to be an inexpensive, simplified method with reduced pathogen detection time. Such characteristics for fluorescent dyes such as Eva Green (Bundidamorn *et al.*, 2018), Acridine Orange (Guo *et al.*, 2017), SYTO 9 (Skerniskyte *et al.*, 2016), CTC-DAPI (Wideman *et al.*, 2021), etc. were used for the detection of common foodborne pathogens. Table 8.1 discusses fluorescent dyes in

Table 8.1: Different Fluorescent Dyes Used in the Detection of Foodborne Pathogens

Dyes	Microorganism Detected	Detection Limit	Reaction	Advantages	References
Fluorescein isothiocyanate (FITC)	*Acidovorax citrulli*	10^6 CFU/mL	FITC emits visible yellow-green light	Helps to locate the pathogens intracellularly, helps to visualize live cells and stains them either violet or blue color, etc.	(Zeng et al., 2018) (Fakruddin et al., 2013)
CTC-DAPI (5-cyano-2,3-ditolyl tetrazolium chloride—4′,6-diamidino-2-phenylindole)	*Listeria monocytogenes*	Not Stated	CTC-DAPI stains help to undergo viable cell count within the bacterial biofilms after antimicrobial treatment	Rapid processing technique helps to estimate viable bacterial cell count	(Wideman et al., 2021)
CTC-DAPI	*Campylobacter jejuni*	Not Stated	CTC stain indicates cellular respiratory activity in bacterial cells, DAPI stain allows the enumeration of viable bacterial cell count. Together as a double staining process, it enables to monitor VBNC of *C. jejuni* easily	Suitable for rapid pathogen detection helps to estimate viable bacterial cell count	(Cappelier et al., 1997)
EVA Green	*Salmonella* spp., *L. monocytogenes*, Shiga toxin-producing *E. coli* (STEC)	1 CFU/25 g	Ability to provide robust PCR signal and strong–sharp DNA melting curve peak, suitable to apply in a closed-tube format such as HRM application, and exhibits reasonable photostability	Very suitable for utilisation in rapid processing detection technique	(Bundidamorn et al., 2018) (Mao et al., 2007)
Acridine Orange	*E.coli*	Not stated	Fluoresce green when it binds to dsDNA and red when it binds to ssDNA or RNA	Actively utilized for cell enumeration in rapid pathogen detection Technique	(Guo et al., 2017)
SYTO 9	*Salmonella* spp., *Yersinia enterocolitica*, *Listeria monocytogenes*, *Campylobacter* spp.	$1–1.9 \times 10^3$ CFU/mL	Cell permeable stain and allow increased quantum yield upon nucleic acid binding	Stains Live/Dead, Gram positive-Gram negative bacterial pathogen and allow rapid detection from the sample	(Skerniskyte et al., 2016)

identifying various pathogenic strains in food samples. A few important aspects are considered for effective detection: 1) the length of time needed to cultivate microorganisms, 2) the lengthening of the incubation period needed for enough microbial growth to create visual colonies, and 3) the length of time needed for the enzyme to produce color after the introduction of a chromogenic substrate. All of these criteria indicate that the culturing approach is a laborious operation (Nehra *et al.*, 2022). For the successful execution of the standard culture method, expensive laboratory tools and skilled workers are needed. However, there are some factors that prevent pathogen isolation, such as insufficient amount of pathogen concentration in the sample, uneven pathogen distribution, presence of native pathogens, variety of food matrices, sluggish determination of microorganisms, restricted accuracy, etc. All of these things indicate the importance of using quick approaches, notably in the food business, to quickly detect viruses in food samples while also being time-efficient, with less error in pathogen detection, minimal labor, etc.

8.3.2 Immune Cell-Based Approaches

Immunoassay relies on the quantitative interaction of antigen and antibody for identification or to check the presence of any pathogen in a food sample. Numerous non-covalent associations between paratopes and the binding location on an antigen, or epitope, are involved in the process of antigen-antibody interaction (Sela-culang *et al.*, 2013). Immunoglobulins, or antibodies, are soluble globulin-class proteins that play a key role in the immunological reactions within the body (Igs). When an antigen or immunogen is present, mature effector B cells release Igs. Antibodies are frequently used as a great biorecognition component for pathogen identification in various food or other environmental samples (Nehra *et al.*, 2022). Enzyme-linked immunosorbent assay (ELISA), lateral flow immunoassays (LFIAs), reserve passive agglutination tests, etc., are popular pathogen detection immunoassays in food matrices. Essential features such as the purity and specificity of an antibody play an important role in pathogen detection immunoassays (Priyanka *et al.*, 2016). The ELISA-based pathogen detection method is widely used as a sandwich ELISA, which is often the most popular pathogen detection method. In this immunoassay, the amount of bound antigen from the enrichment culture is determined using two primary antibodies, a capture antibody, and a detection antibody (Zhao *et al.*, 2017). Nearly all ELISA techniques employ chromogenic substrates that produce some sort of perceptible color change to indicate the presence of antigen. Sandwich ELISA is a sensitive and reliable method that uses both primary and secondary antibodies. Typically, the walls of the microtiter plate wells are used to immobilize the primary antibody (capture antibody). The immobilized capture primary antibody binds to the food sample's target antigen, and any leftover unbound antigens are rinsed off the microtiter wells. The remaining unbound secondary antibodies are eliminated when the bound antigen is sandwiched with an enzyme-conjugated secondary antibody to produce a complex antigen-antibody structure. The target antigen is sandwiched between the capture primary antibody and the secondary antibody that has been enzyme-conjugated to generate the complex structure. When a colorless substrate is added, an enzymatic reaction takes place with the substrate, which causes the colorless substrate to change into a colored form. The amount of antigen present in the supplied sample is thus visualized by the color intensity, which is recorded as a signal (Figure 8.1) (Law *et al.*, 2015). β-galactosidase, alkaline phosphatase, and horseradish peroxidase enzymes (HRP) are frequently used in the ELISA method (Yeni *et al.*, 2014). The use of HRP enzymes has successfully conducted the detection of pathogens from various food samples such as contaminated vegetables, meat (Shen *et al.*, 2014), non-fat dairy products (Capo *et al.*, 2020), and cattle feces (Zhang *et al.*, 2016). In order to meet the new, emerging issues, there has been an ongoing effort to improve the ELISA approach. As monoclonal antibodies (mAbs) have an affinity for monovalency, they have become a potent tool for pathogen identification. A study was conducted comparing pathogen detection's effectiveness using direct ELISA and conventional culture results. It was shown that indirect ELISA, which uses monoclonal antibodies, outperformed traditional culture, which detected 23% of *Salmonella* spp. in the contaminated chicken meal sample, with a sensitivity, specificity, and accuracy rate of 94% (Schneid *et al.*, 2006). In another study, mAbs were generated to detect the presence of *Vibrio parahaemolyticus* pathogen in seafood. mAbs produced against recombinant thermostable-related hemolysin (TRH) protein allowed detection of all the pathogenic strains of *V. parahaemolyticus* using the sandwich ELISA technique. Results obtained from the PCR assay did not show any differentiation between live and dead cells, while a sandwich ELISA assay successfully detected the live cells in the enrichment culture. A combination of monoclonal antibodies-based methods that included both sandwich ELISA and PCR techniques showed an accurate and enhanced sensitive

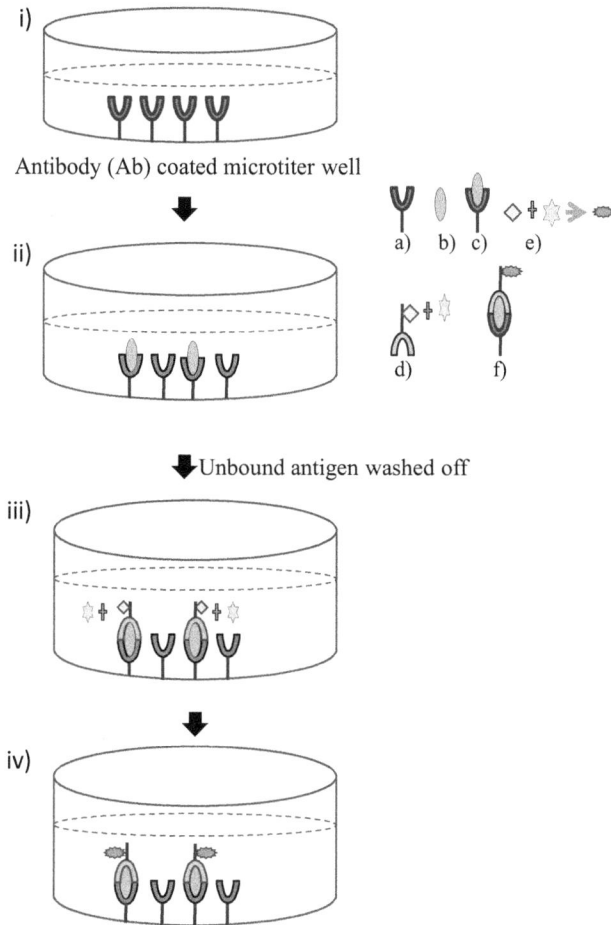

Figure 8.1 Schematic representation of sandwich ELISA technique for pathogen detection: i) primary capture Ab is made immobilized on a solid surface; ii) target Ag binds to the primary Ab and left unbound Abs are washed off; iii) on incubation, primary Ab-target Ag-enzyme conjugated-secondary Ab complex structure is formed; iv) on enzymatic reaction with the colorless substrate, the coloured signal is imparted.

detection of *Vibrio parahaemolyticus* in the seafood enrichment culture (Kumar *et al.*, 2011). Similar to such above-mentioned studies, the ELISA detection technique has further been used in numerous experiments to identify the pathogenic strains in the food sample; a few significant ones were discussed in Table 8.2. Immunological identification of pathogenic strains from food samples has improved in sensitivity, specificity, and reliability with quick results. *Bacillus cereus* can be found in research and clinical laboratories using the ELISA technique with effectiveness. In a study, entire *Bacillus cereus* cells were used as immunogens to make mouse and rabbit polyclonal antibodies (pAbs) and monoclonal antibodies (mAbs) in preparation for the double-antibody sandwich ELISA technique's detection of *B. cereus* pathogen in food samples containing minced meat. This investigation demonstrated that the use of polyclonal antibodies increases sensitivity as they can recognize more epitope sites than monoclonal antibodies (mAbs) and can even detect pathogenic *B. cereus* strains in extremely low concentrations (0.9×10^3 cells/mL in phosphate-buffered saline) (Zhu 2016). The use of nanobody-fused reported proteins in a sandwich ELISA immunoassay has successfully detected *Salmonella* foodborne pathogens in the food samples. Such improved immunoassay techniques exhibit high-affinity outcomes within a reduced time and minimum use of the reagents (Gu *et al.*, 2022).

Table 8.2: Detection of Foodborne Pathogens Using ELISA Detection Technique

Pathogen Detected	Food Sample	Substrate Utilised	Enzyme Employed	Detection Limit	Assay Time (hrs)	References
Escherichia coli	Artificial contamination of milk, vegetables, ground beef, etc.	3,3′,5,5′-tetramethylbenzidine-hydrogen peroxide (TMB-H_2O_2),	Horseradish peroxide (HRP)	68 CFU/mL in PBS (Phosphate- Buffered Saline) 6.8×10^3 CFU/mL detected in the food sample	3	(Shen *et al.*, 2014)
Listeria monocytogenes	Non-fat dried milk	3,3′,5,5′-tetramethylbenzidine (TMB)	Horseradish peroxide (HRP)	1×10^2 CFU/mL	8	(Capo *et al.*, 2020)
Enterohemorrhagic *Escherichia coli* (EHEC O157:H7)	Artificial contamination of vegetables (lettuce, spinach, sprouts, etc.), grounded beef, raw milk, and cattle feces	Tetramethylbenzidine (TMB)	Horseradish peroxide (HRP)	1×10^3 CFU/mL for *E.coli* O157:H7 culture, 1×10^4 CFU/g before enrichment of contaminated food samples, 1×10^2 CFU/g after enrichment of contaminated food samples.	Variable	(Zhang *et al.*, 2016)
Vibrio parahaemolyticus	Seafood homogenate	TMB-H_2O_2	Horseradish peroxide (HRP)	1×10^3 cells/mL after enrichment	16	(Kumar *et al.*, 2011)
Bacillus cereus	Artificially contaminated comminuted meat	TMB	HRP	0.9×10 cells/mL	Variable	(Zhu *et al.*, 2016)
Salmonella enteritidis	Artificially contaminated milk sample	TMB	HRP	5×10^4 CFU/mL in milk, 10 CFU/mL after the enrichment step	8	(Gu *et al.*, 2022)

The ELISA method demonstrates traits including a high rate of sensitivity and specificity of pathogen detection, a high-efficiency rate due to the ability to conduct simultaneous analysis without the need for time-consuming sample pre-treatment, a high accuracy rate, a safe, eco-friendly method, and a cost-effective assay. Although ELISA has numerous advantages as a pathogen detection method, it has some disadvantages, including the need for specialist equipment, antibody instability, and contamination at the intermediate phase of chemical-conjugate specific binding that can result in false positive results.

8.3.3 Polymerase Chain Reaction (PCR)-Based Approach

The polymerase chain reaction is one of the most frequently used pathogen detection methods (PCR). The detection of specific target DNA or RNA sequences of the pathogen present in the sample drives the PCR techniques for pathogen detection (Law *et al.*, 2015). It permits *in-vitro* amplification of the specific target nucleic acid sequences and also amplifies targeted either natural or synthetic nucleic acid sequences present in low concentration in the sample (Nehra *et al.*, 2022). PCR method undergoes amplification of specific target DNA fragment following a three-step cyclic process (Mandal, 2011): a) Specific target dsDNA fragment is denatured into ssDNA at high-temperature conditions; b) two specific primers (synthetic oligonucleotides) are annealed to the target ssDNA sequence that acts as a DNA template for hybridisation at specific high temperatures; c) the primers complementary to the ssDNA fragments are polymerized in the presence of deoxyribonucleotides and thermostable DNA polymerase enzymes in the final step of the primer polymerization process. In order to increase the quantity of target DNA fragments, these three processes are performed repeatedly. Once the PCR-amplified products have been stained with EtBr (ethidium bromide), they can easily be seen as bands on the electrophoresis gel (Lin, Wang, and Oh 2013). Because of their dependable and quick detection results, PCR-based techniques have been employed to identify foodborne pathogens from food samples in addition to the benefits they offer, such as speed, rate of high accuracy, sensitivity, and specificity. By amplifying target DNA fragments and specific genes that encode bacterial toxins, PCR techniques are very helpful for detecting the toxic genes produced by pathogens like *Vibrio cholera, E. coli, Staphylococcus aureus, Campylobacter jejuni, Salmonella* spp., *Shigella* spp., *Listeria monocytogenes*, etc. Compared to procedures based on culture and immunoassay, this method is quicker. The amplified product from a PCR technique can be obtained in 30 minutes, and as the technology uses several primers, it has become easier to distinguish between different pathogenic strains (Priyanka *et al.*, 2016). Despite being useful, the typical PCR approach gives inaccurate estimates of the viability of cells when used for pathogen identification. This occurs because the DNA could not be distinguished as coming from either live or dead cells using the conventional PCR technique (Foddai and Grant, 2020). Viability PCR is used to identify pathogens based on an unbroken cell wall or cell membrane, which has been observed to overcome the restriction of correct distinction among live or dead cells by the conventional PCR approach. Biological dyes are used in conjunction with PCR technology in viability PCR. Testing for viability by PCR uses dyes like EMA (ethidium monoazide) or PMA (propidium monoazide, which is a derivative of ethidium bromide). The food samples are first stained with EMA or PMA dye because they have the capacity to penetrate damaged cell membranes and attach to DNA both in free form and in cells with damaged cell membranes or cell walls. After the cells are exposed to intense light, the dye creates a covalent link with the nucleic acids, resulting in irreparable damage such as breaking the nucleic acid strands. The cells with an intact cell membrane would only be amplified after DNA amplification, while the cells with a compromised cell membrane or extracellular cells with degraded DNA would produce inadequate templates (Emerson *et al.*, 2017). As a result, PCR shows the ability to distinguish between live and dead cells rapidly and effectively, and it has developed into one of the faster methods for identifying foodborne pathogens. However, the presence of an intact cell membrane in every cell does not guarantee that the cell is alive. Evidence suggests that some cells have intact cell membranes but are metabolically inactive, producing false results (Ayrapetyan and Oliver, 2016). According to a study, when bacteria including *Escherichia coli* 0157:H7, *Micrococcus* species, *Staphylococcus aureus, Streptococcus* species, etc., were exposed to EMA dye, the dye not only eliminated DNA from dead cells but also partially decreased the DNA content of living cells (Nocker *et al.*, 2006). It has been discovered that messenger RNA (mRNA) is a more accurate predictor of cell viability than DNA. This is because the mRNA molecule is only found in cells that are metabolically active. Reverse transcriptase enzyme is used in reverse-transcriptase PCR (RT-PCR), sometimes referred to as quantitative PCR (qPCR), to transform extracted messenger RNA (mRNA) into complementary DNA (cDNA). The freshly synthesized complementary DNA is then employed as the template strand for exponential amplification using traditional or quantitative PCR if quantification is necessary (qPCR). Since bacterial transcripts are readily destroyed by intracellular and extracellular RNases and the detectable mRNA becomes restricted to the active and live cells within the sample, the fundamental concept of RT-PCR demonstrates rapid reduction of mRNA levels following cellular death (Foddai and Grant, 2020). Because RT-PCR does not use gel electrophoresis to identify PCR products, it differs from simple PCR. The produced PCR product is quantified using the fluorescent signal generated by intercalating dyes or probes (Law et al., 2015). Even when the pathogen content in the mixture if initially remains very low, RT-PCR could precisely accurately measure the target nucleic acids. By using fluorescent reporter molecules, which enable real-time evaluation of PCR products by watching the amplification of a specific target sequence based on fluorescent technology, this method accelerates the process of detecting infectious microbes in the provided sample (Vizzini et al., 2017). Fluorescent dye or DNA probes, along with thermostable DNA polymerase, nucleotides, and a sample, allow for the quantification of amplified DNA in qPCR (Nehra et al., 2022). Several DNA probes, like TaqMan probes, molecular beacon probes, FRET probes, etc., are utilized to evaluate the amplified products. Taq DNA polymerase enzyme cleaves particular DNA fragments in the 5' to 3' direction, allowing TaqMan probe to perform quantification and mutation detection functions (Vizzini et al., 2017). A molecular beacon, on the other hand, is an oligonucleotide probe with a stem-loop or hair-pin shape that contains a quencher and a fluorophore at the 5' and 3' ends, respectively (C. Wang and Yang, 2013). The loop section and the stem, which are created by annealing two complementary arm pieces, contain a complementary sequence to a particular target sequence. The molecular beacon probe has dyes connected to both ends of a reporter dye and a quencher dye. No fluorescence is produced because both dyes are kept in close proximity by the hybrid stem. The voluntary conformational changes that occur during hybridization separate the two dyes, causing emission of a fluorescence signal at the exact target sequence (Law et al., 2015). In the PCR method, various dyes, such as SYBRGreen, SYBERGold, ethidium bromide, etc., are used to detect amplicons. In Table 8.3, a few significant studies are detailed about the use of different types of dyes utilized in various PCR-induced studies. SYBRGreen dye was discovered, which could attach to freshly produced copies of the target DNA fragments during DNA amplification using PCR, considerably enhancing the fluorescent signal. This explains why the amplicons generated by the PCR directly correlate with the fluorescent strength signal (Priyanka et al., 2016). The amplicons produced would be higher the more intense the fluorescent signal detected. SYBERGreen dye has often been applied in the detection of various pathogens from food samples viz., fresh meat, raw milk and pasteurized cheese (Bastam et al., 2021), neonate's milk (Carvalho et al., 2013), dairy products (Singh et al., 2012), etc. In a study employing qPCR test, SYBERGreen dye was used to identify the pathogenic strains of *Vibrio parahaemolyticus* in tropical shellfish, and *V. parahaemolyticus* was found in the culture at a concentration of about 1×10^2 CFU/mL (Tyagi et al., 2009). Multiplex PCR (mPCR) allows simultaneous amplification of numerous loci in a single reaction for the quick detection of pathogens from various food samples like ready-to-eat takeout foods (Lee, 2014), contaminated tomatoes, ground beef (Yang et al., 2013), lettuce (Y. Mao et al., 2016), minced meat (Boukharouba, 2022) etc. Different primer sets are amalgamated into a single PCR assay (Lin, Wang, and Oh 2013). The development of multiplex PCR depends heavily on the primer design since good mPCR assays require primer sets with identical annealing temperatures (Law et al., 2015). In mPCR assays, primer concentration is a crucial variable. It helps to alter the primer concentration to enable the synthesis of accurate PCR products because interactions between different primer sets may lead to the creation of primer dimers (Law et al., 2015). Buffer concentration, cycle temperatures, DNA template strands, calibration between deoxynucleotide concentration and $MgCl_2$, DNA polymerase enzyme, and other important parameters are taken into consideration. In one of the earliest applications of PCR, *Mycobacterium tuberculosis* was found in sputum samples. The assay's results demonstrate sensitivity, specificity, and cost-effectiveness, considerably easing the tuberculosis laboratory diagnosis process (Eisenach et al., 1991). Since then, practically, almost, all infections can be detected rapidly and with easy thanks to the discovery and use of various PCR techniques.

The polymerase chain reaction method holds a number of potential advantages, including a high rate of positivity and sensitivity compared to other detection methods, the need for a small amount of sample to detect the microbes present in it and the ability to amplify their copies thousands to millions of times, improved virus detection abilities, quick detection, low cost, etc. (Liu et al., 2019). The PCR process provides a lot of benefits, but it also has certain drawbacks. Due to the

Table 8.3: Detection of Foodborne Pathogens Using Different PCR Techniques

Types of PCR Techniques	Food Sample	Pathogen Detected	Dye Used	Detection Limit	Reference
Multiplex PCR	Artificially contaminated food products viz., tomato, lettuce, and ground beef	Salmonella typhimurium, Escherichia coli O157: H7, Listeria monocytogenes	PMA	5.1×10^3 CFU/g for Salmonella typhimurium, 7.5×10^3 CFU/g for E. coli O157:H7 and 8.4×10^3 CFU/g for Listeria monocytogenes	(Yang et al., 2013)
Multiplex PCR	Korean ready-to-eat food	Escherichia coli O157:H7, Bacillus cereus, Salmonella spp., Vibrio parahaemolyticus, Staphylococcus aureus, Listeria monocytogenes	Not Stated	Variable	(Lee et al., 2014)
Multiplex PCR	Artificially contaminated live shrimp sample	Vibrio vulnificus	Not stated	1.3×10^5 CFU/mL	(Roig et al., 2022)
Multiplex PCR	Artificially contaminated organic lettuce and minced meat	E. coli, Salmonella enterica, Listeria monocytogenes	Not Stated	10^3 CFU/mL	(Boukharouba et al., 2022)
Multiplex PCR	Artificially contaminated vegetable (lettuce)	Listeria monocytogenes, Listeria ioanooi	Not Stated	10 CFU/g	(Mao et al., 2016)
RT-PCR	Artificially contaminated raw milk, pasteurised milk and cheese	Staphylococcus aureus, Listeria monocytogenes, Salmonella typhi	SYBER Green	Not Stated	(Bastam et al., 2021)
RT-PCR	Artificially contaminate Modified-Atmosphere-Packed (MAP) Salmon Steaks	Photobacterium phosphoreum	Propidium monoazide (PMA), SYBR green	3–8 log CFU/g	(Mace et al. 2013)

(Continued)

TABLE 8.3 (*continued*)

Types of PCR Techniques	Food Sample	Pathogen Detected	Dye Used	Detection Limit	Reference
RT-PCR	Fresh pork meat	*Salmonella* spp., *Shigella* spp., *Staphylococcus aureus*	Not Stated	2.0 CFU/g for *Salmonella*, 6.8 CFU/g for *Shigella* spp., and 9.6 CFU/g for *Staphylococcus aureus*	(Ma et al., 2014)
RT-PCR	Infant milk formula	*Salmonella* spp. *Listeria monocytogenes*, *E.coli* O157	SYBER Green	1.7 CFU/25 g	(Azinheiro, Carvalho, and Prado 2020)
RT-PCR	Dairy products	*Listeria monocytogenes*, *Salmonella* spp.	SYBER Green	3 log CFU/ml in non-fat dried milk	(Singh et al., 2012)

extraordinary sensitivity of PCR, any contamination of the material, even in trace amounts of DNA, can produce false findings. Another issue is that the primers employed in the PCR process can lead to the non-specific annealing of somewhat similar sequences but not exactly the same as the target DNA (Garibyan, 2013).

8.3.4 Biosensing Approach: An Efficient Rapid Foodborne Pathogen Detection Technique

An analytical tool called a biosensor device has two primary parts: a bioreceptor and a transducer. While the transducer transforms biological connections into detectable electrical signals, the bio-receptor detects biological materials such as antibodies, enzymes, nucleic acids, binding proteins, aptamers, etc. (Law *et al.*, 2015). Utilizing a biosensing technique for pathogen detection offers dependability, sensitivity, accuracy with specificity, quick detection, and cost-effectiveness (Nehra *et al.*, 2022). For pathogen detection, biosensors need a small amount of material. Unlike nucleic acid or immunology-based techniques, this technology does not require a pre-enrichment phase for proper pathogen detection. The important property of a biosensor is that it enables quick and portable pathogen detection for both laboratory and field experimentation. Due to its extremely quick pathogen detection capabilities, it allows interactive information on food ingredients and gives opportunity to take prompt corrective action. The following section discusses a few types of biosensors that can quickly identify foodborne pathogens.

8.3.4.1 *Optical Biosensors*

An optical biosensor compact analytical device consists of a biological sensing component connected to an optical transducer system. This makes use of many transduction hypotheses to find foodborne pathogens. The most frequently employed method is surface plasmon resonance (SPR). Momentary waves are used in the SPR optical system to measure the modifications that occur in the relative index (RI) relatively close to the sensor surface. When incident light contacts a free electron at a specific angle (α), known as the SPR angle, waves are produced. Any modifications at the interface of two media surface significantly impact the angle (Poltronieri *et al.*, 2014). Antibodies are first fixed on a thin metal surface, delivering an optically transparent waveguide, regulating the capture of target pathogens. When near-infrared (IR) or visible light passes through a waveguide, the interaction of the light with the electrons cloud on the metal surface, allows the production of a powerful vibration. The binding of a pathogen causes a shift in the resonance to generate longer wavelengths, and the resulting shift shows how many pathogens are bound to the metal's surface. According to a study, the pathogens *E. coli* O157:H7, *Salmonella* spp., *Campylobacter jejuni*, and *Listeria monocytogenes* were detected simultaneously fin artificially contaminated apple juice using the SPR optical biosensor technique, which is based on wavelength division multiplexing. Each bacterial species detected in the study has a detection limit that varies from 3.4×10^3 to 1.2×10^5 CFU/mL (Taylor *et al.*, 2006). Another study used the SPR biosensing technology and ground beef and cucumber that had been intentionally contaminated with *E.coli* O157:H7, having a detection limit of 3×10^3 CFU/mL (Wang *et al.*, 2013). SPREETA Biosensor and BIOCORE 3000 Biosensor are two commercially available optical biosensors for detecting foodborne pathogens. *E. coli* O157:H7 can be found in apple juice, milk, and ground beef, according to a study employing the SPREETA biosensor with a detection limit of around 10^2–10^3 CFU/mL (Waswa *et al.*, 2007). In a different investigation, a BIOCORE biosensor with a detection limit of 10^5 cells/mL was used to identify *Listeria monocytogenes* (Leonard *et al.*, 2004). Optical biosensor exhibits beneficial properties like high sensitivity, real-time, or label-free immunosensing detection system etc., its high-cost rate that limits the usage of this method.

8.3.4.2 *Electrochemical Biosensors*

Due to the specific identification function of the bio-recognition element within the test sample, the biosensor selectively recognizes the target molecule and captures it on the surface of the electrode. The electrode, as the primary signal converter, has the capacity to collect the identifying signal produced on its surface and convert it into an electrical signal that is measurable and can be processed to produce a qualitative or quantitative result for the particular target (Zhang *et al.*, 2019). Amperometric, potentiometric, conductometric, and impedimetric biosensors are some additional categories for electrochemical biosensing techniques. The electrochemical biosensing technique was successful in detecting several pathogens from various samples. *Staphylococcus aureus* was detected using an amperometric biosensor with a detection limit of approximately 1 CFU/mL (Ávila *et al.*, 2012). In an experimental investigation, a contaminated food meal that included sprouts, cooked corn, fried rice, tomatoes, strawberries, and lettuces were utilized to detect *Bacillus cereus*; the detection limit was determined to be between 35–88 CFU/mL (Pal *et al.*, 2008). Another study described using an electrochemical biosensor based on Cerium(IV) oxide/chitosan-modified electrodes to detect the DNA of *Clostridium perfringens* from dairy foods while monitoring impedance change during the experiment (Qian *et al.*, 2018).

Nevertheless, of the huge studies and publications on biosensors based on pathogen detections, there is very minimal knowledge about the commercial availability of biosensors and their applications in different areas. Some of the commercial biosensors used in different sectors are mentioned. For instance, glucose testing biosensor kits, pregnancy test kits, insulin detectors, hemoglobin testing kits, etc., are most commonly known and used in the field of medicine (Bahadir and Sezgintürk, 2015), VitFast systems, BIOFISH 300 Sulphite, BIOFISH 700 SUL, BIOWINE 700 MUST & WINE, RIDACUBE SCAN etc. used in food additives and quality control (Di Nardo and Anfossi, 2020), BOD (Biological Oxygen Demand) biosensors in aquaculture (Bahadir and Sezgintürk, 2015), Metrohm Dropsens, PalmSens, GAMRY instruments, etc. are utilized for pathogen detections (Akgönüllü *et al.*, 2020), etc. Biosensors provided specificity, sensitivity, and rapid monitoring of the targeted analyte in different sectors and proved to be a useful technology in the present time, and its applications are expected to increase in the coming days. The biosensor technology is being studied, developed, and focused upon to progress in its advancement and potential by overcoming limitations. There is a need to find alternate ways to mitigate the life span limitation of the biological constituents, easy shifting of the prototype to the site of mass production, proper handling of the tool, minimize the analysis cost, and find ways that are capable of enhancing the ability of multiplex testing (Di Nardo and Anfossi 2020).

8.4 CONCLUSION

Although traditional methods for detecting microorganisms in various sources yield positive findings, they are time- and labor-intensive. Therefore, various quick pathogen detection techniques are being developed to address the shortcomings of traditional approaches. Rapid detection techniques deliver more precise, sensitive, accurate, and dependable results than traditional procedures. These quick detection methods also deliver quick and effective results, which have been valuable in preventing foodborne outbreaks and containing any epidemics. Utilizing nucleic-based techniques, immunoassays, and other effective detection methods necessitates having a thorough understanding of the specialized equipment and the working principles. Effective training is also highly important. Furthermore, there is not just one important strategy that favors or offers a quick, sensitive, and highly accurate outcome for the infections found. Therefore, combining various quick pathogen detection techniques improves the consistency of positive and confirmed pathogen results.

REFERENCES

Akgönüllü, S., D. Çimen, M. Bakhshpour et al., 2020. Commercial sensors for pathogen detection. In *Commercial Biosensors and Their Applications*, ed. M. K. Sezgintürk, 89–106. Torino, Italy: Elsevier.

Argudín, María Ángeles, María Carmen Mendoza, and María Rosario Rodicio. 2010. Food poisoning and Staphylococcus aureus enterotoxins. *Toxins* 2, no. 7: 1751–1773.

Ayrapetyan, Mesrop. and James D. Oliver. 2016. The viable but non-culturable state and its relevance in food safety. *Current Opinion in Food Science* 8: 127–133.

Azinheiro, Sarah, Carvalho, Joana, Prado, Marta, & Garrido-Maestu, Alejandro (2020). Multiplex Detection of Salmonella spp., E. coli O157 and L. monocytogenes by qPCR Melt Curve Analysis in Spiked Infant Formula. *Microorganisms*,8, no.9:1359 https://doi.org/10.3390/microorganisms8091359.

Bajaj, Swati and Puja Dudeja. 2019. Food poisoning outbreak in a religious mass gathering. *Medical Journal Armed Forces India* 75: 339–343.

Bantawa, Kamana, Kalyan Rai, Dhiren Subba Limbu, and Hemanta Khanal. 2018. Food-borne bacterial pathogens in marketed raw meat of Dharan, eastern Nepal. *BMC Research Notes* 11: 1–5.

Baraketi, Amina, Stephane Salmieri and Monique Lacroix. 2018. Foodborne pathogens detection: persevering worldwide challenge. *Biosensing Technologies for the Detection of Pathogens—A Prospective Way for Rapid Analysis*, ed. T. Rinken and K. Kirivand. United Kingdom: Intechopen, 53–72.

Barker, J., M. Naeeni and S. F. Bloomfield. 2003. The effects of cleaning and disinfection in reducing Salmonella contamination in a laboratory model kitchen. *Journal of Applied Microbiology* 95: 1351–1360.

Bastam, Mahsa Morovati, Mahsa Jalili, Iraj Pakzad, Abbas Maleki, and Sobhan Ghafourian. 2021. Pathogenic bacteria in cheese, raw and pasteurised milk. *Veterinary Medicine and Science* 7: 2445–2449.

Bhattacharya, Alex, Saran Shantikumar, Damon Beaufoy, Adrian Allman, Deborah Fenelon, Karen Reynolds, Andrea Normington et al., 2020 Outbreak of Clostridium perfringens food poisoning linked to leeks in cheese sauce: an unusual source. *Epidemiology & Infection* 148: 1–7.

Bahadır, Elif Burcu, and Mustafa Kemal Sezgintürk. 2015. Applications of commercial biosensors in clinical, food, environmental, and biothreat/biowarfare analyses. *Analytical Biochemistry* 478: 107–120.

Bintsis, Thomas. 2017. Foodborne pathogens. *AIMS Microbiology* 3: 529–563.

Boukharouba, Aya, Ana González, Miguel García-Ferrús, María Antonia Ferrús, and Salut Botella. 2022. Simultaneous detection of four main foodborne pathogens in ready-to-eat food by using a simple and rapid multiplex PCR (mPCR) assay. *International Journal of Environmental Research and Public Health* 19: 1–18.

Bundidamorn, Damkerng, Wannakarn Supawasit, and Sudsai Trevanich. 2018. A new single-tube platform of melting temperature curve analysis based on multiplex real-time PCR using Eva Green for simultaneous screening detection of Shiga toxin-producing *Escherichia* coli, *Salmonella* spp. and *Listeria monocytogenes* in food. *Food Control* 94: 195–204.

Capo, Alessandro, Sabato D'Auria, and Monique Lacroix. 2020. A fluorescence immunoassay for a rapid detection of Listeria monocytogenes on working surfaces. *Scientific Reports* 10: 1–12. 10.1038/s41598-020-77747-y.

Cappelier, J. M., B. Lazaro, A. Rossero, A. Fernandez-Astorga, and Michel M. Federighi. 1997. Double staining (CTC-DAPI) for detection and enumeration of viable but non-culturable Campylobacter jejuni cells. *Veterinary Research* 28: 547–555. HAL Id: hal-00902502.

Carvalho, Aline Feola de, Daniela Martins da Silva, Sergio Santos Azevedo, Rosa Maria Piatti, Margareth Elide Genovez, and Eliana Scarcelli. 2013. Detection of CDT toxin genes in Campylobacter spp. strains isolated from broiler carcasses and vegetables in São Paulo, Brazil. *Brazilian Journal of Microbiology* 44: 693–699. 10.1590/s1517-83822013000300005.

Di Nardo, Fabio, and Laura Anfossi. 2020. Commercial biosensors for detection of food additives, contaminants, and pathogens. In *Commercial Biosensors and Their Applications*, ed. M. K. Sezgintürk, 183–215. Torino, Italy: Elsevier.

Eisenach, Kathleen D., Mark D. Sifford, M. Donald Cave, Joseph H. Bates, and Jack T. Crawford. 1991. Detection of *Mycobacterium tuberculosis* in sputum samples using a polymerase chain reaction. *American Review of Respiratory Disease* 144: 1160–1163.

Emerson, Joanne B., Rachel I. Adams, Clarisse M. Betancourt Román, Brandon Brooks, David A. Coil, Katherine Dahlhausen, Holly H. Ganz et al. 2017. Schrödinger's microbes: tools for distinguishing the living from the dead in microbial ecosystems. *Microbiome* 5: 1–23.

145

Esteban-Fernandez de Ávila, Berta, María Pedrero, Susana Campuzano, Vanessa Escamilla-Gómez, and José M. Pingarrón. 2012. Sensitive and rapid amperometric magnetoimmunosensor for the determination of Staphylococcus aureus. *Analytical and Bioanalytical Chemistry* 403: 917–925.

Fakruddin, Md, Khanjada Shahnewaj Bin Mannan, and Stewart Andrews. 2013. Viable but nonculturable bacteria: food safety and public health perspective. *International Scholarly Research Notices* 2013: 1–7.

Foddai, Antonio CG., and Irene R. Grant. 2020. Methods for detection of viable foodborne pathogens: Current state-of-art and future prospects. *Applied Microbiology and Biotechnology* 104: 4281–4288.

Food and Drug Administration. 2017. Food Code: 2017 Recommendations of the United States Public Health Service Food and Drug Administration. URL: https://www.fda.gov/food/fda-food-code/food-code-2017 (accessed: October 8, 2022).

Frank, Christina, M. S. Faber, Mona Askar, Helen Bernard, Angelika Fruth, Andreas Gilsdorf, Michael Höhle et al. 2011. Large and ongoing outbreak of haemolytic uraemic syndrome, Germany, May 2011. *Euro Surveillance* 16: 1–3.

Garibyan, Lilit, and Nidhi Avashia. 2013. Research techniques made simple: polymerase chain reaction (PCR). *The Journal of Investigative Dermatology* 133: 1–8.

Gouali, Malika, Corinne Ruckly, Isabelle Carle, Monique Lejay-Collin, and François-Xavier Weill. 2013. Evaluation of CHROMagar STEC and STEC O104 chromogenic agar media for detection of Shiga toxin-producing *Escherichia coli* in stool specimens. *Journal of Clinical Microbiology* 51: 894–900.

Gu, Kui, Zengxu Song, Changyu Zhou, Peng Ma, Chao Li, Qizhong Lu, Ziwei Liao et al. 2022. Development of nanobody-horseradish peroxidase-based sandwich ELISA to detect *Salmonella Enteritidis* in milk and in vivo colonization in chicken. *Journal of Nanobiotechnology* 20: 1–18.

Guo, Rachel, Cushla McGoverin, Simon Swift, and Frederique Vanholsbeeck. 2017. A rapid and low-cost estimation of bacteria counts in solution using fluorescence spectroscopy. *Analytical and Bioanalytical Chemistry* 409: 3959–3967.

Hardstaff, Joanne L., Helen E. Clough, Vittoria Lutje, K. Marie McIntyre, John P. Harris, Paul Garner, and Sarah J. O'Brien. 2018. Foodborne and food-handler Norovirus outbreaks: a systematic review. *Foodborne Pathogens and Disease* 15: 589–597.

Hennekinne, Jacques-Antoine, De Buyser, Marie-Laure, & Dragacci, Sylviane (2012). *Staphylococcus aureus* and its food poisoning toxins: characterization and outbreak investigation. *FEMS Microbiology Reviews*, 36:815–836 .

Hirvonen, Jari J., Anja Siitonen and Suvi-Sirkku Kaukoranta. 2012. Usability and performance of CHROMagar STEC medium in detection of Shiga toxin-producing Escherichia coli strains. *Journal of Clinical Microbiology* 50: 3586–3590.

Igwaran, Aboi, and Anthony Ifeanyi Okoh. 2019. Human campylobacteriosis: a public health concern of global importance. *Heliyon* 5: 1–14.

Kadariya, Jhalka, Tara C. Smith, and Dipendra Thapaliya. 2014. Staphylococcus aureus and staphylococcal food-borne disease: an ongoing challenge in public health. *BioMed Research International* 2014: 1–9.

Kumar, Ballamoole Krishna, Pendru Raghunath, Devananda Devegowda, Vijay Kumar Deekshit, Moleyur Nagarajappa Venugopal, Iddya Karunasagar, and Indrani Karunasagar. 2011.

Development of monoclonal antibody based sandwich ELISA for the rapid detection of pathogenic Vibrio parahaemolyticus in seafood. *International Journal of Food Microbiology* 145: 244–249.

Kunwar, R., Harpreet Singh, Vipra Mangla, and R. Hiremath. 2013. Outbreak investigation: Salmonella food poisoning. *Medical Journal Armed Forces India* 69: 388–391.

Law, Jodi Woan-Fei, Nurul-Syakima Ab Mutalib, Kok-Gan Chan, and Learn-Han Lee. 2015. Rapid methods for the detection of foodborne bacterial pathogens: principles, applications, advantages and limitations. *Frontiers in Microbiology* 5: 1–20.

Lee, Heeyoung, and Yohan Yoon. 2021. Etiological agents implicated in foodborne illness worldwide. *Food Science of Animal Resources* 41: 1–7.

Lee, Nari, Kyung Yoon Kwon, Su Kyung Oh, Hyun-Joo Chang, Hyang Sook Chun, and Sung-Wook Choi. 2014. A multiplex PCR assay for simultaneous detection of *Escherichia coli* O157: H7, *Bacillus cereus*, *Vibrio parahaemolyticus*, *Salmonella* spp., *Listeria monocytogenes*, and *Staphylococcus aureus* in Korean ready-to-eat food. *Foodborne Pathogens and Disease* 11: 574–580.

Leonard, Paul, Stephen Hearty, John Quinn, and Richard O'Kennedy. 2004. A generic approach for the detection of whole *Listeria monocytogenes* cells in contaminated samples using surface plasmon resonance. *Biosensors and Bioelectronics* 19: 1331–1335.

Liu, Harry Y.,Grant C. Hopping, Uma Vaidyanathan, Yasmyne C. Ronquillo, Phillip C. Hoopes, and Majid Moshirfar. 2019. Polymerase chain reaction and its application in the diagnosis of infectious keratitis. *Medical Hypothesis, Discovery and Innovation in Ophthalmology* 8: 152–155.

Lozano-Torres, Beatriz, Juan F. Blandez, Félix Sancenón, and Ramón Martínez-Máñez. 2021. Chromo-fluorogenic probes for β-galactosidase detection. *Analytical and Bioanalytical Chemistry* 413: 2361–2388.

Ma, Kai, Yi Deng, Yu Bai, Dixin Xu, Erning Chen, Huijuan Wu, Baoming Li, and Lijuan Gao. 2014. Rapid and simultaneous detection of *Salmonella, Shigella*, and *Staphylococcus aureus* in fresh pork using a multiplex real-time PCR assay based on immunomagnetic separation. *Food Control* 42: 87–93.

Manafi, M. 1996. Fluorogenic and chromogenic enzyme substrates in culture media and identification tests. *International Journal of Food Microbiology* 31: 45–58.

Mandal, P. K., A. K. Biswas, K. Choi, and U. K. Pal. 2011. Methods for rapid detection of foodborne pathogens: an overview. *American Journal of Food Technology* 6: 87–102.

Mao, Fei, Wai-Yee Leung, and Xing Xin. 2007. Characterization of EvaGreen and the implication of its physicochemical properties for qPCR applications. *BMC Biotechnology* 7: 1–16.

Mao, Yan, Xiaolin Huang, Sicheng Xiong, Hengyi Xu, Zoraida P. Aguilar, and Yonghua Xiong. 2016. Large-volume immunomagnetic separation combined with multiplex PCR assay for simultaneous detection of *Listeria monocytogenes* and *Listeria ivanovii* in lettuce. *Food Control* 59: 601–608.

March, Sandra B., and Samuel Ratnam. 1986. Sorbitol-MacConkey medium for detection of *Escherichia coli* O157: H7 associated with hemorrhagic colitis. *Journal of Clinical Microbiology* 23: 869–872.

Méndez-Olvera, Estela T., Jaime A. Bustos-Martínez, Yolanda López-Vidal, Antonio Verdugo-Rodríguez, and Daniel Martínez-Gómez. 2016. Cytolethal distending toxin from *Campylobacter jejuni* requires the cytoskeleton for toxic activity. *Jundishapur Journal of Microbiology* 9: 1–10.

Mustafa, M. S., S. Jain, and V. K. Agrawal. 2009. Food poisoning outbreak in a military establishment. *Medical Journal Armed Forces India* 65: 240–243.

Nehra, Monika., Virendra Kumar, Rajesh Kumar, Neeraj Dilbaghi, and Sandeep Kumar. 2022. Current scenario of pathogen detection techniques in agro-food sector. *Biosensors* 12: 1–17.

Nocker, Andreas, Ching-Ying Cheung, and Anne K. Camper. 2006. Comparison of propidium monoazide with ethidium monoazide for differentiation of live vs. dead bacteria by selective removal of DNA from dead cells. *Journal of Microbiological Methods* 67: 310–320.

Onyeneho, Sylvester N., and Craig W. Hedberg. 2013. An assessment of food safety needs of restaurants in Owerri, Imo State, Nigeria. *International Journal of Environmental Research And Public Health* 10: 3296–3309.

Pal, Sudeshna., Wendy Ying, Evangelyn C. Alocilja, and Frances P. Downes. 2008. Sensitivity and specificity performance of a direct-charge transfer biosensor for detecting *Bacillus cereus* in selected food matrices. *Biosystems Engineering* 99: 461–468.

Parrón, Ignacio, J. Álvarez, Mireia Jané, T. Cornejo Sánchez, E. Razquin, S. Guix, G. Camps, C. Pérez, À.Domínguez, and Working Group for the Study of Outbreaks of Acute Gastroenteritis in Catalonia. 2019. A foodborne norovirus outbreak in a nursing home and spread to staff and their household contacts. *Epidemiology & Infection* 147: 1–7.

Perry, John D. 2017. A decade of development of chromogenic culture media for clinical microbiology in an era of molecular diagnostics. *Clinical Microbiology Reviews* 30: 449–479.

Poltronieri, Palmiro, Valeria Mezzolla, Elisabetta Primiceri, and Giuseppe Maruccio. 2014. Biosensors for the detection of food pathogens. *Foods* 3: 511–526.

Priyanka, B., Rajashekhar K. Patil, and Sulatha Dwarakanath. 2016. A review on detection methods used for foodborne pathogens. *The Indian Journal of Medical Research* 144, no. 3 (2016): 327–338. 10.41 03/0971-5916.198677.

Qian, Xingcan., Qing Qu, Lei Li, Xin Ran, Limei Zuo, Rui Huang, and Qiang Wang. 2018. Ultrasensitive electrochemical detection of *Clostridium perfringens* DNA based morphology-dependent DNA adsorption properties of CeO_2 nanorods in dairy products. *Sensors* 18: 1–15.

Radosavljevic, V., Finke, E.-J., & Belojevic, G. (2014). Escherichia coli O104:H4 outbreak in Germany--clarification of the origin of the epidemic. *The European Journal of Public Health*, 25, no. 1:125–129 .

Roig, Arnau Pérez., Héctor Carmona-Salido, Eva Sanjuán, Belén Fouz, and Carmen Amaro. 2022. A multiplex PCR for the detection of *Vibrio vulnificus* hazardous to human and/or animal health from seafood. *International Journal of Food Microbiology* 377: 1–7.

Scallan, Elaine, Robert M. Hoekstra, Frederick J. Angulo, Robert V. Tauxe, Marc-Alain Widdowson, Sharon L. Roy, Jeffery L. Jones, and Patricia M. Griffin. 2011. Foodborne illness acquired in the United States—Major Pathogens. *Emerging Infectious Diseases* 17: 1–21.

Scharff, Robert L., John Besser, Donald J. Sharp, Timothy F. Jones, Gerner-Smidt Peter, and Craig W. Hedberg. 2016. An economic evaluation of PulseNet: a network for foodborne disease surveillance. *American Journal of Preventive Medicine* 50: S66–S73.

Schlievert, Patrick M., Lynn M. Jablonski, Manuela Roggiani, Ingrid Sadler, Scott Callantine, David T. Mitchell, Douglas H. Ohlendorf, and Gregory A. Bohach. 2000. Pyrogenic toxin superantigen site specificity in toxic shock syndrome and food poisoning in animals. *Infection and Immunity* 68: 3630–3634.

Schneid, Andréa dos Santos., Kelly Lameiro Rodrigues, Davi Chemello, Eduardo Cesar Tondo, Marco Antônio Zacchia Ayub, and José Antonio Guimarães Aleixo. 2006. Evaluation of an indirect ELISA for the detection of Salmonella in chicken meat. *Brazilian Journal of Microbiology* 37: 350–355.

Seitz, Scot R., Juan S. Leon, Kellogg J. Schwab, G. Marshall Lyon, Melissa Dowd, Marisa McDaniels, Gwen Abdulhafid et al. 2011. Norovirus infectivity in humans and persistence in water. *Applied and Environmental Microbiology* 77: 6884–6888.

Sela-Culang, Inbal, Vered Kunik, and Yanay Ofran. 2013. The structural basis of antibody-antigen recognition. *Frontiers in Immunology* 4: 1–13.

Shen, Zhiqiang., Nannan Hou, Min Jin, Zhigang Qiu, Jingfeng Wang, Bin Zhang, Xinwei Wang, Jie Wang, Dongsheng Zhou, and Junwen Li. 2014. A novel enzyme-linked immunosorbent assay for detection of Escherichia coli O157: H7 using immunomagnetic and beacon gold nanoparticles. *Gut Pathogens* 6: 1-8.

Singh, Jitender, Virender K. Batish, and Sunita Grover.2012. Simultaneous detection *of Listeria monocytogenes* and *Salmonella* spp. in dairy products using real time PCR-melt curve analysis. *Journal of Food Science and Technology* 49: 234–239.

Skerniškytė, Jūratė, Julija Armalytė, Raimonda Kvietkauskaitė, Vaida Šeputienė, Justas Povilonis, and Edita Sužiedėlienė. 2016. Detection of *Salmonella* spp., *Yersinia enterocolitica*, *Listeria monocytogenes* and *Campylobacter* spp. by real-time multiplex PCR using amplicon DNA melting analysis and probe-based assay. *International Journal of Food Science & Technology* 51: 519–529.

Taylor, Allen D., Jon Ladd, Qiuming Yu, Shengfu Chen, Jiří Homola, and Shaoyi Jiang. 2006. Quantitative and simultaneous detection of four foodborne bacterial pathogens with a multi-channel SPR sensor. *Biosensors and Bioelectronics* 22: 752–758.

Tyagi, Anuj, Vasudevan Saravanan, Iddya Karunasagar, and Indrani Karunasagar. 2009. Detection of *Vibrio parahaemolyticus* in tropical shellfish by SYBR green real-time PCR and evaluation of three enrichment media. *International Journal of Food Microbiology* 129: 124–130.

Tzschoppe, Markus, Annett Martin, and Lothar Beutin. 2012. A rapid procedure for the detection and isolation of enterohaemorrhagic *Escherichia coli* (EHEC) serogroup O26, O103, O111, O118, O121, O145 and O157 strains and the aggregative EHEC O104: H4 strain from ready-to-eat vegetables. *International Journal of Food Microbiology* 152: 19–30.

Vizzini, Priya., Lucilla Iacumin, Giuseppe Comi, and Marisa Manzano. 2017. Development and application of DNA molecular probes. *Aims Bioengineering* 4: 113–132.

Wang, Chunming, and Chaoyong James Yang. 2013. Application of molecular beacons in real-time PCR. In *Molecular Beacons*, ed. C. J. Yang, and W. Tan, 45–59. Heidelberg, Berlin: Springer.

Wang, Yixian, Zunzhong Ye, Chengyan Si, and Yibin Ying. 2013. Monitoring of Escherichia coli O157: H7 in food samples using lectin based surface plasmon resonance biosensor. *Food Chemistry* 136: 1303–1308.

Waswa, John, Joseph Irudayaraj, and Chitrita Deb Roy. 2007. Direct detection of *E. coli* O157: H7 in selected food systems by a surface plasmon resonance biosensor. *LWT-Food Science and Technology* 40: 187–192.

World Health Organization. 2016. Burden of foodborne diseases in the South-East Asia Region. URL: https://apps.who.int/iris/bitstream/handle/10665/332224/9789290225034-eng.pdf (accessed October 8, 2022).

World Health Organization. 2015."*WHO estimates of the Global Burden of Foodborne Diseases: Foodborne Disease Burden Epidemiology Reference Group 2007-2015."* World Health Organization. URL: https://apps.who.int/iris/bitstream/handle/10665/199350/?sequence=1. (accessed: October 8, 2022).

Wideman, Nathan E., James D. Oliver, Philip Glen Crandall, and Nathan A. Jarvis. 2021. Detection and potential virulence of viable but non-culturable (VBNC) *Listeria monocytogenes*: a review. *Microorganisms* 9: 1–11.

Yang, Youjun, Feng Xu, Hengyi Xu, Zoraida P. Aguilar, Ruijiang Niu, Yong Yuan, Jichang Sun et al. 2013. Magnetic nano-beads based separation combined with propidium monoazide treatment and multiplex PCR assay for simultaneous detection of viable *Salmonella Typhimurium*, *Escherichia coli* O157: H7 and *Listeria monocytogenes* in food products. *Food Microbiology* 34: 418–424.

Yeni, F., Sibel Acar, Ö. G. Polat, Y. Soyer, and H. Alpas. 2014. Rapid and standardized methods for detection of foodborne pathogens and mycotoxins on fresh produce. *Food Control* 40: 359–367.

Zeng, Haijuan, Xuzhao Zhai, Manman Xie, and Qing Liu. 2018. Fluorescein isothiocyanate labeling antigen-based immunoassay strip for rapid detection of *Acidovorax citrulli*. *Plant Disease* 102: 527–532.

Zhang, Xuehan, Meng Li, Bicheng Zhang, Kangming Chen, and Kongwang He. 2016. Development of a sandwich ELISA for EHEC O157: H7 Intimin γ1. *Plos One* 11: 1–14.

Zhang, Zhenguo, Jun Zhou, and Xin Du. 2019. Electrochemical biosensors for detection of foodborne pathogens. *Micromachines* 10: 1–16.

Zhao, Xihong, Junliang Zhong, Caijiao Wei, Chii-Wann Lin, and Tian Ding. 2017. Current perspectives on viable but non-culturable state in foodborne pathogens. *Frontiers in Microbiology* 8: 1–16.

Zhu, Longjiao, Jing He, Xiaohan Cao, Kunlun Huang, Yunbo Luo, and Wentao Xu. 2016. Development of a double-antibody sandwich ELISA for rapid detection of *Bacillus Cereus* in food. *Scientific Reports* 6: 1–10.

9 Intelligent Point-of-Care Testing for Food Safety

Mycotoxins

Xiaofeng Hu, Shenling Wang, Zhaowei Zhang, and Peiwu Li

9.1 INTRODUCTION: Typical Mycotoxins in Grain and Oil

The word "mycotoxin" originates from the combination of the Greek "mykes" and the Latin "toxicum" (Van der Zijden *et al.* 1962). Mycotoxins are the secondary metabolites of small molecules with different chemical structures and complex toxicity produced by fungi in the metabolic process. It is one of food's most harmful natural risk factors. At present, more than 400 kinds have been isolated and identified. The mycotoxins produced by *Aspergillus*, *Fusarium*, and *Alternaria* fungi, including aflatoxin (AFT), ochratoxin A (OTA), and fusarium toxin, are the most harmful ones. Mycotoxins easily pollute most agricultural products such as grain and oil, threatening people's safety and health. Research results from the International Food and Agriculture Organization show that mycotoxins contaminate about 25% of the world's grain and oil products. According to incomplete statistics from China's Grain Administration, the annual grain loss caused by mixed mycotoxin pollution accounts for 6.2% of the total grain production, which is more than six times the yearly increase of grain needed to ensure national food security.

9.1.1 Aflatoxin

Aflatoxins are toxic secondary metabolites produced by *Aspergillus flavus*, *Aspergillus parasiticus*, *Aspergillus nomius*, etc. It has carcinogenic, teratogenic, and mutagenic hazards (Li *et al.* 2009). Aflatoxin was first identified in the early 1960s when more than 100,000 turkeys died on British farms, and the disease was known as "Turkey X disease" because of its unknown cause (Wannop 1961). Studies have shown that the cause of Turkey's mortality was aflatoxin-contaminated feed (Lancaster *et al.* 1961). Aflatoxins are a class of difuran coumarin derivatives catalyzed by polyketone compound synthetase. The basic structure is composed of a difuran ring and a coumarin. The difuran ring is the basic toxin structure, while the coumarin is associated with carcinogenesis. Since the discovery of aflatoxin, more than 20 kinds of aflatoxin and its derivatives have been isolated, including aflatoxin B_1 (AFB$_1$), B_2 (AFB$_2$), G_1 (AFG$_1$), G_2 (AFG$_2$), M_1 (AFM$_1$), M_2 (AFM$_2$), etc. The chemical structure formula is shown in Figure 9.1.

In 1993, the World Health Organization's (WHO) International Agency for Research on Cancer (IARC) classified AFB1 as a Group I carcinogen, that is, substances with confirmed carcinogenicity in humans (Ostry *et al.* 2017). The toxicity of aflatoxins varies significantly among different chemical structures. The toxicity of main aflatoxins was ranked as AFB$_1$>AFM$_1$>AFG$_1$>AFB$_2$>AFG$_2$ (Tahir *et al.* 2018), aflatoxin with double-bond structure at the end of difuran rings is more toxic, while aflatoxin without double-bond structure is relatively less toxic. Among them, AFB1 is the most widely distributed, the most poisonous, and the most carcinogenic. Aflatoxin has cytotoxicity, hepatotoxicity, genotoxicity, reproductive toxicity, and immunotoxicity. It is easy to cause cancer, teratogenicity, and mutagenicity and has acute and chronic toxicity (Thomas *et al.* 2003). Large doses of aflatoxin can cause acute toxicity and even death. In contrast, small amounts of aflatoxin are chronically toxic, causing growth disorders, nutritional deficiencies, chronic liver damage, and immune system impairment in animals (Bedard *et al.* 2006). Aflatoxin-contaminated food can easily lead to aflatoxin-contaminated food poisoning. The severity depends on the amount and duration of exposure. Acute symptoms include vomiting, bleeding, abdominal pain, jaundice, pulmonary edema, cerebral edema, coma, convulsions, and even death (Mwanda *et al.* 2005). In 2004, an outbreak of aflatoxin disease in Kenya resulted in 125 deaths from consuming aflatoxin-contaminated corn food. The study found aflatoxin contamination levels in 55% of corn samples tested>20 µg/kg, 35% aflatoxin contamination level>100 µg/kg, 7% aflatoxin contamination level>1,000 µg/kg (Khlangwiset *et al.* 2011). The chronic toxicity of aflatoxin can induce tumors or induce other diseases and suppress immune function. Long-term aflatoxin exposure is strongly associated with growth and cognitive impairment in children (Khlangwiset *et al.* 2011). Therefore, most countries and regions set a limited level of aflatoxin in various agricultural products and food. Aflatoxin contamination in human food is strictly regulated by the Food and Drug Administration (FDA). Aflatoxin levels in grains and products should be below 20 µg/kg. The European Union (EU) has adopted stricter limits (0.1~2 µg/kg) for aflatoxin in cereals or cereal products intended for human consumption (Ismail *et al.* 2018).

DOI: 10.1201/9781003334859-9

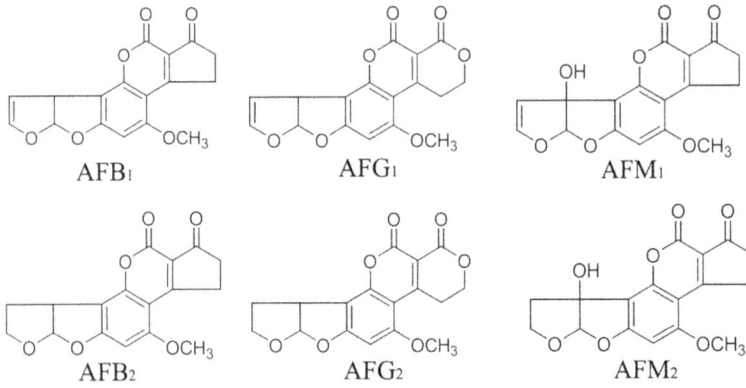

Figure 9.1 Chemical structures of six main aflatoxins.

9.1.2 Zearalenone

Zearalenone (ZEN), also known as F-2 toxin, is a harmful secondary metabolite with estrogen-like effects produced primarily by Fusarium. The *Fusarium spp.* that can produce ZEN mainly includes *Fusarium graminis, F. culmorum, F. equiseti, F. verticillioides,* and *F. semitectum.* In 1962, Stob et al., Purdue University isolated ZEN from corn infected with ear rot (Stob *et al.* 1962). In the same year, Christensen et al. also reported that ZEN was isolated from corn culture infected with scab and named F-2 toxin. In 1966, Dr. Urry et al. used NMR and mass spectrometry to identify the chemical structure of F-2 toxin as a sihydroxybenzoic acid lactone compound, which was named zearalenone (Ryu *et al.* 1999). The chemical formula of ZEN is $C_{18}H_{22}O_5$, and the relative molecular weight is 318. ZEN can be reduced to zearalenol (ZOL) and zeranol (ZAL) in animals (Figure 9.2). Both reduction forms have two diastereoscopic stereoisomers, α-ZOL and β-ZOL, and four metabolic structures α-ZAL and β-ZAL.

In temperate climates, ZEN mainly contaminates corn, barley, wheat, oats, rice, rye, and sorghum. ZEN production occurs mainly in the field but can also be synthesized under harsh grain storage conditions, such as humidity greater than 30% to 40% (Zinedine *et al.* 2007). That's because ZEN and deoxynivalenol (DON) are metabolized by the same fungus and often contaminate food and oil at the same time. Like most mycotoxins, ZEN is difficult to remove completely during food processing because of its high thermal stability. Cereal products such as breakfast cereals, baked snacks, bread, pasta, and even cooking oils (such as corn and wheat germ oil) are considered significant sources of human exposure to ZEN (Mally *et al.* 2016). Due to its strong estrogen activity, ZEN can bind to estrogen receptors and mainly acts on the reproductive system (Marin *et al.* 2013). Sporadic epidemics suggest that ZEN triggers central precocious puberty and may lead to early puberty in children (Yang *et al.* 2016). ZEN has been classified as a Group III carcinogen by IARC, and the European Union regulates the content of ZEN in food products to be no higher than 75–350 µg/kg in different foods (Pinotti *et al.* 2016).

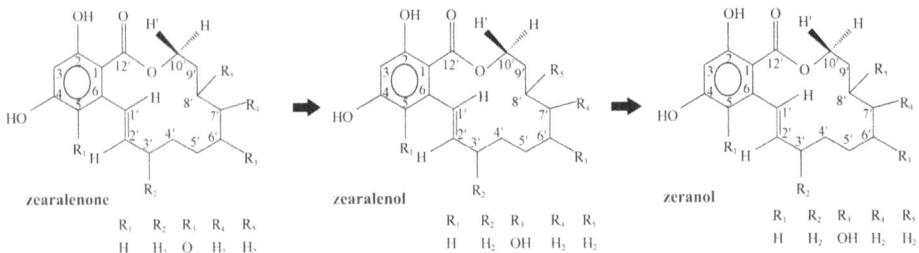

Figure 9.2 Chemical structures of mian zearalenones.

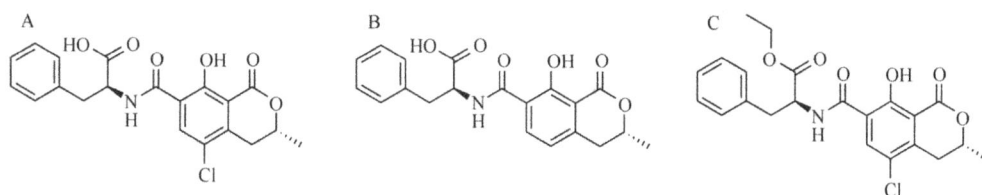

Figure 9.3 Chemical structures of (A) OTA, (B) OTB, and (C) OTC.

9.1.3 Ochratoxin

Ochratoxin, also known as brown *Aspergillus*, is a toxic secondary metabolite produced by *Aspergillus* and *Penicillium*. Ochratoxin was first discovered in 1965, Van der Merwe published a paper in the journal *Nature* entitled "Ochratoxin A, a toxic metabolite produced by *Aspergillus ochraceus Wilh*" (Van der Merwe *et al.* 1965). The toxic secondary metabolites produced by the strain were named ochratoxin A (OTA). Subsequently, scientists identified ochratoxin B (OTB) and ochratoxin C (OTC). OTA often contaminates grain, oil, and products.

Ochratoxin is a para-chlorophenolic group containing a dihydroisocoumarin moiety that is amide-linked to L-phenylalanine. The chlorine atoms of OTAs are replaced by hydrogen atoms to produce OTBs, and OTC is the ethyl compound of OTAs (Figure 9.3). Based on dietary intake studies, OTA can bind to plasma proteins through the gastrointestinal tract and accumulate in the kidneys with a long half-life (about 35 days) (Ringot *et al.* 2006). OTA can compete with phenylalanine hydroxylase in the kidney and liver and inhibit certain protein synthesis as well as ribonucleic acid (RNA) and deoxyribonucleic acid (deoxyribonucleic acid, DNA) synthesis (Studer-Rohr *et al.* 2000), with strong nephrotoxicity, hepatotoxicity, neurotoxicity, and immunotoxicity. For this reason, the IARC classifies OTA as a Group II B carcinogen. It is important to note that high-dose OTA exposure is only found in limited areas. For example, OTA levels of 8.91 and 148 ng/mL were detected in human urine in Egypt and Sierra Leone, respectively (Bui-Klimke *et al.* 2015). The EU has set OTA limits of 5 mg/kg for raw foods and 3 mg/kg for cereals.

9.1.4 Deoxyniverenol

Deoxynivalenol (DON), also known as vomitoxin/emetic toxin, belongs to the B-type mono-trichosporene toxin. DON was first identified in 1970 in barley infected with scab in Kagawa Prefecture, Japan, and was named Rd toxin (Morooka *et al.* 1972). In the following year, Japanese scholars Yoshizawa and Morooka clarified that the structure of the toxin was a 4-deoxy derivative of nivalenol (NIV), which was renamed deoxynivalenol (Yoshizawa *et al.* 1973). Subsequently, DON-contaminated agricultural products were found both in Ohio, USA, and in Shanghai (Vesonder *et al.* 1973). DON is a tetracyclic sesquiterpenoid compound whose chemical names are $3\alpha,7\alpha$, 15-trihydroxy-12, 13-epoxy-monotrophorus 9-ene-8-one, and molecular formula is $C_{15}H_{20}O_6$. Its chemical structure formula is shown in Figure 9.4. There is one ketogroup at C8 and three hydroxyl groups at C3, C7, and C15. It has been shown that the epoxy groups at positions C12 and C13 in the chemical structure of DON are the main toxic groups, which can inhibit protein biosynthesis, and the C3 hydroxyl group is related to its toxicity. The DON pollution of Chinese crops showed a trend of high pollution rate and low pollution degree. In terms of acute toxicity, ingestion of food contaminated by DON may cause nausea, vomiting, diarrhea, headache, dizziness, abdominal pain, and fever (Sobrova *et al.* 2010). In terms of chronic toxicity, it has been

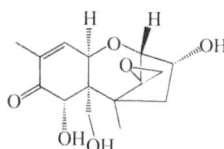

Figure 9.4 Chemical structures of DON.

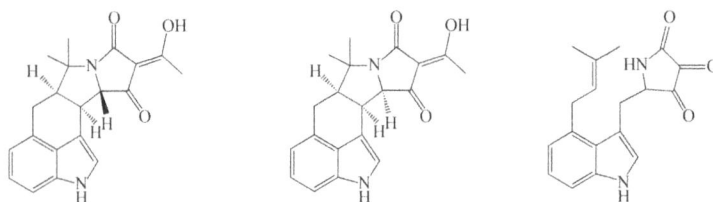

Figure 9.5 Chemical structures of CPA and its derivatives.

found in animal experimental studies that long-term exposure to low-dose DON-contaminated food can damage intestinal morphology and function (Pestka 2010). Epidemiological studies in animal models have been widely used to establish limit standards around the world, and DON has been classified as a Group III carcinogen by IARC (Ostry *et al.* 2017). The U.S. FDA has set the limit of DON in finished grains as 1 mg/kg.

9.1.5 Cyclopianic Acid

Cyclopiazonic acid (CPA) is a toxic metabolite produced by some fungi of *Aspergillus* and *Penicillium* (Lalitha Rao *et al.* 1985). Discovered in 1973, CPA was isolated from a liquid culture medium of *P. cyclopium* and its structure was identified (Sweeney *et al.* 1998). In addition to the two fungi, *A. flavus, A. Aspergillus parasiticus, A. oryzae, A. tamarii, P. camembertii,* and other fungi can also produce CPA. Studies have shown that CPA is synthesized from one molecule of L-tryptophan, one molecule of mevalonic acid, and two molecules of acetic acid, and its precursor compound is β-CPA. Iso-α-CPA is a new CPA analog discovered in 2009 that is presumed to be derived from D-tryptophan (Figure 9.5). CPA is a neurotoxin that acts as a specific sarcoplasmic reticulum/endoplasmic reticulum Ca^{2+} pump inhibitor to induce nervous system disorders (Selli *et al.* 2016). CPA interferes with the differentiation of human monocytes into macrophages and has immunosuppressive activity in humans (Hymery *et al.* 2014). CPA can also cause Koudua poisoning, with symptoms mainly manifested as dizziness and vomiting (Lalitha Rao *et al.* 1985). Moreover, since both CPA and AFB_1 can be produced by *Aspergillus parasiticus*, mixed contamination of the two mycotoxins often occurs (Cole 1986).

9.2 RESEARCH PROGRESS OF INTELLIGENT POCT METHOD

9.2.1 Optical Intelligent POCT Method

Optical sensing detection technology is an analytical method of molecular recognition detection using optical signals, which plays an important role in the field of analysis and detection. Optical detection can provide information such as concentration, molecular structure, microstructure, and binding dynamics of the object to be measured. It relies on detecting changes in optical signals and makes it highly compatible with various spectral measurements by detecting changes in wavelength, phase, time, intensity, and polarity of light. For example, fluorescence, chemiluminescence, near infrared spectroscopy, hyperspectral technology, surface plasmon resonance, surface-enhanced Raman scattering, and so on. It is widely used in disease marker detection, water quality detection, food safety and dangerous goods detection, and other fields.

9.2.1.1 Fluorescence

Fluorescence (FL) refers to the emitted light, for example, when a substance is irradiated by the light of a certain wavelength, absorbs light energy, enters an excited state, and then returns to the ground state. Fluorescent molecules can be used as labeling materials and have a wide range of applications in smart sensing fields such as biochemistry and medicine. Hu et al. developed a typical AIEgen (bromo-modified tetraphenylene) with high luminous efficiency, uniform size, and good biocompatibility (Hu *et al.* 2021). AIEgen nanospheres were coupled with AFB1 and CPA monoclonal antibodies, respectively, and an AIEgen synchronous intelliSensing method for AFB1 and CPA was established. The experimental results show that the lowest detection limits (LOD) of AFB_1 and CPA are 0.003 ng/mL and 0.01 ng/mL, respectively, the linear ranges are 0.05–1.2 ng/mL and 0.8–50.0 ng/mL, respectively, and the recovery ranges from 90.3% to 110.0%. The method has excellent specificity, repeatability (coefficient of variation CV less than 4.6%), and reproducibility (CV less than 6.7%). The AFLB is stable for 180 days at 4°C. This provided a new method for highly intelliSensing typical harmful metabolites of *Aspergillus* flavus in grain and oil. Hu et al. also developed silica template with uniform particle size and labyrinth inside (Hu 2021). The loading method for QDs-loaded silica nanosphere was established, and the LQDB with high luminous efficiency was constructed. QDs are efficiently loaded by the large specific surface area of the labyrinth-like inner wall. Intelligent POCT equipment and apps are designed. The synchronous pretreatment method of mycotoxins in grain and oil was established. The time for detecting a single sample is shortened, and the shortest time for pretreatment and detection is 7 minutes. The LOD of AFB_1 and ZEN in corn and urine was as low as 0.002 ng/mL and 0.02 ng/g, respectively. The CV values of repeatability and reproducibility are less than 5%, and the specificity is satisfied. The stability test shows that the LQDB test strip can be stored at 4°C for 180 days, and the recoveries range from 97.3 to 108.8%. The comparison of corn and urine samples showed that the result of this method was consistent with that of the UPLC-MS/MS method, which showed that the LQDB strip had good accuracy, which provided a new idea for the research of multi-toxin intelliSensing and highly sensitive fluorescent labeling materials.

9.2.1.2 Chemiluminescence

Chemiluminescence (CL) is a kind of light radiation phenomenon in a chemical reaction. According to the linear relationship between the concentration of the substance and the chemiluminescence intensity under certain conditions, the content of the substance can be analyzed. Zangheri et al. have proposed a chemiluminescent detection-based biosensor that can be used for intelligent sensing of OTA (Zangheri *et al.* 2021). The OTA and horseradish peroxidase OTA conjugate (HRP-OTA) in the sample were competitively reacted with OTA antibody on the cellulose nitrate strip, and then chemiluminescence reaction was generated after the addition of luminol /H_2O_2. The smartphone camera was used as the photodetector to perform intelligent sensing for OTA in the sample. Shahvar et al. developed an intelligent sensing detection method for morphine detection based on chemiluminescence (Shahvar *et al.* 2018). They dripped potassium permanganate onto a thin layer of samples, captured the chemiluminescence via a smartphone, and calculated the results. The detection limit can reach 0.5 mg/L. The CL intelligent method is simple and low cost, which is suitable for on-the-spot detection of morphine.

9.2.1.3 Near Infrared Spectrum

Near-infrared spectroscopy (NIR spectroscopy) studies the interaction of matter and light in the near-infrared region of the electromagnetic spectrum between 750 nm and 2,500 nm (Zeng *et al.* 2021). When infrared light interacts with molecules in the sample, the bonds of those molecules vibrate at different frequencies depending on the type of bond. In the near infrared region, C-H, N-H, and O-H vibrational bonds are most prevalent and determine the spectral shape of a given sample. Since it analyzes functional groups, NIR cannot only analyze information about the chemical composition of a substance, but also provide information about its function. Khan et al., using NIR, developed a prediction model using partial least squares (PLS) regression to analyze and predict the parameters that have a significant impact on the quality of milk powder, such as fine particle size fraction, dispersion, and packing density of various milk powder samples, with an accuracy of 88%–90% (Khan *et al.* 2021). As a nondestructive technique, NIR technology also shows good potential in adulteration identification. For example, Zaukuu et al. used NIR's LDA model to identify and quantify different degrees of chili powder mixed with corn meal, and the identification accuracy and prediction accuracy of corn meal adulteration were 95.55% and 95.02% (Zaukuu *et al.* 2019). In addition, Kamboj et al. used NIR analysis to establish a stoichiometric model, and used multivariate analysis to analyze the content of sugar in milk qualitatively and quantitatively. The correlation coefficient of the regression model is higher than 0.9, and the root mean square error of the verification is 0.04. Therefore, NIR technology can provide the dairy industry with a simple, efficient, fast, green, and non-destructive technology for detecting and quantifying milk adulteration (Kamboj *et al.* 2020).

9.2.1.4 Hyperspectral Imaging

Hyperspectral image (HSI) combines spectral and imaging techniques in an imaging mode to obtain spectral and spatial information simultaneously (Senthilkumar *et al.* 2015). HSI can quickly and nondestructively provide valuable information about the external physical and internal chemical properties of agricultural and food products. It has a wide range of applications in quality and safety assessment of different agricultural products and food products such as fruits

(Chen *et al.* 2015), vegetables (Trong *et al.* 2011), poultry carcasses (Nakariyakul *et al.* 2007), grains (Fox *et al.* 2014), dairy products (Forchetti *et al.* 2017), etc. First of all, HIS can be used for adulteration identification analysis. Fu et al. used near-infrared hyperspectral imaging to distinguish wheat flour from low levels (< 5.0%) talcum powder and benzoyl peroxide (BPO) granules. The spectra of wheat talc mixture samples were analyzed by the first derivative band difference method, and the spectra of wheat BPO mixture samples were analyzed by spectral correlation measurement and band ratio method, and adulterants were successfully identified from wheat talc/BPO mixture (Fu *et al.* 2020). Secondly, HIS can be used for quality identification. Liu et al. collected HIS images of healthy, damaged, and moldy peanuts of 1,066 peanut samples by means of spectrograph for comparison, and used peanut recognition index as data feature pre-extraction and fusion into HIS images. After feature pre-extraction, peanuts were identified by deep learning technology. This method can improve the accuracy of peanut identification (Liu et al. 2020b). Thirdly, HIS can identify the pollution level of heavy metals. Sun et al. use the HIS system to obtain HIS images of lettuce leaves, and then use ENVI software to extract hyperspectral data from all samples. Based on the full spectral data and the spectral data selected by CARS, and using the established SVM and PLS-DA discriminative models, a method for identifying lead pollution levels in vegetables based on HIS technology was established, which was successfully applied to identify the degree of lead pollution in lettuce leaves (Sun *et al.* 2021).

9.2.1.5 Surface Plasmon Resonance Technology

Surface plasmon resonance (SPR) technology is based on optical phenomena occurring on conductive films at the interface of media with different refractive indices. SPR has the advantages of no labeling of samples, no purification of analyte, real-time monitoring and reflection of a dynamic process, high sensitivity, and low cost, and can avoid false positive signals caused by labeling materials. SPR is an efficient method to detect surface affinity interaction, and has been widely used in biosensors and chemical sensors and has become a powerful analytical technology for biological, medical diagnostics, and food and animal feed risk assessment monitoring. SPR technology has been applied to the detection of mycotoxin in grain and oil. Xu et al. synthesized gold nanoparticles of about 25 nm, coated them on the cross section of the end of optical fiber, then modified ZEN nucleic acid aptamers, and developed a low-cost, portable, and reusable fiber-based ZEN local surface plasmon resonance biosensor, which can be used for the detection of ZEN. The minimum limit of detection (LOD) was 0.102 ng/mL (Xu *et al.* 2021). Wei using self-assembly monomolecular membrane preparation of SPR sensor chips, established at the same time detecting AFB1 in the corn and wheat, OTA, ZEN, and DON SPR method, LOD range of 0.59~3.26 ng/mL, relative standard deviation less than 10%. In addition, the results obtained by this SPR method were consistent with those obtained by the HPLC-MS/MS method, indicating that this SPR method can be applied to grain and oil safety detection (Wei *et al.* 2019). Rehmat et al. used the spin coating technique to coat chitosan and carboxymethyl chitosan on SPR chips to analyze OTA in a complex coffee matrix, and developed a fast and sensitive immunoassay method for competitive inhibition of SPR, with LODs up to 3.8 ng/mL, respectively. A highly sensitive and low-cost immunoassay was achieved (Rehmat *et al.* 2019).

9.2.1.6 Surface Enhanced Raman Spectroscopy

Surface enhanced Raman scattering (SERS) is the signal amplification generated by electromagnetic interaction between light and metal nanoparticles, which is achieved by a laser field generated by excitation of plasma resonance. SERS combines Raman spectroscopy and nanotechnology, and is a spectral method based on light scattering. The signal is generated by an inelastic collision between a sample and incident photon emitted by a monochromatic light source (such as laser beam) (Yan *et al.* 2021). It has been widely used in food detection and cancer diagnosis as an effective tool for detecting interface properties and molecular interactions, and characterizing surface molecular adsorption behavior and molecular structure. The advantages of SERS include ultra-sensitive detection, fast turnover, on-site sampling, on-site monitoring, low cost, portable sensor, and suitable for large-scale screening. Combining biological cognitive events with SERS can significantly improve the analytical performance of such a method, but also increase the complexity and cost of the method. Duan reported a method using interface assembly of plasma nanoparticles and current displacement reaction to prepare Ag-gold bimetallic nanowire arrays as SERS substrates for the detection of Fumei Shuang in juice samples and melamine in milk samples, with LODs up to 1 and 10 nM, respectively (Duan *et al.* 2021). Pan et al. developed a rapid SERS method for the rapid detection of acetamidine residue on cabbage leaves based on the synergistic combination of AuNPs stability and AgNPs optical properties. The detection limit of this method was 0.14 mg/kg, which was lower than the minimum residue limit set in China. This method is simple in sample pretreatment and has great potential in field and nondestructive testing (Pan *et al.* 2021). Wang et al. successfully developed a high-sensitivity SERS method for detecting nitrite, adopted the signal conversion strategy based on the one-step nitrite-mertan reaction between nitrite and 2-thiobarbituric acid to improve the sensitivity of SERS, and used the portable Raman spectrometer system and self-made gold nanoparticles as the enhanced matrix. The resultant S-nitroso-mercaptan compound was detected in the solution (Wang *et al.* 2021).

9.2.2 Electrochemical Immune Intelligent POCT Method

The principle of electrochemical immune intelligent POCT is to use chemical identification system to convert the information of the object to be tested into electrical signals that can be detected and output. It has the advantages of simple equipment, low price, easy-to-realize automation, high sensitivity, wide measuring range, instant, economic, and so on. Wu et al. constructed an electrochemical aptamer sensor based on CdTe/CdS/ZnS quantum dots and Luminol for intelligent sensing of AFB_1 in food with LOD of 0.43~0.12 p mol/L and linear range of 5.0~10 nM (Wu *et al.* 2017). Kudr et al. used an inkjet printed electrochemical reduction GO microelectrode to detect HT-2 toxins with a LOD of 1.6 ng/mL and a linear range of 6.3 to 100 ng/mL (Kudr *et al.* 2020). Malvano et al. developed an unlabeled electrochemical impedance biosensor for detecting *Escherichia coli* O157:H7 in food. By adding an electron transfer medium to activate ferrocene, the electrical performance of the system was improved and the sensitivity of the electrochemical impedance biosensor to *Escherichia coli* O157:H7 was improved. LOD was 3 CFU/mL (Malvano *et al.* 2018).

9.2.3 Magnetic Intelligent POCT Method

Nuclear magnetic resonance (NMR) measures the magnetic properties of a spin, related to the physical or chemical properties of the object under test. NMR is a physical process in which atomic nuclei with non-zero magnetic moments absorb radiation of a certain frequency in resonance under the action of an external magnetic field. The detectors detect and receive NMR signals emitted as electromagnetic radiation, which can be sent to a computer and converted into images through data processing. NMR technology can be used for water analysis. Bertam et al. used low-pulse field NMR to study the changes in water activity and distribution of white muscle and black dried meat during freezing storage. The results showed that with the increase in freezing time, the content of free water in pork also increased significantly, and the degeneration and structural changes of protein were closely related to the changes in water activity in meat. White muscle was more prone to water migration and deterioration under freezing conditions than black dry meat (Bertram *et al.* 2007). Cornillon et al. used NMR to study the water mobility and distribution in corn chips, chocolate cookies, soft caramel, corn starch/water, and other low-moisture cereal and biscuit systems, and pointed out that various chemical interactions in the system lead to changes in water mobility (Cornillon *et al.* 2000). MacMillan et al. determined oil and water content in French fries using magnetic resonance imaging (Macmillan *et al.* 2008).

9.2.4 Bionics Intelligent POCT Method
9.2.4.1 Olfactory Intelligent POCT Method

Electronic nose is a kind of analytical equipment that is used to detect and identify odor mixture quickly. It can imitate the working principle of human smell and carry out olfactory intelligent perception of the measured object. Specific chemical sensors are used in the device, which produce a characteristic odor distribution based on the interaction with the gas mixture and identify the composition of the mixture by comparing the resulting odor curve with the odor standard (Aouadi *et al.* 2020). The electronic nose is similar to the human nose in that it is based on the same operating principle, in which the volatile components of the study sample are analyzed by chemical sensors that mimic the olfactory cells in the nose and then the signals are sent to a data recognition system that mimics brain function. Triyana et al. reported on a portable electronic nose based on low-cost dynamic headspace and metal oxide gas sensor arrays for identification analysis of vegetable oils (sunflower, grape seed) and animal fats (chicken, lamb, pig). This portable electronic nose, consisting of ten field-effect tube-type sensors, has been used to evaluate lipid oxidation in various edible oils (olive oil, peanut oil, soybean, rapeseed, camellia oil, corn, sunflower, flaxseed, and walnut) with good results. The recognition rate of calibration and verification models was 100%

(Triyana *et al.* 2015). In recent years, electronic noses have also been used to detect and identify moldy (*Penicillium expansus, Aspergillus Niger*) apples, with predictive accuracy between 72% and 96.3% (Jia et al. 2019). In addition, the electronic nose can also be used to monitor the damage and ripening degree of lychee fruits (Xu *et al.*, 2021) and the damage degree of yellow peach fruits (Yang *et al.* 2020), with an accuracy of 93.33%.

9.2.4.2 Taste Intelligent POCT Method

Electronic tongue, also known as artificial tongue or taste sensor, is mainly used for taste classification of various chemical substances in liquid samples, which can imitate human taste for taste intelligent perception. Electronic tongues can be used to apply fingerprinting, which compares the profile of a mixture with that of a standard substance, to identify, classify, and analyze multicomponent mixtures in a qualitative and quantitative manner (Ciosek *et al.* 2007; Deisingh *et al.* 2004). As an intelligent perception technology close to human taste perception ability, the electronic tongue has completely changed the traditional food evaluation system, and can also solve the problem of food adulteration. Dias et al.'s identification model of goat, cow, and goat/cow mixture using a potential electronic tongue can detect adulteration of goat milk and cow milk. The results showed that the overall classification recognition and prediction abilities of the electronic tongue were 97% and 87%, respectively (Dias *et al.* 2009). In another study, voltammetric electronic tongues were used to detect antibiotic residues in milk, allowing for quantitative detection of antibiotics. Wei et al. investigated the ability to monitor quality attributes of yogurt samples during fermentation, ripening, and storage by deploying a voltammetric electronic tongue. The results suggest that the electronic tongue can predict the sensory properties of cheddar cheese during a 12-month storage period (Wei *et al.* 2011).

9.2.4.3 Visual Intelligent POCT Method

Machine vision perception is an intelligent perception technology that uses cameras and computers to process images and imitate human eyes for image recognition, tracking, and measurement. Visual intelligent perception systems can achieve high quality, high efficiency, complete automation, and can replace the manual inspection method, so as to eliminate the error and inconsistency of human judgment results, but also reduce the tedious operation of manual inspection. Some researchers have explored the scope of applying machine vision systems to grain quality detection and classification. Visual intelligent perception has been applied to the detection of meat and meat products (Li *et al.* 2001), fish (Mery *et al.* 2011), fruits and vegetables (Cubero *et al.* 2011), and cooked food (Pedreschi *et al.* 2004).

9.2.4.4 Tactile Intelligent POCT Method

Tactile intelligent perception is the intelligent perception of taste, brittleness, texture, thickness, and viscosity of the measured object. So far, various types of tactile intelligent perception detection devices have been developed (Kuppuswamy *et al.* 2020; Shimonomura 2019). Tactile intellisthesia detection devices typically consist of a camera and tactile skin that converts physical contact into light signals that can be captured by the camera, depending on how the contacts are converted into optical information. The marker displacement method is one of the most widely used research methods, in which cameras measure the displacement of visual markers embedded in soft materials (Hofer *et al.* 2021). This method is suitable for detecting the magnitude and direction of forces. In the reflective film method, the camera captures small deformation on the surface of a flexible material sheet covered with a reflective film (Yuan *et al.* 2017). Because small, irregular changes in the sensor's surface are accentuated by light applied from the side, even very small irregularities in the contact area, such as human fingerprints, can be visualized. The haptic intelligent perception technology can be used to obtain a three-dimensional touch feeling that is closer to human feeling and measure the characteristics of the substance that cannot be measured by traditional methods.

9.2.5 Other Intelligent POCT Method

Terahertz spectrum refers to electromagnetic waves with frequencies ranging from 0.1 to 10 THz (wavelength 30 μm to 3 mm). Studies have shown that the vibrational and rotational energy levels of most biomolecules (DNA, proteins, and amino acids) are in this band (Bernier *et al.* 2013; Redo-Sanchez *et al.* 2013); therefore, terahertz technology has intelligent sensing ability to detect and distinguish biological samples. Terahertz spectroscopy is considered to be the most promising detection method for its low energy, high resolution, and high penetration, and has shown extensive application potential in many areas of food process monitoring and quality control, such as the identification of transgenic cotton, rice, and soybean seeds (Liu *et al.* 2016a; Liu *et al.* 2016b). For the detection of melamine in food (Baek *et al.* 2014), Liu et al. quickly determined AFB1 in soybean oil by terahertz spectroscopy with LOD of 2 μg/kg, and the accuracy was above 90% (Liu et al. 2020a).

Biological speckle is an optical and non-destructive intelligent sensing technology used to analyze biological materials. When the target object is irradiated by coherent light such as a laser, biological speckle phenomenon will occur (Zdunek *et al.* 2014). The wide range of interactions between coherent light and complex biological materials produce changes in speckle patterns (bright and dark areas) that are captured by the camera and transmitted to the processing system. There are two ways to measure biological speckle activity: backscatter and forward scatter speckle. In a backscatter device, the laser reaches the surface of the sample and the reflected light is captured. In a forward scatter device, the laser passes through the sample and reaches the camera. At present, some agricultural studies have involved biospeckle intelligent sensing applications, such as animal reproduction and parasite monitoring (Rabal and Braga 2010), detection of wheat seed germination process and simulation of germination damage (Sutton *et al.* 2017), and monitoring of food emulsions (Silva, 2010). The meat quality was analyzed (Amaral *et al.* 2013), and the biospeckle activity was correlated with the chlorophyll content of apples (Zdunek *et al.* 2012). Biological speckle can provide knowledge about the biological and physical properties of tissues. The method is fast, simple, and economical, and has a wide range of applications in food quality and safety.

9.3 CONCLUSION

In this chapter, harm of typical mycotoxins and recent research studies of the intelligent POCT method are discussed. These research works demonstrate that intelligent POCT provides rapid and accurate detection and analysis for a variety of analytes in food quality monitoring.

More importantly, intelligent POCT is portable, inexpensive, and easy to use compared with conventional detection devices, offering sustainable detection technology for resource-limited areas of the world.

REFERENCES

Amaral, I. C., Braga, R. A. Jr., Ramos, E. M., Souza Ramos, A. L., Rezende Roxael, E. A. 2013. Application of biospeckle laser technique for determining biological phenomena related to beef aging. *Journal of Food Engineering* 119:135–139.

Aouadi, B., Zaukuu, J.-L.Z., Vitalis, F., Bodor, Z., Feher, O., Gillay, Z., Bazar, G., Kovacs, Z. 2020. Historical evolution and food control achievements of near infrared spectroscopy, electronic nose, and electronic tongue-critical overview. *Sensors* 20:5479.

Baek, S. H., Lim, H. B., Chun, H. S. 2014. Detection of melamine in foods using terahertz time-domain spectroscopy. *Journal of Agricultural and Food Chemistry* 62:5403–5407.

Bedard, L. L., Massey, T. E. 2006. Aflatoxin B-1-induced DNA damage and its repair. *Cancer Letters* 241:174–183.

Bernier, M., Garet, F., Coutaz, J. L. 2013. Precise determination of the refractive index of samples showing low transmission bands by THz time-domain spectroscopy. *Ieee Transactions on Terahertz Science and Technology* 3:295–301.

Bertram, H. C., Andersen, R. H., Andersen, H. J. 2007. Development in myofibrillar water distribution of two pork qualities during 10-month freezer storage. *Meat Science* 75:128–133.

Bui-Klimke, T. R., Wu, F. 2015. Ochratoxin A and human health risk: A review of the evidence. *Critical Reviews In Food Science And Nutrition* 55:1860–1869.

Chen, S., Zhang, F., Ning, J., Liu, X., Zhang, Z., Yang, S. 2015. Predicting the anthocyanin content of wine grapes by NIR hyperspectral imaging. *Food Chemistry* 172:788–793.

Ciosek, P., Wroblewski, W. 2007. Sensor arrays for liquid sensing - electronic tongue systems. *Analyst* 132:963–978.

Cole, B. J. 1986. The social behavior of Leptothorax allardycei (Hymenoptera, Formicidae): Time budgets and the evolution of worker reproduction. *Behavioral Ecology and Sociobiology* 18:165–173.

Cornillon, P., Salim, L. C. 2000. Characterization of water mobility and distribution in low- and intermediate-moisture food systems. *Magnetic Resonance Imaging* 18:335–341.

Cubero, S., Aleixos, N., Molto, E., Gomez-Sanchis, J., Blasco, J. 2011. Advances in machine vision applications for automatic inspection and quality evaluation of fruits and vegetables (vol 4, pg 487, 2011). *Food and Bioprocess Technology* 4:829–830.

da Silva, E. Jr., Teixeira da Silva, E. R., Muramatsu, M., da Silva Lannes, S. C. 2010. Transient process in ice creams evaluated by laser speckles. *Food Research International* 43:1470–1475.

Deisingh, A. K., Stone, D. C., Thompson, M. 2004. Applications of electronic noses and tongues in food analysis. *International Journal of Food Science and Technology* 39:587–604.

Dias, L. A., Peres, A. M., Veloso, A. C. A., Reis, F. S., Vilas-Boas, M., Machado, A. A. S. C. 2009. An electronic tongue taste evaluation: Identification of goat milk adulteration with bovine milk. *Sensors and Actuators B-Chemical* 136:209–217.

Duan, B., Hou, S., Wang, P., Chen, Y., Xiong, Q., Das, P., Duan, H. 2021. Chemical processing of interfacially assembled metal nanowires for surface-enhanced Raman scattering detection of food contaminants. *Journal of Raman Spectroscopy* 52:532–540.

Forchetti, D. A. P., Poppi, R. J. 2017. Use of NIR hyperspectral imaging and multivariate curve resolution (MCR) for detection and quantification of adulterants in milk powder. *Lwt-Food Science And Technology* 76:337–343.

Fox, G., Manley, M. 2014. Applications of single kernel conventional and hyperspectral imaging near infrared spectroscopy in cereals. *Journal Of The Science of Food And Agriculture* 94:174–179.

Fu, X., Chen, J., Fu, F., Wu, C. 2020. Discrimination of talcum powder and benzoyl peroxide in wheat flour by near-infrared hyperspectral imaging. *Biosystems Engineering* 190:120–130.

Hofer, M., Sferrazza, C., D'Andrea, R. 2021. A vision-based sensing approach for a spherical soft robotic arm. *Frontiers in Robotics And Ai* 8:630935.

Hu X., Huang L., Wang S., Ahmed R., Li P., Demirci U., Zhang Z. 2023, Color-selective labyrinth-like quantum dot nanobeads enable point-of-care dual assay of Mycotoxins. *Sensors And Actuators B-Chemical* 376: 132956.

Hu, X., Zhang, P., Wang, D., Jiang, J., Chen, X., Liu, Y., Zhang, Z., Tang, B. Z., Li, P. 2021 AIEgens enabled ultrasensitive point-of-care test for multiple targets of food safety: Aflatoxin B_1 and cyclopiazonic acid as an example. *Biosensors and Bioelectronics* 182: 113188.

Hymery, N., Masson, F., Barbier, G., Coton, E. 2014. Cytotoxicity and immunotoxicity of cyclopiazonic acid on human cells. *Toxicology In Vitro* 28:940–947.

Ismail, A., Goncalves, B. L., de Neff, D. V., Ponzilacqua, B., Coppa, C. F. S. C., Hintzsche, H., Sajid, M., Cruz, A. G., Corassin, C. H., Oliveira, C. A. F. 2018. Aflatoxin in foodstuffs: Occurrence and recent advances in decontamination. *Food Research International* 113:74–85.

Jia, W., Liang, G., Tian, H., Sun, J., Wan, C. 2019. Electronic nose-based technique for rapid detection and recognition of moldy apples. *Sensors* 19:1526.

Kamboj, U., Kaushal, N., Mishra, S., Munjal, N. 2020. Application of selective near infrared spectroscopy for qualitative and quantitative prediction of water adulteration in milk. *Materials Today-Proceedings* 24:2449–2456.

Khan, A., Munir, M. T., Yu, W., Young, B. R. 2021. Near-infrared spectroscopy and data analysis for predicting milk powder quality attributes. *International Journal of Dairy Technology* 74:235–245.

Khlangwiset, P., Shephard, G. S., Wu, F. 2011. Aflatoxins and growth impairment: A review. *Critical Reviews in Toxicology* 41:740–755.

Kudr, J., Zhao, L., Nguyen, E. P., Arola, H., Nevanen, T. K., Adam, V., Zitka, O., Merkoci, A. 2020. Inkjet-printed electrochemically reduced graphene oxide microelectrode as a platform for HT-2 mycotoxin immunoenzymatic biosensing. *Biosensors & Bioelectronics* 156:112109.

Kuppuswamy, N., Alspach, A., Uttamchandani, A., Creasey, S., Ikeda, T., Tedrake, R., Ieee, 2020. Soft-bubble grippers for robust and perceptive manipulation. *Ieee/Rsj International Conference on Intelligent Robots and Systems* 9917–9924

Lalitha Rao, B., Husain, A. 1985. Presence of cyclopiazonic acid in kodo millet (Paspalum scrobiculatum) causing "kodua poisoning" in man and its production by associated fungi. *Mycopathologia* 89:177–180.

Lancaster, M. C., Jenkins, F. P., Philp, J. M. 1961. Toxicity associated with certain samples of groundnuts. *Nature, London* 192:1095–1096.

Li, J., Tan, J., Shatadal, P. 2001. Classification of tough and tender beef by image texture analysis. *Meat Science* 57:341–346.

Li, P., Zhang, Q., Zhang, W., Zhang, J., Chen, X., Jiang, J., Xie, L., Zhang, D. 2009. Development of a class-specific monoclonal antibody-based ELISA for aflatoxins in peanut. *Food Chemistry* 115:313–317.

Liu, W., Liu, C. H., Chen, F., Yang, J. B., Zheng, L. 2016a. Discrimination of transgenic soybean seeds by terahertz spectroscopy. *Scientific Reports* 6:35799.

Liu, W., Liu, C. H., Hu, X. H., Yang, J. B., Zheng, L. 2016b. Application of terahertz spectroscopy imaging for discrimination of transgenic rice seeds with chemometrics. *Food Chemistry* 210:415–421.

Liu, W., Zhao, P. G., Wu, C. S., Liu, C. H., Yang, J. B., Zheng, L. 2020a. Rapid determination of aflatoxin B-1 concentration in soybean oil using terahertz spectroscopy with chemometric methods. *Food Chemistry* 293:213–219.

Liu, Z., Jiang, J., Qiao, X., Qi, X., Pan, Y., Pan, X. 2020b. Using convolution neural network and hyperspectral image to identify moldy peanut kernels. *LWT* 132:109815.

MacMillan, B., Hickey, H., Newling, B., Ramesh, M., Balcom, B. 2008. Magnetic resonance measurements of French fries to determine spatially resolved oil and water content. *Food Research International* 41:676–681.

Mally, A., Solfrizzo, M., Degen, G. H. 2016. Biomonitoring of the mycotoxin Zearalenone: current state-of-the-art and application to human exposure assessment. *Archives of Toxicology* 90:1281–1292.

Malvano, F., Pilloton, R., Albanese, D. 2018. Sensitive detection of escherichia coli O157:H7 in food products by impedimetric immunosensors. *Sensors* 18:2168.

Marin, S., Ramos, A. J., Cano-Sancho, G., Sanchis, V. 2013. Mycotoxins: Occurrence, toxicology, and exposure assessment. *Food and Chemical Toxicology* 60:218–237.

Mery, D., Lillo, I., Loebel, H., Riffo, V., Soto, A., Cipriano, A., Aguilera, J. M. 2011. Automated fish bone detection using X-ray imaging. *Journal of Food Engineering* 105:485–492.

Morooka, N., Uratsuji, N., Yoshizawa, T., Yamamoto, H. 1972. Studies on the toxic substances in barley infected with Fusarium spp. *Journal – Food Hygienic Society of Japan* 13:368–375.

Mwanda, O. W., Otieno, C. F., Omonge, E. 2005. Acute aflatoxicosis: Case report. *East African Medical Journal* 82:320–324.

Nakariyakul, S., Casasent, D. 2007. Fusion algorithm for poultry skin tumor detection using hyperspectral data. *Applied Optics* 46:357–364.

Ostry, V., Malir, F., Toman, J., Grosse, Y. 2017. Mycotoxins as human carcinogens-the IARC Monographs classification. *Mycotoxin Research* 33:65–73.

Pan, T.-t., Guo, W., Lu, P., Hu, D. 2021. In situ and rapid determination of acetamiprid residue on cabbage leaf using surface-enhanced Raman scattering. *Journal of The Science of Food and Agriculture* 101:3595–3604.

Pedreschi, F., Mery, D., Mendoza, F., Aguilera, J. M. 2004. Classification of potato chips using pattern recognition. *Journal of Food Science* 69:E264–E270.

Pestka, J. J. 2010. Deoxynivalenol: mechanisms of action, human exposure, and toxicological relevance. *Archives of Toxicology* 84:663–679.

Pinotti, L., Ottoboni, M., Giromini, C., Dell'Orto, V., Cheli, F. 2016. Mycotoxin contamination in the EU feed supply chain: A focus on cereal byproducts. *Toxins* 8:45.

Rabal, H. J., & Braga Jr, R. A. (Eds.). 2018. *Dynamic laser speckle and applications.* CRC press.

Redo-Sanchez, A., Laman, N., Schulkin, GelSight: High-Resolution Robot Tactile Sensors for Estimating Geometry and Force B., Tongue, T. 2013. Review of terahertz technology readiness assessment and applications. *Journal of Infrared Millimeter and Terahertz Waves* 34:500–518.

Rehmat, Z., Mohammed, W. S., Sadiq, M. B., Somarapalli, M., Anal, A. K. 2019. Ochratoxin A detection in coffee by competitive inhibition assay using chitosan-based surface plasmon resonance compact system. *Colloids and Surfaces B-Biointerfaces* 174:569–574.

Ringot, D., Chango, A., Schneider, Y. J., Larondelle, Y. 2006. Toxicokinetics and toxicodynamics of ochratoxin A, an update. *Chemico-Biological Interactions* 159:18–46.

Ryu, D., Hanna, M. A., Bullerman, L. B. 1999. Stability of zearalenone during extrusion of corn grits. *Journal of Food Protection* 62:1482–1484.

Selli, C., Tosun, M. 2016. Effects of cyclopiazonic acid and dexamethasone on serotonin-induced calcium responses in vascular smooth muscle cells. *Journal Of Physiology And Biochemistry* 72:245–253.

Senthilkumar, T., Jayas, D. S., White, N. D. G. 2015. Detection of different stages of fungal infection in stored canola using near-infrared hyperspectral imaging. *Journal of Stored Products Research* 63:80–88.

Shahvar, A., Saraji, M., Shamsaei, D. 2018. Smartphone-based chemiluminescence sensing for TLC imaging. *Sensors And Actuators B-Chemical* 255:891–894.

Shimonomura, K. 2019. Tactile image sensors employing camera: A review. *Sensors* 19:3933.

Sobrova, P., Adam, V., Vasatkova, A., Beklova, M., Zeman, L., Kizek, R. 2010. Deoxynivalenol and its toxicity. *Interdisciplinary Toxicology* 3:94–99.

Stob, M., Baldwin, R. S., Tuite, J., Andrews, F. N., Gillette, K. G. 1962. Isolation of an anabolic, uterotrophic compound from corn infected with Gibberella zeae. *Nature* 196:1318-1318.

Studer-Rohr, I., Schlatter, J., Dietrich, D. R. 2000. Kinetic parameters and intraindividual fluctuations of ochratoxin A plasma levels in humans. *Archives Of Toxicology* 74:499–510.

Sun, J., Cao, Y., Zhou, X., Wu, M., Sun, Y., Hu, Y. 2021. Detection for lead pollution level of lettuce leaves based on deep belief network combined with hyperspectral image technology. *Journal of Food Safety* 41:e12866.

Sutton, D. B., Punja, Z. K. 2017. Investigating biospeckle laser analysis as a diagnostic method to assess sprouting damage in wheat seeds. *Computers and Electronics in Agriculture* 141:238–247.

Sweeney, M. J., Dobson, A. D. W. 1998. Mycotoxin production by Aspergillus, Fusarium and Penicillium species. *International Journal Of Food Microbiology* 43:141–158.

Tahir, N. I., Hussain, S., Javed, M., Rehman, H., Shahzady, T. G., Parveen, B., Ali, K. G. 2018. Nature of aflatoxins: Their extraction, analysis, and control. *Journal of Food Safety* 38:e12561.

Thomas, A. E., Coker, H. A. B., Odukoya, O. A., Isamah, G. K., Adepoju-Bello, A. 2003. Aflatoxin contamination of Arachis hypogaea (groundnuts) in Lagos area of Nigeria. *Bulletin of Environmental Contamination and Toxicology* 71:42–45.

Triyana, K., Subekti, M. T., Aji, P., Hidayat, S. N., Rohman, A. J. A. M. 2015. Development of electronic nose with low-cost dynamic headspace for classifying vegetable oils and animal fats. *Materials* 771:50–54.

Trong, N. N. D., Tsuta, M., Nicolai, B. M., De Baerdemaeker, J., Saeys, W. 2011. Prediction of optimal cooking time for boiled potatoes by hyperspectral imaging. *Journal of Food Engineering* 105:617–624.

van der Merwe, K. J., Steyn, P. S., Fourie, L., Scott, D. B., Theron, J. J. 1965. Ochratoxin A, a toxic metabolite produced by Aspergillus ochraceus Wilh. *Nature* 205:1112–1113.

Van Der Zijden, A. S. M., Koelensmid, W. A. A. B., Boldixgh, J., Barrett, C. B., Ord, W. O., Philp, J. 1962. Isolation in crystalline form of a toxin responsible for Turkey X disease. *Nature, London* 195:1060–1062.

Vesonder, R. F., Ciegler, A., Jensen, A. H. 1973. Isolation of the emetic principle from Fusarium-infected corn. *Applied Microbiology* 26:1008–1010.

Wang, P., Sun, Y., Li, X., Shan, J., Xu, Y., Li, G. 2021. One-step chemical reaction triggered surface enhanced Raman scattering signal conversion strategy for highly sensitive detection of nitrite. *Vibrational Spectroscopy* 113:103221.

Wannop, C. C. 1961. Turkey "X" disease. *Veterinary Record* 73:310–311.

Wei, T., Ren, P., Huang, L., Ouyang, Z., Wang, Z., Kong, X., Li, T., Yin, Y., Wu, Y., He, Q. 2019. Simultaneous detection of aflatoxin B1, ochratoxin A, zearalenone and deoxynivalenol in corn and wheat using surface plasmon resonance. *Food Chemistry* 300:125176.

Wei, Z., Wang, J. 2011. Detection of antibiotic residues in bovine milk by a voltammetric electronic tongue system. *Analytica Chimica Acta* 694:46–56.

Wu, L., Ding, F., Yin, W., Ma, J., Wang, B., Nie, A., Han, H. 2017. From Electrochemistry to electroluminescence: Development and application in a ratiometric aptasensor for aflatoxin B1. *Analytical Chemistry* 89:7578–7585.

Xu, Y., Xiong, M., Yan, H. 2021. A portable optical fiber biosensor for the detection of zearalenone based on the localized surface plasmon resonance. *Sensors and Actuators B-Chemical* 336:129752.

Yan, M., Li, H., Li, M., Cao, X., She, Y., Chen, Z. 2021. Advances in surface-enhanced raman scattering-based aptasensors for food safety detection. *Journal of Agricultural and Food Chemistry* 69:14049–14064.

Yang, R., Wang, Y.-M., Zhang, L., Zhao, Z.-M., Zhao, J., Peng, S.-Q. 2016. Prepubertal exposure to an oestrogenic mycotoxin zearalenone induces central precocious puberty in immature female rats through the mechanism of premature activation of hypothalamic kisspeptin-GPR54 signaling. *Molecular and Cellular Endocrinology* 437:62–74.

Yang, X., Chen, J., Jia, L., Yu, W., Wang, D., Wei, W., Li, S., Tian, S., Wu, D. 2020. Rapid and non-destructive detection of compression damage of yellow peach using an electronic nose and chemometrics. *Sensors* 20:1866.

Yoshizawa, T., Morooka, N. 1973. Deoxynivalenol and its monoacetate: New mycotoxins from Fusarium roseum and moldy barley. *Agricultural and Biological Chemistry* 37:2933–2934.

Yuan, W. Z., Dong, S. Y., Adelson, E. H. 2017. GelSight: High-resolution robot tactile sensors for estimating geometry and force. *Sensors* 17:2762.

Zangheri, M., Di Nardo, F., Calabria, D., Marchegiani, E., Anfossi, L., Guardigli, M., Mirasoli, M., Baggiani, C., Roda, A. 2021. Smartphone biosensor for point-of-need chemiluminescence detection of ochratoxin A in wine and coffee. *Analytica Chimica Acta* 1163: 338515.

Zaukuu, J. L. Z., Bodor, Z., Vitalis, F., Zsom-Muha, V., Kovacs, Z. 2019. Near infrared spectroscopy as a rapid method for detecting paprika powder adulteration with corn flour. *Acta Periodica Technologica* 50:346–352.

Zdunek, A., Adamiak, A., Pieczywek, P. M., Kurenda, A. 2014. The biospeckle method for the investigation of agricultural crops: A review. *Optics and Lasers in Engineering* 52:276–285.

Zdunek, A., Herppich, W. B. 2012. Relation of biospeckle activity with chlorophyll content in apples. *Postharvest Biology and Technology* 64:58–63.

Zeng, J., Guo, Y., Han, Y., Li, Z., Yang, Z., Chai, Q., Wang, W., Zhang, Y., Fu, C. 2021. A review of the discriminant analysis methods for food quality based on near-infrared spectroscopy and pattern recognition. *Molecules* 26:749.

Zinedine, A., Soriano, J. M., Molto, J. C., Manes, J. 2007. Review on the toxicity, occurrence, metabolism, detoxification, regulations and intake of zearalenone: An oestrogenic mycotoxin. *Food and Chemical Toxicology* 45:1–18.

10 Mycotoxin Degradation Methods in Food

*Jaqueline Garda-Buffon, Francine Kerstner de Oliveira, Juliane Lima da Silva,
Wesclen Vilar Nogueira, and Eliana Badiale-Furlong*

10.1 INTRODUCTION

Mycotoxins are a group of toxic compounds produced by filamentous fungi, such as some species of the genera *Aspergillus*, *Penicillium* and *Fusarium*, during their growth, when subjected to stress conditions. Several studies have been carried out to evaluate the mycotoxin occurrence in different foods, such as cereals (Khodaei *et al.*, 2021), nuts (Narváez *et al.*, 2020), coffee (Oueslati *et al.*, 2022), spices (Potortì *et al.*, 2020), oilseeds (Kholif *et al.*, 2021), fruits and juice (Ji *et al.*, 2022), beer (Schabo *et al.*, 2021), wines (Kochman *et al.*, 2021), salami (Parussolo *et al.*, 2019), dry-cured meat (Peromingo *et al.*, 2019) and eggs (Osaili *et al.*, 2022). These studies even reported concentrations higher than those established as maximum limits by regulatory agencies. Increased awareness of food safety and public health has influenced the global concern to prevent, minimize and control mycotoxins in food and feed. This mycotoxin occurrence in food proves the consumption risk of these contaminants by humans and animals since they are detected frequently in most different food types. Some of the effects of mycotoxin contamination include immune and neurological system damage, gastrointestinal and kidney disorders and even cancer (Luz *et al.*, 2022). In addition, mycotoxin contamination may cause huge economic losses, since it is estimated that up to 25% of the world's food is contaminated with mycotoxins every year (FAO - Food and Agriculture Organization of the United Nations, 2019). Researchers worldwide have focused their objectives on exploring approaches for the decontamination and/or degradation of these substances. Strategies include the use of chemical, physical and biological methods (Luo *et al.*, 2018). The chemical methods consist of the application of agents that degrade the mycotoxins structure (Chandravarnan *et al.*, 2022). Physical methods are based on the use of washing, heating or irradiation to reduce contamination by degradation and/or leaching (Massarolo *et al.*, 2022). Finally, the biological methods are based mainly on microorganisms (with emphasis on bacteria, filamentous fungi and yeasts) or enzymes capable of altering their structure or adsorbing these contaminants (Nešić *et al.*, 2021). In this way, the focus of the chapter is to provide information on the different methods of mycotoxin degradation (Figure 10.1) related to food production and guaranteeing food safety.

10.2 PHYSICAL METHODS

Physical methods aimed at reducing mycotoxin concentrations in food matrices can be divided into traditional and non-thermal methods. Traditional methods include sorting, washing, peeling, separating the visibly contaminated portion (e.g., by density), stripping (Hojnik *et al.*, 2017) and grinding (e.g., wet and dry) (Massarolo *et al.*, 2022). Even if the initial techniques are rigorous, they may not be sufficient to reduce mycotoxin levels below legislated limits (Agriopoulou *et al.*, 2016). Therefore, the techniques for reducing the mycotoxin concentration must be associated with others during the subsequent stages of the production chain of a given product. Drying raw materials to a critical safe water content for raw material storage is one of the alternatives for reducing fungal proliferation and possible mycotoxin synthesis during storage (Schmidt *et al.*, 2019). In addition, the management and logistics of batches of raw materials and processed products in storage facilities is an indicated alternative (Moncini *et al.*, 2020). However, the reduction of mycotoxin levels cannot depend only on the initial stages of conservation and post-harvest processing, as initial concentrations can be high and not only be associated with matrices that have a visibly contaminated portion (Schaarschmidt and Fauhl-Hassek, 2018). Furthermore, the use of such methods can be laborious, inefficient and often impractical, especially on an industrial scale (Udomkun *et al.*, 2017). In addition to the methods already mentioned, the use of thermal treatment in the industrialization of raw materials usually favors the control of mycotoxins by reducing their concentrations (Table 10.1). When evaluating the studies, the baking has greater efficiency in the degradation of a variety of mycotoxins: deoxynivalenol (94.2%), beauvericin (90%), fumonisin B_1 (82.6%), fumonisin B_2 (82.6%), aflatoxin B_1 (78.6%), aflatoxin G_2 (82.2%) and moniliformin (71%). Promising results are also observed when microwave radiation, extrusion and pressure cooking are employed.

Mycotoxin reduction depends on several factors such as chemical structure, mycotoxin concentration, moisture content, pH and ionic concentration of the matrix during treatment. In addition, two factors stand out: temperature and treatment time. The time for mycotoxin

DOI: 10.1201/9781003334859-10

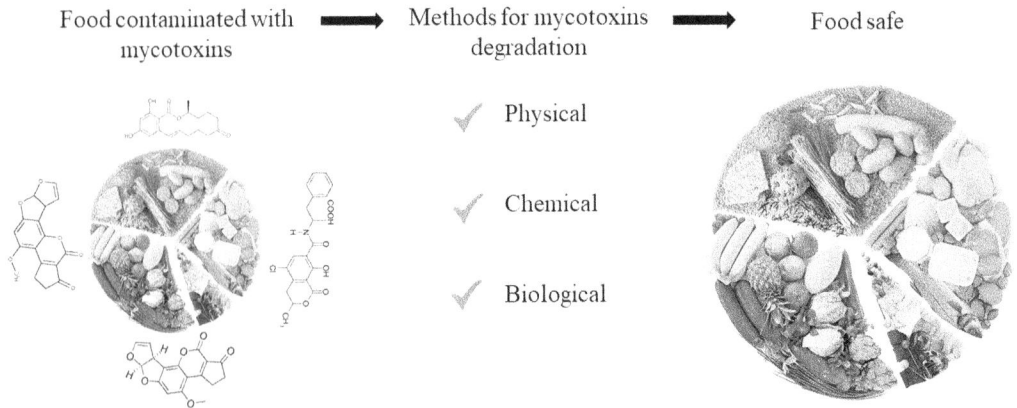

Figure 10.1 Methods used to degrade mycotoxins and obtain safe food.

degradation can be long, requiring high energy consumption, making the process costly, in addition to affecting the quality of treated food products (Rastegar *et al.*, 2017). This last fact was observed by Schmidt *et al.* (2019), who, when evaluating the reduction of mycotoxins in dried grains using microwaves, observed severe damage to enzyme activity and gluten proteins, resulting in low-quality flour and bread. On the other hand, even if the processing time is short, such as those used in extrusion, its process characteristics (e.g., high temperature, high pressure and short duration) do not necessarily imply the degradation of mycotoxins, as these can combine with biopolymers during processing (e.g., proteins and carbohydrates), forming so-called modified mycotoxins. These mycotoxins cannot be detected by the usual methods (Rychlik *et al.*, 2014). In addition, they may pose an additional risk to consumers, as they are hydrolyzed during digestion and can return to their free form (Kovač *et al.*, 2018). Modified mycotoxins can also occur naturally in plants, through hydrolysis, reduction or oxidation processes, resulting in the formation of reactive groups in the chemical structures of mycotoxins, allowing their connection with biopolymers (Broekaert *et al.*, 2015). Examples are the occurrence in wheat of zearalenone-14-glucoside, modified form of zearalenone (Schneweis *et al.*, 2002), deoxynivalenol-glutathione (Kluger *et al.*, 2013), deoxynivalenol-3-sulfate and deoxynivalenol-15-sulfate, modified forms of deoxynivalenol (Warth *et al.*, 2015), the occurrence in wheat and oats of glycoside derivatives of T-2 toxin and HT-2 toxin toxins (Lattanzio *et al.*, 2012), the occurrence in wheat flour N-(1-deoxy-d-fructos-1-yl) and N-(1-deoxy-d-fructos-1-yl), fumonisin B_2 and B_3 glucose conjugates, respectively (Matsuo *et al.*, 2015). For these reasons, other approaches began to be evaluated in order to reduce the mycotoxin concentration in different food matrices. Among the new approaches, non-thermal methods stand out (Table 10.2). The use of these methods can affect the chemical structure of mycotoxins by contact (e.g., direct and indirect) or oxidative, leading to their degradation. However, its effectiveness will depend on some specific factors such as moisture in the food matrix, mycotoxin concentration and exposure intensity (Luo *et al.*, 2017). Among these factors, humidity is the main factor responsible for assisting in the degradation of mycotoxins. This may be related to the fact that a higher moisture content allows the formation of free radicals by water ionization (Jalili *et al.* 2012).

Jalili *et al.* (2012) evaluated the effect of gamma radiation (5 to 30 kGy) on the concentrations of aflatoxins and ochratoxin A in pepper. The authors observed that peppers with the highest moisture content (18%) showed the greatest reductions in aflatoxins and ochratoxin A, around 35 and 55%, respectively. Kumar *et al.* (2012) evaluated this same method (10 kGy) in reducing the concentration of ochratoxin A in green coffee beans. The authors found that mycotoxin degradation was inversely proportional to the moisture content of the samples. Reductions in the initial concentration of ochratoxin A were 5, 9, 20, 90 and 100% for samples with 9, 10, 12, 23 and 58% moisture, respectively. Similar behavior was observed in the degradation for degradation of ochratoxin A and zearalenone and in maize by electron beam irradiation (10 kGy) (Luo *et al.*, 2017), of alternariol, alternariol monomethyl ether and tentoxin in wheat flour by cold plasma

Table 10.1: Degradation of Mycotoxins in Food Matrices by Thermal Processes

Matrice	Treatment	Condition for MR	Mycotoxin (MR - %)	Reference
Corn tortillas	Alkaline cooking	365 °C for 20 min	MON (71)	Pineda-Valdes et al. (2002)
Apple juice	Evaporation	80 °C for 20 min	PAT (14.06)	Kadakal et al. (2003)
Rice	Conventional cooking	160 °C for 20 min	AFB$_1$ (36)	Park et al. (2006)
Rice	Pressure cooking	160 °C for 20 min	AFB$_1$ (88)	Park et al. (2006)
Bread	Baking	200 °C for 20 min	BEA (90)	Meca et al. (2012)
Peanut	Roasting	200 °C for 25 min	AFB$_1$ (89.7)	Martins et al. (2017)
Pistachio nuts	Roasting	120 °C for 60 min	AFB$_1$ (93.1)	Rastegar et al. (2017)
Bread	Baking	180 °C for 35 min	DON (49)	Tibola et al. (2018)
Cookies	Baking	200 °C for 30 min	T-2 (45) and HT-2 (20)	Kuchenbuch et al. (2019)
Crunchy muesli	Baking	170 °C for 20 min	T-2 (15) and HT-2 (19.2)	Kuchenbuch et al. (2019)
Maize bread	Steam cooking	150 °C for 30 min	AFB$_1$ (33.9), AFB$_2$ (39), AFG$_1$ (34.4) and AFG$_2$ (37.2)	Lin et al. (2019)
Chicken breast	Baking	200 °C for 5 min	AFB$_1$ (78.6), AFG$_1$ (82.2), FB$_1$ (82.6) and FB$_2$ (82.6)	Sobral et al. (2019)
Chicken breast	Microwave	350 W for 0.45 s	AFB$_1$ (81.6), AFG$_2$ (84.6), FB$_1$ (70.1) and FB$_2$ (70.1)	Sobral et al. (2019)
Grain cookies	Baking	190 °C for 12 min	DON (94.2)	Devos et al. (2020)
Cornmeal	Extrusion	160 °C	AFB$_1$ (83.7), AFB$_2$ (80.5), AFG$_1$ (74.7) and AFG$_2$ (87.1)	Massarolo et al. (2021)

AFB$_1$ = aflatoxin B$_1$, AFB$_2$ = aflatoxin B$_2$, AFG$_1$ = aflatoxin G$_1$, AFG$_2$ = aflatoxin G$_2$, BEA = beauvericin, DON = deoxynivalenol, FB$_1$ = fumonisin B$_1$, FB$_2$ = fumonisin B$_2$, HT-2 = HT-2 toxin, MON = moniliformin, OTA = ochratoxin A, MR = maximum reduction, T-2 = T-2 toxin.

Table 10.2: Mycotoxin Degradation in Food by Non-Thermal Processes

Treatment	Matrice	Condition	Mycotoxin (MR - %)	Reference
High hydrostatic pressure	Olives	5 min at 250 MPa and 35 °C	CIT (100)	Tokuşoğlu et al. (2010)
Pulsed electric field	Milk	Maximum voltage and frequency of 80 kV and 1 kHz respectively for 4-32 μs pulse duration	AFM₁ (72.2)	Khoori et al. (2020)
Gamma irradiation	Corn	10 kGy	OTA (67.9) and ZEN (71.1)	Luo et al. (2017)
	pepper	30 kGy	OTA (25)	Woldemariam et al. (2021)
	Sorghum	46.596 Gy per min	AFB₁ (59) and OTA (32)	Amara et al. (2022)
UV light radiation	Chili powder	365 nm wavelength for 15 and 60 min	AFB₁ (87.8)	Tripathi et al. (2010)
	Grape juice	254 nm wavelength from 14.2 to 99.4 mJ/cm2	PAT (43.4)	Dong et al. (2010)
	Milk	Wavelength of 360 nm at 0, 2,5 e 5 J/cm² for 30 min	AFM₁ (74.26)	Khoori et al. (2020)
Pulsed light	Rice and rice bran	9 cm from the light source, 3 pulses per s for 20 s	AFB₁ (90.3) and AFB₂ (86.7)	Wang et al. (2016)
	Peanut	7 cm from the light source, 3 pulses per s for 4 min	AFB₁ (98.9) and AFB₂ (98.1)	Abuagela et al. (2019)
Cold plasma	Hazelnut	50 mm away from the plasma source (1,000 W) with duration of 1, 2, 4 and 12 min	AFB₁ (79.1) and AFs (74.2)	Siciliano et al. (2016)
	Wheat flour	6 mm away from the plasma source with a duration of 180 s	AOH (60.6), AME (73.8) and TEN (54.5)	Hajnal et al. (2019)
	Oat	2 mm away from the plasma source with duration of 0, 2, 4, 6, 8 and 10 min	DON (54.4)	Feizollahi et al. (2020)

AFB₁ = aflatoxin B₁, AFB₂ = aflatoxin B₂, AFM₁ = aflatoxin M₁, AFs = sum of aflatoxins B₁, B₂, G₁ e G₂, AME = alternariol monomethyl ether, AOH = alternariol, CIT = citrinin, OTA = ochratoxin A, PAT = patulin, MR = maximum reduction, TEN = tentoxin, ZEN = zearalenone.

(Hajnal *et al.*, 2019). However, the moisture content did not influence the degradation of deoxynivalenol in barley when treated by cold plasma (Feizollahi *et al.*, 2020). In addition, the application of these methods may have some peculiarities, with a negative impact on the quality of food matrices. The use of pulsed light used to degrade aflatoxin B_1 and B_2, promoted significant changes ($p < 0.05$) in the surface color of peanuts when treated for 4 min, observed in the values of L* (brightness), a* (redness) and b* (yellowing) (Abuagela *et al.*, 2019). The use of gamma radiation, despite not having as objective the evaluation of degrading mycotoxins, significantly affected ($p < 0.05$) germination (Kottapalli *et al.*, 2003), soluble protein content, must color, alpha-amylase and diastatic power of barley, decreasing with increasing radiation dose (Kottapalli *et al.*, 2006). Zhu *et al.* (2014) used ultraviolet light irradiation in apple juice to degrade patulin. The authors observed an increase in the value of L* and a decrease in the values of a* and b*. Furthermore, the loss of ascorbic acid was 36.5% under the conditions used to degrade 90% of the mycotoxin. Therefore, this is a disadvantage for the food industry.

The use of high hydrostatic pressure used in olives, in addition to degrading citrinin (100% degradation), increased the content of total phenolic compounds and hydroxytyrosol. However, the concentration of oleuropein, the compound responsible for preventing bone loss in humans, was lower after high-pressure treatment for 5 min at 250 MPa and 35°C (Tokuşoğlu *et al.*, 2010). Thus, although there are contrasting reports on the effect of different non-thermal methods to degrade mycotoxins in different food matrices, these can be considered promising for the industry. However, the effect on the physical-chemical composition of the food matrices must be considered. Furthermore, any method should be used in combination with good manufacturing and storage practices to prevent toxigenic mold growth and mycotoxin synthesis.

10.3 CHEMICAL METHODS

Chemical treatments for mycotoxin degradation in food and feed products are based on the use of chemical compounds, such as ammonia, hydrogen peroxide, sodium bisulfite, organic acids and ozone, among others, to degrade the mycotoxin structure. Although there is no regulation regarding the application of any chemical treatment for degrading mycotoxins in foods for human consumption, many studies report the effectiveness of these treatments (Table 10.3). Additionally, chemical decontamination techniques for feedstuff were approved in the European Union in 2015, provided that the treatment does not adversely affect the characteristics and the nature of the feed, is effective and irreversible and does not result in harmful residues for the feed or environment (European Union, 2015). The first chemical method studied for the degradation of mycotoxins was ammonia (Norred, 1982). However, treatments with ammonia are not practical, regardless of its efficiency, because of the high potential for toxic derivates formations and the modifications in nutrient content and sensory properties of the product (Ismail *et al.*, 2018; Yagen *et al.*, 1989). On the other hand, hydrogen peroxide is an efficient oxidizing agent that can destroy mycotoxin structure and it is safe to be used in certain food processes. The advantage of hydrogen peroxide application is that it can be easily removed after degradation treatment. Also, studies have indicated that irradiating UV associated with hydrogen peroxide can generate more free radicals than hydrogen peroxide alone. This process, known as "advanced oxidation processes", is widely used in oxidating organic compounds in environmental and food matrices (Shen, 2021). However, there are no recent studies that apply these processes in mycotoxins. Once sodium bisulfite has a low cost and is already commonly added to several types of food and beverages, including wine, fruit juice, jellies and dried fruit for acting as an inhibitor of enzymatic degradation, as an antioxidant and as a bacteriostatic agent, the residual bisulfite is an approved food additive and studies indicate may be a possibility to mycotoxin degradation, mainly DON and AFB_1 (Dänicke *et al.*, 2012; Yagen *et al.*, 1989). Sodium bisulfite may destroy aflatoxin B_1 when is inserted into the double bond of the furan ring in the mycotoxin structure, eliminating the main binding site in the DNA and consequently the main toxic effect of the molecule (Yagen *et al.*, 1989).

Organic acids such as lactic acid and citric acid may be used for mycotoxin control with a focus on preventing the growth of the fungal population or on degradation. The inactivation with the hydrolysis provocated by these acids leads to the conversion of the degraded compound into products with much lower toxicity (Doyle *et al.*, 1982). However, drastic conditions like high temperatures are necessary to convert mycotoxins into compounds less toxic to acids, which limits its application in the food industry (Nunes *et al.*, 2021). Ozone is a strong oxidant used to disinfect food processing and packaging equipment. Also, has been used to inhibit fungal growth and decontaminate different types of food and beverages (Chandravarnan *et al.*, 2022). Ozone may

Table 10.3: Studies on Chemical Approaches for Mycotoxin Degradation in Food Products

Matrice	Treatment	Mycotoxin	Spiked Concentration ($ug\ kg^{-1}$)	Maximum Degradation	Reference
Corn	Ammonia	AFB_1	1000	99%	Norred (1982)
Corn	Ammonia	AFB_1	1000	88%	Nyandieka et al. (2009)
		AFB_2		85%	
		AFG_1		96%	
		AFG_2		93%	
Chili poder	Hydrogen peroxide	AFB_1	100	58%	Tripathi and Mishra (2009)
Corn	Hydrogen peroxide	AFB_1	*	100%	Tabata et al. (1994)
		AFB_2		100%	
		AFG_1		100%	
		AFG_2		100%	
Black pepper	Ammonia	AFB_1	60 ug/kg^{-1} for OTA, AFB_1 and AFG_1, and 18 ug/kg^{-1} for AFB_2 and AFG_2	44%	Jalili et al. (2011)
		AFB_2		40%	
		AFG_1		45%	
		AFG_2		41%	
		OTA		52%	
Black pepper	Hydrogen peroxide	AFB_1		44%	
		AFB_2		32%	
		AFG_1		43%	
		AFG_2		37%	
		OTA		38%	
	Citric Acid	AFB_1		29%	
		AFB_2		25%	
		AFG_1		23%	
		AFG_2		24%	
		OTA		26%	
	Sodium bisulfite	$AFB1$		44%	
		$AFB2$		23%	
		$AFG1$		35%	
		$AFG2$		25%	
		OTA		41%	
Cereal grains	Sodium metabisulfite	DON	*	> 95%	Dänicke et al. (2012)
Sorghum	Citric acid	$\Sigma(AFB_1 + AFB_2)$	140	92%	Méndez-Albores et al. (2009)
Soybean	Lactic acid	$AFB1$	10	94%	Lee et al. (2015)
Parboiled rice	Ozone	AFB_1	46	81%	Luz et al. (2022)
		AFB_2	48	62%	
		AFG_1	259	59%	
		AFG_2	625	48%	

Food	Method	Mycotoxin	Concentration	Degradation	Reference
Apple juice	Ozone	DON	164	56%	Diao et al. (2019)
		OTA	647	88%	
		ZEA	1,536	76%	
		PAT	202**	75%	
Corn	Ozone	AFB$_1$	50	57%	Porto et al. (2019)
		AFG$_1$	50	55%	
		AFB$_2$	50	30%	
		AFG$_2$	50	36%	
Peanuts	Ozone	AFB1	500	79%	Li et al. (2019)
Milk	Ozone	AFM$_1$	0.56	50%	Mohammadi et al. (2017)
Wheat	Ozone	DON	1,000	64%	Trombete et al. (2017)
		Σ(AFB$_1$ + AFB$_2$ + AFG$_1$ + AFG$_2$)	200	48%	
Red pepper	Ozone	AFB$_1$	25	74%	Kamber et al. (2017)
Corn	Ozone	OTA	67	64%	Qi et al. (2016)
		ZEA	2932	86%	
Corn	Ozone	AFB$_1$	53.6	79%	Luo et al. (2014)
		AFB$_2$	12.08	72%	
		AFG$_2$	2.42	71%	
Wheat (pericarp)	Ozone	DON	1,065	100%	Savi et al. (2014)
Wheat (endosperm)		DON	534	100%	

Notes

* Not mentioned.

** Spiked concentration unit: ug L^{-1}. AFB$_1$ = aflatoxin B$_1$; AFB$_2$ = aflatoxin B$_2$; AFG$_1$ = aflatoxin G$_1$; AFG$_2$ = aflatoxin G$_2$; AFM$_1$ = aflatoxin M$_1$; DON = deoxynivalenol; OTA = ochratoxin A; PAT = patulin; ZEA = zearalenone.

attack the double bond in organic compounds and produce lower molecular weight and less toxic compounds without affecting the quality properties of food minimally, being this the main advantage of its use (Mir *et al.*, 2021). Studies indicate that the gaseous form of ozone is more useful in the decontamination of fungal and mycotoxins than the aqueous form (Freitas-Silva and Venâncio, 2010). Besides, the treatment effectiveness depends on the gas concentration, exposure time, moisture, food matrice and mycotoxin structure (Afsah-Hejri *et al.*, 2020). Although ozone is safe, efficient and environmentally friendly, it is an unstable compound that has a short half-life that quickly decomposes to form oxygen (Chandravarnan *et al.*, 2022). As it needs to be kept constant in the degradation process, this requirement can present some cost and logistical disadvantages (Pandiselvam *et al.*, 2019).

Another interesting approach that is unprecedented in the literature for the degradation or reduction of mycotoxins is the application of plant extracts and essential oils. Aqueous plant extracts and essential oils have the potential to prevent and control mycotoxin contamination because of its antifungal activities (Chandravarnan *et al.*, 2022; Hamad *et al.*, 2023). Therefore, an interesting approach is the application of these extracts and different essential oils in food and the evaluation of their relation with the synthesis or degradation of mycotoxin, since the impact of these oils and their chemical constituents on mycotoxin degradation remains unknown. Although some were published more than 30 years ago, the research contains valuable information to be discussed because of their contribution to mycotoxins detoxification (Table 10.3). The fact that more data exists with aflatoxins is probably because aflatoxin B_1 is considered the most toxic naturally occurring mycotoxin, followed by aflatoxins G_1, B_2 and G_2 (Luz *et al.*, 2022).

Comparing the research, applying ammonia, hydrogen peroxide, sodium bisulfite and organic acids have been of less interest probably because of these approaches' disadvantages. The degradation differences between the studies show that the same method can be effective for different matrices and/or different mycotoxins. Ammonia, for example, was used to degrade aflatoxins by Norred (1982) and Nyandieka *et al.* (2009) in corn and both studies have a maximum degradation higher than 90%. Jalili *et al.* (2011) studied different methods for degradate aflatoxins and OTA in black pepper and, although the best result was applying ammonia, all methods degraded the mycotoxins. Sodium metabisulfite was an excellent treatment to degrade DON in cereal grains in the study made by Dänicke *et al.* (2012). Treatments with citric and lactic acids were studied by Méndez-Albores *et al.* (2009) and Lee *et al.* (2015) to degrade aflatoxins in sorghum and soybean, respectively, and all the maximum degraded results were bigger than 92%, showing the efficiency of these approaches. Although these chemical methods can effectively detoxify mycotoxins, the safety of the degraded compounds and the removal of residual chemicals after treatments largely limit their application, besides that, some degradation processes can be easily reversible, such as some acid treatments (Chandravarnan *et al.*, 2022).

Studies using ozone are more recent and frequent and there is also a wider variety of mycotoxins studied using this chemical approach. The degradation differences between the studies show that mycotoxins have a different sensitivity to treatments using ozone. In addition, the food matrix is also an important parameter that has different sensitivities. Ozone treatment was more effective when applied to wheat than parboiled rice for DON. Furthermore, the application condition is also very important, as the degradation was higher for DON in the Savi *et al.* (2014) study than in the Trombete *et al.* (2017) (both in wheat). In the first study, Savi *et al.* (2014) applied a higher concentration of O_3 and a lower mass of wheat. Among the aflatoxins, AFB_1 appears to be the most sensitive among ozone treatments because it has the best degradation (Table 10.3). As much as all the presented approaches are effective in decontamination, each chemical treatment has its own limitations and application conditions, so there is a need to identify environmentally and biologically safe and cost-effective mycotoxin decontamination agents that are in accordance with the aim of the proposed study.

10.4 BIOLOGICAL METHODS

The biological control of mycotoxins based on the use of fungi, bacteria and enzymes capable of degrading and/or adsorbing mycotoxins is a biological approach for reducing mycotoxin levels, promising in terms of efficiency and specificity, constituting a positive impact for the production of safe foods for consumption. Furthermore, it is established as an ecologically correct and safe strategy for the elimination of mycotoxins in food production (Nešić *et al.*, 2021). Fungi (in particular, yeasts) and bacteria (mainly lactic acid) can act to reduce mycotoxin levels due to the adsorption phenomenon or the mycotoxin degradation caused by microbial secretions

(Wang *et al.*, 2019). Adsorption uses cell surfaces of active or inactive microorganisms to bind mycotoxins (Luo *et al.*, 2020). Differently, the enzymes act in the degradation and/or transformation of the chemical structure of these contaminants, resulting in compounds with less toxicity (Oliveira *et al.*, 2021). The mycotoxin reduction by biological methods is affected by different factors, such as the type of biological agent and mycotoxin, adsorbent and contaminant concentration, pH and temperature of the medium (Luo et al. 2019). Thus, the optimization of these factors is fundamental to guarantee efficiency in the reduction of mycotoxin levels using biological methods. In this context, the main microorganisms and enzymes used in the biological control of mycotoxins will be discussed below, focusing on their application in food.

10.4.1 Lactic Acid Bacteria

Lactic acid bacteria are microorganisms widely studied for application in mycotoxin degradation by biological methods, mainly due to their safety history in food applications and cultivation and maintenance conditions. Lactic acid bacteria can act to reduce the mycotoxin concentration in foods using two different mechanisms: the use of cells and/or enzymes produced by some strains (Muhialdin *et al.*, 2020). The main structure of the cell wall of lactic acid bacteria is made up of peptidoglycan encrusted with teichoic and lipoteichoic acids, a protein and polysaccharide layer. Among these compounds, peptidoglycan and polysaccharides are responsible for the adsorption of mycotoxins (Luo et al. 2019). Wang *et al.* (2015b) reported the reduction of patulin concentration in a culture medium attributed to adsorption by selected lactic acid bacteria. Furthermore, the authors describe the positive relationship between increased adsorption and greater surface area and cell wall volume, an effect attributed to the high number of binding sites. In addition to adsorption, it is important to note that viable cells of lactic acid bacteria are responsible for the production of several bioactive metabolites that can prevent mycotoxin production in food. These bioactive compounds include acids, carbon dioxide, hydrogen peroxide, phenyllactic acid and low molecular weight bioactive peptides. It is estimated that these metabolites can bind to mycotoxins, resulting in reduced toxicity (Muhialdin *et al.* 2020). However, the mechanisms of adsorption and/or degradation of mycotoxins by lactic acid bacteria are still unclear.

10.4.2 Yeasts

Yeasts are microorganisms widely used in biotechnological processes to produce bread, beer, wine and spirits, among others. As in lactic acid bacteria, the cell wall in the yeast structure makes the cells capable of adsorbing a wide range of compounds, including mycotoxins. Viable cells, non-viable cells and yeast products are able to reduce the bioavailability of mycotoxins in food. In the literature, there are reports that show the ability of yeast to remove different mycotoxins from raw materials and foods in fermentation (Piotrowska, 2021). Traditionally, different strains of the yeast *Saccharomyces cerevisiae* are studied for their ability to degrade mycotoxins (Luo *et al.*, 2020). Zhang *et al.* (2019) evaluated the adsorption of patulin by the yeast *Saccharomyces cerevisiae* CCTCC 93161 in cultivation in an aqueous solution. The authors concluded that the proteins and polysaccharides in the yeast cell walls interacted with patulin and were responsible for physical adsorption, reducing the patulin concentration by about 22% at the beginning of fermentation (0h). After 24 hours of fermentation, the percentage of patulin removal reached approximately 85%. In another study, conducted by Li *et al.* (2018), the interaction between *Saccharomyces cerevisiae* CITCC 93161 and the mycotoxin patulin during cultivation were evaluated. The authors observed the complete degradation of patulin and attributed the action of endo and exoenzymes produced during fermentation. The mitigation potential of the mycotoxin nivalenol by the alcoholic fermentation of *Saccharomyces cerevisiae* US-05 with the application of magnetic fields was evaluated by Boeira *et al.* (2021). The authors concluded that alcoholic fermentation mitigated contamination by nivalenol by up to approximately 56% after 96 hours, showing a promising method for mitigating this mycotoxin. Despite the wide emphasis given to the genus Saccharomyces, other yeasts have been studied, among them, *Kluveromyces marxianus* for aflatoxin M_1 (Martínez *et al.*, 2019) and *Meyerozyma guilliermondii* for patulin (Fu *et al.*, 2021).

10.4.3 Enzymes

In recent years, efforts have been made to identify enzymes capable of degrading mycotoxins. Enzymes are proteins that have catalytic activity, acting as biocatalysts that increase the speed of reactions, accelerating the conversion of substrates into products (Nelson *et al.*, 2014). Due to their catalytic nature and detoxification abilities, enzymes are efficient decontamination agents

(Manubolu *et al.*, 2018). Enzymes can come from the product of microorganisms or be intentionally added to the medium with a view to decontamination. In the first case, it may be difficult to distinguish the decontamination mechanism, due to the possibility of adsorption by the micro-organism cells and the action of enzymes in the degradation of the physical structure, or even due to the possibility of a synergistic effect (Luo et al. 2019). Enzymes can act in the biotransformation of mycotoxins forming less toxic or non-toxic compounds. The main conversion pathways are hydroxylation, hydrogenation, hydrolysis, oxidation, esterification, glucuronidation and glycosylation, deep oxidation, methylation, sulphation, demethylation and deamination (Nešić *et al.*, 2021). However, the enzymatic biotransformation mechanism of mycotoxins is variable and difficult to elucidate, depending on the type of mycotoxin and enzyme used. Different factors can influence mycotoxin degradation (e.g., enzyme activity, pH, temperature and incubation time). The enzyme activity applied in the mycotoxin degradation is a limiting factor for the industry application, mainly due to its high cost (Oliveira *et al.*, 2021). Thus, the optimization of these factors is crucial regarding the potential application of enzymes in mycotoxin degradation.

10.4.4 Food Application of Biological Methods

In the last decade, several bacteria, yeasts and enzymes have been proposed for use as decontaminating agents in foods, however, not all of them have been effectively evaluated in food matrices. The difficulty in implementing biological methods is mainly due to the need to be non-pathogenic, specific, effective and not result in toxic effects. Table 10.4 shows some literature reports on the application of biological agents to reduce mycotoxin levels in food. Milk is the highlight of research carried out in recent years, possibly due to its wide consumption in different forms in the human diet and its fundamental role in the diet of children. Thus, guaranteeing the quality of milk is essential to

Table 10.4: Mycotoxin Degradation by Biological Agents in Food

Treatment	Mycotoxin	Matrice	Reduction (%)	Reference
Bacteria				
Lactobacillus kefiri	AFB_1	Milk	82	Wang *et al.* (2015a)
Lactobacillus kefiri	OTA	Milk	94	Wang *et al.* (2015a)
Lactobacillus kefiri	ZEA	Milk	100	Wang *et al.* (2015a)
Lactobacillus delbrueckii spp. *bulgaricus* LB340 + *Lactobacillus rhamnosus* HOWARU + *Bifidobacterium lactis* FLORA-FIT BI07	AFM_1	Milk	11.7	Corassin *et al.* (2013)
S. thermophilus sub. *Lactis* + *L. bulgaricus*	AFM_1	Yogurt	69.8	Elsanhoty *et al.* (2014)
S. thermophilus sub. *Lactis* + *L. bulgaricus* + *L. plantrium*	AFM_1	Yogurt	87.8	Elsanhoty *et al.* (2014)
S. thermophilus sub. *Lactis* + *L. bulgaricus* + *L. acidophilus*	AFM_1	Yogurt	72.8	Elsanhoty *et al.* (2014)
Lactobacillus rhamnosus RC007	AFM_1	Milk	61	Martínez *et al.* (2019)
Pediococcus acidilactici RC005	AFM_1	Milk	34	Martínez *et al.* (2019)
Pediococcus pentosaceus RC006	AFM_1	Milk	26	Martínez *et al.* (2019)
Yeast				
Saccharomyces cerevisiae	AFM_1	Milk	92.7	Corassin *et al.* (2013)
Saccharomyces cerevisiae S6u	OTA	Wine	51.6	Cecchini *et al.* (2018)
Saccharomyces cerevisiae S10c	OTA	Wine	45.4	Cecchini *et al.* (2018)
Saccharomyces cerevisiae RC016	AFM_1	Milk	19	Martínez *et al.* (2019)
Kluveromyces marxianus VM003	AFM_1	Milk	36	Martínez *et al.* (2019)
Saccharomyces boulardii RC009	AFM_1	Milk	25	Martínez *et al.* (2019)
Enzymes				
Peroxidase	AFB_1	Beer	24	Sibaja *et al.* (2019)
Peroxidase	AFB_1	Milk	97	Sibaja *et al.* (2019)
Peroxidase	AFM_1	Milk	65	Sibaja *et al.* (2019)
Peroxidase	OTA	Grape juice	17	Nora *et al.* (2019)
Peroxidase	OTA	Beer	4.8	Garcia *et al.* (2020)
Peroxidase	ZEA	Beer	10.9	Garcia *et al.* (2020)
Lacase CotA	ZEA	Corn meal	90	Guo *et al.* (2022)

AFB_1 = aflatoxin B_1; AFM_1 = aflatoxin M_1; OTA = ochratoxin A; ZEA = zearalenone.

provide safe products to consumers. Based on the data presented in Table 10.4, it can be observed that bacteria, yeasts and enzymes have been evaluated for the reduction of the main mycotoxins (aflatoxins B_1 and M_1) associated with this food. The yeast *Saccharomyces cerevisiae* showed prominence in the degradation of aflatoxin M_1, resulting in the mitigation of up to 92.7%, while the peroxidase enzyme resulted in the mitigation of up to 97% of aflatoxin B_1 in that same matrix. The high levels of reduction observed demonstrate the potential application of biological methods for the degradation of mycotoxins. However, studies that clarify the degradation products, as well as their non-toxicity, are crucial for the industrial application of these biological degradation methods.

10.5 FINAL CONSIDERATIONS

Due to the frequent occurrence of mycotoxins and their toxicity, these compounds are a matter of concern for health control agencies. In view of these factors, there is a need for studies that have as their scope the degradation of mycotoxins, which is the alternative to avoid the great disposal of raw materials or food when aiming at the health of the consumer. Among the methods described, the microbiological methods stand out due to their applicability and innovation in the food production chain. However, the use of combined methods, mainly physical and biological, is promising, enabling the increase of degradation and the availability of safe and quality food for humans and animals.

REFERENCES

Abuagela, M. O., B. M. Iqdiam., H. Mostafa., S. M. Marshall., Y. Yagiz., M. R. Marshall., L. Gu., and P. Sarnoski. 2019. Combined effects of citric acid and pulsed light treatments to degrade B-aflatoxins in peanut. *Food and Bioproducts Processing* 117: 396–403. 10.1016/j.fbp.2019.08.011

Afsah-Hejri, L., P. Hajeb, and R. J. Ehsani. 2020. Application of ozone for degradation of mycotoxins in food: A review. *Comprehensive Reviews in Food Science and Food Safety* 19(4): 1777–1808. 10.1111/1541-4337.12594

Agriopoulou. S., A. Koliadima., G. Karaiskakis., and J. Kapolos. 2016. Kinetic study of aflatoxins' degradation in the presence of ozone. *Food Control* 61: 221–226. 10.1016/j.foodcont.2015.09.013

Amara, A. B., A. Mehrez., C. Ragoubi., R. Romero-González., A. G. Frenich., A. Landoulsi., and I. Maatouk. 2022. Fungal mycotoxins reduction by gamma irradiation in naturally contaminated sorghum. *Journal of Food Processing and Preservation* 46(3): e16345. 10.1111/jfpp.16345

Boeira, C. Z., M. A. C. Silvello, R. D. Remedi, A. C. P. Feltrin, L. O. Santos, and J. Garda-Buffon. 2021. Mitigation of nivalenol using alcoholic fermentation and magnetic field application. *Food Chemistry*, 340: 127935. 10.1016/j.foodchem.2020.127935

Broekaert, N., N. Devreese., S. Baere., P. Backer., and S. Croubels. 2015. Modified *Fusarium* mycotoxins unmasked: From occurrence in cereals to animal and human excretion. *Food and Chemical Toxicology* 80: 17–31. 10.1016/j.fct.2015.02.015

Cecchini, F., M. Morassut, J. C. Saiz, and E. Garcia-Moruno. 2018. Anthocyanins enhance yeast's adsorption of ochratoxin a during the alcoholic fermentation. *European Food Research and Technology* 245: 309–314. 10.1007/s00217-018-3162-9

Chandravarnan, P., D. Agyei, and A. Ali. 2022. Green and sustainable technologies for the decontamination of fungi and mycotoxins in rice: A review. *Trends in Food Science & Technology* 124 (June): 278–295. 10.1016/j.tifs.2022.04.020

Corassin, C. H., F. Bovo, R. E. Rosim, and C. A. F. Oliveira. 2013. Efficiency of *Saccharomyces cerevisiae* and lactic acid bacteria strains to bind aflatoxin M_1 in uht skim milk. *Food Control*, 31: 80–83. 10.1016/j.foodcont.2012.09.033

Dänicke, S., S. Kersten, H. Valenta, and G. Breves. 2012. Inactivation of deoxynivalenol-contaminated cereal grains with sodium metabisulfite: A review of procedures and toxicological aspects. *Mycotoxin Research* 28(4): 199–218. 10.1007/s12550-012-0139-6

Devos, R. J. B., C. S. Tibola., B. Biduski., L. M. Colla., and L. C. Gutkoski. 2020. Deoxynivalenol reduction through the processing of whole grain cookies. *Research, Society and Development* 9(12): e39991211098. 10.33448/rsd-v9i12.11098

Diao, E., J. Wang, X. Li, X. Wang, H. Song, and D. Gao. 2019. Effects of Ozone Processing on Patulin, Phenolic Compoundspatulin, phenolic compounds and Organic Acidsorganic acids in Apple Juiceapple juice. *Journal of Food Science and Technology* 56(2): 957–965. 10.1007/s13197-018-03561-0

Dong, Q., D. C. Manns., G. Feng., T. Yue., J. J. Churey, and R. W. Worobo. 2010. Reduction of patulin in apple cider by UV radiation. *Journal of Food Protection* 73(1): 69–74. 10.4315/0362-028X-73.1.69

Doyle, M. P., R. S. Applebaum, R. E. Brackett, and E. H. Marth. 1982. Physical, chemical and biological degradation of mycotoxins in foods and agricultural commodities, 8.

Elsanhoty, R. M., S. A. Salam, M. F. Ramadan, and F. H. Badr. 2014. Detoxification of aflatoxin M_1 in yogurt using probiotics and lactic acid bacteria. *Food Control* 43: 129-13. 10.1016/j.foodcont.2014.03.002

European Union. 2015. European Union. COMMISSION REGULATION (EU) 2015/ 786 – of 19 May 2015 – Defining acceptability criteria for detoxification processes applied to products intended for animal feed as provided for in directive 2002/ 32/ EC of the European Parliament and of the council, 5.

FAO – Food and Agriculture Organization of the United Nations. 2019. *Market Monitor. Agricultural Market Information System (AMIS), Rome.*

Feizollahi, E., B. Iqdiam., T. Vasanthan., M. S. Thilakarathna., and M. S. Roopesh. 2020. Effects of atmospheric-pressure cold plasma treatment on deoxynivalenol degradation, quality parameters, and germination of barley grains. *Applied Sciences* 10: 3530. 10.3390/app10103530

Freitas-Silva, O., and A. Venâncio. 2010. Ozone applications to prevent and degrade mycotoxins: A review. *Drug Metabolism Reviews* 42(4): 612–620. 10.3109/03602532.2010.484461

Fu, Y., Q. Yang, D. Solairaj, E. A. Godana, M. N. Routledge, and H. Zhang. 2021. Biodegradation of mycotoxin patulin by the yeast *Meyerozyma guilliermondii. Biological Control*, 160: 104692. 10.1016/j.biocontrol.2021.104692

Garcia, S. O., K. V. M. Sibaja, W. V. Nogueira, A. C. P. Feltrin, D. F. A. Pinheiro, M. B. R. Cerqueira, E. Badiale-Furlong, and J. Garda-Buffon. 2020. Peroxidase as a simultaneous degradation agent of ochratoxin A and zearalenone applied to model solution and beer. *Food Research International*, 131: 109039. 10.1016/j.foodres.2020.109039

Guo, Y., Y. Wang, Y. Liu, Q. Ma, C. Ji, and L. Zhao. 2022. Detoxification of the mycoestrogen zearalenone by *Bacillus licheniformis* spore CotA laccase and application of immobilized laccase in contaminated corn meal. *LWT*, 163: 113548. 10.1016/j.lwt.2022.113548

Hajnal, E. J., M. Vukić., L. Pezo., D. Orčić., N. Puač., N. Škoro., A. Milidrag., and D. Š. Simović. 2019. Effect of atmospheric cold plasma treatments on reduction of *Alternaria* toxins content in wheat flour. *Toxins* 11(12): 704. 10.3390/toxins11120704

Hamad, G. M., T. Mehany, J. Simal-Gandara, S. Abou-Alella, O. J. Esua, M. A. Abdel-Wahhab, and E. E. Hafez. 2023. A review of recent innovative strategies for controlling mycotoxins in foods. *Food Control* 144(February): 109350. 10.1016/j.foodcont.2022.109350

Hojnik N., U. Cvelbar, G. Tavčar-Kalcher., J. L. Walsh., and I. Križaj. 2017. Mycotoxin decontamination of food: Cold atmospheric pressure plasma versus "classic" decontamination. *Toxins* 9(5): 151. 10.3390/toxins9050151

Ismail, A., B. L. Gonçalves, D. V. Neeff, B. Ponzilacqua, C. F. S. C. Coppa, H. Hintzsche, M. Sajid, A. G. Cruz, C. H. Corassin, and C. A. F. Oliveira. 2018. Aflatoxin in foodstuffs: Occurrence and recent advances in decontamination. *Food Research International* 113(November): 74–85. 10.1016/j.foodres.2018.06.067

Jalili, M., S. Jinap, and R. Son. 2011. The effect of chemical treatment on reduction of aflatoxins and ochratoxin A in black and white pepper during washing. *Food Additives & Contaminants: Part A* 28(4): 485–493. 10.1080/19440049.2010.551300

Jalili, M., S. Jinap, and M. A. Noranizan. 2012. Aflatoxins and ochratoxin A reduction in black and white pepper by gamma radiation. *Radiation Physics and Chemistry* 81(11): 1786–1788. 10.1016/j.radphyschem.2012.06.001

Ji, X., Y. Xiao, C. Jin, W. Wang, W. Lyu, B. Tang, and H. Yang. 2022. Alternaria mycotoxins in food commodities marketed through e-commerce stores in China: Occurrence and risk assessment. *Food Control* 140(October): 109125. 10.1016/j.foodcont.2022.109125

Kadakal, Ç., and S. Nas. 2003. Effect of heat treatment and evaporation on patulin and some other properties of apple juice. *Journal of the Science of Food and Agriculture* 83(9): 987–990. 10.1002/jsfa.1339

Kamber, U., G. Gülbaz, P. Aksu, and A. Doğan. 2017. Detoxification of Aflatoxinaflatoxin B_1 in Red Pepperred pepper (*Capsicum Annuum L.*) by Ozone Treatmentozone treatment and Its Effectits effect on Microbiologicalmicrobiological and Sensory Quality: DETOXIFICATION OF AFLATOXINsensory quality: Detoxification of aflatoxin B_1 IN RED PEPPERin red pepper. *Journal of Food Processing and Preservation* 41(5): e13102. 10.1111/jfpp.13102

Khodaei, D., F. Javanmardi, and A. M. Khaneghah. 2021. The global overview of the occurrence of mycotoxins in cereals: A three-year survey. *Current Opinion in Food Science* 39(June): 36–42. 10.1016/j.cofs.2020.12.012

Kholif, O. T., A. S. Sebaei, F. I. Eissa, and O. H. Elhamalawy. 2021. Size-exclusion chromatography selective cleanup of aflatoxins in oilseeds followed by HPLC determination to assess the potential health risk. *Toxicon* 200(September): 110–117. 10.1016/j.toxicon.2021.07.009

Khoori, E., V. Hakimzadeh., A. M. Sani., and H. Rashidi. 2020. Effect of ozonation, UV light radiation, and pulsed electric field processes on the reduction of total aflatoxin and aflatoxin M_1 in acidophilus milk. *Journal of Food Processing and Preservation* 44: e14729. 10.1111/jfpp.14729

Kluger, B., C. Bueschl., M. Lemmens., F. Berthiller., G. Häubl., G. Jaunecker., G. Adam., R. Krska., and R. Schuhmacher. 2013. Stable isotopic labelling-assisted untargeted metabolic profiling reveals novel conjugates of the mycotoxin deoxynivalenol in wheat. *Analytical and Bioanalytical Chemistry*, 405: 5031–5036. 10.1007/s00216-012-6483-8

Kochman, J., K. Jakubczyk, and K. Janda. 2021. Mycotoxins in red wine: Occurrence and risk assessment. *Food Control* 129(November): 108229. 10.1016/j.foodcont.2021.108229

Kottapalli, B., C. E. Wolf-Hall., P. Schwarz., J. Schwarz., and J. Gillespie. 2003. Evaluation of hot water and electron beam irradiation for reducing fusarium infection in malting barley. *Journal of Food Protection* 66(7): 1241–1246. 10.4315/0362-028X-66.7.1241

Kottapalli, B., C. E. Wolf-Hall., and P. Schwarz. 2006. Effect of electron-beam irradiation on the safety and quality of *Fusarium*-infected malting barley. *International Journal of Food Microbiology* 110(3): 224–231. 10.1016/j.ijfoodmicro.2006.04.007

Kovač, M., D. Šubarić., M. Bulaić., T. Kovač., and B. Šarkanj. 2018. Yesterday masked, today modified; what do mycotoxins bring next? *Archives of Industrial Hygiene and Toxicology* 69(3): 196–214. 10.2478/aiht-2018-69-3108

Kuchenbuch, H. S., M. Schulz., S. Becker., B. Cramer., and H. U. Humpf. 2019. Thermal reactions and the formation of degradation products of T-2 and HT-2 toxin during processing of oats. *Food-Borne Toxicants: Formation, Analysis, and Toxicology* 1306: 109–122. 10.1021/bk-2019-1306.ch007

Kumar, S., A. Kunwar., S. Gautam., and A. Sharma. 2012. Inactivation of A. ochraceus spores and detoxification of ochratoxin A in coffee beans by gamma irradiation. *Journal of Food Science* 77(2): T44–T51. 10.1111/j.1750-3841.2011.02572.x

Lattanzio, V. M. T., A. Visconti., M. Haidukowski., and M. Pascale. 2012. Identification and characterization of new Fusarium masked mycotoxins, T2 and HT2 glycosyl derivatives, in naturally contaminated wheat and oats by liquid chromatography–high-resolution mass spectrometry. *Journal of Mass Spectrometry* 47(4): 466–475. 10.1002/jms.2980

Lee, J., J. Her, and K. Lee. 2015. Reduction of Aflatoxins (B1, B2, G1, and G2) in Soybean-Based Model Systems. *Food Chemistry* 189: 45–51. 10.1016/j.foodchem.2015.02.013

Li, M., W. Chen, Z. Zhang, Z. Zhang, and B. Peng. 2018. Fermentative degradation of patulin by *Saccharomyces cerevisiae* in aqueous solution *LWT*, 97: 427–7338. 10.1016/j.lwt.2018.07.040

Li, H., Z. Xiong, D. Gui, Y. Pan, M. Xu, Y. Guo, J. Leng, and X. Li. 2019. Effect of Ozonation and UV Irradiationirradiation on Aflatoxin Degradationaflatoxin degradation of Peanutspeanuts. *Journal of Food Processing and Preservation* 43(4): e13914. 10. https://doi.org/1111/jfpp.13914

Lin, X., X. Hu., Y. Zhang, Y. Xia., and M. Zhang. 2019. Bioaccessibility in daily diet and bioavailability in vitro of aflatoxins from maize after cooking. *World Mycotoxin Journal* 12(2): 173–181. 10.3920/WMJ2018.2350

Luo, X., R. Wang, L. Wang, Y. Li, Y. Wang, and Z. Chen. 2014. Detoxification of Aflatoxinaflatoxin in Corn Flourcorn flour by Ozone:ozone: Detoxification of Aflatoxinaflatoxin in Corn Flourcorn flour by Ozoneozone. *Journal of the Science of Food and Agriculture* 94(11): 2253–2258. 10.1002/jsfa.6550

Luo, X., L. Qi., Y. Liu., R. Wang., D. Yang., K. Li., L. Wang., Y. Li., Y. Zhang., and Z. Chen. 2017. Effects of electron beam irradiation on zearalenone and ochratoxin A in naturally contaminated corn and corn quality parameters. *Toxins* 9(3): 84. 10.3390/toxins9030084

Luo, Y., Liu, X., and Li, J. 2018. Updating techniques on controlling mycotoxins - A review. *Food Control*, 89: 123–132. 10.1016/j.foodcont.2018.01.016

Luo, Y., X. Liu, L. Yuan, and J. Li. 2020. Complicated interactions Between bio-adsorbents and mycotoxins during mycotoxin adsorption: Current research and future prospects. *Trends in Food Science & Technology*, 96: 127–134. 10.1016/j.tifs.2019.12.012

Luz, S. R., F. A. Villanova, C. T. Rockembach, C. D. Ferreira, L. J. Dallagnol, J. L. F. Monks, and M. Oliveira. 2022. Reduced of mycotoxin levels in parboiled rice by using ozone and its effects on technological and chemical properties. *Food Chemistry* 372(March): 131174. 10.1016/j.foodchem.2021.131174

Manubolu, M., L. Goodla, K. Pathakoti, and K. Malmlof. 2018. Enzymes as direct decontaminating agents – Mycotoxins. In *Enzymes in Human and Animal Nutrition*: 313–330. 10.1016/B978-0-12-80541 9-2.00016-2

Martínez, M. P., A. P. Magnoli, G. Pereyra, and L. Cavaglieri. 2019. Probiotic bacteria and yeasts adsorb aflatoxin M_1 in milk and degrade it to less toxic AFM_1-metabolites. *Toxicon* 172: 1–7. 10.1016/j.toxicon.2019.10.001

Martins, L. M., A. S. Sant'Ana, B. T. Iamanaka., M. I. Berto, J. I. Pitt, and M. H. Taniwaki. 2017. Kinetics of aflatoxin degradation during peanut roasting. *Food Research International* 97: 178–183. 10.1016/j.foodres.2017.03.052

Massarolo, K. C., J. R. Mendoza., T. Verma, L. Kupski., E. Badiale-Furlong., and A. Bianchini. 2021. Fate of aflatoxins in cornmeal during single-screw extrusion: A bioaccessibility approach. *LWT* 138: 110734. 10.1016/j.lwt.2020.110734

Massarolo, K. C., P. Rodrigues., C. F. J. Ferreira., L. Kupski., and E. Badiale-Furlong. 2022. Simultaneous distribution of aflatoxins B_1 and B_2, and fumonisin B_1 in corn fractions during dry and wet-milling. *Journal of Food Science and Technology* 59: 3192–3200. 10.1007/s13197-022-05373-9

Matsuo, Y., K. Takahara., Y. Sago., M. Kushiro., H. Nagashima., and H. Nakagawa. 2015. Detection of N-(1-deoxy-d-fructos-1-yl) fumonisins B_2 and B_3 in corn by high-resolution LC-Orbitrap MS. *Toxins* 7(9): 3700–3714. 10.3390/toxins7093700

Meca, G., A. Ritieni., and J. Mañes. 2012. Influence of the heat treatment on the degradation of the minor Fusarium mycotoxin beauvericin. *Food Control* 28(1): 13–18. 10.1016/j.foodcont.2012.04.016

Méndez-Albores, A., J. Veles-Medina, E. Urbina-Álvarez, F. Martínez-Bustos, and E. Moreno-Martínez. 2009. Effect of citric acid on aflatoxin degradation and on functional and textural properties of extruded sorghum. *Animal Feed Science and Technology* 150(3–4): 316–329. 10.1016/j.anifeedsci.2008.10.007

Mir, S. A., B. N. Dar, M. A. Shah, S. A. Sofi, A. M. Hamdani, C. A. F. Oliveira, M. H. Moosavi, M. H. A. M. Khaneghah, and A. S. Sant'Ana. 2021. Application of new technologies in decontamination of mycotoxins in cereal grains: Challenges, and perspectives. *Food and Chemical Toxicology* 148(February): 111976. 10.1016/j.fct.2021.111976

Mohammadi, H., S. M. Mazloomi, M. H. Eskandari, M. Aminlari, and M. Niakousari. 2017. The Effecteffect of Ozoneozone on Aflatoxinaflatoxin M_1, Oxidative Stability, Carotenoid Contentoxidative stability, carotenoid content and the Microbial Countmicrobial count of Milkmilk. *Ozone: Science & Engineering* 39(6): 447–453. 10.1080/01919512.2017.1329647

Moncini, L., S. Sarrocco, G. Pachetti, A. Moretti, M. Haidukowski, and G. Vannacci. 2020. N_2 controlled atmosphere reduces postharvest mycotoxins risk and pests attack on cereal grains. *Phytoparasitica* 48: 555–565. 10.1007/s12600-020-00818-3

Muhialdin, B. J., N. Saari, A. S. M. Hussin. 2020. Review on the biological detoxification of mycotoxins using lactic acid bacteria to enhance the sustainability of foods supply. *Molecules*, 25(11):2655. 10.3390/molecules25112655

Narváez, A., Y. Rodríguez-Carrasco, L. Castaldo, L. Izzo, G. Graziani, and A. Ritieni. 2020. Occurrence and exposure assessment of mycotoxins in ready-to-eat tree nut products through ultra-high performance liquid chromatography coupled with high resolution q-orbitrap mass spectrometry. *Metabolites* 10(9): 344. 10.3390/metabo10090344

Nelson, D. L., and M. M. Cox. 2014. *Princípios de Bioquímica de Lehninger*. Porto Alegre: Artmed.

Nešić, K., K. Habschied, and K. Mastanjević. 2021. Possibilities for the biological control of mycotoxins in food and feed. *Toxins*, 13(3): 198. 10.3390/toxins13030198

Nora, N. S., A. C. P. Feltrin, K. V. M. Sibaja, E. Badiale-Furlong, and J. Garda-Buffon. 2019. Ochratoxin A reduction by peroxidase in a model system and grape juice. *Brazilian Journal of Microbiology*, 50: 1075–1082. 10.1007/s42770-019-00112-3

Norred, W. P. 1982. Ammonia treatment to destroy aflatoxins in corn. *Journal Of Food Protection* 45: 6.

Nunes, V. M. R., M. Moosavi, A. M. Khaneghah, and C. A. F. Oliveira. 2021. Innovative modifications in food processing to reduce the levels of mycotoxins. *Current Opinion in Food Science* 38(April): 155–161. 10.1016/j.cofs.2020.11.010

Nyandieka, H. S., J. O. Maina, and C. Nyamwange. 2009. Destruction of aflatoxins in contaminated maize samples using ammoniation procedures, 5.

Oliveira, F. K., L. O. Santos, and J. Garda-Buffon. 2021. Mechanism of action, sources, and application of peroxidases. *Food Research International*, 143: 110266. 10.1016/j.foodres.2021.110266

Osaili, T. M., A. R. Al-Abboodi, M. A. L. Awawdeh, and S. A. M. A. L. Jbour. 2022. Assessment of mycotoxins (deoxynivalenol, zearalenone, aflatoxin B1 and fumonisin B1) in hen's eggs in Jordan. *Heliyon* 8(10): e11017. 10.1016/j.heliyon.2022.e11017

Oueslati, S., S. B. Yakhlef, P. Vila-Donat, N. Pallarés, E. Ferrer, F. J. Barba, and H. Berrada. 2022. Multi-mycotoxin determination in coffee beans marketed in Tunisia and the associated dietary exposure assessment. *Food Control* 140(October): 109127. 10.1016/j.foodcont.2022.109127

Pandiselvam, R., M. R. Manikantan, V. Divya, C. Ashokkumar, R. Kaavya, A. Kothakota, and S. V. Ramesh. 2019. Ozone: An advanced oxidation technology for starch modification. *Ozone: Science & Engineering* 41(6): 491–507. 10.1080/01919512.2019.1577128

Park, J. W., and Y. B. Kim. 2006. Effect of pressure cooking on aflatoxin B_1 in rice. *Journal of Agricultural and Food Chemistry* 54(6): 2431–2435. 10.1021/jf053007e

Parussolo, G., M. S. Oliveira, M. V. Garcia, A. O. Bernardi, J. G. Lemos, A. Stefanello, C. A. Mallmann, and M. V. Copetti. 2019. Ochratoxin A production by aspergillus westerdijkiae in Italian-type salami. *Food Microbiology* 83(October): 134–140. 10.1016/j.fm.2019.05.007

Peromingo, B., M. Sulyok, M. Lemmens, A. Rodríguez, and M. Rodríguez. 2019. Diffusion of mycotoxins and secondary metabolites in dry-cured meat products. *Food Control* 101(July): 144–150. 10.1016/j.foodcont.2019.02.032

Pineda-Valdes, G., D. Ryu, D. S. Jackson, and L. B. Bullerman. 2002. Reduction of moniliformin during alkaline cooking of corn. *Cereal Chemistry* 79(6): 779–782 10.1094/CCHEM.2002.79.6.779

Piotrowska, M. 2021. Microbiological decontamination of mycotoxins: Opportunities and limitations. *Toxins*, 13(11):819. 10.3390/toxins13110819

Porto, Y. D., F. M. Trombete, O. Freitas-Silva, I. M. Castro, G. M. Direito, and J. L. R. Ascheri. 2019. Gaseous Ozonationozonation to Reduce Aflatoxins Levelsreduce aflatoxins levels and Microbial Contaminationmicrobial contamination in Corn Gritscorn grits. *Microorganisms* 7(8): 220. 10.3390/microorganisms7080220

Potortì, A. G., A. Tropea, V. L. Turco, V. Pellizzeri, A. Belfita, G. Dugo, and G. Bella. 2020. Mycotoxins in spices and culinary herbs from Italy and Tunisia. *Natural Product Research* 34(1): 167–171. 10.1080/14786419.2019.1598995

Qi, L., Y. Li, X. Luo, R. Wang, R. Zheng, L. Wang, Y. Li, D. Yang, W. Fang, and Z. Chen. 2016. Detoxification of Zearalenonezearalenone and Ochratoxinochratoxin A by Ozoneozone and Quality Evaluationquality evaluation of Ozonised Cornozonized corn. *Food Additives & Contaminants: Part A* 33(11): 1700–1710. 10.1080/19440049.2016.1232863

Rastegar, H., S. Shoeibi, H. Yazdanpanah, M. Amirahmadi, A. M. Khaneghah, F. B. Campagnollo, and A. S. Sant'Ana. 2017. Removal of aflatoxin B_1 by roasting with lemon juice and/or citric acid in contaminated pistachio nuts. *Food Control* 71: 279–284. 10.1016/j.foodcont.2016.06.045

Rychlik, M., H. Humpf, D. Marko, S. Dänicke, A. Mally, F. Berthiller, H. Klaffke, and N. Lorenz. 2014. Proposal of a comprehensive definition of modified and other forms of mycotoxins including "masked" mycotoxins. *Mycotoxin Research* 30: 197–205. 10.1007/s12550-014-0203-5

Savi, G. D., K. C. Piacentini, K. O. Bittencourt, and V. M. Scussel. 2014. Ozone Treatment efficiency on fusarium graminearum and deoxynivalenol degradation and its effects on whole wheat grains (triticum aestivum l.) quality and germination. *Journal of Stored Products Research* 59(October): 245–253. 10.1016/j.jspr.2014.03.008

Schaarschmidt, S., and C. Fauhl-Hassek. 2018. The fate of mycotoxins during the processing of wheat for human consumption. *Comprehensive Reviews in Food Science and Food Safety* 17(3): 556–593. 10.1111/1541-4337.12338

Schabo, D. C., L. Freire, A. S. Sant'Ana, D. W. Schaffner, and M. Magnani. 2021. Mycotoxins in artisanal beers: An overview of relevant aspects of the raw material, manufacturing steps and regulatory issues involved. *Food Research International* 141 (March): 110114. 10.1016/j.foodres.2021.110114

Schmidt, M., E. Zannini, and E. K. Arendt. 2019. Screening of post-harvest decontamination methods for cereal grains and their impact on grain quality and technological performance. *European Food Research and Technology* 245: 1061–1074. 10.1007/s00217-018-3210-5

Schneweis, I., K. Meyer., G. Engelhardt., and J. Bauer. 2002. Occurrence of zearalenone-4-β-d-glucopyranoside in wheat. *Journal of Agricultural and Food Chemistry* 50(6): 1736–1738. 10.1021/jf010802t

Shen, M. 2021. Detoxification of aflatoxins in foods by ultraviolet irradiation, Hydrogen Peroxide, and Their Combination - A Review, 6.

Sibaja, K. V. M., S. O. Garcia, A. C. P. Feltrin, R. D. Remedi, M. B. R. Cerqueira, E. Badiale-Furlong, and J. Garda-Buffon. 2019. Aflatoxin biotransformation by commercial peroxidase and its application in contaminated food. *Chemical Technology and Biotechnology* 94(4): 1187–1194. 10.1002/jctb.5865

Siciliano, I., D. Spadaro, A. Prelle, D. Vallauri, M. C. Cavallero, A. Garibaldi, and M. L. Gullino. 2016. Use of cold atmospheric plasma to detoxify hazelnuts from aflatoxins. *Toxins* 8(5): 125. 10.3390/toxins8050125

Sobral, M. M. C., S. C. Cunha, M. A. Faria, Z. E. Martins, and I. Ferreira. 2019. Influence of oven and microwave cooking with the addition of herbs on the exposure to multi-mycotoxins from chicken breast muscle. *Food Chemistry* 276: 274–284. 10.1016/j.foodchem.2018.10.021

Tabata, S., H. Kamimura, A. Ibe, H. Hashimoto, and Y. Tamura. 1994. Degradation of aflatoxins by food additives. *Journal of Food Protection* 57(1): 42–47. 10.4315/0362-028X-57.1.42

Tibola, G. S., M. Z. Miranda, F. F. Paiva, J. M. C. Fernandes, E. M. Guarienti., and M. Nicolau. 2018. Effect of breadmaking process on mycotoxin content in white and whole wheat breads. *Cereal Chemistry* 95(5): 660–665. 10.1002/cche.10079

Tokuşoğlu, Ö., H. Alpas, and F. Bozoğlu. 2010. High hydrostatic pressure effects on mold flora, citrinin mycotoxin, hydroxytyrosol, oleuropein phenolics and antioxidant activity of black table olives. *Innovative Food Science & Emerging Technologies* 11(2): 250–258. 10.1016/j.ifset.2009.11.005

Tripathi, S., and H. N. Mishra. 2009. Studies on the efficacy of physical, chemical and biological aflatoxin B1 detoxification approaches in red chili powder. *International Journal of Food Safety, Nutrition and Public Health* 2(1): 69. 10.1504/IJFSNPH.2009.026920

Tripathi, S., and H. N. Mishra. 2010. Enzymatic coupled with UV degradation of aflatoxin B_1 in red chili powder. *Journal of Food Quality*, 33(1): 186–203. 10.1111/j.1745-4557.2010.00334.x

Trombete, F. M., Y. D. Porto, O. Freitas-Silva, R. V. Pereira, G. M. Direito, T. Saldanha, and M. E. Fraga. 2017. Efficacy of ozone treatment on mycotoxins and fungal reduction in artificially contaminated soft wheat grains: Efficacy of O_3 on mycotoxins and fungi. *Journal of Food Processing and Preservation* 41(3): e12927. 10.1111/jfpp.12927

Udomkun, P., A. N. Wiredu, M. Nagle, J. Müller, B. Vanlauwe, and R. Bandyopadhyay. 2017. Innovative technologies to manage aflatoxins in foods and feeds and the profitability of application – A review. *Food Control* 76: 127–138. 10.1016/j.foodcont.2017.01.008

Wang, L., Z. Wang, Y. Yuan, R. Cai, C. Niu, and T. Yue. 2015a. Identification of key factors involved in the biosorption of patulin by inactivated lactic acid bacteria (LAB) cells. *PlosOne*. 10.1371/journal.pone.0143431

Wang, L., T. Yue, Y. Yuan, Z. Wang, M. Ye, and R. Cai. 2015b. A new insight into the adsorption mechanism of patulin by the heat-inactive lactic acid bacteria cells. *Food Control* 50:104–110. 10.1016/j.foodcont.2014.08.041

Wang, B., N. E. Mahoney, Z. Pan, R. Khir, B. Wu, H. Ma, and L. Zhao. 2016. Effectiveness of pulsed light treatment for degradation and detoxification of aflatoxin B_1 and B_2 in rough rice and rice bran. *Food Control* 59: 461–467. 10.1016/j.foodcont.2015.06.030

Wang, N., W. Wu, J. Pan, and M. Long. 2019. Detoxification strategies for zearalenone using microorganisms: A review. *Microorganisms* 7(7):208. 10.3390/microorganisms7070208

Warth, B., P. Fruhmann, G. Wiesenberger, B. Kluger, B. Sarkanj, M. Lemmens, C. Hametner, J. Fröhlich, G. Adam, R. Krska, and R. Schuhmacher. 2015. Deoxynivalenol-sulfates: Identification and quantification of novel conjugated (masked) mycotoxins in wheat. *Analytical and Bioanalytical Chemistry* 407: 1033–1039. 10.1007/s00216-014-8340-4

Woldemariam, H. W., M. Kießling, S. A. Emire, P. G. Teshome, S. Töpfl, and K. Aganovic. 2021. Influence of electron beam treatment on naturally contaminated red pepper (*Capsicum annuum* L.) powder: Kinetics of microbial inactivation and physicochemical quality changes. *Innovative Food Science & Emerging Technologies* 67: 102588. 10.1016/j.ifset.2020.102588

Yagen, B., J. E. Hutchins, R. H. Cox, W. M. Hagler, and P. B. Hamilton. 1989. Aflatoxin B1S: Revised structure for the sodium sulfonate formed by destruction of aflatoxin B1 with sodium bisulfite, 4.

Zhang, Z., M. Li, C. Wu, and B. Peng. 2019. Physical absorption of patulin by *Saccharomyces cerevisiae* during fermentation. *Journal of Food Science and Technology* 56(4):2326–2331. 10.1007/s13197-019-03681-1

Zhu, Y., T. Koutchma, K. Warriner, and T. Zhou. 2014. Reduction of patulin in apple juice products by UV light of different wavelengths in the UVC range. *Journal of Food Protection* 77(6): 963–971. 10.4315/0362-028X.JFP-13-429

11 Food Irradiation for Food Safety

Md. Hasan Tarek Mondal and Md. Akhtaruzzaman

11.1 INTRODUCTION

Food safety is a serious issue of the current world with dominant human as well as environmental significances if ineffectually processed. With increasing the globalization of the food supply, ensuring the safety of supply for the consumers has become a worldwide collaborative effort. The major concern for confirming food safety can be demonstrated by the extent of foodborne illness around the globe. Food irradiation technology has been widely used to control the foodborne pathogen in food products during industrial processing. The International Atomic Energy Agency claims that ionizing radiation is used commercially in 33 nations across the world to process various food products (International Atomic Energy Agency, 2007). The commonly irradiated food items are fresh fruits, vegetables, poultry, beef and pork, shellfish and eggs. In recent decades, the safety issue of irradiated food products has been a considerable debate. The concerns of irradiated food over a period of time are toxicity compounds produced and their adequate nutritional characteristics of food that are processed by irradiation treatments. Food products that are treated by ionizing radiation not only meet the quarantine requirements but also promise an alternative method in food processing. The food material is transiently exposed to radioactivity in a radioactivity chamber at a predetermined speed to control the quantity of dosage engrossed by the particular food substances. According to the WHO, FAO and IAEA's prior experimental findings, the severe ionizing energy level should not exceed 10 kGy, and the energy of X-rays and the electron produced by appliance bases should be worked at 5 MeV to 10 MeV in order to halt radioactivity of treated food. Food is subjected to an exact amount of powerful radiant energy, often known as ionizing radiation, through a process termed "food irradiation." Chemical bonds can be broken by ionizing radiation. In beef, poultry and shellfish, irradiation can destroy harmful bacteria and other germs. Additionally, it can prevent the sprouting of potatoes, onions and other tubers and bulbs improve the shelf life of fresh fruits and vegetables, and de-infest spices. A versatile technique for food preservation is food irradiation. It is a secure food preservation technique that the U.S. Food and Drug Administration (FDA) and more than 60 other national food safety authorities have approved for use with a range of foods. Ionizing radiation that can be used to irradiate food includes gamma rays from cesium 137 (^{137}Cs) or cobalt 60 (^{60}Co), X-rays from machine sources operating at or below a 5 MeV energy level, and electrons from machine sources operating at or below a 10 MeV energy level (also known as E-Beam). Typically, the dosages of radiation applied to food are expressed in kilograys (kGy; 1 kGy = 1,000 Gy). Microorganisms' DNA reacts to radiation very quickly. Consequently, food irradiation damages microbial cells by reacting either directly or indirectly on DNA molecules. To ensure food preservation and safety, the precise dose of food irradiation is essential.

11.2 FOOD IRRADIATION

Irradiating food is a novel method for processing of food materials by using precise quantities of radiation (X-rays, γ-rays and faster electron) for destroying the pathogens that are present in food with maintaining safety and suitability. Food irradiation is a non-thermal food treating method for the disruption of microbial cell that achieves microbial safety with extending shelf life of a food product. With this method of processing, food is not rendered radioactive. Irradiation kills pathogens and makes them unable to reproduction. The following list of ionizing radiation that has been approved for use in food processing materials:

i. Cobalt-60 (^{60}Co) emits gamma radiation of minimum and maximum energy of 1.17 MeV to 1.33 MeV.

ii. Cesium-137 (^{137}Cs) provides maximum energy of 0.662 MeV.

iii. Electrons that have been accelerated to the highest energy of 10 MeV.

iv. X-rays with a maximum energy of 5 MeV.

An ionizing radiation dose of less than 1 kGy applications can preclude decay in onion, bulbs and tubers. It can also be used for preventing spoilage of fresh fruits, vegetables, dried fish, beef and also delay the ripening process of fruits. The applications of intermediary dose of 1–10 kGy with combination of refrigeration also prevent spoilage of stored products such as poultry, meat, egg,

DOI: 10.1201/9781003334859-11

seafood and strawberries. High ionizing energy doses in the diversity of 10–71 kGy are also applied in packaged food products such as meat, seafood, poultry and prepared foods and it is also applied for sterilizing some spices and food additives. These food products treated by radiation technology preserve their excellence features such as color, texture, taste and nutrient content (Ehlermann, 2016). Though ionizing radiation retains the quality of foods, chemical changes of food are also occurring during processing. The chemical alterations in food due to the application of radiation treatments are tiny and the possibility of toxic effects on patron is also insignificant. In the food irradiation technique, the aesthetical and nutritional features of foods are conserved to a higher and healthier rank rather than other methods of processing. Food irradiation is mainly applied for preservation purposes of food as well as sterilization application of various types of food items in processing industry. The main benefits of food irradiations are summarized below:

i. Destroying pathogenic microorganisms that are associated with food spoilage.

ii. Sprouting inhibition of potatoes and other bulbs during storage.

iii. Delay fruits ripening process.

iv. Disinfesting pests and parasites in horticultural produces.

v. Sterilizing of packed food items.

vi. Improves keeping qualities of various food products.

"Irradiated food" refers to food that has received ionizing radiation treatment. In the presence of ionizing radiation, food absorbs energy. The term "absorbed dosage" refers to the amount of energy absorbed, and it is expressed in units of Gray (Gy). Food absorbs energy from the environment, creating free radicals, which mix with other food molecules and generate short-lived compounds that kill microorganisms. Bacteria are killed by irradiation, which damages their DNA and proteins. The food irradiation process cannot be used for all food products. This drawback arises from the fact that the ionizing radiation significantly alters the flavor or texture of the product. Look for dairy products that have undergone pasteurization methods to ensure lower levels of exposure danger if you want to encourage high levels of food safety.

11.3 PRINCIPLES OF FOOD IRRADIATION

When particular amount of radiation passes through food substances, the energy is engrossed by the food substances and leads to excitation of the atoms and constituents of food molecules resulting in biological and some chemical changes recognized to happen during irradiation of food. The reaction of isomerization and dissociation (loss of hydrogen atom) happens within the molecules of the irradiated substances and produces free radicals. Usually the free radicals that are produced during irradiation require a very squat period of time and dried food products comprising bone will have inadequate mobility and persevere for a long time. The prime radiation technique is used in food processing industries. In food processing industries, gamma (γ) rays are mostly used for irradiation purposes that are obtained from the radioactive sources in the forms of cobalt (^{60}Co) or cesium (^{137}Cs). Hydrolysis and oxidative degradation are the principal impacts of irradiation on carbohydrates. The breakdown of complex carbohydrates produces simple compounds. Ionizing radiation may alter the physical and chemical composition of foods that are high in carbohydrates, although such alterations have no lasting nutritional impact. However, certain foods with very high sugar content lose their nutritional value and become unpleasant as a result of high ionizing radiation dosages. Due to the protective qualities of the food's constituents, vitamins in food are less ionizing radiation resistant than vitamins in pure solutions (Chauhan et al., 2009; Zanardi et al., 2018). It may be required to maintain low temperatures and exclude oxygen in order to protect vitamins A, E, K, and thiamin, particularly from medium and high dosage irradiation. The impact of radiation on folic acid is poorly known. Radiation's effects on the vitamin C content in fruits and vegetables have been the subject of conflicting reports. These differences may result from conversion to the physiologically active dehydroascorbic acid. Antithiamine and antivitamin B_6 tests on high-dose irradiation chicken and meat produced no results.

11.4 DEGREES OF FOOD IRRADIATION

Radiation energy secretes from one source to another in space in a quantifiable environment. Different measures of radiant energy are emanated from an electromagnetic spectrum. In the irradiation process, the ionizing energy is applied to materials such as food substance for achieving the sterilization before preservation of food by abolishing microbes, pests and parasites. Electromagnetic gamma (γ) and corpuscular beta (β) rays, which are produced by radioactive versions of materials like cobalt (^{60}Co) and cesium (^{137}Cs), are the main radiations employed in the food industry. Three procedures can be used to treat food products in the food processing industry using ionizing energy.

11.4.1 Radappertization

In this method, the food materials are treated with necessary amounts of doses to stop decay, poisonousness caused by organisms in food. This is also termed "sterilization." The obligatory dose is usually maintained in the range of 25–45 kGy.

11.4.2 Radicidation

In this method, the food substance is treated with a sufficient dose of ionizing radiation in order to decrease the quantity of feasible, pathogenic, non-spore-forming, bacteria and also inactivates foodborne parasites. The recommended dose is used in the range of 2–8 kGy.

11.4.3 Radurization

In this method, the food substance is treated with a sufficient dose of ionizing radiation to enhance its keeping quality by eliminating the numbers of substantial viable spoilage organisms. The essential dose is used in the range of 0.4–10 kGy.

For the goal of preserving food, ionizing radiation is frequently utilized. According to previous research findings, the main benefits of using ionizing radiation for food processing include preventing sprouting, delay of fruit and vegetable ripening and senescence, reducing pest infestation in agricultural products, destroying microbes responsible for spoilage and removing parasites and pathogens that are dangerous to human health. Through a significant decrease in the number of bacteria that cause rotting, radiation improves the keeping quality of some goods (Fan *et al.*, 2008; International Atomic Energy Agency, 2005). The process of irradiation is dependent on the amount of radiation exposure. The range of modest radiation doses of 0.02–0.2 kGy can suppress the sprouting of yams, potatoes, onions, garlic and other commodities. The dose range of 0.2–1 kGy is employed for physiological processes that take place over time, such as the ripening of fruits, and the 1–7 kGy dose range is employed in such treatments to expose fresh meat and seafood, depending on the product (Gautam and Tripathi, 2016).

11.5 MECHANISM OF MICROBIAL DESTRUCTION OF IRRADIATED FOOD

The mechanism regarding microbial destruction during irradiation of food happens primarily on the basis of interruption of the cell membrane or cellular bustle through damaging the DNA of organisms (Erkmen and Bozoglu, 2016). During ionizing radiation, the electron deletion occurs from a radioactive source, resulting in energy deposition transpired on mark molecules, for example, DNA directly. Water in nearby molecules is available in the cell of organisms. When ionizing treatment is applied in food, the molecules of water misplace an electron and produce H_2O^+ and e^-. The product formed by radiolysis of water reacts with the additional molecules to yield a variety of chemical compounds such as hydrogen, oxygen, hydroxyl radicals ($^{•}$OH), hydrogen peroxide (H_2O_2) and molecular hydrogen. The formed chemical compounds are very much reactive and cause breakdown at the sugar-phosphate bonds of single-strand DNA and also collapse the bonds that are connected with the base pair towards a reverse in double-strand DNA (Erkmen and Bozoglu, 2016). As a result, bacterial cell death occurs. Generally, bacteria, yeasts and viruses are less sensitive to irradiation than molds. The success of inactivating microorganisms depends on a number of variables, including the microorganisms' physiological state, bacterial strain, the composition of the food, the dose utilized and capacity to repair radiation-induced DNA damage (Farkas *et al.*, 2014). The figure of the microbial destruction mechanisms of gamma rays on food samples is illustrated in Figure 11.1.

A change in the membrane's lipid composition, particularly in the polyunsaturated fatty acids, disturbs the membrane's normal structure and has an impact on the permeability of the membrane, among other membrane activities. Additionally, membrane enzyme activity may be impacted (Erkmen and Bozoglu, 2016). Most predominant foodborne pathogens and the main food spoilage microorganisms are usually sensitive to irradiation and can be rendered inactive by low and medium doses of radiation between 1 and 7 kGy. At the same radiation dose, cocci are more

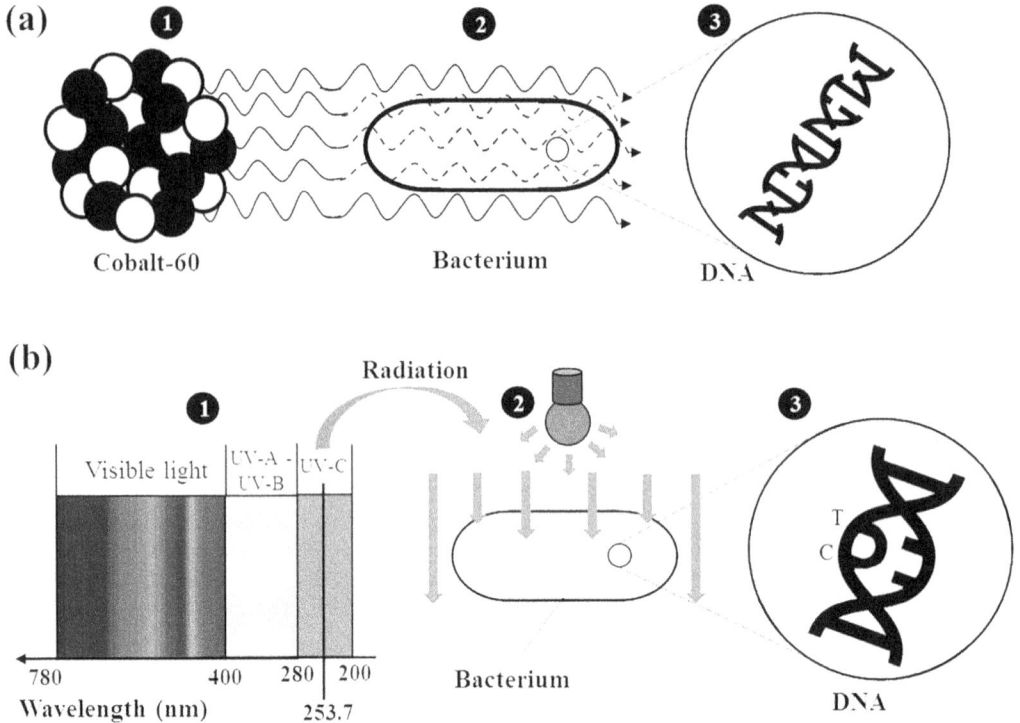

Figure 11.1 The ^{60}Co irradiation source that generates gamma rays (Rosario *et al.*, 2020) [(1) with the ability to penetrate structures and bacteria is the microbial inactivation source (a) (2) Gamma rays harm the DNA of the microbe, making the cell inert (3) The inactivation of microbes by UV-C light (b): Mercury lamps emit UV-C photons with a maximum efficacy of 253.7 nm (1) Influence the organisms (2) The production of pyrimidine dimers (Thymine-Cytosine) is primarily responsible for the inactivation and slowing of microbial growth (3)].

resistant to radiation than rods, and Gram-positive bacteria are more radiation resistant than Gram-negative bacteria. Different isolates and strains of the same species of bacteria may have different irradiation sensitivity. Microbes' ability to adapt to stressful environments may lead to an increase in their resistance to radiation. Some of the harmful bacteria may be able to endure higher radiation doses due to certain resistance genes to specific medications (Skowron *et al.*, 2018). The matrix composition, the length of time used to apply the method, the intensity of application, inactivation temperature and the resistance of the bacteria all affect how effectively microorganisms are inactivated (Figure 11.2). Viruses and yeasts have high radiation tolerance as well (Calado *et al.*, 2014). Additionally, vegetative microbial cells are more sensitive than spores.

11.6 IRRADIATION FOOD PRODUCTS

Food processing with ionizing radiation has some beneficial effects, such as extending shelf life of product, inactivation of pathogenic bacteria, insects, parasites, yeast and molds, delaying fruit and vegetable ripening and sprouting inhabitation of tuber crops. The effects can be achieved by using lower radiation exposures. The potential application of different doses of radiation on food products, their effects are summarized in Table 11.1.

11.7 SAFETY ISSUE OF IRRADIATED FOOD

The safety issue of irradiated food is an important aspect for the present century. Evidence has proven the safety issue of irradiated food. Over 50 years of research on the food safety of irradiated food has proven their evidence against safety approach of irradiated food. The Food and Drug Administration (FDA) has claimed that the irradiation processing is safe by conducting research for

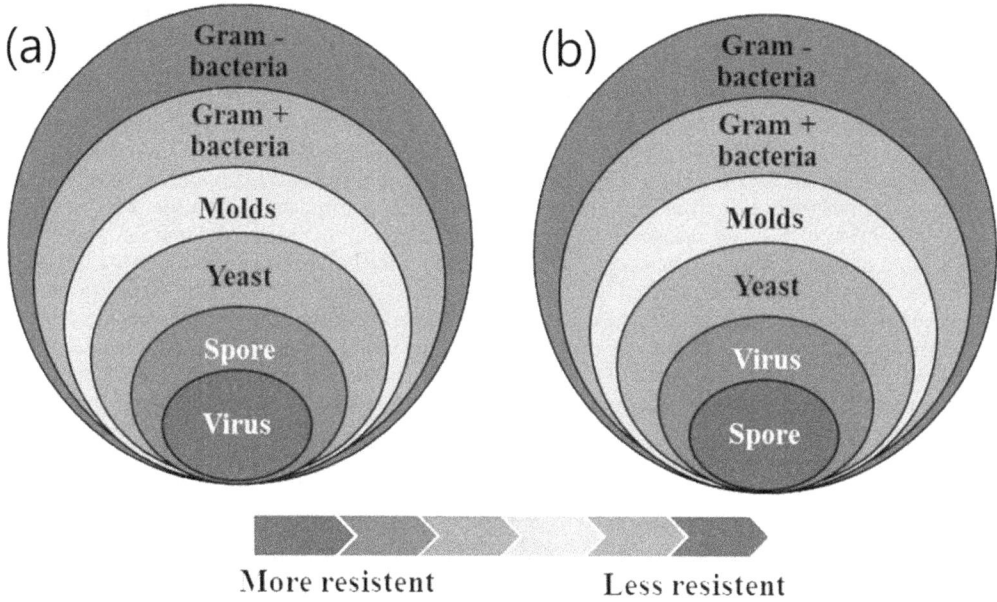

Figure 11.2 Microbial resistance levels (a) Gamma Radiation (Mahapatra *et al.*, 2005), (b) UV-C Radiation.

Table 11.1: A Variety of Radiation Doses and Their Potential Applications

Dose Range	Effects/Purpose	Foodstuff
Lower Dose (around 1 kGy)	Avoid sprouting (potatoes, onions, garlic, etc.) After harvesting; eradicate insects and larvae in wheat, flour, fruits and vegetables. Slow process of ripening. Kill several harmful food-related parasites.	Bananas, mangoes, other non-citrus fruits, grains and pulses, dehydrated vegetables, dry fish and meat, fresh pork, potatoes, onions, garlic and ginger.
Moderate dose (1 to 10 kGy)	Severe reduction in or elimination of specific parasites and microorganisms that cause food deterioration. Decrease or eradication of numerous pathogenic bacteria.	Berries, grapes, dried vegetables, fish, poultry, raw or frozen seafood and meat.
Sterilization at a high level (10 to 50 kGy)	Food sterilization for specific applications, such as meals for patients with immune system compromises. Several disease-causing viruses are eliminated.	Sterilized food for immunocompromised patients.

more than 30 years. A combined report prepared by FAO, IAEA and WHO stated that food treated with any dose is acceptable, palatable and healthy. This statement guides strong evidence that any food exposed to incorrect irradiation treatment may have lost its vital assets but is not risky for consumption (Filho *et al.*, 2014). According to a 1981 report, the WHO, FAO and IAEA team on the wholesomeness of irradiated food foodstuffs established a maximum radiation level of 10 kGy as being safe and wholesome. When energy was delivered up to 10 kGy, toxicological risk was absent, and there were no obvious negative impacts on nutrition or microbiology. According to a different conclusion from the same panel, irradiated foodstuffs should continue to be calculated separately while taking into account the scientific requirement and their safety. It is appropriate to set a maximum dosage level for the treatment of some food products by ionizing radiation. Without significantly harming nutrition or microbiology, the energy carried up to 10 kGy did not pose a toxicological risk. It is immobile considered to be a problem and defensive principles should be used until such documents are obtainable because there are insufficient documents on the effects of eating a diet consisting of irradiated foods on long-term health impacts and on the effects of long-term consumption of irradiated foods on human health. However, in vitro research has shown that 2-ACBs have tumor-promoting properties. According to a recent scientific view from the European Food Safety Authority on chemical safety of irradiated food along with their inherent treatments, the most radiolytic compounds are 2-ACBs, hydrocarbons and cholesterol. In food products, oxides and furans are also generated that have undergone various food processing procedures and are not just formed by irradiation. Regardless of the intended purposes for the treatment, evidence supports the safety of food and food products or raw materials used as ingredients exposed to ionizing radiation (Vaz et al., 2012). Additionally, the amounts that they increase in food that has been exposed to radiation were not noticeably higher than those that result from heat treatments. The majority of the information from the most recent literature regarding various biological risks supports the food classes and radiation levels mentioned, as per prior assessments by the Scientific Committee on Food of the European Commission. Another crucial topic of the 21st century is the toxicological perspective of foods that have been exposed to ionizing radiation. Radiolytic compounds including short-chain hydrocarbons and formaldehyde are formed as a result of the food irradiation process, which also creates reactive radicals. 2-Alkyl-Cyclobutanones are produced when triglycerides are radioactively lysed (2 ACBs). Because these molecules are unique, they are frequently used as indicators to distinguish foodstuffs that have undergone radioactivity.

11.7.1 Radioactivity Safety

As a result of exposure to common ambient radiation, foods are normally radioactive to variable degrees. Food products exposed to gamma radiation (cobalt-60) do not become radioactive since no neutrons are secreted during the treatment and no atomic variations occur in the food molecules' nucleus matrix (Loaharanu, 2007). It is highlighted that the redox potential is simply one of the variables that significantly affects how reactive an irradiated aqueous system, such as food, is when it comes to radiolysis of water. The electron concentration at the reaction site and the activation energy play further roles in determining the rates of reactions between the radiolysis product of water and reactive moieties in substrates. IAEA studies revealed that the increase in background radiation exposure from eating food exposed to gamma rays from cobalt-60 or cesium-137, 10 MeV electrons, or X-rays produced by electron beams with energy below 5 MeV is negligible and is best described as zero.

11.7.2 Toxicological Safety

Safety issue of irradiated food based on toxicological evidence is the most important because during irradiation, series of radiolytic product called cyclic ketones (2-alkylcyclobutanes) are formed. These cyclic chemicals are created when irradiating fat-counting food products. Short-term studies have been started to date to look into probable toxicological compounds and their health risks. In the cell line of bacteria and humans, Hartwig et al. (2007) examined the toxic and colon carcinogenic potential of several pure synthesized 2-alkylcyclobutanones. Various pure synthetic 2-alkylcyclobutanones were examined by Hartwig et al. (2007) in bacterial and human cell lines for their potential to be harmful and to cause colon cancer. Although the Ames test did not indicate any mutagenic action in Salmonella strains, they discovered that significant cytotoxicity was observed in bacteria. Genotoxicity was suggested by the alkaline unwinding process' detection of DNA base damages in mammalian cells but not by the comet assay. The makeup of triacylglycerol's fatty acid profile was also discovered to affect cytotoxicity and genotoxicity. The radiolytic products found in radiation-treated beef were studied by the Federation of American Societies for Experimental Biology between 1976 and 1979, and they came to the conclusion that the quantities that would be consumed would not have any toxicological effects. All of the detected radiolytic compounds are also present in non-irradiated food products. After 35 years of investigation, no compounds that can only be detected by the food irradiation process have been found. There is no conclusive incontestable proof that irradiated foods contain radiation-formed carcinogens or other hazardous elements, according to extensive toxicological research. Last but not least, a World Health Organization expert committee formed in 1980 examined all the data from 1,200 studies and suggested that its member countries allow any food irradiated up to an average dose of 10 kGy without the need for additional testing. These studies, which were carried out in the middle of the 1980s, showed no negative outcomes. Animals fed diets that were irradiated with doses between 25 and 50 kGy, which is far greater than the level used for human food, did not experience any mutagenic, teratogenic, or oncogenic harmful effects that may be related to eating irradiated food for several generations (Kava, 2007).

11.7.3 Microbiological Safety

The important effect of radiation on microorganisms depends on the interaction of free radicals with their DNA and RNA. Depending on the strains and physiological condition of the strains employed, different microorganisms have varying radiation resistance. Cells under stress characteristically exhibit amplified levels of radiation resistance. The pathogenic and decay organisms like *L. monocytogenes, Staphylococcus aureus, Salmonella, E. coli* O157:H7, *Campylobacter, Yersinia enterocolitica*, yeast and mold can be successfully removed from food by irradiation. Raw meats' sanitary quality can be improved by food irradiation by avoiding potential health risks. The use of radiation activity during the processing of foodstuffs depends on sensitivity of the microbes, the species, size, spores and various microenvironments, as well as the physical condition of the food, the oxygen content and the presence of chemical substances such as proteins, sulfites, nitriles and sulfhydryl compounds. Low and medium dose (1 to 7 kGy) radiation can inactivate Gram-negative and spoilage bacteria like *Enterobacteriaceae* and *Pseudomonas* spp. since they are overly susceptible to it. Ionizing radiation greatly lengthens the life of food with acceptable range through eliminating germs (1 to 3 years). Additionally, if the food was contaminated after irradiation, it has been predicted that irradiated foods might be more favorable to the proliferation of foodborne diseases.

11.7.4 Nutritional Safety of Food Irradiation

Food chemical composition can be altered by irradiation, which alters the nutritional value of the product. With a dosage of 1 kGy, irradiation does not result in any appreciable loss of macronutrients (Indiarto *et al.*, 2020). Even at doses above 10 kGy, irradiation has no effect on the nutritional value of proteins, lipids and carbohydrates; however, there may be some sensory effects (Ehlermann, 2014). According to animal feeding experiments, irradiating food at any practical interest dosage level will not have an adverse effect on the nutritional content of these ingredients (Nair and Sharma, 2016). The nutritional value and quality of processed food items can be preserved with the right treatment, such as radiation conditions paired with packing methods (Indiarto *et al.*, 2020). From a nutritional point of view, irradiation has little effect on the digestibility and amino acid composition of proteins, so meat, poultry and fish are good choices for irradiation (Prakash, 2020). In a similar manner, chicken, mackerel and cod post-irradiation do not appear to have an impact on the biological value and digestibility of dietary proteins or amino acid pattern (Woodside, 2015). The impact of radiation on proteins depends on their state, structure and content, as well as whether they are native or denatured, dry or in solution, liquid or frozen and the presence or absence of certain other chemicals. However, long-term feeding studies found that irradiating raw and cooked beef (up to 70 kGy) does not affect their digestion or nutritional content and only serves to extend the shelf life (Pedreschi and Mariotti-Celis, 2020; Woodside, 2015).

Carbohydrates are substantially less vulnerable to radiation when consumed as dietary ingredients than when utilized in pure form. Carbohydrates are protected by the protein (Harder *et al.*, 2016). There were no discernible modifications in the bulk density or composition of wheat flour or starch following gamma irradiation. It's interesting to note that, depending on the dosage, the flour's amylose content rose by 25% to 36% (Bashir *et al.*, 2017). Based on the fruit type, the stage of ripening, and the radiation dose, irradiation can either increase or decrease the amount of sugars in fruits. Individual sugar alterations caused by radiation are similarly minimal. As an illustration, modest doses of irradiation result in minor but considerable increases in the glucose content of Custard apples and decreases in the glucose and sucrose contents of mangoes and Imperial mandarins (Prakash and Ornelas-Paz, 2019). Fruits' viscosity or hardness may alter as a result of the hydrolysis or depolymerization of carbohydrates like pectin and starches (Prakash, 2020). The most vulnerable macromolecule to food radiation is lipids (Prakash, 2020). Depending on the composition, irradiation causes fatty acids and lipids to undergo oxidation, polymerization, decarboxylation, and dehydration and releases a variety of chemicals

(Ravindran and Jaiswal, 2019). Foods with high lipid content (52%–70%), and particularly those with a high amount of unsaturated fatty acids, are especially vulnerable to radiation because the formation of free radicals during radiation accelerates the oxidation of lipids. The susceptibility to oxidation and, thus, to radiation treatment increases with the extent of unsaturation (Sajilata and Singhal, 2006; Štajner *et al.*, 2007). Irradiating at freezing temperatures and packaging to block the effects of light and oxygen can prevent the alterations that occur in lipids (Ravindran and Jaiswal, 2019). The results of several investigations on the effects of meat irradiation on lipids have demonstrated that, at low doses of radiation, lipids in the presence of their endogenous antioxidants are not especially vulnerable to radiation-induced peroxidation (Roberts, 2016). Additionally, rather than only affecting fat, when food is exposed to radiation as a whole, all of its contents are affected (Harder *et al.*, 2016). Irradiation has little to no effect on minerals, just like it has little to no effect on macronutrients. Food irradiation can lower the amount of minerals that are easily obtainable, but for those who eat a balanced diet, the reduction does not result in a significant loss of nutrition (Bevelacqua and Javad Mortazavi, 2020; Woodside, 2015).

Foods containing the very sensitive vitamins A, B1, C and E can be altered by irradiation. Free radicals produced by irradiation are mostly to blame for these changes (Prakash and Ornelas-Paz, 2019). The amount of radiation used; the makeup of the food; how it is packaged, stored, and processed; as well as those circumstances' temperature and oxygen content will all affect how much vitamin is actually lost (Woodside, 2015). Vitamins B1 and E, which are the most vulnerable to radiation, can be preserved not only by excluding oxygen, but also by low-temperature irradiation. The kind of product and irradiation dose have an impact on vitamin C losses as well (Nair and Sharma, 2016). Vitamin C cannot be changed by low doses of radiation (1 kGy) (Prakash and Ornelas-Paz, 2019). When exposed to radiation, vitamin C partially transforms into dehydroascorbic acid, which also possesses vitamin C action in people (Nair and Sharma, 2016). To reduce impacts on organoleptic changes and vitamin losses, food is frequently irradiated at dosages lower than 10 kGy (Woodside, 2015). There are certain less sensitive vitamins in diet. However, not all vitamins are equally sensitive to irradiation when properly handled during irradiation (Roberts, 2016). Table 11.2 lists the relative sensitivity to irradiation of certain vitamins. Vitamin radiation sensitivity decreases in the order shown below (Stefanova *et al.*, 2010):

Fat-soluble vitamins: vitamin E > carotene > vitamin A > vitamin D > vitamin K

Water-soluble vitamins: vitamin B1 > vitamin C > vitamin B6 > vitamin B2 > folate, niacin, vitamin B12

The nutritional value of irradiated foods is often on pace with or higher than that of non-irradiated foods that have undergone standard processing. Irradiation can be harmful to the nutritional

Table 11.2: Vitamins' Relative Susceptibility to Radiation Treatment (Pedreschi and Mariotti-Celis, 2020; Woodside, 2015)

High Sensitivity	Low Sensitivity
Vitamin C [W]	Carotene [F]
Vitamin B1 (thiamin) [W]	Vitamin D [F]
Vitamin E [F]	Vitamin K [F]
Vitamin A (pyridoxine) [F]	Vitamin B6 (pyridoxine) [W]
	Vitamin B2 (riboflavin) [W]
	Vitamin B12 (cobalamin) [W]
	Vitamin B3 (niacin) [W]
	Vitamin B9 (folate) [W]
	Vitamin B5 (pantothenic acid) [W]

Notes
[W] Water-soluble vitamin
[F] Fat-soluble vitamin

content of food, but conventional heating and drying techniques can be far worse. Nutrient losses from irradiation won't be significant enough to have a negative impact on people's or populations' nutritional state. Due to the irradiation process, irradiated food manufactured in accordance with recognized good manufacturing practice is recognized safe and nutritionally appropriate (Nair and Sharma, 2016; Nishihira, 2020).

11.8 HEALTH CONCERN OF IRRADIATED FOOD

Ionizing energy is primarily used to treat food with the intention of enhancing or protecting human health by eliminating disease-causing microorganisms and avoiding the use of chemicals that are frequently used to preserve or disinfect food products and may potentially leave behind toxic or cancer-causing residues. One of the biggest health risks in the world, according to the World Health Organization, is disease-causing organisms contaminating food. By extending the shelf lives of food while keeping their nutritional value and organoleptic (sensory) qualities (such as taste, odor, color and texture), using ionizing radiation to preserve food can significantly improve the health of humans by eliminating malnutrition at a global scale. Food that is prevented from spoiling has a financial value and benefits for human health that far outweigh the cost of irradiation. Although food irradiation is thought to be safe, concerns have been raised about the effects of irradiated food on health. Numerous scientific inquiries have yielded both favorable and unfavorable results. People who have impaired immune systems are particularly vulnerable to infections spread through their food. Patients with AIDS or receiving chemotherapy may be given treatment to sanitize the digestive tract and placed on a sterile diet in order to prevent infection problems. Foods may be sterilized with irradiation and then kept for several months without refrigeration (Eustice, 2020; Mohácsi-Farkas, 2016). High-dose irradiation has been used recently to prepare meals for infants and individuals with impaired immune systems (Feliciano, 2018).

One of the main health issues in the world today is food allergies, which have been rising alarmingly in recent years. Food allergens are typically proteins, which have a complicated spatial structure and are made up of several amino acids. Allergenicity may be decreased by changing the proteins' spatial arrangement. Alternately, free radicals such $\bullet OH$, $\bullet H$, and e_{aq}^- may oxidize and deoxidize the amino acids in allergy proteins, affecting them. Thus, peptide bonds experience breakage and crosslinking, which results in the elimination of allergen native constructions. The majority of food allergens have been successfully eliminated using various ways by irradiation, which has been demonstrated to be effective in practice (Pan et al., 2021). The use of irradiation technology to combat lectins, the most prevalent cause of food intolerance and their immunological and allergy consequences has been proven to be both safe and efficient. As a result, it has been suggested as a means to lower or completely eradicate food allergenicity (Vaz et al., 2011). Most nations' usual diets include nitrates, nitrites and nitrosamines. Food additives called nitrates and nitrites give processed meat its color and taste. These additives have been found to have the potential to cause cancer. The majority of nitrosamines can cause DNA adductions and gene mutations in animals, which can lead to carcinogenesis (Song et al., 2015). Numerous studies have looked into the possible application of radiation as a food preservation technique to reduce unwanted and hazardous chemicals in food products like N-nitrosamine, residual nitrite and biogenic amines (Shalaby et al., 2016; Yousefi and Razdari, 2015). The production of nitrosamine and nitrite-related compounds in cured meat can be managed by irradiation. To preserve the flavor and color, sterilizing dosages of irradiation on cured meat dramatically reduces or eradicates the concentrations of nitrates and nitrites (Ravindran and Jaiswal, 2019). Also, ionizing radiations have proven to be effective in plummeting biogenic amines in matured cheese during the period of six months of storage (Shalaby et al., 2016; Yousefi and Razdari, 2015).

During the irradiation process, harmful bacteria in food are killed or destroyed by the production of reactive molecules such free radicals and hydrated electrons. When electrons hit water molecules, these chemicals are created. However, in addition to microbial disinfection, other chemical processes are also started, which have the potential to produce a number of compounds and alter the characteristics of irradiated foods. The new chemical compounds that are produced are directly related to the composition of the foodstuff. Among all the substances produced during irradiation, benzene, toluene, formaldehyde, and malonaldehyde have prompted the most safety concerns (Ravindran and Jaiswal, 2018). There have been reports of formaldehyde and malondialdehyde being powerful mutagenic and skin tumor-causing agents (in mice). According to a study, formaldehyde-exposed mice develop chromosomal abnormalities while going through the spermatogenic process (Ravindran and Jaiswal, 2018). Among dietary pollutants, benzene is reported to possess one of the maximum carcinogenicity values. Leukemia can develop after consuming benzene over a long period of time (Johnson *et al.*, 2007). According to studies, excessive toluene exposure in mice inhibited hippocampal neurogenesis while having no effect on the organs' other functions (such as the lungs, liver, and kidney) (Kim *et al.*, 2020). Cancer incidence will rise as a result of a rise in the amount of a mutagen in food caused by irradiation. There will be an increase in the incidence of cancer for several decades after food irradiation is eventually prohibited. 2-alkylcyclobutanones (ACBs) are triglyceride radiolytic derivatives that are produced only by irradiating foods that contain fat. Some hazardous effects were to be predicted due to their unique chemical structure. These ingredients are present in extremely low amounts in irradiated food (Ehlermann, 2014). Over the past ten years, a number of experiments have been carried out to ascertain the toxicological and mutagenesis consequences linked to 2-ACB ingestion. These investigations have shown that at low doses, 2-ACBs have no mutagenic or genotoxic effects on cell lines of mammals. However, rat and human colon cells have experienced cytotoxicity and genetic damage when exposed to these compounds at greater levels. The researchers came to the conclusion that while these substances have little toxicological potential, eating irradiated fattening food won't have any negative effects on people (Arvanitoyannis and Dionisopoulou, 2010; Ravindran and Jaiswal, 2019). Due to ascorbic acid, fructose, sucrose and glucose, ionizing radiation caused the production of furan in apple and orange juices. After the first three days of storage, furan levels in both juices continued to rise as the radiation dosage rose from 0 to 5 kGy. The increase in furan during the earlier period of storage may have been caused by the after-effects of irradiation (Vranova and Ciesarova, 2009). Furan is toxic to the liver in rats and mice. It causes cholangiofibrosis in rats and hepatocellular adenomas or cancers in mice. There is little evidence of chromosomal damage occurring in living organisms, and the underlying process is not well understood (EFSA Panel on Contaminants in the Food Chain (CONTAM) et al., 2017). Irradiation is never an appropriate solution for fixing a contamination problem that already contains mold or mycotoxins. Food contaminated with mycotoxins and reduced in quality and quantity due to the presence of mold can have serious negative effects on health (Calado *et al.*, 2014).

11.9 RISK ASSESSMENT OF IRRADIATED FOOD

Risk assessment is a methodical process used to find potential risks to human health or the environment posed by certain activity. It is basically a method of detecting potential health risks that could arise from a specific set of activities and evaluating the impact. Two categories exist within the process of risk analysis: Firstly, a qualitative process (process of identifying major hazards to which an initiative is unprotected) and secondly, quantitative (offers a direct correlation to the assessment of the hazards that require safety). Risk assessment covers a wide range of actions, such as pest control, genetically modified organisms, and preservation methods (i.e., migration in packaging and irradiation). It is crucial to emphasize that few research have been conducted in this field, and those that have are of limited validity regarding the risk assessment of irradiation (X-rays, e-beam, irradiation and microwave). Therefore, conducting epidemiological research with reference to risk assessment of consuming irradiated food for both animals and humans is a priority. The goal of these studies is to shed more light on the risks associated with irradiation exposure and eating irradiated feed and food.

Assessments of the effect and potential danger of radioactive contamination are typically based on data about the source term or contamination density ($Bq\ m^{-2}$) of specific radionuclides, transport in soils, transport to vegetation or animals, and biological uptake and accumulation (for example, for fish, the concentration ratio is the Becquerel/kg tissue per Becquerel/L water). The average bulk mass or surface concentration is typically used to estimate ambient radioactivity because radionuclides in sample matrices are considered to be uniformly distributed as simple atomic, molecular or ionic species. Risk assessments ought to be based on the greatest available science, according to experts and decision-makers from all sectors. Using a variety of nuclide-specific measurement results of activity concentrations in sand, air, flora and products as well as outdoor dose rates, radio-ecological models were employed to determine the upper level of present and future absorbed radiation doses for populations in various regions of the German Federal Republic. Any food that is irradiated or that contains an irradiated ingredient must be labeled with the word "irradiated" prominently, either as part of the main label or right next to the irradiated ingredient. Additionally, it might (optionally) display the "Radura" symbol, a universal symbol indicating radioactive food.

11.10 CONCLUSIONS

Ionizing radiation is used in the food business as a method of controlling harmful bacteria and parasites in foods, preventing postharvest food losses and extending the shelf life of perishable foods. Numerous studies have been accomplished to assess the nutritional value and safety issue of irradiated foods. Despite the fact that there is evidence to suggest that irradiation processing causes chemical changes, irradiated foods are just as safe as those that have undergone other common processing techniques, like heating, pasteurization and canning. In order to increase the safety of food products, ionizing radiation has been explored for the processing of food for many years. The process of food irradiation is currently a recognized technique for preventing harmful bacteria causing human foodborne disorders. Processing different food products using this method is both safe and efficient. According to evidence from numerous studies, the impact of ionizing radiation on approved food and food products eradicated harmful microorganisms like *E. coli* and *Salmonella*, two bacteria that are responsible for food spoilage, as well as numerous types of dangerous foodborne illnesses. Irradiation tries to stop the process of sprouting of tubers and ripening of fruits so that food can be maintained for a lengthier period of time without suffering a momentous loss of nutrients. Food is irradiated in an enclosed procedure that can be returned for further processing or discarded. The process of irradiating food also has a respectable safety record, unlike radioactive resources. It has no negative effects on the environment or on people's bodies. In fact, there haven't been enough reports of potential health problems associated with eating irradiated food, and the food doesn't turn radioactive. The major issue with irradiation technology is how consumers feel about using ionizing radiation in the food processing sector. To dispel the myth that food irradiation uses nuclear energy, the acceptance of irradiation technology depends on adequate education, support and communication. The majority of people hold the belief that eating food and eating materials exposed to ionizing radiation poses harm to their health and increases the possibility of creating malformed organisms. Some people are also scared of the potentially disastrous incidents that might take place at the food radiation factories and think that if more food irradiators are permitted, the possibility of accidents will rise. Yet, compared to other foods, the cost of irradiation foods is notably higher. Another issue is how customers react to the irradiated food products; various studies have been conducted on this subject. Research outcomes have consistently revealed that a number of peoples have fallacies about food irradiation process and they think that irradiation makes food unsafe. According to the findings of various studies, the buying rate of irradiated food and food items will increase if people are conscious about the irradiation process and its control on food pathogens. The chief difficulty to acquiring irradiation technique is the deficiency of adequate knowledge and its supporting technologies, which led to several disagreements. In order for consumers to make an informed decision, it is vital to supply exact data and employ the food irradiation technique as a substitute technology that is safe for food and food products preservation. In conclusion, the myth that irradiation technology is used for fraud and death should be dispelled at this time, and it is urgent to use this wonderful source of ionizing radiation as irradiation technology for food processing and preservation in order to increase the security and safety of the world's food supply.

REFERENCES

Arvanitoyannis, I. S., and Dionisopoulou, N. K. (2010). Risk assessment of irradiated foods. In *Irradiation of Food Commodities* (pp. 141–168). Academic Press.

Bashir, K., Swer, T. L., Prakash, K. S., and Aggarwal, M. (2017). Physico-chemical and functional properties of gamma irradiated whole wheat flour and starch. *LWT-Food Science and Technology* 76(Part A): 131–139.

Bevelacqua, J. J., and Javad Mortazavi, S. M. (2020). Can irradiated food have an influence on people's health? In *Genetically Modified and Irradiated Food: Controversial Issues: Facts versus Perceptions* (pp. 243–257). Academic Press.

Calado, T., Venâncio, A., and Abrunhosa, L. (2014). Irradiation for mold and mycotoxin control: A review. *Comprehensive Reviews in Food Science and Food Safety* 13(5): 1049–1061.

Chauhan, S. K., Kumar, R., Nadanasabapathy, S., and Bawa, A. S. (2009) Detection methods for irradiated foods. *Comprehensive Reviews in Food Science and Food Safety* 8(1): 4–16.

EFSA Panel on Contaminants in the Food Chain (CONTAM), Knutsen, H. K., Alexander, J., Barregård, L., Bignami, M., Brüschweiler, B., Ceccatelli, S., Cottrill, B., Dinovi, M., Edler, L., Grasl-Kraupp, B., Hogstrand, C., Hoogenboom, L., Nebbia, C. S., Oswald, I. P., Petersen, A., Rose, M., Roudot, A. C., Schwerdtle, T., …Wallace, H. (2017). Risks for public health related to the presence of furan and methylfurans in food. *EFSA Journal* 15(10): e05005.

Ehlermann, D. A. E. (2014). Safety of food and beverages: Safety of irradiated foods. In *Encyclopedia of Food Safety* (Vol. 3, pp. 447–452). Academic Press.

Ehlermann, D. A. E. (2016). The early history of food irradiation. *Radiation Physics and Chemistry* 129: 10–12.

Erkmen, O., and Bozoglu, F. (2016). Food preservation by irradiation. In *Food Microbiology: Principles into Practice* 1st ed. (Vol. 2, pp. 106–127). John Wiley & Sons, Ltd. Published.

Eustice, R. F. (2020). Novel processing technologies: Facts about irradiation and other technologies. In *Genetically Modified and Irradiated Food: Controversial Issues: Facts versus Perceptions* (pp. 269–286). Academic Press.

Fan, X., Niemira, B. A., and Prakash, A. (2008). Irradiation of fresh and fresh-cut fruits and vegetables. *Food Technology* 3: 36–43.

Farkas, J., Ehlermann, D. A. E., and Mohacsi-Farkas, C. (2014). Food technologies: Food irradiation. In *Encyclopedia of Food Safety* (pp. 178–186). Elsevier.

Feliciano, C. P. (2018). High-dose irradiated food: Current progress, applications, and prospects. *Radiation Physics and Chemistry* 144: 34–36.

Filho, L. T., Lucia, S. M. D., Limaa, R. M., Scolforoa, C. Z., Carneiroa, J. C. Z., Pinheirob, C. J. G., and Passamai Jr., J. L., (2014). Irradiation of strawberries: influence of information regarding preservation technology on consumer sensory acceptance. *Innovative Food Science and Emerging Technologies* 26: 242e247.

Gautam, S., and Tripathi, J. (2016) Food processing by irradiation: An effective technology for food safety and security. *Indian Journal of Experimental Biology* 54(11): 700–707.

Harder, M. N. C., Arthur, V., and Arthur, P. B. (2016). Irradiation of foods: Processing technology and effects on nutrients: Effect of ionizing radiation on food components. In *Encyclopedia of Food and Health* (pp. 476–481). Academic Press.

Hartwig, A., Pelzer, A., Burnouf, D., Tite´ca, H., Delince´e, H., Briviba, K., Soika, C., Hodapp, C., Raul, F., Miesch, M., Werner, D., Horvatovich, P., and Marchioni, E. (2007). Toxicological potential of 2-alkylcyclobutanonesdSpecific radiolytic products in irradiated fat-containing food in bacteria and human cell lines. *Food and Chemical Toxicology* 45: 2581–2591.

Indiarto, R., Pratama, A. W., Sari, T. I., and Theodora, H. C. (2020). Food irradiation technology: A review of the uses and their capabilities. *International Journal of Engineering Trends and Technology* 68(12): 91–98.

International Atomic Energy Agency (IAEA) (2005). *Gamma Irradiators for Radiation Processing (IAEA Brochure)*. IAEA.

International Atomic Energy Agency (IAEA) (2007). Food irradiation clearances database [cited 3 March 2008] Available from: URL: International Atomic Energy Agency (IAEA). IAEA safety standards [cited 2 April 2008] Available from: URL: http://www ns.iaea.org/standards/

Johnson, E. S., Langård, S., and Lin, Y. S. (2007). A critique of benzene exposure in the general population. *Science of the Total Environment* 374(2–3): 183–198.

Kava, R. (2007). *Irradiated Foods* 6th ed. American Council on Science and Health.

Kim, J., Lim, J., Moon, S. H., Liu, K. H., and Choi, H. J. (2020). Toluene Inhalation causes early anxiety and delayed depression with regulation of Dopamine turnover, 5-HT1A receptor, and adult neurogenesis in mice. *Biomolecules & Therapeutics* 28(3): 282–291.

Loaharanu, P. (2007). In Kava R. ed. *Irradiated Foods* 6th ed. American Council on Science and Health.

Mahapatra, A. K., Muthukumarappan, K., and Julson, J. L. (2005). Applications of ozone, bacteriocins and irradiation in food processing: A review. *Critical Reviews in Food Science and Nutrition* 45(6): 447–461.

Mohácsi-Farkas, C. (2016). Food irradiation: Special solutions for the immuno-compromised. *Radiation Physics and Chemistry* 129: 58–60.

Nair, P. M., and Sharma, A. (2016). Food irradiation. In *Innovative Food Processing Technologies* (pp. 19–29). Elsevier. 10.1016/B978-0-12-815781-7.02950-4

Nishihira, J. (2020). Safety of irradiated food. In *Genetically Modified and Irradiated Food: Controversial Issues: Facts versus Perceptions* (pp. 259–267). Academic Press.

Pan, M., Yang, J., Liu, K., Xie, X., Hong, L., Wang, S., and Wang, S. (2021). Irradiation technology: An effective and promising strategy for eliminating food allergens. *Food Research International* 148: 110578.

Pedreschi, F., and Mariotti-Celis, M. S. (2020). Irradiation kills microbes: Can it do anything harmful to the food? In *Genetically Modified and Irradiated Food: Controversial Issues: Facts versus Perceptions* (pp. 233–242). Academic Press.

Prakash, A. (2020). What is the benefit of irradiation compared to other methods of food preservation? In *Genetically Modified and Irradiated Food: Controversial Issues: Facts versus Perceptions* (pp. 217–231). Academic Press.

Prakash, A., and Ornelas-Paz, J. de J. (2019). Irradiation of fruits and vegetables. In *Postharvest Technology of Perishable Horticultural Commodities* (pp. 563–589). Woodhead Publishing.

Ravindran, R., and Jaiswal, A. K. (2018). Toxicological aspects of irradiated foods. In *Food Irradiation Technologies: Concepts Applications and Outcomes* (Issue 4, pp. 337–351). Royal Society of Chemistry.

Ravindran, R., and Jaiswal, A. K. (2019). Wholesomeness and safety aspects of irradiated foods. *Food Chemistry* 285: 363–368.

Roberts, P. B. (2016). Food irradiation: Standards, regulations and world-wide trade. *Radiation Physics and Chemistry* 129: 30–34.

Rosario, D. K. A., Rodrigues, B. L., Bernardes, P. C., and Juniora, C. A. C. (2020). Principles and applications of non-thermal technologies and alternative chemical compounds in meat and fish. *Critical Reviews in Food Science and Nutrition* 61:1163.

Sajilata, M. G., and Singhal, R. S. (2006). Effect of irradiation and storage on the antioxidative activity of cashew nuts. *Radiation Physics and Chemistry* 75(2): 297–300.

Shalaby, A. R., Anwar, M. M., Sallam, E. M., and Emam, W. H. (2016). Quality and safety of irradiated food regarding biogenic amines: Ras cheese. *International Journal of Food Science & Technology* 51(4): 1048–1054.

Skowron, K., Grudlewska, K., Gryn, G.; Skowron, K. J., Swieca, A., Paluszak, Z., Zimek, Z., Rafalski, A., GospodarekKomkowska E. (2018). Effect of electron beam and gamma radiation on drug-susceptible and drug-resistant Listeria monocytogenes strains in salmon under different temperature. *Journal of Applied Microbiology* 125: 828–842.

Song, P., Wu, L., and Guan, W. (2015). Dietary nitrates, nitrites, and nitrosamines intake and the risk of gastric cancer: A meta-analysis. *Nutrients* 7(12): 9872–9895. 10.3390/NU7125505

Stefanova, R., Vasilev, N. V., and Spassov, S. L. (2010). Irradiation of food, current legislation framework, and detection of irradiated foods. In *Food Analytical Methods* (Vol. 3, Issue 3, pp. 225–252). Springer.

Štajner, D., Milošević, M., and Popović, B. M. (2007). Irradiation effects on phenolic content, lipid and protein oxidation and scavenger ability of soybean seeds. *International Journal of Molecular Sciences* 8(7): 618–627.

Vaz A. F., Costa R. M., Coelho L. C., Oliva M. L., Santana L. A., et al. (2011). Gamma irradiation as an alternative treatment to abolish allergenicity of lectins in food. *Food Chemistry* 124(4): 1289–1295.

Vaz A. F., Souza M. P., Carneiro-da-Cunha M. G., Medeiros P. L., Melo A. M., et al. (2012). Molecular fragmentation of wheat-germ agglutinin induced by food irradiation reduces its allergenicity in sensitised mice. *Food Chemistry* 132(2): 1033–1039.

Vranova, J., and Ciesarova, Z. (2009). Furan in food – a review. *Czech Journal of Food Sciences* 27(1): 1–10.

Woodside, J. V. (2015). Nutritional aspects of irradiated food. *Stewart Postharvest Review* 11(3): 1–6.

Yousefi, M. R., and Razdari, A. M. (2015). Irradiation' and its potential to food preservation. *International Journal of Advanced Biological and Biomedical Research* 3(1): 51–54.

Zanardi, E., Caligiani, A., and Novelli, E. (2018). New insights to detect irradiated food: an overview. *Food Analytical Methods* 11(1): 224–235.

12 Using Inorganic Nanoparticles for Sustainable Food Safety and Quality Control

Neela Badrie

12.1 INTRODUCTION

Securing food safety and quality is the goal towards achieving food and nutrition security, but could be challenging due to the rapid growth of the global food supply, the diversity of food companies and public health sectors and changing consumer food habits (Tertis *et al.*, 2021). A food control system safeguards the safety, quality, wholesomeness and fitness for consumer consumption, begins on the farm and ends at the consumer plate and includes foods which are traded domestically, regionally and internationally (FAO, 2022). According to the World Health Organization (WHO, 2020), about 600 million people become ill after eating contaminated food every year. The effects of unsafe foods cost about US$95 billion in lost productivity each year for low- and middle-income economies (UN, 2022).

The global nanotechnology industry has shown overall standing and development and is projected to exceed US$125 billion by 2024 (Liu and Xia, 2020) and to reach US$126.8 billion by 2027 (Machado *et al.*, 2022). The novel uses of nanoparticles (NPs) technologies in the agri-food industry have transformed traditional foods and improved on food security, safety and quality (Ashraf *et al.*, 2021). The growth of nanotechnological applications has enabled functional food development, facilitated nutrition enrichment, advanced food smart and active nanopackaging systems, detected and monitored foodborne pathogens and contaminants, increased the shelf-life of agroprocessed foods and have guided regulations (Abbas *et al.*, 2019; Siddiqui and Alrumman, 2021; Ungureanu *et al.*, 2022). However, despite these advances, there are various safety and health hazards linked to the consumption of nanofoods, accumulations of nanomaterials (NMs) in human bodies and environmental pollutant hazards (Siddiqui and Alrumman, 2021). In this chapter, the focus is on the characteristics and antimicrobial activity of 'green' synthesized NPs, their applications in various nanobiosensors for detection and monitoring of food quality and their incorporation and functions in nanobiosensors for smart food packaging. The scientific issues related to the biosafety and toxicity of inorganic/metallic NPs and their potential negative health safety effects on workers and the food processing environment are included.

12.2 USING NANOPARTICLES FOR SUSTAINABLE FOOD QUALITY AND SAFETY CONTROL

Nanotechnology is an evolving driven technology of the 21st century due to its multi-functions involving nanomaterials (NMs) ranging in size from 1–100 nanometers (nm) (Lugani *et al.*, 2021; Machado *et al.*, 2022). The NPs are nanosized organic, inorganic or hybrid materials with at least one component ranging from 1 to 100 nm and can be defined by their size and diameter (Shang *et al.*, 2019; Baig *et al.*, 2021; Ndwandwe *et al.*, 2021). Metallic nanoparticles (NPs) have distinctive chemical, physical and biological properties in comparison to the pure natural bulk materials related to high surface-area-to-volume ratio on the reduction of NP size to the nanometer scale (Ponce *et al.*, 2018; Singh *et al.*, 2021; Ndwandwe *et al.*, 2021). The NPs have served as carriers for antimicrobial polypeptides to protect against deterioration of foods, as nanopesticides and nanobiosensors to detect mycotoxins, viruses and bacteria (Cerqueira and Pastrana, 2019; Das *et al.*, 2019; Lugani *et al.*, 2021). The structural alterations of metallic NPs result in NP-induced toxicity, related to the principles of oxidative stress and levels of reactive oxygen species (ROS) that hinder cell proliferation and differentiation and cause cell death (Dayem *et al.*, 2017). The NPs, which show reactivity and biological activity, possess intense affinity to their targets such as proteins (Zehra *et al.*, 2021). The antimicrobial effectiveness of NPs is related to the driving force of photons, the structural changes in microbial cellular permeability, disulfide bond formation associated to the functional thiol group of proteins, the ROS generation and hyperoxidation of DNA and RNA nucleotides (Guzmán-Altamirano *et al.*, 2022).

12.3 'GREEN' SYNTHESIS OF NANOPARTICLES FOR DETECTION/MONITORING OF FOOD SAFETY AND QUALITY

The production of metal and metal oxide NPs can occur by the 'bottom-up' and 'top-down' methods (Basavegowda *et al.*, 2020). In general, the NPs are most often formed by the 'bottom up'

DOI: 10.1201/9781003334859-12

by processing of atoms or molecules to construct complex nanostructures followed by bioreduction such as 'green' synthesis (Anirudhan *et al.*, 2018; Rane *et al.*, 2018; Rovera *et al.*, 2020; Adeyeye and Ashaolu, 2021). The 'top-down' approach involves the reduction of the larger inorganic particles into nanosized particles (Anirudhan *et al.*, 2018; Basavegowda *et al.*, 2020; Rovera *et al.*, 2020). In an effort to promote environmental sustainability, the in-vivo 'green' synthesis of NPs by plants, bacteria, fungi and algae (Singh and Singh, 2019) adds to simplicity, sustainability, biocompatibility, cost-effectiveness, and minimizes the potential of environmental risks (Ahmad *et al.*, 2019; Nasrollahzadeh *et al.*, 2019; Singh and Singh, 2019; Castillo-Henríquez *et al.*, 2020; Rahman *et al.*, 2020; Cassani *et al.*, 2021; Couto and Almeida, 2022; Malini *et al.*, 2022; Maťátková et al. 2022 Maťátková 2022Maťátková 2022; Mustapha *et al.*, 2022). The NPs when employed in 'green' nanosynthesis are devoid of harsh operating chemical and physical methods common to the conventional methods (Carrillo-Inungaray *et al.*, 2018; Saratale *et al.*, 2018; Sanjay. 2019; Dikshit *et al.*, 2021). Metal NPs have been synthesized and applied to food safety and quality analysis, crop development and environmental monitoring (Devra, 2022). Generally, the NPs are engineered and possess electronic, catalytic and unique optical properties which are applied to the field in special sensing and Surface-Enhanced Raman spectroscopy (SERS) (Li *et al.*, 2020; Khan *et al.*, 2022). Magnetic NPs have been designed and synthesized to include functional groups and with high surface to mass ratios facilitating the mechanisms for preconcentration of trace levels of molecules, heavy metals and foodborne pathogens in complex food matrices (Yu *et al.*, 2022). Hence, the focus in this section is on the sustainable development of bio-inspired NPs for food safety and quality control uses.

12.3.1 Green Synthesis of Metallic Nanoparticles from Plants

The usage of plant extracts for the manufacture of metal NPs offers several advantages (Ungureanu *et al.*, 2022), such as a wide range of biomolecular variability, cheap material, low toxicity, simplicity, short production time, suitability for scale-up production, can act as blocking/stabilizing agents and reducing agents, thereby reducing the use of synthetic NPs (Sanjay, 2019; Tsekhmistrenko *et al.*, 2021; Khan *et al.*, 2022). As reported by Rafique *et al.* (2017), the synthesis of Ag-NPs using plants could overcome the challenges of the slow route of using microorganisms and the required stringent control in sustaining the microbial culture.

There is a growing trend of utilizing agri-food waste extract as reducing agents for NPs synthesis (Rodríguez-Félix *et al.*, 2021). Table 12.1 shows the utilization of various plant waste for NPs synthesis. The AgNPs were synthesized from onion (*Allium cepa*) peel collected as domestic waste from households in India (Santhosh *et al.*, 2021). These fabricated AgNPs showed active antibacterial effects. Omran et al. (2021) recognized the importance of upcycling of mandarin (*Citrus reticulum*) waste peels into functional compounds. The mandarin peel extract demonstrated excellent performance as bioreductant, biostabilizer and biocapping agent for the biological fabrication of AgNPs. The waste banana stem, *Musa paradisiaca* Linn aqueous extract (Doan *et al.*, 2021) served in both AgNPs and AuNPs synthesis and banana waste peduncles used for AgNPs synthesis for antibacterial activities (El-Desouky et al., 2021). Safflower (*Carthamus tinctorius* L.) waste extract aided in the synthesis of AgNPs and possesses antibacterial activity (Rodríguez-Félix *et al.*, 2021). Among all the NPs, the AgNPs were the most commercially made and applied due to their antimicrobial activity while AuNPs have been studied as a sensor/detector (He *et al.*, 2019).

Table 12.1 shows the various applications of 'green' synthesized metallic/inorganic NPs as antimicrobials in food safety and quality control. These NPs have been synthesized mainly for antimicrobial activities from black currant (*Ribes nigrum*) leaf extract (Hovhannisyan *et al.*, 2022), raspberry (*Rubus ellipticus*) root extract, waste banana stem, *Musa paradisiaca* Linn. (WBS) aqueous extract and banana peduncle (Doan *et al.*, 2021, El-Desouky et al., 2021), algae *Padina pavonica* (El-Zamkan *et al.*, 2021), carboxyl methyl cellulose and cellulose extract (He *et al.*, 2021), blueberries *Vaccinium arctostaphylos* aqueous extract (Khodadadi *et al.*, 2021), *Ocimum tenuiflorum* fresh leaves, *Mentha* (mint) leaves, *Murrayakoenigii* (curry leaves), *Aloe vera* (Kumari *et al.*, 2021), *Citrus limon* (lemon peel), *Citrus sinensis* (orange peel) and *Citrus tangerina* (tangerine peel) (Niluxsshun *et al.*, 2021), *Citrus limetta* (sweet lemon peel) extract (Dutta *et al.*, 2020), *Allium cepa* (onion) peel extract (Santhosh *et al.*, 2021), Reishi mushroom *Ganoderma lucidum*) extract (Aygün *et al.*, 2020), *Moringa oleifera* flower extract (Bindhu *et al.*, 2020; Adrianto *et al.*, 2022), *Parkia speciosa* leaf aqueous extract (Ravichandran *et al.*, 2020), *Capparis zeylanica* leaf broth (Ravindran *et al.*, 2020), globe amaranth *Gomphrena globosa* aqueous extract of fresh leaves (Tamilarasi and Meena, 2020) and *suspensa* fruit water extract (Du *et al.*, 2020). The CuNPS were fabricated from walnut, *Juglans regia* green husk aqueous extract (Hassan *et al.*, 2022), AuNPs, from Peacock flower or Barbados Pride, *Caesalpinia pulcherrima*

Table 12.1: Applications of 'Green' Synthesized Metallic/Inorganic Nanoparticles as Antimicrobials in Food Safety and Quality Control

Nanoparticles (NPs)	Green Synthesis Substrate	Particle Size	Applications/ Methodology	Findings	References
AgNPs	Black currant (*Ribes nigrum*) leaf extract	AgNPs size, 1 to 50 nm	Demonstrate AgNPs for antibacterial activities against *Escherichia coli* and kanamycin-resistant *E. coli*	Composite film with AgNPs had lower bacterial contamination compared to polyethylene films	Hovhannisyan *et al.*, 2022
AgNPs	(*Carthamus tinctorius* L.), safflower waste aqueous extract	AgNPs size, 8.67 ± 4.7 nm	Synthesize AgNPs from waste and test antibacterial activity	Effective antibacterial activity for *Staphylococcus aureus* and *Pseudomonas fluorescens*	Rodriguez-Felix *et al.*, 2021
AgNPs	Raspberry (*Rubus ellipticus*) root extract	AgNPs size 13.85 to 34.30 nm	Bioactive components in root extract reduced Ag+ ion into AgNPs	Compared to root extract, AgNPs resulted in higher antibacterial and antioxidant properties	Khanal *et al.*, 2022
AgNPs and AuNPs	Waste banana stem, *Musa paradisiaca* Linn (WBS) aqueous extract	WBS-AgNPs size, of 7–13 nm and AuNPs of 11–14 nm	Fabricate of AgNPs and AuNPs using WBS for antimicrobial effects	WBS-AgNPs have antibacterial activity	Doan *et al.*, 2021
AgNPs	Banana waste peduncles (BWP)	Plasmonic AgNPs size, ~14.1 nm	Fabricate of AgNPs and AgNPs/Degussa nanocomposite fabricated from banana waste	Nanocomposites exhibited synergistic antibacterial activities	El-Desouky et al., 2021
AgNPs	Algae *Padina paonica* petroleum ether extract	AgNPs diameter size, 46.21 nm	Synthesize AgNPs and screen dairy milk and environmental samples for virulence and disinfectant-resistance genes of *Listeria* spp.	Deactivation of *L. monocytogenes* in cheese milk curd and whey by AgNPs and extract after 14 days and deactivated after 28 days of storage	El-Zamkan *et. al.*, 2021

(*Continued*)

TABLE 12.1 (continued)

Nanoparticles (NPs)	Green Synthesis Substrate	Particle Size	Applications/ Methodology	Findings	References
AgNPs	Blueberries *Vaccinium arctostaphylos* aqueous extract	AgNPs size range of 7–16 nm	Synthesize AgNPs and evaluate antibacterial properties	AgNPs showed broad antibacterial activities	Khodadadi *et al.*, 2021
AgNPs	Curry leaves, *Aloe vera*, *Ocimum tenuiflorum* (tulasi) fesh leaves, *Mentha* (mint)	*Aloe vera* mediated AgNPs, diameter size range of 27–31 nm Cellulosic fibers AgNPs, average size of about 20 nm	Mix plant extract with silver nitrate and coat on blotting paper as packaging	Antibacterial activities of packaging for *E. coli* and *S, aureus* which extend storage of tomatoes to about 30 days and coriander leaf to 15 days	Kumari *et al.*, 2021
AgNPs	*Citrus limon, Citrus sinensis, Citrus tangerina*, lemon, orange, and tangerine peels respectively	AgNPs size 5–80 nm	Phytochemicals in peel extracts served as reducing and stabilizing agents for AgNPs and to investigate antibacterial activity using well-diffusion method	Broad spectrum of antibacterial activity	Niluxsshun *et al.*, 2021
AgNPs	*Allium cepa* (onion) peels extract	AgNPs size, 2–80 nm	Synthesis and characterize of AgNPs	AgNPs demonstrated antibacterial activity	Santhosh *et al.*, 2021
AgNPs	Reishi mushroom *Ganoderma lucidum* extract	AgNPs diameter size, 15–22 nm	Microwave assisted green synthesis of Ag NPs	AgNPs showed DNA cleavage activity against many bacteria and fungus *C. albicans*	Aygün *et al.*, 2020
AgNPs	*Moringa oleifera* flower extract (MOF)	AgNPs size, 22 nm	Synthesis of AgNPs from MOF extract, assess antimicrobial and sensing properties	AgNPs inhibited bacteria and detected the presence of Cu by optical sensor-based surface plasmon resonance	Bindhu *et al.*, 2020

AgNPs	Lemon peel extract, *Citrus limetta*	AgNPs size, of 18 nm	Synthesize and investigate AgNPs for antibacterial activities	AgNPs demonstrated antibacterial and antifungal activities cell membrane permeability	Dutta *et al.*, 2020
AgNPs	*Parkiaspeciosa* leaf aqueous extract	AgNPs, size, 31 nm by SEM, 35 nm by TEM, and 155.3 d. nm by Dynamic Light Scattering	Reduce silver nitrate leaf aqueous extract to fabricate AgNPs	AgNPs showed antimicrobial and antioxidant radical scavenging activities	Ravichandran *et al.*, 2020
AgNPs	*Capparis zeylanica* leaf broth	–	Synthesize AgNPs using leaf broth and mix with PVA/PEG biopolymer and investigate antimicrobial activity	AgNPs showed excellent antibacterial activity	Ravindran *et al.*, 2020
Ag NPs	*Gomphrena globose*, globe amaranth Aqueous fresh leaves	–	Reduce Ag+ to metallic Ag0 by phytochemicals present in the leaves and test for antibacterial effects	AgNPs showed excellent antibacterial activity	Tamilarasi and Meena, 2020
Ag NPs	*Forsythia suspensa* fruit extract	AgNPs size, 47.3 ± 2.6 nm by Dynamic Light Scattering	Evaluate AgNPs antibacterial activities against foodborne pathogens	AgNPs showed morphological alterations and antibacterial activities against foodborne pathogens resulting in damage to cell membrane integrity, induce release of nucleic acids and disruption of cell reproduction	Du *et al.*, 2020

(Continued)

TABLE 12.1 (*continued*)

Nanoparticles (NPs)	Green Synthesis Substrate	Particle Size	Applications/ Methodology	Findings	References
CuNPs	Walnut, *Juglans regia*green husk aqueous extract	CuNPs core diameter size, 53–28 nm	Investigate CuNPs for antibacterial, antifungal and antibiofilm properties	CuNPs exhibited free radical DPPH scavenging capacity, antibacterial, antifungal, antibiofilm properties, exhibited photocatalytic activity against methyl orange	Hassan *et al.*, 2022
AuNPs	Peacock flower or Barbados Pride, *Caesalpinia pulcherrima* extract	AuNPs particle diameter size, 15.2 ±1.1 nm	Conjugate AuNPs with gallic acid (GA-AuNPs) and investigate antimicrobial and antioxidant properties	GA-AuNPs suitable for food packaging lining	Mehmood *et al.*, 2022

extract (Mehmood *et al.*, 2022), Damask rose petals, *Rosa damascene* aqueous extract (Du *et al.*, 2020) and SeNPs from blue-green microalgae cyanobacterium, *Spirulina platensis* (Alipour *et al.*, 2021) showed antimicrobial effects and antioxidant activities (Table 12.1).

12.3.2 'Green' Synthesis of Nanoparticles from Fungi

The biosynthesis of NPs from the fungi is an eco-friendly technology. Fungi produce enzymes and proteins as reducing and capping agents to create metallic NPs from metal salts (Abdel-Aziz *et al.*, 2018). Various fungal genera have served for the synthesis of NPs, among them are *Aspergillus* and *Fusarium* with Ag being explored extensively for their antimicrobial activity (Roy *et al.*, 2018; Chippa, 2019). The biogenic synthesis of NPs through the enzymatic reduction of metal ions into their zerovalent nanoforms using macrofungi was investigated for mushroom antimicrobials and alternative nutraceuticals (Pandey *et al.*, 2020). Table 12.1 indicates that AgNPs were synthesized using Reishi mushroom (*Ganoderma lucidum*) extract and demonstrated antibacterial effects and antifungal effects against *Candida albicans* (Aygün *et al.*, 2020).

12.4 USE OF NANOPARTICLES IN NANOBIOSENSORS

Nanobionsensors have gained attention for food safety and quality analysis being driven by the requirement to comply with various standards and regulations (Ragavan and Neethirajan, 2019, Hari *et al.*, 2020). The advances in nanotechnology and NPs have improved the specificity, reproducibility, limit of detection and detection range of biosensors (Li *et al.*, 2020). Nanobiosensors have at least one sensing element less than 100 nm (Fraceto *et al.*, 2016) can detect specific analytes, molecules, biological agents or surrounding conditions (Yang and Duncan, 2021). For food analysis, nanobiosensors measure the physicochemical and biological quality of foods such as to detect foodborne pathogens, toxins, contaminants, monitor quality changes of food packaging to minimize food waste (Yu *et al.*, 2018; Cerqueira, *et al.*, 2019; Das *et al.*, 2019; Joshi *et al.*, 2019). The recent diagnostic point of care (POC) devices could be paper-based or chip-based for quick food safety and quality study (Neethirajan *et al.*, 2018; Choi *et al.*, 2019). These biosensors could detect and record every change occurring in cells and identify test substances at very low concentrations (Khan *et al.*, 2022). The NPs have enabled higher sensitivity by signal amplification and introduce several enhanced transduction principles to the aptasensors as detection tools for contaminants (Sharma *et al.*, 2015). Surface-Enhanced Raman Spectroscopy (SERS) has offered improvements over traditional analytical techniques being rapid, sensitive 'structural fingerprinting' of low concentration of analytes and being utilizable as an analytical field tool (Han *et al.*, 2022). In food analysis, NMs have advanced the potential of smart electrochemical biosensors with device miniaturization, automatization, high sensitivity and specificity, and connectivity-enabled through the internet (Garrido-Maestu *et al.*, 2018; Antje and Baeumner, 2020; Curulli, 2020) to facilitate faster real-time qualitative and quantitative food analysis. In some complex designs of nanobiosensors, the quality changes of commercial food products over time could be monitored by smartphones (Yang and Duncan, 2021) and the analyte can be quantified by a smartphone app, with enabling signal analysis (Choi *et al.*, 2019). The coupling of nanotechnology-enhanced fluorescent immunochromatographic test strips has improved sensitivity and quantitative analysis in food safety (Wu *et al.*, 2021). The AuNPs have been widely used in different sensors due to their optical, electronic, catalytic and chemical properties (Li *et al.*, 2020). The NMs, such as AuNPs and gold nanorods (AuNRs), have improved the sensitivity and multiplexing capabilities in DNA detection due to their optical properties (Cerqueira and Pastrana, 2019). Dextrin-capped AuNPs(d-AuNPs) in concentrated ionic conditions were used in the development of an unamplified genomic DNA nanobiosensor for sequence-specific detection (Baetsen-Young et al., 2018). The d-AuNPs detected the unamplified sequence from *Pseudoperonospora cubensis*, the causal agent of cucurbit downy mildew (Baetsen-Young *et al.*, 2018). Table 12.2 shows that CuO NPs-based smartphone-combined digital colorant biosensor from *Camellia sinensis* polyphenols was selective for ammonia to monitor for food spoilage (Karakuş *et al.*, 2022).

12.4.1 Nanobiosensor Applications to Food Safety and Quality

Nano/biosensors have gained attention among other food-analysis techniques due their promising properties such as quick response, sensitive, less labor-intensive procedures, enhanced accuracy and being able to conduct remotely (Castillo *et al.*, 2017). Table 12.2 shows some recent applications of nanomaterials/nanoparticles as nanosensors for food safety and quality analysis.

Table 12.2: Applications of Metallic Nanoparticles as Nanobiosensors in Food Safety and Quality Analysis

NPs-Based Biosensors	Mode of Action/Technology/Objectives	Food/Samples	Main Findings/Applications	References
AgNPs	A label-free SERS using rough flower AgNPs to sense three mycotoxins in rice	Rice	Limit of detection (LOD): aflatoxin B_1 (AFB_1), 1.145, ochratoxin A (OTA), 1.133, and ochratoxin B (OTB) 1.180 µg. Kg	He et al., 2023
AgNPs	A SERS-based 3-aminobenzeneboronic acid labeled AgNPs to detect total arsenic (TA) in black tea leaves	Black tea leaves	LOD of 0.028 µg/g TA Good recoveries of 83.84–109.53% in tea	Barimah et al., 2022
CuONPs	Camellia sinensis polyphenols, CuONPs biosensors as a smartphone-integrated digital colorimetric ammonia indicator for food spoilage	Food samples	Ammonia concentration range of 12.5–100 µM, and a LOD of 40.6 nM	Karakuş et al., 2022
Ag and Ag-Ni NPs	NPs synthesized from plant extract of Bunya, Araucaria bidwilli and use of spectrophotometric probe to determine Cu^{+2}	Real water samples	AgNP size, 42 nm and Ag-Ni NP, 25 nm with optimum response to cupric ions of contact time of 12.5 and 7.5 min, respectively	Shahzadi et al., 2022
AuNPs	SERS based on plasmonic metal AuNPs decorated ZnO/ZnFe$_2$O$_4$ SERS-active with electronic and optical features for nanocomposite for melamine detection	Milk samples	Decorated AuNPs selected quick sensing of melamine between 0.39 –7.92 µM	Tiwari et al., 2022
NiHCF-NPs and AuNPs	Zn/Fe bimetallic ZIF derived nanoporous carbon facilitated electron transfer nickel hexacyanoferrate NiHCF-NPs as signal probe in electrochemical aptasensor for detection of paraquat	Lettuce and cabbage	LOD of 0.34 µg.L. Recoveries ranged from 96.20% to 104.02%	Wu et al., 2022
AgNPs	Orange (Citrus sinensis) peel AgNPs extract to detect Hg^{2+} ions drinking water by visual colorimetry.	Drinking water	AgNPs size of 55 nm, selective for colorimetric detection of Hg^{2+} ions, LOD (mol. L) of Hg^{2+} 1.24 × 10^{-6} (0.25 ppm)	Aminu and Oladepo, 2021
CuSNPs	Copper monosulfide (CuSNPs) and a Cu^{2+} fluorescent probe for OTA and conjugated with anti-OTA antibodies with OTA antigens	Soybean, coffee, corn samples	LOD of 0.01 ng. mL^{-} Recoveries from 94–110%	Chen et al., 2021
AuNPs	Hibiscus rosa sinensis extract and AuNPs coelectrospun with polycaprolactone and polyethylene oxide into fiber mats which were sensitive to quality changes	Shrimp	Sensors showed color changes to pH and trimethylamine nitrogen, indole, and total microbial counts for the prediction of shrimp quality	Jovanska et al., 2022
AuNPs	Color changes of AuNPs-based colorimetric sensor induced by pesticides were captured by camera of smartphone and processed by software	Tea	LOD of phosalone, 90 nMLOD of thiram, 13.8 nM	Ma et al., 2021

AuNPs	Curcumin mediated poly (ethylene glycol thiol acid conjugated AuNPs for optical determination of melamine	Turmeric	Flourescence intensity in the range of 0–10 mM with 33 nM detection limit	Shehab et al., 2020
CoPcNPs	Use cobalt phthalocyanine NPs to synthesize iron-based metal-organic framework nanocomposites	CoPc nanocomposites for food packaging	LOD of OTA is 0.063 fg/mL^{-1}	Song et al., 2021
AgNPs	AgNPs fabricated from onion extract with localized plasmon resonance to detect Hg^{2+} ions	Water samples	Increase in Hg^{2+} ions confirmed by reduction of Ag with over 92% recovery	Alzahrani, 2020
Ag$_2$ONPs	Ag$_2$ONPs synthesized by reducing AgNO$_3$ precursor with *Brassica rapa* L. subsp. *Pekinensis* Nappa cabbage extract and used as sensing electrode to detect p-nitrophenol	Water treatment & cabbage samples	Ag$_2$ONP-carbon black/nickel foam electrode responded to p-nitrophenol concentrations from 1.0 to 0.1 pM	Banua and Han, 2020
AuNPs	AuNPs for fabrication surface plasmon resonance (SPR) biosensor for AFB$_1$ detection	Wheat samples	LOD of AFB$_1$, 0.19 nM Average recoveries, 93–90.1%	Bhardwaj et al., 2020
AgNPs	Localized SPR enhanced sensing of Hg2+ by a probe with chitosan functionalized AgNPs for selective detection of Hg2+ in aqueous medium	Aqueous medium	LOD of Hg^{2+} 1.5 ppb lower than WHO permissible limit of 2 ppb	Boruah et al., 2020
AuNPs	AuNPs-multiplex PCR multiplex test for colorimetric real-time detection of specific foodborne pathogens	Food samples	LOD (pg/μL^{-1}) for both *Listeria monocytogenes*, and *Salmonella typhimurium*, 10 and for *Escherichia coli* O157: H7:50	Du et al., 2020
AuNPs	Synthesize salt tolerant AuNP using *Rosa damascene* Aqueous extract and detect *L. monocytogenes* by AuNPs-based lateral flow immunoassay	Pork tenderloin	LOD(CFU/mL) of pure *L. monocytogenes*, 2.5 × 10^5 and in pork tenderloin 2.85 × 10^5	Du et al., 2020
AgNPs	AgNPs-based SERS sensor linked to chemometric algorithms for detect OTA and AFB$_1$ in cocoa beans	Cocoa beans	LOD (pg/mL) of OTA, 2.63 and for AFB$_1$, 4.15, and LOD for spiked-cocoa-beans, 0.002 μg/ mL	Kutsanedzie et al., 2020
AuNPs	Carbon nanotube stabilized AuNPs decorated polymeric nanocomposite to estimate xanthine (XN), and hypoxanthine (HX) in the detection of food spoilage	Fish samples	LOD for XN, 24.1 nM, HX 90.5 nM	Sen and Sarkar, 2020
AuNPs	AuNPs-based time-temperature indicators (TTIs) to relate to the peroxide value of muffins	Muffins	Changes in colors of TTIs from yellow to pink to deep purple to indicate spoilage and inedibility of muffins	Zhang et al., 2020

12.4.1.1 Allergens

The advances of combination of nanomaterials and electrochemical biosensors for food allergen detection have been reviewed (Sheng *et al.*, 2022). A mast RBL-2H3 cell-based biosensor analysed shrimp allergen tropomyosin and fish allergen parvalbumin in which the cationic magnetic fluorescent NPs were activated by the allergen antigen in the electrochemical analysis (Jiang *et al.*, 2015).

12.4.1.2 Antibiotic Residues

In a review, various design and fabrication of various NMs-based aptasensors to detect antibiotic residues in animal-derived foods were summarized (Liu *et al.*, 2022). Sensitive signal labels were developed to detect chloramphenicol in contaminated milk samples using aptamer-conjugated magnetic NPs that serve as recognition and concentration elements and up-conversion NPs as signal labels (Wu *et al.*, 2022). In another research, the limit of detection (LOD) for chloramphenicol in raw milk was low at 18.3 pM using an apta-sensing colorimetric platform developed by changing AuNPs by short-sequence aptamers (Javidi *et al.*, 2018).

12.4.1.3 Melamine

Table 12.2 shows that an AuNPs decorated $ZnO/ZnFe_2O_4$ composite SERS substrate identified melamine between 0.39 μM–7.92 μM in milk samples (Tiwari *et al.*, 2022). Methoxy polyethylene glycol thiol-coated AuNPs were employed for optical determination with LOD of 33 nM for melamine (Shehab *et al.*, 2020). A colorimetric detection method for melamine in vegetables, fruit, milk and water samples utilized differently sized citrate-capped AuNPs and showed an LOD of 2.37×10^{-8} M from the aggregated and disaggregated behavior of metal NPs (Paul *et al.*, 2017).

12.4.1.4 Heavy Metals

Different phytochemical extracts of *Araucaria bidwilli* acted as reducing, capping and stabilizing agents in the 'green' synthesis of Ag and Ag-Ni bimetallic NPs for sensing of cupric ions (Shahzadi *et al.*, 2022). Table 12.2 indicates that a colorimetric detector applying 'green' synthesized AgNPs from onion extract with localized surface plasmon resonances (SPR) detected Hg^{2+} ions in water samples (Alzahrani, 2020).

12.4.1.5 Mycotoxins

The development of innovative aptasensors for simultaneous analysis of multi-mycotoxins is crucial for food safety (Guo *et al.*, 2020). Mycotoxins are often present in low concentrations in complex food matrices and may be produced by one or several fungal species, making detection difficult (Mishra *et al.*, 2018). A review was presented on the various SERS methodologies for both qualitative and quantitative analysis of mycotoxins using different SERS substrates (Martinez and He, 2021). Table 12.2 has indicated a label-free SERS sensor using rough AgNPs for rapid sensing of three mycotoxins in rice (He *et al.*, 2023). The LODs (μg/Kg) were aflatoxin B_1 1.145, ochratoxin A, 1.133 (OTA) and ochratoxin B 1.180μg/Kg (OTB) for rice and were lower than the value given by the European Commission. In another research, a SERS sensor quantified the OTA and AFB_1 levels for cocoa beans using synthesized AgNP@pH-11 to fabricate a sensor applied to two chemometric algorithms to predict the two mycotoxins (Kutsanedzie *et al.*, 2020). A solvent-mediated extraction method improved on the extraction of ochratoxin A (OTA) from wheat and wine and facilitated the distribution of the NPs for improved detection signal using SERS (Rojas *et al.*, 2020).

Table 12.2 shows that functionalised AuNPs were used to fabricate SPR Au chip for AFB_1 detection in wheat samples with the LOD of 0.003 nM (Bhardwaj *et al.*, 2020). In another research, a SERS colorimetric aptasensor was fabricated for OTA detection using $Au@Fe_3O_4$ NPs (Shao *et al.*, 2018). NPs function as signal indicator and magnetic separator with LOD of the aptasensor being 0.004 ng/mL^{-1}. Copper monosulfide NPs coupled to an anti-OTA antibody with signal from Cu^2 fluorescent probe detected OTA in foodstuff such as corn, soybean and coffee (Chen *et al.*, 2021). An impedimetric aptasensor based on an Fe-based organic platform is embedded with cobalt phthalocyanine NPs detected OTA (Song *et al.*, 2021). In another research, the fabrication of 1-hexyl-3-methylimidazolium hexafluorophosphate ZnO nanoflowers had a high infinity for aflatoxins (AFs) in wheat and peanut samples (Zhu *et al.*, 2022).

12.4.1.6 Pesticides

A review summarized the modifications of metal NPs with specific ligands for the low detection of pesticides using limited samples (Mehta *et al.*, 2022). Table 12.2 shows that an AuNPs-based colorimetric sensor for the quantitative analysis of pesticides was captured by color-induced changes of the aggregated AuNPs and the images were captured by the camera of a smartphone (Ma *et al.*, 2021). An electrochemical aptasensor with the deposition of AuNPs on nanoporous carbon and nickel hexacyanoferrate NPs was developed to detect paraquat assay in lettuce, cabbage and agriculture irrigation water samples (Wu *et al.*, 2022).

12.4.1.7 Xanthine/Hypoxanthine

Dervisevic *et al.* (2019) reviewed the nanotechnology-based electrochemical and optical sensors for hypoxanthine and xanthine as markers of food spoilage in meat. The incorporated NMs served as fluorescence emitters and quenchers in optical sensing instruments such as H_2O_2 mimicking materials. There is an issue in respect to the selectivity of xanthine oxidase towards hypoxanthine and xanthine as xanthine oxidase is not the only source of H_2O_2 in foods.

A multi-walled carbon nanotube decorated with AuNP film was synthesized by electropolymerization with enhanced conductivity for the detection of xanthine and hypoxanthine in fish (Sen and Sarkar, 2020). This sensor demonstrated high recovery rate between 95.03%–104.8% (Table 12.2). A bionanocomposite film fabricated by embedding graphene oxide sheets and decorated with Fe_3O_4 NPs into poly(glycidyl methacrylate-co-vinylferrocene) and by covalent immobilization of xanthine oxidase. This developed platform served as a xanthinebiosensor for fish (Dervisevic *et al.*, 2015). The LOD was 0.17 µM.

12.5 NANOPARTICLES APPLICATIONS TO SMART FOOD PACKAGING

This section relates to the uses of engineered NPs as antimicrobial additives, the incorporation of NPs in materials for smart and active food packaging, and nanobiosensors in the analysis and monitor of changes in food packaging. The functioning of nanocomposite films and packaging is related to the size of the nanofillers and the uniformity of NPs distribution and dispersion in the matrix (Jafarzadeh and Jafari, 2020). Many nanotechnological applications to food packaging have led to advancements (Tabari, 2018) such as inclusion of functionalities to food packaging that are beyond those of the conventional packaging (Dudefoi *et al.*, 2018). The inclusion of NPs into food packaging materials has promoted the effective attachment to biological molecules related to their higher ratio of surface to volume (Basavegowda *et al.*, 2020). Considering the safety issues of inorganic/metallic NPs, the phytosynthesized NPs have tremendous advantages such as lower toxicity and increased antimicrobial effects in comparison to other types of synthesized NPs (Ungureanu *et al.*, 2022). 'Smart packaging' is the term most often used but active or intelligent packaging is sometimes used (Green, 2020). Active and smart packaging sense changes in food packages, which signal those changes to consumers and release active functional ingredients for preservation (Naseer *et al.*, 2018). Smart packaging tracks product quality and environmental conditions (Green, 2020, Chelliah *et al.*, 2021) using oxygen scavengers, moisture absorbers and barrier packing product (Chellaram *et al.*, 2014). Active packaging is based on active NMs that react by releasing antimicrobial and antioxidant agents, reduce/absorb residual oxygen concentration and removal of undesirables (Hutapea *et al.*, 2022). A review paper highlighted the applications of SeNPs synthesized from plant extracts for active packaging of foods (Ndwandwe *et al.*, 2021). The safety and health risks of smart and active packaging, and nanoprocessed foods and the regulatory policies require further investigation (Hutapea *et al.*, 2022).

Smart NMs like nanobiosensors and nanobarcodes have been incorporated sensing devices for tracking the integrity of food products and food packages, checking temperature changes, leaks, reporting deviations and remote control of quality of food products (Castillo *et al.*, 2017, Pandhi *et al.*, 2021). Smart and active packaging incorporate NPs as nanobiosensors and nanodevices in the detection of harmful contamination and toxic materials very rapidly, to detect gases, odors, pathogens, freshness in foods and to monitor changes in packaging conditions or integrity (Kuswandi, 2016; Colica *et al.* 2018 Colica 2018; Yu *et al.*, 2018; Cerqueira and Pastrana, 2019; He *et al.*, 2019; Shawon *et al.*, 2020; Mohammadpour and Naghib, 2021; Hutapea *et al.*, 2022). A nanobarcode detection system for food and biological samples employed fluorescence under ultraviolet light with the color being read by a computer scanner (Aigbogun *et al.*, 2017). Time and temperature indicators are simple, cost-effective and simple to use in real-time monitoring such as the effect of temperature on quality and safety of foods (Chelliah *et al.*, 2021). The quality of the food along the food chain could be tracked by radio frequency identification (RFID) labels, indicators and tags. A wireless biosensor design was developed using dextrin-capped gold AuNP as markers that could be adapted to existing RFID to monitor the quality of milk in a real-time limit (Karuppuswami *et al.*, 2018). A DNA and antibody-based sensor can detect growth of food pathogens bacteria, determine the food authenticity during transportation and storage and give warnings to consumers as to the freshness of foods (Ötles and Sahyar, 2017; Cerqueira and Pastrana, 2019). Table 12.2 shows that an ethanol extract of *Hibiscus rosa* sinensis was coelectrospun into fiber mats with polycaprolactone, polyethylene oxide and AgNPs to detect which were sensitive to the quality changes of shrimp for pH, trimethylamine nitrogen, indole and total microbial counts changes for shrimp (Jovanska *et al.*, 2022). In another research, a sensor detection limit for bisphenol A(BPA) was 6.63 ± 0.77 nM using dendritic platinum NPs coated on AuNPs and deposited on a carbon electrode (Shim *et al.*, 2018).

12.5.1 Biodegradable Nanocomposite Films

The incorporation of NPs in biodegradable polymer packaging has increased the mechanical strength and barrier properties to extend the storage of various food products (Basavegowda *et al.*, 2020; Ningthoujam *et al.*, 2022). The incorporation of TiO_2NPs in chitosan film has improved the mechanical properties of the film and sensory quality for packaging of minced beef (Hosseinzadeh *et al.*, 2020). A biodegradable nanocomposite active film was synthesized from jackfruit polysaccharide, incorporating TiO_2NPs by photocatalysis, which showed reduced moisture content and transparency and exhibited antimicrobial activity applied for active food packaging (Jinn *et al.*, 2017). The incorporation of AgNPs into edible films promoted their antimicrobial properties (Krásniewska *et al.*, 2020). The ZnONPs-based (bio)polymer composites increased the mechanical, gas barrier and antimicrobial properties (Abbas *et al.*, 2019)

12.6 NANOSAFETY AND TOXICITY

Natural NPs have always existed in foods that are consumed without noticeable safety risks. Despite the advantages of NPs in nanotechnology, there are safety and health concerns while processing for intelligent and active packaging and the consumption of nanoprocessed food regarding NPs accumulation in human bodies and environment risks and pollution (Bajpai *et al.*, 2018). The thrust has been to identify novel NP applications and to study the potential effects of NPs on human health by researchers. The physicochemical and structural properties of NPs when dispersed in food matrices change resulting in the cellular activities such as ROS generation related to NP-related toxicity, which cause on toxicity and health problems in the body (McClements and Xiao, 2017; Adeyeye *et al.*, 2021). Different mechanisms such as excess ROS generation have been key in metallic NP-induced toxicity in cellular signaling pathways to cell death (Daye *et al.*, 2017), oxidative stress, genotoxic and the potential of being carcinogenic (Jain *et al.*, 2018). The entry routes of NPs are mainly respiratory, dermal and gastrointestinal (Jain, 2018; Karimi *et al.*, 2018). There are concerns as to the environmental and occupational exposure properties of NPs for workers at nanotechnology labs who are prone to unknown risks and hazards (Ahn *et al.*, 2016). The toxicity levels of NPs are still indefinable, due to limited information on risk assessments and effects on human health. The NPs' toxicity challenges have demanded changes to regulatory policies (Bajpai *et al.*, 2018; Ashraf *et al.*, 2021). There are guidance documents on risk assessment of NPs in food, animal and human health and on the technical requirements for regulated food applications to establish the presence of NPs (EFSA, 2021). The U.S. FDA has several guidance documents as to whether FDA-regulated products would involve the application of nanotechnology and as to the regulation approach by the FDA of nanotechnology products (FDA, 2014, 2018).

12.7 CONCLUSION AND FUTURE OUTLOOK

The inorganic NPs have several roles in food safety and quality control. By their properties, NPs have contributed to the antimicrobial and antioxidants effects on foods, as nanobiosensors aided in the analysis of foods for safety and quality, served for detection and monitoring for food spoilage, foodborne pathogens, mycotoxins, allergens, antibiotics and heavy metals melamine and pesticides, have enhanced the mechanical strength of biopolymers, and have functional roles in active and smart packaging. The main metallic NPs applied as antimicrobials were the AgNPs in food preservation. In the safeguard of consumer food safety, the NMs/NPs during the food product life cycle should be measured (Gondal and Tayyiba, 2022) and the interactions of NPs with the food systems need to be evaluated as to the effects on the digestibility of the food constituent, consumer health and on environmental risks (Hossain *et al.*, 2021). It is critical to develop new tests to examine the toxic effects of NPs on the health of human beings and effects of risk exposures. The migration of metal and metal oxide NPs into food is related to the physicochemical and structural properties, dosage, route of administration and duration of exposure. Hence, it is critical to develop new engineering techniques for NPs, based on the toxicity profiles of known metal/metal oxides NPs to address possible food safety consequences and health hazards associated with the consumption of nanofoods.

Plants have been the major sources for 'green' synthesis of NPs in the mitigation of health and environmental risks towards a sustainable future. The phytochemicals of the several plant extracts have served as bioreductants, biostabilizer and biocapping agents for NPs synthesis. Considering the safety issues, phytosynthesized NPs have advantages such as their lower toxicity and antimicrobial effects in comparison with other synthesized NPs (Fierascu *et al.*, 2020). However, these 'green' technologies present several constraints such as in the effective stabilization of NPs, easy-to-use analyses, viability for lengthy time, up-scale of technology and the shift in NMs development toward eco-friendlier options such as the use of vegetable wastes.

The incorporation of NPs has been beneficial in the fabrication of packaging materials to add tensile strength, resistance, and thermal performance to biopolymers. For smart and active packaging structures, NPs are applied to detect changes in food packages, alert consumers of any changes and release of functional components to preserve foods. However, the incorporation of NPs and antigen-specific biomarkers for preparing nanocomposite polymeric films could be a future development (Vijayakumar *et al.*, 2022). The metallic/metal oxides NPs have significant roles in the identification and quantification of food pathogens and contaminants. However, the assessment of NPs presents several analytical challenges, such as to the definition of identity, quantification and size of NPs, which may exist in different morphologies and vary in dispersed states of a particle (Linsinger *et al.*, 2013). The key advantages that have been identified for nanobiosensors are sensitivity, speed, specificity, small size and diagnostic point of care. The AgNPs, AuNPs, CuONPs, CuSNPs and cobalt phthalocyanine NPs are some NPs that have served as detecting sensors in various applications such as for foodborne pathogens, various mycotoxins such as AFB1, ochratoxin A and B (OTA), arsenic, ammonia for spoilage, cupric and mercury (II) ions, melamine, pesticides such as paraquat, xanthine/hypoxanthine mainly by SERS sensors and optical and colorimetric detectors.

The commercialization of nanobiosensor has been slowly driven which could be linked to the low transformation of university-driven nanosensor research into commercial laboratory-ready detection tools (Yang and Duncan, 2021; Thakur *et al.*, 2022). Also, it was reported that most modern food manufacturers may be reluctant to reveal their use due to the negative perception of nanobiosensors (Ile *et al.*, 2019). Other challenges in the development of nanobiosensors are in the regulatory compliance, ethical concerns of use, consumer acceptance and whether the food should be nano-labeled as nonlabeled (Yang and Duncan, 2021).

REFERENCES

Abbas, M., M. Buntinx, W. Deferme, and R. Peeters. 2019. (Bio)polymer/Zno nanocomposites for packaging applications: A review of gas barrier and mechanical properties. *Nanomaterials*, 9: 1494. 10.3390/nano9101494.

Abdel-Aziz, S. M., R. Prasad, A. A. Hamed, and M. Abdelraof. 2018. Fungal nanoparticles: A novel tool for a green biotechnology? In *Fungal Nanobionics: Principles and Applications*, eds. R. Prasad, V. Kumar, M. Kumar, and S. Wang, 61–87. Singapore: Springer. 10.1007/978-981-10-8666-3_3.

Adeyeye, S. A. O., and T. J. Ashaolu. 2021. Applications of nano-materials in food packaging: A review. *Journal of Food Process Engineering*, 44: 13708. 10.1111/jfpe.13708.

Adrianto, N., A. M.. Panre, N. I. Istiqomah, M. Riswan, F. Apriliani, and E. Suharyadi. 2022. Localized surface plasmon resonance properties of green synthesized silver nanoparticles. *Nano-Structures and Nano-Objects*, 31: 100895. 10.1016/j.nanoso.2022.100895.

Ahmad, S., S. Munir, N. Zeb et al. 2019. Green nanotechnology: A review on green synthesis of silver nanoparticles-An ecofriendly approach. *International Journal of Nanomedicine*, 14: 5087–5107. 10.2147/IJN.S200254.

Ahn, J. J., Y.. Kim, E. A.. Corley et al. 2016. Laboratory safety and nanotechnology workers: An analysis of current guidelines in the USA. *Nanoethics*, 10: 5–23. 10.1007/s11569-016-0250-9.

Aigbogun, I. E., S. S. D. Mohammed, A. A. Orukotan, and J. D. Tanko. 2017. The role of nanotechnology in food industries - A review. *Journal of Advances in Microbiology*, 7 (4): 1–9. 10.9734/JAMB/2017/38175.

Alipour, S., S. Kalari, M. H. Morowvat, Z. Sabahi, and A. Dehshahri. 2021. Green synthesis of selenium nanoparticles by cyanobacterium *Spirulina platensis* (abdf2224): Cultivation Condition Quality Controls. *Hindawi BioMed Research International*, 2021: Article ID 6635297. 10.1155/2021/6635297.

Alzahrani, E. 2020. Colorimetric detection based on localized surface plasmon resonance optical characteristics for sensing of mercury using green-synthesized silver nanoparticles. *Hindawi Journal of Analytical Methods in Chemistry*, Article ID 6026312. 10.1155/2020/6026312.

Aminu, A., and S. A., Oladepo 2021. Fast orange peel-mediated synthesis of silver nanoparticles and use as visual colorimetric sensor in the selective detection of mercury(II) ions. *Arabian Journal for Science and Engineering*, 46: 5477–5487. 10.1007/s13369-020-05030-3.

Anirudhan, T. S., V. S. Athira, and V. C. Sekhar. 2018. Electrochemical sensing and nano molar level detection of Bisphenol-A with molecularly imprinted polymer tailored on multiwalled carbon nanotubes. *Polymer*, 146: 312–320. 10.1016/j.polymer.2018.05.052.

Antje, C. G., and J. Baeumner. 2020. Biosensors to support sustainable agriculture and food safety. *TrAC Trends in Analytical Chemistry*, 128: 115906. 10.1016/j.trac.2020.115906.

Ashraf, S. A., A. J. Siddiqui, A. E. O. Elkhalifa et al. 2021. Innovations in nanoscience for the sustainable development of food and agriculture with implications on health and environment. *Science of the Total Environment*, 768: 144990. 10.1016/j.scitotenv.2021.144990.

Aygün, A., S. Özdemir, M. Gülcan, K. Cellat, and F. Şen. 2020. Synthesis and characterization of reishi mushroom-mediated green synthesis of silver nanoparticles for the biochemical applications. *Journal of Pharmaceutical and Biomedical Analysis*, 178: 112970.

Baetsen-Young, A. M., M. Vasher, L. L. Matta, P. Colgan, E.C. Alocilja, and B. Day. 2018. Direct colorimetric detection of unamplified pathogen DNA by dextrin-capped gold nanoparticles. *Biosensors and Bioelectronics*, 101, 29–36. 10.1016/j.bios.2017.10.011

Baig, N., I. Kammakakam, and W. Falath. 2021. Nanomaterials: A review of synthesis methods, properties, recent progress, and challenges. *Materials Advances*, 2: 1821–1871. 10.1039/ d0ma00807arsc.li/materials-advances.

Bajpai, V. K., M. Kamle, S. Shukla et al. 2018. Prospects of using nanotechnology for food preservation, safety, and security. *Journal of Food and Drug Analysis*, 26 (4): 1201–1214. 10.1016/ j.jfda.2018.06.011.

Banua, J., and J. I. Han. 2020. Biogenesis of prism-like silver oxide nanoparticles using nappa cabbage extract and their p-nitrophenol sensing activity. *Molecules*, 25: 2298. 10.3390/molecules25102298.

Barimah, A. O., P. Chen, L. Yin, H. R. El-Seedi, X. Zou, and Z. Guo. 2022. SERS nanosensor of 3-aminobenzeneboronic acid labeled Ag for detecting total arsenic in black tea combined with chemometric algorithms. *Journal of Food Composition and Analysis*, 110, 104588. 10.1016/j.jfca.2022 .104588.

Basavegowda, N., T. K. Mandal, and K-H Baek. 2020. Bimetallic and trimetallic nanoparticles for active food packaging applications: A review. *Food and Bioprocess Technology*, 13: 30–44. 10.1007/s11 947-019-02370-3.

Bhardwaj, H., G. Sumana, and C. A. Marquette. 2020. A label-free ultrasensitive microfluidic surface plasmon resonance biosensor for aflatoxin B_1 detection using nanoparticles integrated gold chip. *Food Chemistry*, 307: 125530. 10.1016/j.foodchem.2019.125530.

Bindhu, M. R., M. Umadevi, G. A. Esmail, N. A. Al-Dhabi, and M. V. Arasu. 2020. Green synthesis and characterization of silver nanoparticles from *Moringa oleifera* flower and assessment of antimicrobial and sensing properties. *Journal of Photochemistry and Photobiology B: Biology*, 205: 111836. 10.1016/j.jphotobiol.2020.111836.

Boruah, B. S.r, N. Ojah, and R. Biswas. 2020. Bio-inspired localized surface plasmon resonance enhanced sensing of mercury through green synthesized silver nanoparticle. *Journal of Lightwave Technology*, 38: 2086–2091. 10.1109/jlt.2020.2971252.

Carrillo-Inungaray, M. L., J. A., Trejo-Ramirez, A. Reyes-Munguia, and C. Carranza-Alvarez. 2018. Use of nanoparticles in the food industry: Advances and perspectives. *Impact of Nanoscience in the Food Industry, Handbook of Food Bioengineering*, 15: 419–444. Academic Press. .10.1016/B978-0-12-811441-4.00015-7.

Cassani, L., N. E. Marcovich, and A. Gomez-Zavaglia. 2021. Seaweed bioactive compounds: Promising and safe inputs for the green synthesis of metal nanoparticles in the food industry. *Critical Reviews in Food Science and Nutrition*. 10.1080/10408398.2021.1965537.

Castillo, G., Z. Garaiova, and T. Hianik. 2017. New technologies for nanoparticles detection in foods. In *Advances in Food Diagnostics*, eds. F. Toldrá and L. M. L. Nolett, 305–341. Chichester: John Wiley and Sons.

Castillo-Henríquez, L., K. Alfaro-Aguilar, J. Ugalde-Álvarez et al. 2020. Green synthesis of gold and silver nanoparticles from plant extracts and their possible applications as antimicrobial agents in the agricultural area. *Nanomaterials*, 10: 1763. 10.3390/nano10091763.

Cerqueira, M. A., and L. M. Pastrana. 2019. Does the future of food pass by using nanotechnologies? *Frontiers in Sustainable Food Systems*, 3: 16. 10.3389/fsufs.2019.00016.

Chellaram, C., G. Murugaboopathi, A. A. John et al. 2014. Significance of nanotechnology in food industry. *APCBEE Procedia*, 8: 109–113. 10.1016/j.apcbee.2014.03.010.

Chelliah, R., S. Wei, E.B.-M, Daliri et al. 2021. Development of nanosensors based intelligent packaging systems: Food quality and medicine. *Nanomaterials*, 11: 515. 10.3390/nano11061515.

Chen, R., Y. Sun, B. Huo, D. Han, S. Li, and Z. Gao. 2021. A CuS-nanoparticle-based fluorescent probe for the sensitive and specific detection of ochratoxin A. *Talanta*, 222: 121678, January 15th, 2021. 10.1016/j.talanta.2020.121678.

Chippa, H. 2019. Mycosynthesis of nanoparticles for smart agricultural practice: A green and eco-friendly approach. In *Green Synthesis, Characterization and Applications of Nanoparticles, Micro and Nano Technologies*, eds. A. K. Shukla and S. Iravani , 5: 87–109. Amsterdam: Elsevier Inc.. 10.1016/B978-0-08-102579-6.00005-8.

Choi, J. R., K. W. Yong, J. Y. Choi, and A. C. Cowie. 2019. Emerging point-of-care technologies for food safety analysis. *Sensors*, 19: 817. 10.3390/s19040817.

Colica, C., V., Aiello, L. Boccuto et al. 2018. The role of nanotechnology in food safety. *Minerva Biotecnologica*, 30 (2): 69–73. http://ir.librarynmu.com/handle/123456789/237.

Couto, C., and A. Almeida. 2022. Metallic nanoparticles in the food sector: A mini-review. *Foods*, 11: 402. 10.3390/foods11030402.

Curulli, A. 2020. Nanomaterials in electrochemical sensing area: Applications and challenges in food analysis. *Molecules*, 25 (23): 5759. 10.3390/molecules25235759.

Das, G., J. K. Patra, S. Paramithiotis, and H-S. Shin. 2019. The sustainability challenge of food and environmental nanotechnology: Current status and imminent perception. *International Journal of Environmental Research and Public Health*, 16 (23): 4848. 10.3390/ijerph16234848.

Dayem, A. A., Hossain, M. K., S. B. Soo Bin Lee et al. 2017. The role of reactive oxygen species (ROS) in the biological activities of metallic nanoparticles. *International Journal of Molecular Sciences*, 18: 120. 10.3390/ijms18010120.

Dervisevic, M., E. Custiuc, E. Çevik, Z. Durmus, M. Şenel, and A. Durmus. 2015. Electrochemical biosensor based on REGO/Fe$_3$O$_4$ bionanocomposite interface for xanthine detection in fish sample. *Food Control*, 57: 402–410, November 2015.

Dervisevic, M., E. Dervisevic, and M. Şenel. 2019. Recent progress in nanomaterial-based electrochemical and optical sensors for hypoxanthine and xanthine. A review. *Microchimica Acta*, 186 (12): 749. 10.1007/s00604-019-3842-6.

Devra, V. 2022. Applications of metal nanoparticles in the agri-food sector. *Egyptian Journal of Agricultural Research*, 2: 163–183. 10.21608/EJAR.2022.102565.1164.

Dikshit, P. K., J. Kumar, A. K. Das et al. 2021. Green synthesis of metallic nanoparticles: Applications and limitations. *Catalysts*, 11: 902. 10.3390/catal11080902.

Doan, V. D., V. T. Le, T. L. Phan, T. L. H. Nguyen, and T. D. Nguyen. 2021. Waste banana stem utilized for biosynthesis of silver and gold nanoparticles and their antibacterial and catalytic properties. *Journal of Cluster Science*, 32 (6): 1673–1682. 10.1007/s10876-020-01930-4.

Du, J., S. Wu, L. Niu, J. Li, D. Zhao, and Y. Bai. 2020. A gold nanoparticles-assisted multiplex PCR assay for simultaneous detection of *Salmonella typhimurium*, *Listeria monocytogenes* and *Escherichia coli* O157:H7. *Analytical Methods*, 12 (2): 212–217.

Dudefoi, W., A. Villares, and S. Peyron. 2018. Nanoscience and nanotechnologies for biobased materials, packaging and food applications: New opportunities and concerns. *Innovative Food Science and Emerging Technologies*, 46: 107–121.

Dutta, T., N. N. Ghosh, M. Das, R. Adhikary, V. Mandal, and A. P. Chattopadhyaye. 2020. Green synthesis of antibacterial and antifungal silver nanoparticles using *Citrus limetta* peel extract: Experimental and theoretical studies. *Journal of Environmental Chemical Engineering*, 8 (4): 104019. 10.1016/j.jece.2020.104019.

EFSA. 2021. Guidance on risk assessment of nanomaterials to be applied in the food and feed chain: human and animal health. https://www.efsa.europa.eu/en/efsajournal/pub/6768. (Accessed November, 16th, 2022).

EFSA. 2022. Nanotechnology. European Food Safety Authority. https://www.efsa.europa.eu/en/topics/topic/nanotechnology. Accessed November, 16th November, 2022).

El-Dasuki, N., K. R. Shoueir, I. El-Mehasseb, and M. El-Kemary. 2021. Bio-inspired green manufacturing of plasmonic silver nanoparticles/Degussa using banana waste peduncles: Photocatalytic, antimicrobial, and cytotoxicity evaluation. *Journal of Materials Research and Technology*, 10: 671–686. 10.1016/j.jmrt.2020.12.035.

El-Desouky, N., K. R. Shoueir, I. El-Mehasseb, and M. El-Kemary. 2021. Bio-inspired green manufacturing of plasmonic silver nanoparticles/Degussa using banana waste peduncles: Photocatalytic, antimicrobial, and cytotoxicity evaluation. *Journal of Materials Research and Technology*, 10: 671–686. 10.1016/j.jmrt.2020.12.035

El-Zamkan, M. A., B. A. Hendy, H. M. Diab et al. 2021. Control of virulent *Listeria monocytogenes* originating from dairy products and cattle environment using marine algal extracts, silver nanoparticles thereof, and quaternary disinfectants. *Infection and Drug Resistance*, 14: 2721–2739.

FAO. 2022. Food safety and quality. Food and Agriculture Organization of the United Nations. Rome, Italy. https://www.fao.org/food-safety/en/(accessed November 20th, 2022).

FDA. 2014. Considering whether an FDA-regulated product involves the application of nanotechnology guidance for industry. https://www.fda.gov/regulatory-information/search-fda-guidance-documents/considering-whether-fda-regulated-product-involves-application-nanotechnology. (Accessed November 18th, 2022)

FDA. 2018. FDA's Approach to regulation of nanotechnology products. https://www.fda.gov/science-research/nanotechnology-programs-fda/fdas-approach-regulation-nanotechnology-products (Accessed November,15th 2022).

Fierascu, I. I., C. Fierascu, R. I. Brazdis, A. M. Baroi, T. Fistos, and R. C. Fierascu. 2020. Phytosynthesized metallic nanoparticles—between nanomedicine and toxicology. A brief review of 2019's findings. *Materials*, 13 (3): 574. 10.3390/ma13030574.

Fraceto, L. F., R. Grillo, G. A. de Medeiros, V. Scognamiglio, G. Rea, and C. Bartolucci. 2016. Nanotechnology in agriculture: Which innovation potential does it have? *Frontiers in Environmental Science*, 4: 20. 10.3389/fenvs.2016.00020.

Garrido-Maestu, A., S. Azinheiro, J. Carvalho, and Prado, M. 2018. Rapid and sensitive detection of viable *Listeria monocytogenes* in food products by a filtration-based protocol and qPCR. *Food Microbiology*, 73: 254–263. 10.1016/j.fm.2018.02.004.

Gondal, A. H., and L. Tayyiba. 2022. Prospects of using nanotechnology in agricultural growth, environment and industrial food products. *Reviews in Agricultural Science*, 10: 68–81. 10.7831/ras.10.0_68.

Green, E. 2020. Smart, active and intelligent. *Europe Packaging Magazine.Packaging Today*. 8 January 2020. https://www.packagingtoday.co.uk/features/featuresmart-active-and-intelligent-7743916/. (Accessed 15th November 2022).

Guo, X., F. Wen, N. Zheng, M. Saive, M-L, Fauconnier, and J. Wang. 2020. Aptamer-based biosensor for detection of mycotoxins. *Frontiers in Chemistry*, 8: 195. 10.3389/fchem.2020.00195.

Guzmán-Altamirano, M. Á., B. Rebollo-Plata, A. de J. Joaquín-Ramos, and M. G. Gómez-Espinoza. 2022. Green synthesis and antimicrobial mechanism of nanoparticles: applications in agricultural and agrifood safety. *Journal of the Science of the Food and Agriculture*, first published August 8th, 2022. 10.1002/jsfa.12162.

Hari, N., S., Radhakrishnan, and A. J. Nair. 2020. Nanosensors as potential multisensor systems to ensure safe and quality food. In *Biotechnological Approaches in Food Adulterants*, ed. M. A. Verma, 204. Imprint, 35. Florida: CRC Press. 10.1201/9780429354557.

Hassan, S. A., P. Ghadam, and A. A. Ali. 2022. One step green synthesis of Cu nanoparticles by the aqueous extract of *Juglans regia* green husk: Assessing its physicochemical, environmental, and biological activities. *Bioprocess and Biosystems Engineering*, 45 (3): 605–618. 10.1007/s00449-022-02691-2.

He, P., M. M., Hassan, W. Yang et al. 2023. Rapid and stable detection of three main mycotoxins in rice using SERS optimized AgNPs@K30 coupled multivariate calibration. *Food Chemistry*, 398: 133883, 1 January 2023. 10.1016/j.foodchem.2022.133883.

He, X., H. Deng, H.-Min, and H. Wang. 2019. The current application of nanotechnology in food and agriculture. *Journal of Food and Drug Analysis*, 27 (1): 21. 10.1016/j.jfda.2018.12.002.

He, Y., H. Li, X. Fei, and L. Peng. 2021. Carboxymethyl cellulose/cellulose nanocrystals immobilized silver nanoparticles as an effective coating to improve barrier and antibacterial properties of paper for food packaging applications. *Carbohydrate Polymers*, 252: 117156. 10.1016/j.carbpol.2020.117156. PMID: 33183607.

Hossain, A., M. Skalicky, M. Brestic et al. 2021. Application of nanomaterials to ensure quality and nutritional safety of food. *Hindawi Journal of Nanomaterials*, Article ID 9336082. 10.1155/2021/9336082.

Hosseinzadeh, S., R. Partovi, F. Talebi, and A., Babaei. 2020. Chitosan/TiO$_2$ nanoparticle/ *Cymbopogon citratus* essential oil film as food packaging material: Physico-mechanical properties and its effects on microbial, chemical, and organoleptic quality of minced meat during refrigeration. *Journal of Food Processing and Preservation*, 44 (7): e14536. 10.1111/jfpp.14536.

Hovhannisyan, Z., M. Timotina, and J. Manoyan et al. 2022. *Ribes nigrum* L. extract-mediated green synthesis and antibacterial action mechanisms of silver nanoparticles. *Antibiotics*, 11: 1415. 10.3390/antibiotics11101415.

Hutapea, S., S. Ghazi Al-Shawi, and T.-C. Chen. 2022. Study on food preservation materials based on nano-particle reagents. *Food Science and Technology, Campinas*, 42: e39721. 10.1590/fst.39721.

Jafarzadeh, S., and S. M. Jafari. 2020. Impact of metal nanoparticles on the mechanical, barrier, optical and thermal properties of biodegradable food packaging materials. *Critical Reviews in Food Science and Nutrition*, 61, 16.2021. 10.1080/10408398.2020.1783200.

Jain, A., S. Ranjan, N. Dasgupta, and C. Ramalingam. 2018. Nanomaterials in food and agriculture: An overview on their safety concerns and regulatory issues. *Critical Reviews in Food Science and Nutrition*, 58 (2). 10.1080/10408398.2016.1160363.

Javidi, M., M. R. Housaindokht, A. Verdian, and B. M. Razavizadeh. 2018. Detection of chloramphenicol using a novel apta-sensing platform based on aptamer terminal-lock in milk samples. *Analytica Chimica Acta*, 1039: 116–123. 10.1016/j.aca.2018.07.041.

Jiang, D., P. Zhu, H. Jiang, J. Ji, X. Sun, W. Gu, and G. Zhang. 2015. Fluorescent magnetic bead-based mast cell biosensor for electrochemical detection of allergens in foodstuffs. *Biosensors and Bioelectronics*, 70 (15): 482–490. 10.1016/j.bios.2015.03.058

Jinn, B., X. Li, X. Zhou et al. 2017. Fabrication and characterization of nanocomposite film made from a jackfruit filum polysaccharide incorporating TiO$_2$ nanoparticles by photocatalysis. *RSC Advance*, 7: 16931–16937. 10.1039/c6ra28648h.

Joshi, H., P. Choudhary, and S. L. Mundra. 2019. Future prospects of nanotechnology in agriculture. *International Journal of Chemical Studies*, 7 (2): 957–963.

Jovanska, L., C.-H. Chiu, Yi-C. Yeh, W-D. Chiang, C.-C. Hsieh, and R. Wang. 2022. Development of a PCL-PEO double network colorimetric pH sensor using electrospun fibers containing *Hibiscus rosa sinensis* extract and silver nanoparticles for food monitoring. *Food Chemistry*, 368: 130813, 30 January 2022.

Karakuş, S. G., C. Ö. Baytemir, and N., Taşaltın. 2022. An ultra-sensitive smartphone-integrated digital colorimetric and electrochemical *Camellia sinensis* polyphenols encapsulated CuO nanoparticles-based ammonia biosensor. *Inorganic Chemistry Communications*, 143: 109733. September 2022. 10.1016/j.inoche.2022.109733,

Karimi, M., R. Sadeghi, and J. Kokini. 2018. Human exposure to nanoparticles through trophic transfer and the biosafety concerns that nanoparticle-contaminated foods pose to consumers. *Trends in Food Science and Technology*, 75: 129–145. 10.1016/j.tifs.2018.03.012.

Karuppuswami, S., L. L. Matta, E. C. Alocilja, and P. Chahal. 2018. A wireless RFID compatible sensor tag using gold nanoparticle markers for pathogen detection in the liquid food supply chain. *IEEE Sensors Letters*, 2 (2): 1–4. June 2018. 10.1109/LSENS.2018.2822305.

Khan, N., S. Ali, S. Latif, and A. Mehmood. 2022. Biological synthesis of nanoparticles and their applications in sustainable agriculture production. *Natural Science*, 14: 226–234. 10.4236/ns.2022.146022.

Khanal, L. N., K. R. Sharma, H. Paudyal, K. Parajuli, B. Dahal, G. C. Ganga, Y. R. Pokharel, and S. K. Kalauni. 2022. Green synthesis of silver nanoparticles from root extracts of *Rubus ellipticus* Sm. and comparison of antioxidant and antibacterial activity. *Journal of Nanomaterials*, 2022: 1–1110.1155/2022/1832587.

Khodadadi, S., N. Mahdinezhad, B. Fazeli-Nasab, M. J. Heidari, B. Fakheri, and A. Miri. 2021. Investigating the possibility of green synthesis of silver nanoparticles using *Vaccinium arctostaphlyos* extract and evaluating its antibacterial properties. *Ecology and Biotechnological Applications of Biofilms*. Article ID 5572252. 10.1155/2021/5572252.

Kraśniewska, K., S. Galus, and M. Gniewosz. 2020. Biopolymers-based materials containing silver nanoparticles as active packaging for food applications–A review. *International Journal of Molecular Sciences*, 21 (3): 698. 10.3390/ijms21030698.

Kumari, S. C., P. N. Padma, and K. Anuradha. 2021. Green silver nanoparticles embedded in cellulosic network for fresh food packaging. *Journal of Pure and Applied Microbiology*. 15 (3): 1236–1244. Article 6942 | . 10.22207/JPAM.15.3.13.

Kuswandi, B. 2016. Nanotechnology in food packaging. In *Nanoscience in Food and Agriculture, 20: 5, Sustainable Agriculture Reviews*, eds. S. Ranjan, N. Dasgupta, E. Lichtfouse, 151–183. Cham: Springer International Publishing. 10.1007/978-3-319-39303-2_6.

Kutsanedzie, F. Y. H., A. A. Agyekum, V. Annavaram, and Q. Chen. 2020. Signal-enhanced SERS-sensors of CAR-PLS and GA-PLS coupled AgNPs for ochratoxin A and aflatoxin B1 detection. *Food Chemistry*, 315: 126231, 15 June 2020.

Li, L., M. Zhang, and W. Chen. 2020. Gold nanoparticle-based colorimetric and electrochemical sensors for the detection of illegal food additives. *Journal of Food and Drug Analysis*, 28 (4): 641–653. 10.38212/2224-6614.3114.

Linsinger, T. P. J., Q. Chaudhry, V. Dehalu et al. 2013. Validation of methods for the detection and quantification of engineered nanoparticles in food. *Food Chemistry*, 138: 1959–1966. 10.1016/j.foodchem.2012.11.074.

Liu, S., and T. Xia. 2020. Continued efforts on nanomaterial-environmental health and safety is critical to maintain sustainable growth of nanoindustry. *Nano-Micro Small*, 16 (21) (April): 2000603. 10.1002/smll.202000603.

Liu, Y., Y. Deng, S. Li, F. W-N, Chow, M. Liu, and N. He. 2022. Monitoring and detection of antibiotic residues in animal derived foods: Solutions using aptamers. *Trends in Food Science and Technology*, 125: 200–235, July 2022. 10.1016/j.tifs.2022.04.008.

Lugani, Y., B. S. Sooch, P. Singh, and S. Kumar. 2021. Nanobiotechnology applications in food sector and future innovations. *Microbial Biotechnology in Food and Health. Applied Biotechnology Reviews*, 2: 197–225. Elsevier. 10.1016/B978-0-12-819813-1.00008-6.

Ma, G., J. Cao, G. Hu, L. Zhu et al. 2021. Porous chitosan/partially reduced graphene oxide/diatomite composite as an efficient adsorbent for quantitative colorimetric detection of pesticides in a complex matrix. *Analyst*, 146 (14): 4576–4584. 10.1039/d1an00621e.

Machado, T. O., J. Grabow, and Sayer, C. et al. 2022. Biopolymer-based nanocarriers for sustained release of agrochemicals: A review on materials and social science perspectives for a sustainable future of agri- and horticulture. *Advances in Colloid and Interface Science*, 303: 102645. 10.1016/j.cis.2022.102645.

Malini, S., K. Raj, S. Madhumathy et al. 2022. Bioinspired advances in nanomaterials for sustainable agriculture. *Journal of Nanomaterials*, 2022: Article ID 8926133. 10.1155/2022/8926133.

Maťátková, O., J. Michailidu, A. Miškovská, I. Kolouchová, J. Masák, and A. Čejková. 2022. Antimicrobial properties and applications of metal nanoparticles biosynthesized by green methods. *Biotechnology Advances*, 58: 107905, January 11th, 2022, 107905. 10.1016/j.biotechadv.2022.107905.

Martinez, L., and L. He 2021. Detection of mycotoxins in food using surface-enhanced raman spectroscopy: A review. *ACS Applied Bio Materials*, 4: 295–310. 10.1021/acsabm.0c01349.

McClements, D. J., and H. Xiao. 2017. Is nano safe in foods? Establishing the factors impacting the gastrointestinal fate and toxicity of organic and inorganic food-grade nanoparticles. *Npj Science of Food*, 1: 6. 10.1038/s41538-017-0005-1.

Mehmood, S., N. K. Janjua, S. Tabassum, S. Faizi, and H. Fenniri. 2022. Cost effective synthesis approach for green food packaging coating by gallic acid conjugated gold nanoparticles from *Caesalpinia pulcherrima* extract. *Results in Chemistry*, 4: 100437. 10.1016/j.rechem.2022.100437.

Mehta, V. N., N. Ghinaiya, J. Rohit, R. K. Singhal, H. Basu, and S. K. Kailasa. 2022. Ligand chemistry of gold, silver and copper nanoparticles for visual read-out assay of pesticides: A review. *TrAC Trends in Analytical Chemistry*, 153: 116607. 10.1016/j.trac.2022.116607.

Mishra, G. K., A. Barfidokht, F. Tehrani, and R. K. Mishra. 2018. Food safety analysis using electrochemical biosensors. *Foods*, 141. 10.3390/foods7090141.

Mohammadpour, Z., and S. M. Naghib. 2021. Smart nanosensors for intelligent packaging. *Nanosensors for Smart Manufacturing. Micro and Nano Technologies*, 15: 323–346. 10.1016/B978-0-12-823358-0.00017-4.

Mustapha, T., N. Misni, N. R. Ithnin, A. M. Daskum, and N. Z. Unyah. 2022. A review on plants and microorganisms mediated synthesis of silver nanoparticles, role of plants metabolites and applications. *International Journal of Environmental Research and Public Health*, 19: 674. 10.3390/ijerph19020674.

Naseer, B., G. Srivastava, O. S. Qadri, S. A. Faridi, R. Ul Islam, and K. Younis. 2018. Importance and health hazards of nanoparticles used in the food industry. *Nanotechnology Reviews*, 7 (6): 623–641. 10.1515/ntrev-2018-0076.

Nasrollahzadeh, M., S. Mahmoudi-GomYek, N. Motahharifar, and M. G. Gorab. 2019. Recent developments in the plant-mediated green synthesis of Ag-based nanoparticles for environmental and catalytic applications. *The Chemical Record*, 19 (12): 2436–2479.

Ndwandwe, B. K., S. P. Malinga, E. Kayitesi, and B. C. Dlamini. 2021. Advances in green synthesis of selenium nanoparticles and their application in food packaging. *International Journal of Food Science and Technology*, 56: 2640–2650. 10.1111/ijfs.14916.

Neethirajan, S., V. Ragavan, X. Weng, and R. Chand. 2018. Biosensors for sustainable food engineering: Challenges and perspectives. *Biosensors*, 8: 23. 10.3390/bios8010023.

Niluxsshun, M. C. D., K. Masilamani, and U. Mathiventhan. 2021. Green synthesis of silver nanoparticles from the extracts of fruit peel of *Citrus tangerina*, *Citrus sinensis*, and *Citrus limon* for Antibacterial Activities. *Hindawi Bioinorganic Chemistry and Applications*. Article ID 6695734. 10.1155/2021/6695734.

Ningthoujam, R., J. Barsarani, S. Pattanayak et al. 2022. Nanotechnology in food science. In *Bio-Nano*, eds. M. Arakha, A. K., Pradhan, and S. Jha, 59–73. Singapore: Springer. 10.1007/978-981-16-2516-9_4.

Ötleş, S., and B. Y. Şahyar. 2017. Nanotechnology and shelf-life of animal foods. In *Nanotechnology*, eds. R. Prasad, V. Kumar, and M. Kumar, 35–43. Singapore: Springer. 10.1007/978-981-10-4678-0_2.

Omran, B. A., O. Aboelazayem H. N. Nassar et al. 2020. Biovalorization of mandarin waste peels into silver nanoparticles and activated carbon. *International Journal of Environmental Science and Technology*, 18, 1119–1134. 10.1007/s13762-020-02873-z.

Pandey, A. T., I. Pandey, Y., Hachenberger et al. 2020. Emerging paradigm against global antimicrobial resistance via bioprospecting of mushroom into novel nanotherapeutics development. *Trends in Food Science and Technology*, 106: 333–344. December 2020. 10.1016/j.tifs.2020.10.025.

Pandhi, S., D. K. Mahato, and A. Kumar. 2021. Overview of green nanofabrication, technologies for food quality and safety applications. *Food Reviews International*, 1: 21. 10.1080/87559129.2021.1904254.

Paul, I. E., D. N. Kumar, A. Rajeshwari et al. 2017. Detection of food contaminants by gold and silver nanoparticles. *Nanobiosensors*, 4: 129–165. 10.1016/B978-0-12-804301-1.00004-7.

Ponce, A. G., J. F. Ayala-Zavala, N. E. Marcovich, F. J. Vazquez, and M. R. Ansorena. 2018. Industry: recent developments, risks. In *Impact of Nanoscience in the Food Industry, Handbook of Food Bioengineering*, eds. A. M. Grumezescu and A. M. Holban, 12:113–141. Cambridge: Academic Press. 10.1016/B978-0-12-811441-4.00005-4

Rafique, M., I. Sadaf, M. S. Rafique, and M. B. Tahir. 2017. A review on green synthesis of silver nanoparticles and their applications. *Artificial Cells, Nanomedicine, and Biotechnology*. 45 (7): 1272–1291. 10.1080/21691401.2016.1241792.

Ragavan, K. V., and S. Neethirajan. 2019. Nanomaterial-based electrochemical biosensors for food safety. *Journal of Electroanalytical Chemistry*, 781: 147–154. 10.1016/B978-0-12-814130-4.00007-5.

Rahman, A., J. Lin, F. E. Jaramillo et al. 2020. In vivo biosynthesis of inorganic nanomaterials using eukaryotes-A review. *Molecules*, 25 (14): 3246. 10.3390/molecules25143246.

Rane, A. V., K. Kanny, V. K. Abitha, and Thomas, S. 2018. Methods for synthesis of nanoparticles and fabrication of nanocomposites. In *Synthesis of Inorganic Nanomaterials, Advances and Key Technologies, Micro and Nano Technologies*, eds. S. M. Bhagyaraj, O. S. Oluwafemi, N. Kalarikkal, and S. Thomas, 121–139. Sawston: Woodhead Publishing, Elsevier. 10.1016/B978-0-08-101975-7.00005-1.

Ravichandran, V., S. Vasanthi, S. Shalini, A. A. Shah, M. Tripathy, and N. Paliwal. 2020. Green synthesis, characterization, antibacterial, antioxidant and photocatalytic activity of *Parkia speciosa* leaves extract mediated silver nanoparticles. *Results in Physics*, 15: 102565. 10.1016/j.rinp.2019.102565.

Ravindran, R. S. E., V. Subha, and R. Ilangovan. 2020. Silver nanoparticles blended PEG/PVA nanocomposites synthesis and characterization for food packaging. *Arabian Journal of Chemistry*, 13: 6056–6060. 10.1016/j.arabjc.2020.05.005.

Rodríguez-Félix, F., A. G. López-Cota, M. J. Moreno-Vásquez et al. 2021. Sustainable-green synthesis of silver nanoparticles using safflower (*Carthamus tinctorius* L.) waste extract and its antibacterial activity. *Heliyon*, 7: e06923. 10.1016/j.heliyon.2021.e06923.

Rojas, L. M., Y. Qu, and L. He. 2020. A facile solvent extraction method facilitating surface-enhanced Raman spectroscopic detection of ochratoxin A in wine and wheat. *Talanta*, 224: 121792. 10.1016/j.talanta.2020.121792.

Rovera, C., M.. Ghaani, and S. Farris. 2020. Nano-inspired oxygen barrier coatings for food packaging applications: An overview. *Trends in Food Science and Technology*, 97: 210–220. 10.1016/j.tifs.2020.01.024

Roy, L., D. Bera and S. Adak. 2018. Mycosynthesized nanoparticles: Role in food processing industries. In *Fungal Nanobionics: Principles and Applications*, eds. R. Prasad, V. Kumar, M. Kumar, and S. Wang, 287–316. Singapore: Springer. 10.1007/978-981-10-8666-3_12

Sanjay, S. S. 2019. Chapter 2—Safe nano is green nano. In *Green Synthesis, Characterization and Applications of Nanoparticles*, eds. A. K. Shukla and S. Iravani, 27–36. Amsterdam: Elsevier.

Santhosh, A., V. Theertha, P. Prakash, and S. S. Chandra. 2021. From waste to a value added product: Green synthesis of silver nanoparticles from onion peels together with its diverse application. *Materials Today Proceedings*, 46 (Part 10): 4460–4463. 10.1016/j.matpr.2020.09.680.

Saratale, R. G., G. D. Saratale, H. S. Shin et al. 2018. New insights on the green synthesis of metallic nanoparticles using plant and waste biomaterials: Current knowledge, their agricultural and environmental applications. *Environmental Science and Pollution Research*, 25: 10164–10183. 10.1007/s11356-017-9912-6.

Sen, S., and P. Sarkar. 2020. A simple electrochemical approach to fabricate functionalized MWCNT-nanogold decorated PEDOT nanohybrid for simultaneous quantification of uric acid, xanthine and hypoxanthine. *Analytical Chimica Acta*, 1114: 15–28. 10.1016/j.aca.2020.03.060.

Shahzadi, T., S. Iqbal, T. Riaz, and M. Zaib. 2022. A comparative study based on localized surface plasmon resonance optical characteristics of green synthesized nanoparticles towards spectrophotometric determination of cupric ions. *Journal of Taibah University for Science*, 16 (1): 912–922. 10.1080/16583655.2022.2123206.

Shang, Y., Md. K. Hasan, G. J. Ahammed et al. 2019. Applications of nanotechnology in plant growth and crop protection: A review. *Molecules*, 24: 2558. 10.3390/molecules24142558.

Shao, B., X. Ma., S. Zhao, Y. Lv et al. 2018. Nanogapped $Au_{(core)}$ @ $Au-Ag_{(shell)}$ structures coupled with Fe_3O_4 magnetic nanoparticles for the detection of ochratoxin A. *Analytica Chimica Acta*, 1033: 165–172. 29 November 2018. 10.1016/j.aca.2018.05.058.

Sharma. R., K. V. Ragavan, M. S. Thakur, and K. S. M. S. Raghavarao. 2015. Recent advances in nanoparticle based aptasensors for food contaminants. *Biosensors and Bioelectronics*, 74: 612–627. 10.1016/j.bios.2015.07.017.

Shawon, Z. B. Z., Md E. Hoque, and S. R. Chowdhury. 2020. Nanosensors and nanobiosensors: Agricultural and food technology aspects. *Nanofabrication for Smart Nanosensor Applications. Micro and Nano Technologies*, 6: 135–161.

Shehab, S. A., R. El Kurdi, and D. Patra. 2020. Curcumin mediated PEG thiol acid conjugated gold nanoparticles for the determination of melamine. *Microchemical Journal*, 153: 104382. March 2020. 10.1016/j.microc.2019.104382.

Sheng, K. H., Jiang, Y. Fang, L. Wang, and D. Jiang. 2022. Emerging electrochemical biosensing approaches for detection of allergen in food samples: A review. *Trends in Food Science and Technology*, 121: 93–104. 10.1016/j.tifs.2022.01.033.

Shim, K., J. Kim, M. Shahabuddin, Y. Yamauchi, Md. S.A. Hossain, and J. H. Kim. 2018. Efficient wide range electrochemical bisphenol-A sensor by self-supported dendritic platinum nanoparticles on screen-printed carbon electrode. *Sensors and Actuators B: Chemical*, 255: 2800–2808. 10.1016/j.snb.2017.09.096.

Siddiqui, S., and S. A. Alrumman. 2021. Influence of nanoparticles on food: An analytical assessment. *Journal of King Saud University – Science*. 33: 101530. 10.1016/j.jksus.2021.101530.

Singh, P., S. Kumar, and P. P. Sharma. 2021. Nanotechnology applications in advancing agriculture and food technology. *Progressive Agriculture*, 21 (1): 31–36. 10.5958/0976-4615.2021.00006.5.

Singh, V. J., and A. K. Singh. 2019. Role of microbially synthesized nanoparticles in sustainable agriculture and environmental management. In *Role of Plant Growth Promoting Microorganisms in Sustainable Agriculture and Nanotechnology*, eds. A. Kumar, A. K. Singh, and K. K. Choudhary, 4: 55–73. Cambridge, USA: Woodhead Publishing. https://doi.org/10.1016/C2018-0-01338-9

Song, Y., L. He, S. Zhang et al. 2021. Novel impedimetric sensing strategy for detecting ochratoxin A based on NH_2-MIL-101(Fe) metal-organic framework doped with cobalt phthalocyanine nanoparticles. *Food Chemistry*, 351: 129248, 30 July 2021. 10.1016/j.foodchem.2021.129248.

Tabari, M. 2018. Characterization of a new biodegradable edible film based on sago starch loaded with carboxymethyl cellulose nanoparticles. *Nanomedicine Research Journal*, 3 (1): 25–30. 10.22034/nmrj.2018.01.004.

Tamilarasi, P., and P. Meena. 2020. Green synthesis of silver nanoparticles (Ag NPs) using *Gomphrena globosa* (Globe amaranth) leaf extract and their characterization. *Materials Today: Proceedings*, 33 (Part 5): 2209–2216. 10.1016/j.matpr.2020.04.025.

Tertis, M., O. Hosu., B. Feier et al. 2021. Electrochemical peptide-based sensors for foodborne pathogens detection. *Molecules*, 26: 3200. 10.3390/ molecules26113200.

Thakur, M., B. Wang, and M. L. Verma. 2022. Development and applications of nanobiosensors for sustainable agricultural and food industries: Recent developments, challenges and perspectives. *Environmental Technology and Innovation*, 26: 10237, May 2022. 10.1016/j.eti.2022.102371.

Thekkethil, A. J., R. Nair, and A. Madhavan. 2019. The role of nanotechnology in food safety: A review. *International Conference on Computational Intelligence and Knowledge Economy (ICCIKE)*, 405–409, 11–12 December 2019. Dubai: ICCIKE. 10.1109/ICCIKE47802.2019.9004412.

Tiwari, M., A. Singh, S. Dureja, S. Basu, and S. Katunayake. 2022. Au nanoparticles decorated ZnO/$ZnFe_2O_4$ composite SERS-active substrate for melamine detection. *Talanta*, 236: 122819, 1 January 2022, 122819. 10.1016/j.talanta.2021.122819.

Tsekhmistrenko, S., V. Bityutskyy, O. Tsekhmistrenko et al. 2021. Bionanotechnologies: synthesis of metals' nanoparticles with using plants and their applications in the food industry: A review. *Journal of Microbiology, Biotechnology and Food Sciences*, 10 (6): 1513. 10.15414/jmbfs.1513.

UN. 2022. World Food Safety Day highlights need to improve health, prevent foodborne risks. June 7th, 2022. https://news.un.org/en/story/2022/06/1119872. (Accessed 20th November 2022).

Ungureanu, C., G. T. Tihan, R. G. Zgârian et al. 2022. Metallic and metal oxides nanoparticles for sensing food pathogens—An overview of recent findings and prospects. *Materials*, 15: 5374. 10.3390/ma15155374.

Vijayakumar, M. D., G. J. Surendhar, L. Natrayan, P. P. Patil, P. M. B. Ram, and P. Paramasivam. 2022. Evolution and recent scenario of nanotechnology in agriculture and food industries. *Journal of Nanomaterials*, Article ID 1280411, 11 Jul 2022. 10.1155/2022/1280411

WHO. 2020. Food safety. World Health Organization. https://www.who.int/health-topics/food-safety/ (accessed November 15th, 2022).

Wu, Q., H. Tao, Y. Wu, X. Wang, Q. Shi, and D. Xiang. 2022. A label-free electrochemical aptasensor based on Zn/Fe bimetallic MOF derived nanoporous carbon for ultra-sensitive and selective determination of paraquat in vegetables. *Foods*, 11 (16): 2405, Aug 10th, 2022. 10.3390/foods11162405.

Wu, Y., J. Sun, X. Huang, W. Lai, and Y. Xiong. 2021. Ensuring food safety using fluorescent nanoparticles-based immunochromatographic test strips. *Trends in Food Science and Technology*, 118: 658–678. Part A, December 10.1016/j.tifs.2021.10.025.

Yang, T., and T. V. Duncan. 2021. Challenges and potential solutions for nanosensors intended for use with foods. *Nature Nanotechnology*, 16: 251–265. https://doi-org.ezproxygateway.sastudents.uwi.tt/10.1038/s41565-021-00867-7.

Yu, H., J.-Y Park, C. W., Kwon, S.-C, Hong, K.-M Park, and P.-S. Chang. 2018. An overview of nanotechnology in food science: preparative methods, practical applications, and safety. *Hindawi Journal of Chemistry*, 2018: Article ID 5427978. 10.1155/2018/5427978.

Yu, X., T. Zhong, Y. Zhang et al. 2022. Design, preparation, and application of magnetic nanoparticles for food safety analysis: A review of recent advances. *Journal of Agricultural and Food Chemistry*, 70 (1): 46–62. 10.1021/acs.jafc.1c03675.

Zhang, L., R. Sun, H. Yu, H. Yu, G. Xu, L. Deng, and J. Qian 2020. A new method for matching gold nanoparticle-based time–temperature indicators with muffins without obtaining activation energy. *Journal of Food Science*, 85: 2589–2595. 10.1111/1750-3841.15348

Zehra, A., A. Rai, S. K. Singh et al. 2021. An overview of nanotechnology in plant disease management, food safety, and sustainable agriculture. *Food Security and Plant Disease Management*, 10: 193–219. 10.1016/B978-0-12-821843-3.00009-X.

Zhu, A., T. Jiao, S. Ali, Y. Xu, Q. Ouyang, and Q. Chen. 2022. Dispersive micro solid phase extraction based ionic liquid functionalized ZnO nanoflowers couple with chromatographic methods for rapid determination of aflatoxins in wheat and peanut samples. *Food Chemistry*, 391: 133277. 15th October 2022. 10.1016/j.foodchem.2022.133277.

13 Nanozymes in Food Safety

Current Applications and Future Challenges

Ankita Kumari, Anikesh Bhardwaj, Mrinal Samtiya, Tejpal Dhewa, and Sanjeev Kumar

13.1 INTRODUCTION

Enzymes are natural biocatalysts catalyzing biochemical reactions without altering their characteristics, like high catalytic efficiency and substrate specificity. Recently, several enzymes have been employed in the agri-food industry for various applications, such as processing, functionalization, and storage of agri-food products (Huang *et al.*, 2019a). In addition to these applications, the role of these natural biocatalysts has been established in the safety of food products with their high selectivity and sensitivity. Enzymatic methods can be employed to detect the contamination of ions, small molecules, and chemical and biological contaminants (Huang *et al.*, 2019a). However, there are numerous shortcomings to using these enzymes in industrial processes because of the high purification cost and low success rate of recycling. In addition, enzymes are prone to denature under harsh temperatures and pH conditions during processing, making them ineffective for industrial applications (Whitaker *et al.*, 2002). Therefore, studies have focused on finding enzyme substitutes with similar activity and catalytic properties. Artificial enzyme mimics have emerged as better alternatives to overcome the drawbacks of natural enzymes (Gong *et al.*, 2015). Amongst artificial enzymes, nanomaterials (catalytically active), known as nanozymes, are studied widely for their applications and considered next-generation artificial enzymes (Lin *et al.*, 2014). Firstly, Yan et al. reported that ferromagnetic nanoparticles (Fe_3O_4 NPs) have an enzyme-like activity similar to peroxidase enzyme; thus, NPs could oxidize the peroxidase substrate in the presence of H_2O_2 (Gao *et al.*, 2007).

Similarly, many nanomaterials with catalytic activity were recognized alone or in combination with biomolecular ligands (Zhou *et al.*, 2017). The term "nanozyme" refers to metallic and non-metallic nanomaterials with catalytic activity similar to natural enzymes (Wei and Wang, 2013). The mass production of nanozymes is easy and low cost compared to their natural counterparts. Also, these are resistant to severe processing conditions, have high efficiency, and are highly stable in nature (Wu *et al.*, 2019). Nanozymes consist of enzyme-mimicking materials such as cyclodextrins, polymers, metalloproteins, supramolecules, micelles, coordination complexes porphyrins, and dendrimers (Kuah *et al.*, 2016). Progression in nanotechnology improved the properties of nanomaterials and enzyme-mimicking properties, which led to several applications of nanozymes in several fields, including sensor development, food quality and safety assessment, environmental pollutant monitoring, biological metabolite measurement, and medical therapeutics (Jiang *et al.*, 2019; Song *et al.*, 2019; Wang *et al.*, 2019; Wu *et al.*, 2017). In the present chapter, the principles of nanozyme functioning and their applications in food safety (Figure 13.1), followed by future challenges of these applications, have been comprehended.

13.2 CATALYTIC MECHANISMS OF NANOZYMES

Since the discovery of nanozymes, several nanomaterials with biocatalytic activities have been uncovered. Research has shown several nanomaterials such as graphene, nanodots, nanotubes, and nanospheres made up of metals, metal oxides (Fe_3O_4, CeO_2, TiO_2, etc.), and metal chalcogenides can mimic enzymes, including peroxidase (POX) (Guo *et al.*, 2019), superoxide dismutase (SOD) (Liu and Qu, 2019), catalase (CAT) (Asati *et al.*, 2009) (Figure 13.2a), and hydrolases such as nuclease and proteases (Huang *et al.*, 2019a). The diverse nature of these enzymes provides applications in several fields. In recent years, the applicability of nanozymes witnessed a boom in the agri-food industry. Generally, catalysis of nanozymes with peroxidase (POD), oxidases (OXD), and catalase (CAT)-like properties and activities are measured in the form of different signals, including colorimetric, fluorescent, and also chemiluminescent signals (Huang *et al.*, 2019). In addition, these enzyme-mimetic nanomaterials can be designed in several ways to utilize the same in the agri-food industry.

13.2.1 Nanozymes as Recognition Receptor

As nanozymes mimic natural enzymes, the reactions catalyzed by these nanomaterials depend on the type and concentration of substrates similar to natural enzymes. The catalytic efficiency of nanozymes can be adjusted through parameters such as substrate concentration and products of the oxidation process, and the most critical are catalytic sites. Most of the nanozymes have peroxidase-like activities

DOI: 10.1201/9781003334859-13

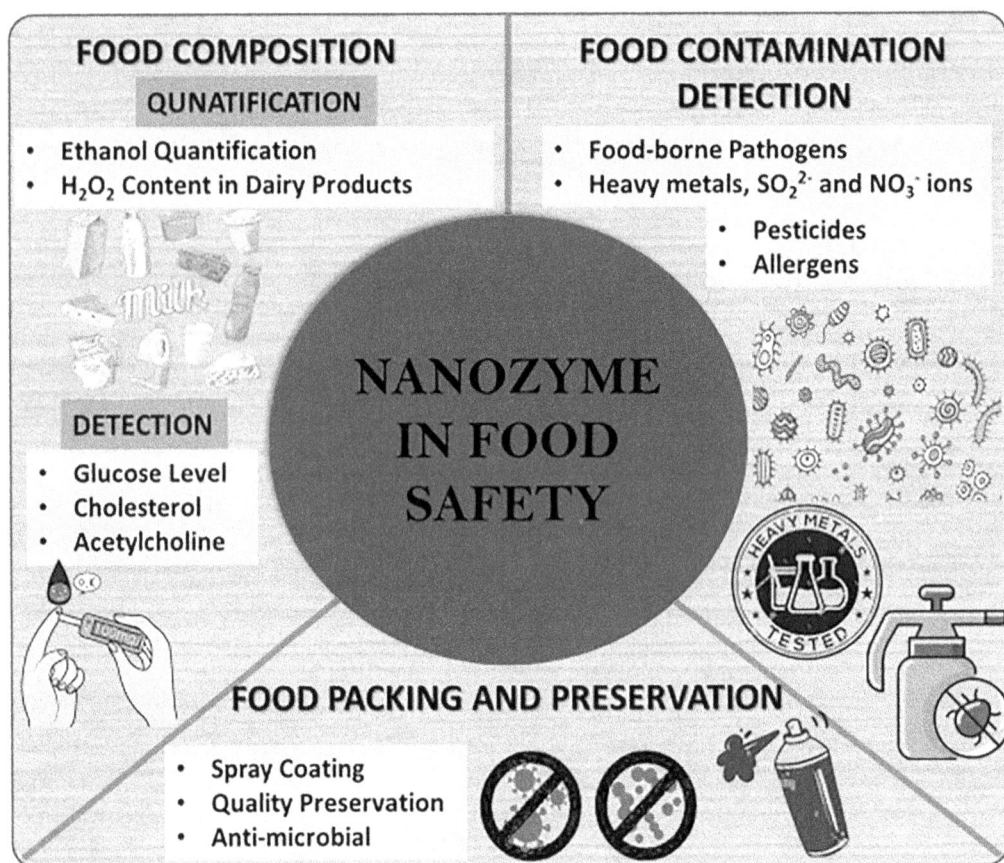

Figure 13.1 Applications of nanozymes in food safety.

that catalyze the oxidation of peroxide substrates such as o-phenylenediamine (OPD), 3,3',5,5'-tetramethylbenzidine (TMB), and 2,2-azinobis (3-ethyl benzothiazoline-6-sulfonic acid) (ABTS) into their corresponding oxidative products, OPDox (orange), TMBox (blue), and oxidized ABTS (green) respectively in the presence of H_2O_2 (Wang et al., 2016) (Figure 13.2b). POD-like enzyme mimics perform similar reactions, and kinetics vary with concentrations of substrates and H_2O_2 as co-substrate and result in varying amounts of product. Oxidative reactions result in a color change that could be used to quantify the concentration of substrates and co-substrates via simple colorimetric methods. Different food analytical methods have been developed for POD-mimic activity based on the amount of H_2O_2 (Zhang et al., 2018). The reactions resulting from the cooperative action of enzymes can also result in analytical detection (Nirala and Prakash, 2018). In addition, nanozyme-based detection systems can be developed using oxidation products that avoid unstable co-substrate like H_2O_2 (Asati et al., 2009). Carbon nanotubes possess peroxidase-like activity on TMB in the presence of H_2O_2, and the reaction can be regulated by multiple factors, including pH, temperature, the concentration of nanotubes, and H_2O_2 (Shamsipur et al., 2014). Metal-organic frameworks (MOFs) consist of organic ligands mutually connected with a cluster of metal ions that have enzyme-mimicking properties (Du et al., 2021). MOFs can mimic several enzymes such as peroxidase (POD), oxidases (OXD), lactase, and hydrolyzes due to the presence of the ion's active site and have attracted attention in applications of food safety (Huang and Sun, 2021).

13.2.2 Nanozymes with Regulatory Mechanisms

Natural enzymes have self-regulatory mechanisms for their catalytic reactions in biological conditions. These regulatory mechanisms involve covalent modifications, spatial organization, and

(a)

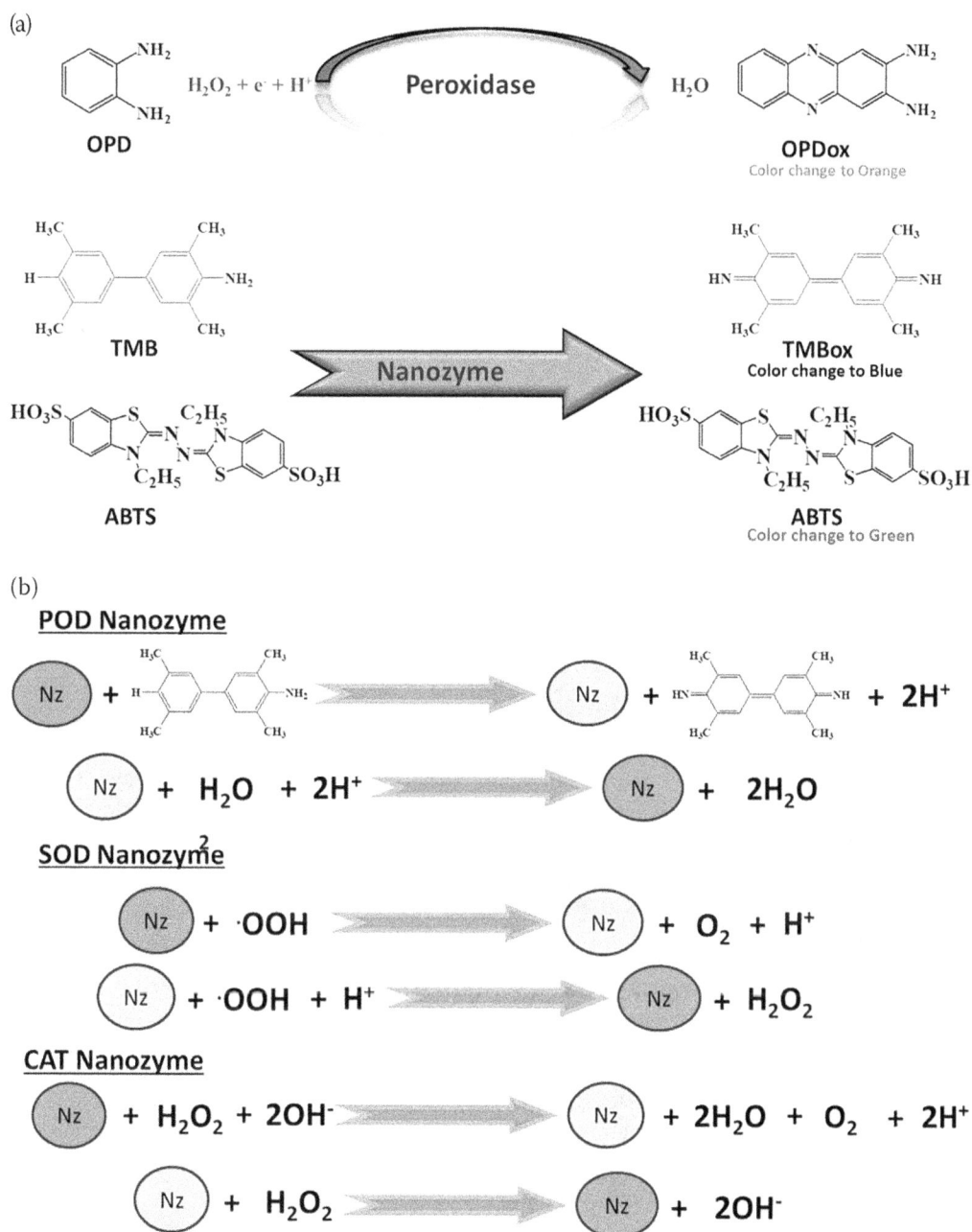

OPD

$H_2O_2 + e^- + H^+$ → Peroxidase → H_2O

OPDox
Color change to Orange

TMB

Nanozyme

TMBox
Color change to Blue

ABTS

ABTS
Color change to Green

(b)

POD Nanozyme

Nz + (TMB) → Nz + (TMBox) + $2H^+$

Nz + H_2O + $2H^+$ → Nz + $2H_2O$

SOD Nanozyme

Nz + $\cdot OOH$ → Nz + O_2 + H^+

Nz + $\cdot OOH$ + H^+ → Nz + H_2O_2

CAT Nanozyme

Nz + H_2O_2 + $2OH^-$ → Nz + $2H_2O$ + O_2 + $2H^+$

Nz + H_2O_2 → Nz + $2OH^-$

Figure 13.2 **(a)** Nanozymes have peroxidase-like activities which catalyze the oxidation of peroxide substrates such as o-phenylenediamine (OPD), 3,3′,5,5′-tetramethylbenzidine (TMB), and 2,2-azinobis(3-ethyl benzothiazoline-6-sulfonic acid) (ABTS) into their corresponding oxidative products, orange OPDox, blue TMBox, and green ABTS respectively in the presence of H_2O_2. **(b)** Mechanisms of peroxidase-like (POD), superoxide dismutase-like (SOD) and catalase-like (CAT) nanozymes. Nz represents nanozyme; Nz in yellow color represents nanozyme in the high oxidative state, and Nz in gray color represents nanozyme in the high oxidative state. All structures are drawn in ChemDraw Pro 8.0 software, PerkinElmer.

external modifications by stimulators or inhibitors. This property of natural enzymes can be utilized in the case of nanozymes. Structural properties of nanozymes, such as particle size and morphology, are important characteristics of their catalytic activities and can be modulated through engineering means (Wu *et al.*, 2019). Multiple modulators such as ions, nucleic acids, peptides, and amino acids are recognized as either stimulators or inhibitors of these artificial enzymes, which can affect the activity of nanozymes. Additionally, the modulation of enzyme surface via adding elements such as antibodies (Ab) for the recognition by molecules modulates covalent modifications or electrostatic absorption, which can alter the interaction of nanozymes with substrates (Huang *et al.*, 2019).

13.2.3 Nanozymes as Signal Tags

In addition to recognition receptors, nanozymes can be used as signal tags to detect analytes. For developing nanozymes as signal tags, antibodies, aptamers, inactive phage antibiotics, and antimicrobial peptides are used as recognition elements and could be used in immunoassays, calorimetric detections with enhanced signal (Loynachan *et al.*, 2018).

13.2.4 Nanozymes as Multifunctional Sensing Elements

Nanozymes can be modulated to perform multi-enzyme functions due to their number of nanostructured modifiable sites (Wu *et al.*, 2019). Researchers have developed several such nanostructures, for example, POD-like iron oxide nanoparticles, which can be separated using magnets, replacing the need for centrifuge due to the ferromagnetism properties of iron (Gao *et al.*, 2007). Nanozyme mimics have another essential characteristic of interfacial adsorption through which analytes can be recognized directly, avoiding using costly specific antibodies. Based on the above, Wang and his associates (2018) developed POD-like hemin-concanavalin, a hybrid nano-flower that directly adsorbed food pathogens on its surface and can be separated using magnets and detected by a simple colorimetric assay.

13.2.5 Nanozymes as a Signal Amplifier

Several nanozymes have been designed to amplify signals in agri-food detection assays. Xie *et al.* (2019) developed a calorimetric assay based on combining immunoreaction with POD-like Au NPs on a chitosan composite membrane to detect *Staphylococcal enterotoxin* B (SE-B). In this system, immunoreaction involved in H_2O_2-based AuNPs formation enables the nanozyme to amplify the signals of AuNPs. Based on AuNPs, other detection systems such as SERS (surface-enhanced Raman scattering) detection of *Salmonella enteritidis* (*S. enteritidis*) via self-growing (Bu *et al.*, 2018); *L. monocytogenes* through colorimetric assay (Liu *et al.*, 2018) were developed.

13.3 APPLICATIONS OF NANOZYMES IN FOOD SAFETY

Nanotechnology is growing at a fast pace, and its applications have become more practical and promising. These applications are not restricted to detecting endogenous ingredients, including glucose, acetylcholine, cholesterol, and allergens. However, contaminants of food, such as H_2O_2, ions, toxins, antibiotics, pesticide residues, and biological contaminants such as food pathogens and mycotoxins, can be detected by nanozyme applications. Nanozymes possess a wide range of applications in safety in the food industry.

13.3.1 Analysis of Food Composition

Food is comprised of both organic (carbohydrates, proteins, lipids, vitamins, etc.) and inorganic components (ions such as Ca, K, Fe, Mg, etc.) (He *et al.*, 2019; Yaseen *et al.*, 2017). A food matrix is a complex assembly of physical and chemical interactions between food components; therefore, the analysis of the composition of food needs high reliability and repeatability. Antioxidants present in fruits and vegetables have a role in protecting against several chronic diseases. Therefore, the assessment of food items for antioxidants is of significance. Nanozymes detect antioxidants based on their capacity to reduce the oxidation products of TMB or ABTS, which are commonly used in oxidation reactions. POD-like Co_3O_4 NPs inhibited the generation of oxABTS through antioxidants, such as ascorbic acid (AA), gallic acid (GA), and tannic acid (TA), with different efficiencies. TA showed the highest antioxidant capability, followed by GA and AA (Jia *et al.*, 2016). In another case, the antioxidant capacity of AA was assessed based on its potential to degrade the nanozyme by converting CoOOH nanoflakes to Co^{2+} with a detection limit of 142 nM within an assay time of 5 min (Ji *et al.*, 2018).

Glucose is an essential component of human energy needs and is an important biomarker of diabetes. Detection of glucose levels in food ingredients is of prime importance to diabetic patients. Several methods of combining natural glucose oxidase (GOX) and POD-mimicking nanozymes to detect glucose have been developed over the years (Wei and Wang, 2008). Huang and team (2018) have developed a glucose detection colorimetric assay based on the same principle with a detection of up to 250 µM with a limit of detection (LOD) of 1.5 µM. But this principle has limited applications because catalysis depends upon pH due to GOX, and sometimes the pH of POD-mimic is incompatible with the pH of GOX. To overcome the above problem, a more advanced system has been developed for glucose detection. Researchers have developed one-pot detection systems which consist of nanomaterials such as Au and MnO_2, which possess dual POD- and GOX-like activities (Han et al., 2018; Huang et al., 2017).

Cholesterol is another important component whose higher levels indicate many diseases; therefore, it can be used as an indicator in the diagnosis of these diseases. Xu and team (2016) have developed a chemiluminescence sensor using POD-like copper nanoclusters and cholesterol oxidase to detect cholesterol within a range of 0.05–10 mM with an LOD of 1.5 µM. This sensor was used to analyze the cholesterol levels in the milk and gave 98% efficiency. Acetylcholine (Ach) has vital importance in brain development, memory improvement, and muscle function, so it is necessary to develop reliable assays to detect the levels of Ach in food. Qian et al. (2014) developed a colorimetric method that measured Ach levels by using POD-mimicking Fe3O4/rGO nanocomposites along with AchE and choline oxidase in a range of 100 to 10 mM and LOD of 39 nM. Detection of choline in milk samples was carried out using MoS_2 nanostructures having POD activity and choline oxidase but had limited sensitivity (Nirala and Prakash, 2018). Compared to these assays, biosensors based on POD-like Fe_3O_4 MNPs and a choline oxidase have a higher sensitivity (0.1 nM) (Zhang et al., 2011). Fe_3O_4 NPs are mainly responsible for higher sensitivity, have higher stability and magnetic properties, and are easy to prepare.

In addition to the above components, ethanol is an important constituent of beer, wine, and other fermented products, which need to be monitored to avoid adulteration and keep the levels according to dietary recommendations. To detect ethanol levels in wine and kefir, an amperometric biosensor was developed by coupling PtRu NPs with alcohol oxidase (Stasyuk et al., 2019). Food allergies are of significant health concern; thus, it is necessary to decipher the presence of allergens to avoid the health issues caused by them. β-lactoglobulin, a milk protein, was used to develop a nanozyme-based ELISA (ELISA) assay to detect cow milk allergy with a lower limit of 1.96 ng/mL (He et al., 2018). Further, the authors developed an improved method utilizing POD-like Pt NPs carrying several antibodies and horseradish peroxidase (HRP) molecules as the signal tag to detect β-lactoglobulin as low as 0.12 ng/mL.

13.3.2 Detection of Food Contaminants

Food contaminants are chemical substances such as pesticides, toxins, drug residues, organic pollutants, etc., or biological agents such as microorganisms, insects, rodents, etc., added to the food either by accident or on purpose, which makes the food unfit for consumption (Kamala and Kumar, 2018). Most methods developed for detecting contaminants in food consider their maximum allowed limit and ignore contaminants' toxicity and negative impact on consumer health. Therefore, developing more advanced methods to detect as much as a low signal from these contaminants is necessary. Nanozymes could be of high advantage to developing ultra-sensitively contaminant detection assays as they have the characteristic to amplify the low signals, one of the principles of their functioning. H_2O_2 is used as a food preservative to keep raw milk fresh by farmers as it is bactericidal in nature. The maximum acceptable amount of H_2O_2 the Food and Agriculture Organization (FAO) set in milk is 0.05% (w/v). The addition of more than the permissive limit reduces the nutritional value of milk and can severely affect the gastrointestinal and neuro system. Several nanozyme-based methods have been developed to check the H_2O_2 content in milk and other dairy products. A colorimetric method developed by Liu et al. (2018) used POD-mimicking iron-doped $CuSn(OH)_6$ microsphere and TMB as a substrate which detected H_2O_2 in a range of 30–1,000 µM with LOD of 9.49 µM. Qi et al. (2017) developed a POD-like nanozyme with a metal-organic framework (PA-Tb-Cu MOF) that produces H_2O_2 concentration-dependent real-time fluorescence signals. These nanostructures can detect the H_2O_2 as low as 0.2 µM with high reproducibility.

The contamination of food and drinking water with ions such as heavy metal, sulfate, and nitrate ions has recently increased to an alarming situation. These ions accumulate in the organs and could pose health dangers. Thus, it is essential to measure the concentration of these ions accurately. Many nanozyme-based techniques have been developed to detect these ions with high sensitivity and accuracy. Deng et al. (2014) developed POD-like Au NPs to detect sulfide ions with a quick assay time of 10 min, and these nanozymes work on the shielding effect of sulfide against Au NPs. Chitosan-functionalized molybdenum (IV) selenide nanospheres (CS-MoSe$_2$ NSs) were developed to detect mercuric ions (Hg^{2+}) through colorimetric sensing (Huang et al., 2019b). In this system, the reduction of chitosan-captured Hg^{2+} ions resulted in the activation of POD- and OXD-like activities of these NSs on the surface with TMB as a substrate producing quantitative signals, which can be measured by spectrophotometer. The authors used this eco-compatible sensor to detect Hg^{2+} ions in drinking water within an assay time of 15 min to the lowest limit of 8.4 nM. In addition to colorimetric methods, electrochemical and chemiluminescent methods were also proposed to analyze these ions' amounts. Zhang et al. (2013) enumerated sulfate ions in white wines using OXD-mimicking CoFe$_2$O$_4$ NPs generating chemiluminescence of aqueous luminol up to a concentration of 2.0×10^{-8} M. An electrochemical method was proposed by Liu et al. (2019) for detection of nitrite ions in sausage samples using a hybrid of histidine-capped gold nanoclusters and rGO (His@Au NCs/rGO) having activities of oxidase. Nitrite can inhibit electrocatalysis of His@Au NCs/rGO during TMB oxidation, which is used as a substrate in most nanozyme reactions. Nanozymes also have been used to detect several other ions, such as Ag^{2+} (Abdolmohammad-Zadeh and Rahimpour, 2015) and Cu^{2+} and Ca^{2+} ions in drinking water milk, respectively (Mu et al., 2014).

The extensive use of antibiotics in food animals, including livestock and poultry, poses a serious risk to human health (Jiang et al., 2019). Zeng et al. (2018) developed a gas pressure-based aptasensing platform using a polyaniline nanowire (NW) functionalized rGO framework and CAT-like Pt nanozyme to detect kanamycin in milk samples. Sharma et al. (2014) proposed an aptamer-based method for detecting kanamycin based on the POD-like Au NPs which has a range of 1–100 nM and high sensitivity. The quantity of antibiotic sulfaquinoxaline was assessed through the electrochemiluminescent method by using POD-like nanozyme activity-based Cu (II)-anchored unzipped COF material (UnZCCTF) (Ma et al., 2018). This system has a high sensitivity for sulfaquinoxaline in the range of 1–20 μM with a LOD of 0.76 pM. However, the complicated process of fabrication of material is a hurdle for this assay for practical applications.

Contamination of food commodities with pesticides used to control pest infestation and crop diseases has become a rising cause of concern (Yaseen et al., 2018). Nanozymes offered highly sensitive and precious approaches to detecting these compounds. Zhang et al. (2016) used OXD-mimicking polyacrylic acid-coated CeO$_2$ NPs (PAA-CeO$_2$) and AChE for the detection of pesticides, especially organophosphates (OP) like dichlorvos and methyl paraoxon. The catalysis of AChE produces tricholine (TCh), which could reduce TMB oxidation. Still, the presence of OPs resulted in the production of more oxidized TMB and can be measured colorimetrically. This reaction proved useful for the quick and accurate measurement of OPs, which are neurotoxic in nature. The paraoxon in cabbage samples was detected using phosphotriesterase-like Co$_3$O$_4$/rGO nanocomposites (Wang et al., 2017). Further, Huang et al. (2019c) designed a portable paper colorimetric sensor to detect an amplified signal produced during the domino reaction of AChE and degradable manganite nanowires (γ-MnOOH NWs). Other toxins in food include highly toxic mycotoxins, which pose health-threatening effects to humans; thus, their accurate detection in food items is necessary. CAT-like rGO/Pt nanocomposites were employed to analyze beer and wine samples to detect the presence of Ochratoxin A and Fumonisin B1 mycotoxins via an "on-the-fly" fluorescent approach (Molinero-Fernandez et al., 2017). Huang and his team (2018) developed a spinel-type manganese cation substituted cobalt oxide (MnCo$_2$O$_4$) nanozyme-based biomolecular sensor. Whose inhibition depends upon the attachment/detachment of aptamer strands on its surface through aptamer-target binding. This method has been proven useful in detecting ochratoxin A in maize with a LOD of 6.5 pg/mL. Recently, Peng et al. (2022) developed a peroxide activities-based porous Iron-Porphyrin-Zr-MOF NanoPCN-223(Fe) nanozyme to develop a MOF-linked immuno-sorbent assay (Ed-MOFLISA) for the detection of aflatoxin B in food samples in the range from 0.05 to 10 ng/mL and LOD of 0.003 ng/mL.

Foodborne pathogens are another contaminant on the list that can create havoc in public health. The development in nanozyme technology also offered the detection of food pathogens, and most assays are based on the immunoassays such as NLISA (nanozyme-based enzyme-linked immunosorbent assay) and N-LIFA (nanozyme with lateral flow immunoassay), a paper-based assay to detect bacteria. Wang et al. (2018) detected *Escherichia coli* O157:H7 (*E. coli* O157:H7), a potent food pathogen using hemin-concanavalin A hybrid nanoflowers-based sandwich structure. This nanostructure functions similarly to ELISA and can detect *E. coli* as low as 4.1 CFU/mL at pH 6.5 and 25°C in milk samples. *E. coli* O157:H7 was also measured using Pt-Au bimetallic nanozyme-based immunochromatographic assay upto 10^2 CFU/mL (Jiang *et al.*, 2016). An NLISA assay was proposed to detect the presence of *E. coli* O157:H7 with Pd-Pt nanozyme up to a range of 10^2–10^6 CFU/mL and LOD of 0.87×10^2 CFU/mL by colorimetric assay (Han *et al.*, 2018). Another common foodborne pathogen, *Listeria monocytogenes* (*L. monocytogenes*), damages neurons and T-cells of the immune system in humans (Faber and Peterkin, 1991) detected using a hybrid of Fe_3O_4 NCs anchored aptamer where Fe_3O_4 NCs acts as a signal amplification ligand (Zhang *et al.*, 2016). *V. parahaemolyticus*, marine bacteria, were detected using highly sensitive nonapeptide pVIII fusion coupled with protein-templated MnO_2 nanospheres. pVIII fusion-*V. parahaemolyticus*-MnO_2 nano-spheres form a sandwich structure to catalyze the TMB in the presence of H_2O_2 to give colorimetric change, thus, displaying its potential use in marine food safety (Liu *et al.*, 2018). Based on N-LIFA, Cheng *et al.* (2017) built a smartphone-integrated dual N-LIFA device to simultaneously detect the presence of *S. enteritidis* and *E. coli* O157:H7 by using POD-mimicking mesoporous core-shell Pa@Pt NPs as a signal amplifier. This device detected *S. enteritidis* and *E. coli* O157:H7 up to 20 CFU/mL and 34 CFU/mL, respectively, in milk samples.

13.3.3 Nanozymes in Food Packaging and Preservation

The contamination of food with contaminants from different sources, including environmental and microbial sources, has increased the concern for food safety. The proper inspection and packaging of food items could help avoid these situations. Nanozymes could contribute to the active packaging of agri-food products and monitor contamination during storage. Naturally ripened fruits differ from artificially ripened and have particular total sugar content, total soluble solids (TSS), and sensory attributes. The glucose content of fruits varies with the ripening stage; thus, it could be the potential marker to determine the ripening stage and can be implemented in lengthening the shelf life. A needle-type electrode biosensor consisting of working, reference, and counter electrodes was developed by Haginoya *et al.* (1994) to determine the sugar content of fruits. In this tri-electrode system, glucose oxidase was immobilized on the working electrode using BSA and glutaraldehyde as a reference to determine glucose. Yousefi *et al.* (2018) developed a biosensor having *E. coli*–specific RNA-cleaving fluorogenic DNAzyme covalently attached to cyclo-olefin polymer packaging film to detect *E. coli* in meat and apple juice as low as 10^3 CFU/mL under a wide range of pH 3–9 for at least 14 days. Recently, Huang *et al.* (2023) used a hybrid of carboxymethyl cellulose nanofibers grafted with enzyme-like metal-organic frameworks (CNF@Ce-MOF) to prepare spray coating for fruits, bananas, and mangoes. This coating is claimed to enhance the quality preservation during the 12-day storage period. Additionally, these nanofibers provide antibacterial protection against common food pathogens: Escherichia *coli* O157: H7, *Staphylococcus aureus,* and methicillin-resistant *Staphylococcus aureus* by generating •O_2 and reducing the levels of ATP via mimicking the oxidase and apyrase activities.

Apart from signal detection, magnetic nanoparticles can amplify the signals and be useful for detecting microbial contamination of food during food storage. Nanoprobes with colorimetric and fluorescence dual-mode were developed to detect *S. aureus* in meat samples. These nanoprobes consist of magnetic nanoparticles (apt-MNPs) labeled with aptamer, which act as capture signal probes and HRP and complementary DNA-functionalized upconversion nanoparticles (HRP-UCNPs-cDNA) as a chromogenic signal probe that forms an immune complex. *S. aureus* binds to apt-MNP, causing the release of HRP-UCNPs-cDNA from the complex, and probes can be detected through fluorescence and colorimetric assays (Ouyang *et al.*, 2021).

13.4 CONCLUSION

The discovery of nanozymes has created a bridge between nanotechnology and biology. Nanozymes provided the platform for tailoring enzymes at the nanoscale and using nanomaterials in biological activities. Most nanozymes consist of nanomaterials with oxidoreductases (e.g., peroxidase, catalase, oxidase, and superoxide dismutase). These enzyme mimics deliver simple, portable, sensitive, selective, and highly producible strategies for food analysis. Moreover, the properties of nanozymes to customization of their properties for precise applications and integration of multifunctional platforms for simultaneous detection of multiple contaminants is a plus for food safety. Additionally, these methods are cheap due to high availability and inexpensive equipment, which enhance their use in the future for food analysis applications.

13.5 FUTURE PERSPECTIVES

Although nanozymes have an exciting future in food safety, but nevertheless have several challenges. Nanozymes have fewer active sites, reducing their activity and still needing improvement. Further, the structure, size, as well as activity of these enzyme mimics need to be optimized to achieve the applications. In applications like analysis of the composition of complex systems such as food, the combination of different signals should be investigated to establish the roles of nanozymes for fast and accurate analysis in food safety. Researchers should focus on developing nanozymes with reusability and biocompatibility while designing new detection approaches.

REFERENCES

Abdolmohammad-Zadeh, Hossein, and Elaheh Rahimpour. "A novel chemosensor for Ag (I) ion based on its inhibitory effect on the luminol–H2O2 chemiluminescence response improved by CoFe$_2$O$_4$ nano-particles." *Sensors and Actuators B: Chemical* 209 (2015): 496–504. 10.1016/J.SNB.2014.11.096.

Asati, Atul, Santimukul Santra, Charalambos Kaittanis, Sudip Nath, and J. Manuel Perez. "Oxidase-like activity of polymer-coated cerium oxide nanoparticles." *Angewandte Chemie* 121, no. 13 (2009): 2344–2348. 10.1002/ANGE.200805279.

Bu, Tong, Qiong Huang, Lingzhi Yan, Lunjie Huang, Mengyue Zhang, Qingfeng Yang, Baowei Yang, Jianlong Wang, and Daohong Zhang. "Ultra technically-simple and sensitive detection for Salmonella enteritidis by immunochromatographic assay based on gold growth." *Food Control* 84 (2018): 536–543. 10.1016/J.FOODCONT.2017.08.036.

Cheng, Nan, Yang Song, Mohamed MA Zeinhom, Yu-Chung Chang, Lina Sheng, Haolin Li, Dan Du et al. "Nanozyme-mediated dual immunoassay integrated with smartphone for use in simultaneous detection of pathogens." *ACS Applied Materials & Interfaces* 9, no. 46 (2017): 40671–40680. 10.1021/ACSAMI.7B12734/SUPPL_FILE/AM7B12734_SI_001.PDF.

Deng, Hao-Hua, Shao-Huang Weng, Shuang-Lu Huang, Ling-Na Zhang, Ai-Lin Liu, Xin-Hua Lin, and Wei Chen. "Colorimetric detection of sulfide based on target-induced shielding against the peroxidase-like activity of gold nanoparticles." *Analytica Chimica Acta* 852 (2014): 218–222. 10.1016/J.ACA.2014.09.023.

Du, Ting, Lunjie Huang, Jing Wang, Jing Sun, Wentao Zhang, and Jianlong Wang. "Luminescent metal-organic frameworks (LMOFs): An emerging sensing platform for food quality and safety control." *Trends in Food Science & Technology* 111 (2021): 716–730. 10.1016/J.TIFS.2021.03.013.

Farber, Jeffrey M., and PI372831 Peterkin. "Listeria monocytogenes, a foodborne pathogen." *Microbiological Reviews* 55, no. 3 (1991): 476–511. 10.1128/MR.55.3.476-511.1991.

Gao, Lizeng, Jie Zhuang, Leng Nie, Jinbin Zhang, Yu Zhang, Ning Gu, Taihong Wang et al. "Intrinsic peroxidase-like activity of ferromagnetic nanoparticles." *Nature Nanotechnology* 2, no. 9 (2007): 577–583. 10.1038/nnano.2007.260.

Gong, Liang, Zilong Zhao, Yi-Fan Lv, Shuang-Yan Huan, Ting Fu, Xiao-Bing Zhang, Guo-Li Shen, and Ru-Qin Yu. "DNAzyme-based biosensors and nanodevices." *Chemical Communications* 51, no. 6 (2015): 979–995. 10.1039/C4CC06855F.

Guo, Wenjing, Mian Zhang, Zhangping Lou, Min Zhou, Peng Wang, and Hui Wei. "Engineering nanoceria for enhanced peroxidase mimics: A solid solution strategy." *ChemCatChem* 11, no. 2 (2019): 737–743. 10.1002/CCTC.201801578.

Haginoya, Ryuichi, Kenji Yokoyama, Masayasu Suzuki, Eiichi Tamiya, and Isao Karube. "Needle type biosensors for analysis of sugars." In *Advanced Materials' 93*, pp. 437–440. Elsevier, 1994. 10.1016/B978-1-4832-8380-7.50107-3.

Han, Jiaojiao, Lei Zhang, Liming Hu, Keyu Xing, Xuefei Lu, Youju Huang, Jiawei Zhang, Weihua Lai, and Tao Chen. "Nanozyme-based lateral flow assay for the sensitive detection of *Escherichia coli* O157: H7 in milk." *Journal of Dairy Science* 101, no. 7 (2018): 5770–5779. 10.3168/JDS.2018-14429.

Han, Lei, Haijiao Zhang, Daoyuan Chen, and Feng Li. "Protein-directed metal oxide nanoflakes with tandem enzyme-like characteristics: colorimetric glucose sensing based on one-pot enzyme-free cascade catalysis." *Advanced Functional Materials* 28, no. 17 (2018): 1800018. 10.1002/ADFM.201800018.

He, Huirong, Da-Wen Sun, Hongbin Pu, Lijun Chen, and Li Lin. "Applications of Raman spectroscopic techniques for quality and safety evaluation of milk: A review of recent developments." *Critical Reviews in Food Science and Nutrition* 59, no. 5 (2019): 770–793. 10.1080/104 08398.2018.1528436.

He, Shengfa, Xin Li, Yong Wu, Shandong Wu, Zhihua Wu, Anshu Yang, Ping Tong, Juanli Yuan, Jinyan Gao, and Hongbing Chen. "Highly sensitive detection of bovine β-Lactoglobulin with wide linear dynamic range based on platinum nanoparticles probe." *Journal of Agricultural and Food Chemistry* 66, no. 44 (2018): 11830–11838. 10.1021/ACS.JAFC.8B04086/SUPPL_FILE/JF8B04086_SI_001.PDF.

Huang, Lunjie, Kai Chen, Wentao Zhang, Wenxin Zhu, Xinnan Liu, Jing Wang, Rong Wang, Na Hu, Yourui Suo, and Jianlong Wang. "ssDNA-tailorable oxidase-mimicking activity of spinel MnCo2O4 for sensitive biomolecular detection in food sample." *Sensors and Actuators B: Chemical* 269 (2018): 79–87. 10.1016/J.SNB.2018.04.150.

Huang, Lunjie, and Da-Wen Sun. "Nanozymes: Advances and applications." In *Nanozymes*, 1st ed. Vol. 1. CRC Press, 2021. 10.1201/9781003109228-15.

Huang, Lunjie, Da-Wen Sun, Hongbin Pu, Cuiyun Zhang, and Daorui Zhang. "Nanocellulose-based polymeric nanozyme as bioinspired spray coating for fruit preservation." *Food Hydrocolloids* 135 (2023): 108138. 10.1016/J.FOODHYD.2022.108138.

Huang, Ying, Zhao, Meiting, Han, Shikui, Lai, Zhuangchai, Yang, Jian, Tan, Chaoliang, Ma, Qinglang, Lu, Qipeng, Chen, Junze, Zhang, Xiao, Zhang, Zhicheng, Li, Bing, Chen, Bo, Zong, Yun, & Zhang, Hua. Growth of Au nanoparticles on 2D metalloporphyrinic metal-organic framework nanosheets used as niomimetic catalysts for cascade reactions. *Advanced Materials*, 29 (2017). 10.1 002/adma.201700102.

Huang, Yanyan, Jinsong Ren, and Xiaogang Qu. "Nanozymes: classification, catalytic mechanisms, activity regulation, and applications." *Chemical Reviews* 119, no. 6 (2019): 4357–4412. 10.1021/ACS.CHEMREV.8B00672/ASSET/IMAGES/MEDIUM/CR-2018-00672R_0038.GIF.

Huang, Lunjie, Da-Wen Sun, Hongbin Pu, and Qingyi Wei. "Development of nanozymes for food quality and safety detection: Principles and recent applications." *Comprehensive Reviews in Food Science and Food Safety* 18, no. 5 (2019a): 1496–1513. 10.1111/1541-4337.12485.

Huang, Lunjie, Qingrui Zhu, Jie Zhu, Linpin Luo, Shuhan Pu, Wentao Zhang, Wenxin Zhu, Jing Sun, and Jianlong Wang. "Portable colorimetric detection of mercury (II) based on a non-noble metal nanozyme with tunable activity." *Inorganic Chemistry* 58, no. 2 (2019b): 1638–1646. 10.1021/ACS.INORGCHEM.8B03193/SUPPL_FILE/IC8B03193_SI_001.PDF.

Huang, Lunjie, Da-Wen Sun, Hongbin Pu, Qingyi Wei, Linpin Luo, and Jianlong Wang. "A colorimetric paper sensor based on the domino reaction of acetylcholinesterase and degradable γ-MnOOH nanozyme for sensitive detection of organophosphorus pesticides." *Sensors and Actuators B: Chemical* 290 (2019c): 573–580. 10.1016/j.snb.2019.04.020.

Ji, Danyang, Yahui Du, Hongmin Meng, Lin Zhang, Zhongming Huang, Yalei Hu, Jianjun Li, Fei Yu, and Zhaohui Li. "A novel colorimetric strategy for sensitive and rapid sensing of ascorbic acid using cobalt oxyhydroxide nanoflakes and 3, 3′, 5, 5′-tetramethylbenzidine." *Sensors and Actuators B: Chemical* 256 (2018): 512–519. 10.1016/J.SNB.2017.10.070.

Jia, Huimin, Dongfang Yang, Xiangna Han, Junhui Cai, Haiying Liu, and Weiwei He. "Peroxidase-like activity of the Co 3 O 4 nanoparticles used for biodetection and evaluation of antioxidant behavior." *Nanoscale* 8, no. 11 (2016): 5938–5945. 10.1039/C6NR00860G.

Jiang, Tao, Yang Song, Tianxiang Wei, He Li, Dan Du, Mei-Jun Zhu, and Yuehe Lin. "Sensitive detection of Escherichia coli O157: H7 using Pt–Au bimetal nanoparticles with peroxidase-like amplification." *Biosensors and Bioelectronics* 77 (2016): 687–694. 10.1016/j.bios.2015.10.017

Jiang, Yingfen, Da-Wen Sun, Hongbin Pu, and Qingyi Wei. "Ultrasensitive analysis of kanamycin residue in milk by SERS-based aptasensor." *Talanta* 197 (2019): 151–158. 10.1016/J.TALANTA.2019.01.015.

Kamala, Katepogu, and Venkobarao Pavan Kumar. "Food products and food contamination." In *Microbial Contamination and Food Degradation*, pp. 1–19. Academic Press, 2018. 10.1016/B978-0-12-811515-2.00001-9.

Kuah, Evelyn, Seraphina Toh, Jessica Yee, Qian Ma, and Zhiqiang Gao. "Enzyme mimics: advances and applications." *Chemistry–A European Journal* 22, no. 25 (2016): 8404–8430. 10.1002/CHEM.201504394.

Lin, Youhui, Jinsong Ren, and Xiaogang Qu. "Catalytically active nanomaterials: a promising candidate for artificial enzymes." *Accounts of chemical research* 47, no. 4 (2014): 1097–1105. 10.1021/AR400250Z/ASSET/IMAGES/MEDIUM/AR-2013-00250Z_0012.GIF.

Liu, Hao, Ya-Nan Ding, Baochan Yang, Zhenxue Liu, Xiao Zhang, and Qingyun Liu. "Iron doped CuSn (OH) 6 microspheres as a peroxidase-mimicking artificial enzyme for H2O2 colorimetric detection." *ACS Sustainable Chemistry & Engineering* 6, no. 11 (2018): 14383–14393. 10.1021/ACSSUSCHEMENG.8B03082/SUPPL_FILE/SC8B03082_SI_001.PDF.

Liu, Lu, Jie Du, Wen-E. Liu, Yongliang Guo, Guofan Wu, Weinan Qi, and Xiaoquan Lu. "Enhanced His@ AuNCs oxidase-like activity by reduced graphene oxide and its application for colorimetric and electrochemical detection of nitrite." *Analytical and Bioanalytical Chemistry* 411 (2019): 2189–2200. 10.1007/S00216-019-01655-Y.

Liu, Yushen, Juan Wang, Xiuling Song, Kun Xu, Huisi Chen, Chao Zhao, and Juan Li. "Colorimetric immunoassay for Listeria monocytogenes by using core gold nanoparticles, silver nanoclusters as oxidase mimetics, and aptamer-conjugated magnetic nanoparticles." *Microchimica Acta* 185 (2018): 1–7. 10.1007/S00604-018-2896-1.

Liu, Zhengwei, and Xiaogang Qu. "New insights into nanomaterials combating bacteria: ROS and beyond." *Science China. Life Sciences* 62, no. 1 (2019): 150–152. 10.1007/S11427-018-9417-1.

Loynachan, Colleen N., Michael R. Thomas, Eleanor R. Gray, Daniel A. Richards, Jeongyun Kim, Benjamin S. Miller, Jennifer C. Brookes et al. "Platinum nanocatalyst amplification: redefining the gold standard for lateral flow immunoassays with ultrabroad dynamic range." *ACS Nano* 12, no. 1 (2018): 279–288. 10.1021/ACSNANO.7B06229/ASSET/IMAGES/LARGE/NN-2017-06229E_0004.JPEG.

Ma, Xionghui, Shuhuai Li, Chaohai Pang, Yuhao Xiong, and Jianping Li. "A Cu (II)-anchored unzipped covalent triazine framework with peroxidase-mimicking properties for molecular imprinting–based electrochemiluminescent detection of sulfaquinoxaline." *Microchimica Acta* 185 (2018): 1–10. 10.1007/s00604-018-3079-9.

Molinero-Fernandez, Agueda, Maria Moreno-Guzman, Miguel Ángel López, and Alberto Escarpa. "Biosensing strategy for simultaneous and accurate quantitative analysis of mycotoxins in food samples using unmodified graphene micromotors." *Analytical Chemistry* 89, no. 20 (2017): 10850–10857. 10.1021/ACS.ANALCHEM.7B02440/SUPPL_FILE/AC7B02440_SI_002.AVI.

Mu, Jianshuai, Li Zhang, Min Zhao, and Yan Wang. "Catalase mimic property of Co3O4 nanomaterials with different morphology and its application as a calcium sensor." *ACS Applied Materials & Interfaces* 6, no. 10 (2014): 7090–7098. 10.1021/AM406033Q/SUPPL_FILE/AM406033Q_SI_001.PDF.

Nirala, Narsingh R., and Rajiv Prakash. "Quick colorimetric determination of choline in milk and serum based on the use of MoS 2 nanosheets as a highly active enzyme mimetic." *Microchimica Acta* 185 (2018): 1–8. 10.1007/S00604-018-2753-2.

Ouyang, Qin, Li Wang, Waqas Ahmad, Yongcun Yang, and Quansheng Chen. "Upconversion nanoprobes based on a horseradish peroxidase-regulated dual-mode strategy for the ultrasensitive detection of Staphylococcus aureus in meat." *Journal of Agricultural and Food Chemistry* 69, no. 34 (2021): 9947–9956. 10.1021/ACS.JAFC.1C03625/SUPPL_FILE/JF1C03625_SI_001.PDF.

Peng, Shuang, Kai Li, Yi-xuan Wang, Lin Li, Yun-Hui Cheng, and Zhou Xu. "Porphyrin NanoMOFs as a catalytic label in nanozyme-linked immunosorbent assay for Aflatoxin B1 detection." *Analytical Biochemistry* 655 (2022): 114829. https://assets.researchsquare.com/files/rs-1641617/v1/0d83c7a9-8888-4374-ab3a-4fda01816764.pdf?c=1652890351.

Qi, Zewan, Li Wang, Qi You, and Yang Chen. "PA-Tb-Cu MOF as luminescent nanoenzyme for catalytic assay of hydrogen peroxide." *Biosensors and Bioelectronics* 96 (2017): 227–232. 10.1016/J.BIOS.2017.05.013.

Qian, Jing, Xingwang Yang, Ling Jiang, Chendan Zhu, Hanping Mao, and Kun Wang. "Facile preparation of Fe3O4 nanospheres/reduced graphene oxide nanocomposites with high peroxidase-like activity for sensitive and selective colorimetric detection of acetylcholine." *Sensors and Actuators B: Chemical* 201 (2014): 160–166. 10.1016/J.SNB.2014.05.020.

Shamsipur, M., A. Safavi, and Z. Mohammadpour. "Indirect colorimetric detection of glutathione based on its radical restoration ability using carbon nanodots as nanozymes." *Sensors and Actuators B: Chemical* 199 (2014): 463–469. 10.1016/J.SNB.2014.04.006.

Sharma, Tarun Kumar, Rajesh Ramanathan, Pabudi Weerathunge, Mahsa Mohammadtaheri, Hemant Kumar Daima, Ravi Shukla, and Vipul Bansal. "Aptamer-mediated 'turn-off/turn-on' nanozyme activity of gold nanoparticles for kanamycin detection." *Chemical Communications* 50, no. 100 (2014): 15856–15859. 10.1039/C4CC07275H.

Song, Wei, Bing Zhao, Ce Wang, Yukihiro Ozaki, and Xiaofeng Lu. "Functional nanomaterials with unique enzyme-like characteristics for sensing applications." *Journal of Materials Chemistry B* 7, no. 6 (2019): 850–875. 10.1039/C8TB02878H.

Stasyuk, Nataliya, Galina Gayda, Andriy Zakalskiy, Oksana Zakalska, Roman Serkiz, and Mykhailo Gonchar. "Amperometric biosensors based on oxidases and PtRu nanoparticles as artificial peroxidase." *Food Chemistry* 285 (2019): 213–220. 10.1016/J.FOODCHEM.2019.01.117.

Wang, Hui, Kaiwei Wan, and Xinghua Shi. "Recent advances in nanozyme research." *Advanced Materials* 31, no. 45 (2019): 1805368. 10.1002/ADMA.201805368.

Wang, Kui-Yu, Sheng-Jun Bu, Chuan-Jing Ju, Chang-Tian Li, Zhong-Yi Li, Ye Han, Cheng-You Ma et al. "Hemin-incorporated nanoflowers as enzyme mimics for colorimetric detection of foodborne pathogenic bacteria." *Bioorganic & Medicinal Chemistry Letters* 28, no. 23-24 (2018): 3802–3807. 10.1016/J.BMCL.2018.07.017.

Wang, Ting, Jiangning Wang, Ye Yang, Ping Su, and Yi Yang. "Co3O4/reduced graphene oxide nanocomposites as effective phosphotriesterase mimetics for degradation and detection of paraoxon." *Industrial & Engineering Chemistry Research* 56, no. 34 (2017): 9762–9769. 10.1021/ACS.IECR.7B02223/SUPPL_FILE/IE7B02223_SI_001.PDF.

Wang, Xiaoyu, Yihui Hu, and Hui Wei. "Nanozymes in bionanotechnology: from sensing to therapeutics and beyond." *Inorganic Chemistry Frontiers* 3, no. 1 (2016): 41–60. 10.1039/C5QI00240K.

Wei, Hui, and Erkang Wang. "Fe3O4 magnetic nanoparticles as peroxidase mimetics and their applications in H_2O_2 and glucose detection." *Analytical Chemistry* 80, no. 6 (2008): 2250–2254. 10.1021/AC702203F.

Wei, Hui, & Erkang Wang. "Nanomaterials with enzyme-like characteristics (nanozymes): next-generation artificial enzymes". *Chemical Society Reviews*, 42 (2013): 6060. 10.1039/c3cs35486e.

Whitaker, John R., Alphons GJ Voragen, and Dominic WS Wong, eds. *Handbook of Food Enzymology*, Vol. 122. CRC Press, 2002.

Wu, Jiangjiexing, Xiaoyu Wang, Quan Wang, Zhangping Lou, Sirong Li, Yunyao Zhu, Li Qin, and Hui Wei. "Nanomaterials with enzyme-like characteristics (nanozymes): next-generation artificial enzymes (II)." *Chemical Society Reviews* 48, no. 4 (2019): 1004–1076. 10.1039/C8CS00457A.

Wu, Shijia, Nuo Duan, Yueting Qiu, Jinghong Li, and Zhouping Wang. "Colorimetric aptasensor for the detection of *Salmonella enterica serovar typhimurium* using $ZnFe_2O_4$-reduced graphene oxide nanostructures as an effective peroxidase mimetics." *International Journal of Food Microbiology* 261 (2017): 42–48. 10.1016/J.IJFOODMICRO.2017.09.002.

Xiaodan, Zhang, He Shaohui, Chen Zhaohui, and Huang Yuming. "$CoFe_2O_4$ nanoparticles as oxidase mimic-mediated chemiluminescence of aqueous luminol for sulfite in white wines." (2013). 10.1021/JF3041269/SUPPL_FILE/JF3041269_SI_001.PDF.

Xie, Xiaoxue, Fang Tan, Aiqing Xu, Keqin Deng, Yulong Zeng, and Haowen Huang. "UV-induced peroxidase-like activity of gold nanoclusters for differentiating pathogenic bacteria and detection of enterotoxin with colorimetric readout." *Sensors and Actuators B: Chemical* 279 (2019): 289–297. 10.1016/J.SNB.2018.10.019.

Xu, Shuangjiao, Yanqin Wang, Dayun Zhou, Meng Kuang, Dan Fang, Weihua Yang, Shoujun Wei, and Lei Ma. "A novel chemiluminescence sensor for sensitive detection of cholesterol based on the peroxidase-like activity of copper nanoclusters." *Scientific Reports* 6, no. 1 (2016): 39157. 10.1038/srep39157.

Yaseen, Tehseen, Da-Wen Sun, and Jun-Hu Cheng. "Raman imaging for food quality and safety evaluation: Fundamentals and applications." *Trends in Food Science & Technology* 62 (2017): 177–189. 10.1016/J.TIFS.2017.01.012.

Yaseen, Tehseen, Da-Wen Sun, Hongbin Pu, and Ting-Tiao Pan. "Detection of omethoate residues in peach with surface-enhanced Raman spectroscopy." *Food Analytical Methods* 11 (2018): 2518–2527. 10.1007/S12161-018-1233-Y.

Yousefi, Hanie, M. Monsur Ali, Hsuan-Ming Su, Carlos DM Filipe, and Tohid F. Didar. "Sentinel wraps: real-time monitoring of food contamination by printing DNAzyme probes on food packaging." *ACS Nano* 12, no. 4 (2018): 3287–3294. 10.1021/ACSNANO.7B08010/SUPPL_FILE/NN7B08010_SI_001.PDF.

Zeng, Ruijin, Zhongbin Luo, Lijia Zhang, and Dianping Tang. "Platinum nanozyme-catalyzed gas generation for pressure-based bioassay using polyaniline nanowires-functionalized graphene oxide framework." *Analytical Chemistry* 90, no. 20 (2018): 12299–12306. 10.1021/ACS.ANALCHEM.8B03889/SUPPL_FILE/AC8B03889_SI_001.PDF.

Zhang, Lisha, Ru Huang, Weipeng Liu, Hongxing Liu, Xiaoming Zhou, and Da Xing. "Rapid and visual detection of Listeria monocytogenes based on nanoparticle cluster catalyzed signal amplification." *Biosensors and Bioelectronics* 86 (2016): 1–7. 10.1016/J.BIOS.2016.05.100.

Zhang, Shi-Xiang, Shi-Fan Xue, Jingjing Deng, Min Zhang, Guoyue Shi, and Tianshu Zhou. "Polyacrylic acid-coated cerium oxide nanoparticles: An oxidase mimic applied for colorimetric assay to organophosphorus pesticides." *Biosensors and Bioelectronics* 85 (2016): 457–463. 10.1016/J.BIOS.2016.05.040.

Zhang, Wenchi, Xiangheng Niu, Xin Li, Yanfang He, Hongwei Song, Yinxian Peng, Jianming Pan, Fengxian Qiu, Hongli Zhao, and Minbo Lan. "A smartphone-integrated ready-to-use paper-based sensor with mesoporous carbon-dispersed Pd nanoparticles as a highly active peroxidase mimic for H_2O_2 detection." *Sensors and Actuators B: Chemical* 265 (2018): 412–420. 10.1016/J.SNB.2018.03.082.

Zhang, Zhanxia, Xiaolei Wang, and Xiurong Yang. "A sensitive choline biosensor using Fe3O4 magnetic nanoparticles as peroxidase mimics." *Analyst* 136, no. 23 (2011): 4960–4965. 10.1039/C1 AN15602K.

Zhang, Xiaodan, He, Shaohui, Chen, Zhaohui, and Huang, Yuming. $CoFe_2O_4$ nanoparticles as oxidase mimic-mediated chemiluminescence of aqueous luminol for sulfite in white wines. *Journal of Agricultural and Food Chemistry*, 61 (2013): 840–847. 10.1021/jf3041269.

Zhou, Yibo, Biwu Liu, Ronghua Yang, and Juewen Liu. "Filling in the gaps between nanozymes and enzymes: challenges and opportunities." *Bioconjugate Chemistry* 28, no. 12 (2017): 2903–2909. 10.1021/ACS.BIOCONJCHEM.7B00673/ASSET/IMAGES/MEDIUM/BC-2017-00673V_0005.GIF.

14 Innovations in Food Safety Technology

A. E. Cedillo-Olivos, R. B. Colorado, S. A. González, and C. Jiménez-Martínez

14.1 INTRODUCTION

This review is divided into two parts; the first part presents the description of the blockchain, its importance for food processing and food safety, advantages and disadvantages of using this technology, and basic principles and characteristics for implementation. It continues with a description of the different blockchain systems and a summary of blockchain applications in the food industry. The second part mentions emerging techniques and technologies that have been used in recent years for the production and packaging of food; it also addresses the main and common problems in food safety, so they develop technologies that approach food safety risks such as shelf life and deterioration in the physicochemical properties; among these techniques, nanozymes, radiation, high hydrostatic pressure, encapsulation, cold plasma, and ultrasound are mentioned; likewise, their classification, advantages, disadvantages, uses of each of the emerging technologies, and their limitations are exposed. It also provides sources and descriptions of the latest food research using one or more of these technologies.

14.2 BLOCKCHAIN TECHNOLOGY

In recent decades, the importation of eating patterns has contributed to modifying the lifestyle as well as the eating habits of the population, which leads to a concern of the consumers to know really what they are ingesting, calling into question the quality characteristics and product safety, along with the reliability of food labeling (Ibarra-Sánchez *et al.*, 2017). On the other hand, the food industry has searched for different technologies that help identify critical points in processes, from the raw material to the finished proceeds, to guarantee safe products arrive quickly and efficiently to the consumer. Recently, the food industry has been implementing blockchain technology for this purpose; this subject will be addressed in this chapter to understand what it is and how it works. Blockchain technology is the latest generation's innovative, decentralized, and distributed technology (Figure 14.1). It is a database type containing a digital record of the history of movements, maintaining the confidentiality, integrity, and security of all transactions and data. The distribution of blockchain systems is mainly

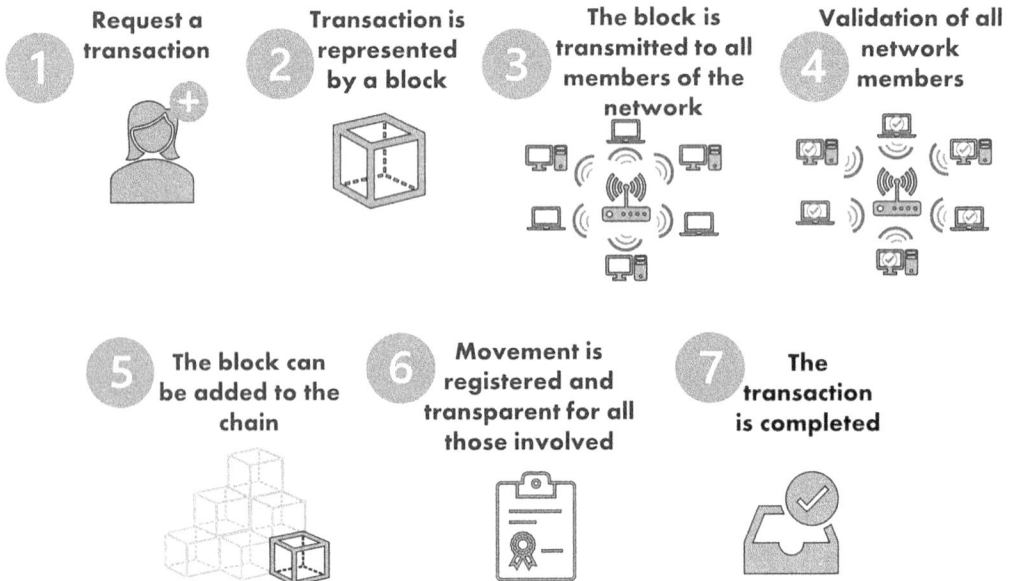

Figure 14.1 Graphical representation of a blockchain system operation (Blasetti, 2017; Zarrin *et al.*, 2021).

DOI: 10.1201/9781003334859-14

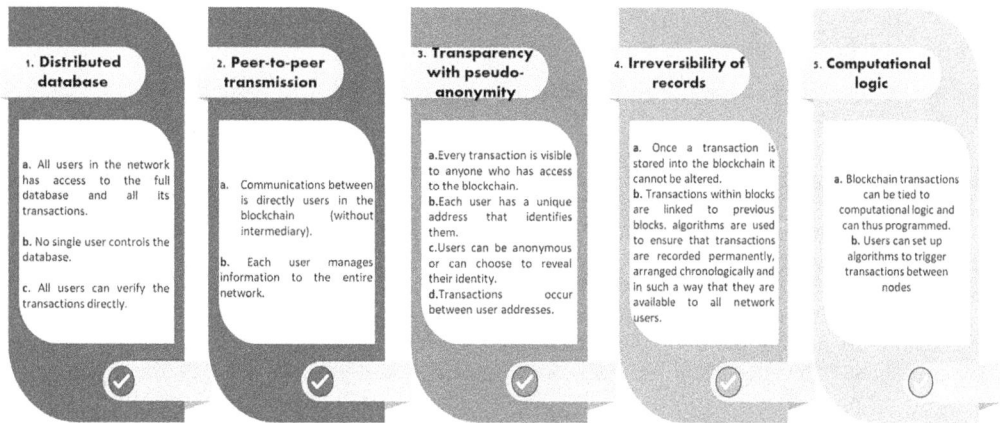

Figure 14.2 Basic principles for the implementation of blockchain technology. Adapted from Lansiti and Lakhani, 2017.

through a network of computers, without centralizing, but sharing the management throughout the whole network. Furthermore, once the information is added, it cannot be edited without modifying the previous records (because it is cryptographically linked to previous transactions); this requires the consent of all/majority of the parties involved (Dutta *et al.*, 2020). The main advantage of a blockchain is that transactions can be traced back to the beginning, so that information about an asset on the blockchain can be reported on how this asset originated and its changes over time. To implement blockchain, it is necessary to consider its five basic principles (Lansiti and Lakhani, 2017). In Figure 14.2, the basic principles of the blockchain are shown, with characteristics of the public blockchain; however, such examples can also occur in the hybrid and private blockchain.

14.2.1 Advantages and Disadvantages of the Blockchain

Before implementing a blockchain, an analysis of this technology's weaknesses and strengths is required, which will help us know if it favors the sector to which it is intended to be applied. Some of the advantages and difficulties of this system are described below (Olsen *et al.*, 2019).

Advantages:

a. Transactions without intermediaries and reliability between those involved are carried out. This increases the speed between the movements to be made and, in turn, the rate of these.

b. Blockchain can be taken as an unalterable record of the history of any asset or industrial good.

c. Easier tracking of products. This can be from the primary source, checking its distribution (place and date), thus helping to withdraw the merchandise if required (withdrawal of defective and dangerous products).

d. Improvements in perishables management and inventory management.

e. Increases efficiency and speed and reduces risks in the supply chain. It detects possible problems in the collection of raw materials, the production capacity, and the capacities of the distribution network with logistics operators since all of these are synchronized.

f. Improvement in the financing, contracting, and international transactions. The blockchain enables the sharing of inventories, information, and financial flows, to improve supply chain finance, procurement, and international business in global value chains.

g. Market creation. We facilitate entry into marketplaces, which allow buyers and farmers to connect directly with companies, increasing the benefits for both parties.

Disadvantages:

a. High energy costs are necessary due to the servers for the calculation capacity of the different consensus systems.

b. Since the transactions in the blockchain are unalterable, some circumstances may produce an error, and no correction can be made in the shared database.

c. The size of the blocks added to the chain is a limiting factor due to performance and efficient usage.

14.2.2 Blockchain Features

For blockchain implementation, the following characteristics are considered: decentralization, persistence, namelessness, and auditability (Zheng *et al.*, 2017; Wang *et al.*, 2018).

14.2.2.1 Decentralization

Transactions within the blockchain system can occur between two users without the need for authentication by an intermediary (for example, banks). In this way, blockchain can reduce server costs and increase the speed and performance of central servers (Xu Jie *et al.*, 2021).

14.2.2.2 Persistence

Each transaction is transmitted through the blockchain network and must therefore be confirmed and recorded in blocks for distribution throughout the network. A node will be obtained on the web, and a copy will be generated in the blockchain system. This also means that any node will be able to validate the encryption of the information and will verify the transactions made on it, making it (almost) impossible to tamper with the data. With existing blockchains, fraud and data falsification can be easily detected.

14.2.2.3 Anonymity

The nodes are the users connected to the blockchain network. Their functions are to store and distribute fragments to update blockchain data. Users interact with the blockchain network using a generated address; this address completely differs from a physical address or an address linked to a specific user account. The nodes have been classified into three, which are as follows:

a. Broadcast nodes carry out transactions and store information on the blockchain through third parties.

b. Full nodes, these types of nodes issue transactions, distribute blockchain information, and compliance with consensus.

c. Mining nodes these nodes solve cryptographic complexes. The mining node attempts to create a new block on the blockchain and demonstrates that this block performs work that helps meet a system need. Once the entire network has verified the proposed transaction, it is added to the existing blockchain as a new block (Meiklejohn *et al.*, 2013; Kosba *et al.*, 2016).

14.2.3 Auditability

Each transaction is validated and recorded with a timestamp. The function of auditing is to check the veracity of the previous records and verify existing ones (since the transaction history, up to the first block of a transaction, is saved and accessible). This feature of the blockchain enhances the traceability and transparency of the data stored on the blockchain by ensuring that information, once recorded, cannot be overwritten or deleted (Bermeo *et al.*, 2018).

14.2.4 Blockchain Systems

There are different ways of building blockchain technology, which is divided into three types: (a) public, (b) private, and (3) consortium (also known as hybrid).

14.2.4.1 Public Blockchain

The public blockchain is based on "proof of work" algorithms; these are open-sourced and unauthorized. These codes allow you to download and start a public node from your device and participate in the verification and consensus process within the blockchain without permission from other users. Anyone can also submit transactions to the blockchain network; if they are valid, anyone can see them permanently stored on the blockchain network. Their nodes' read and write permissions in a public blockchain are free. Some examples are Bitcoin, Ethereum, Monero, Dash,

Litecoin, and Dogecoin. For a blockchain construct to be considered decentralized, three basic principles must be met (Johnston *et al.*, 2014):

a. It must be open-sourced and operate autonomously without intermediaries controlling most of its tokens; transaction data and records should be stored cryptographically on public, decentralized blockchain technology.

b. The application must generate tokens according to the algorithm or standard criteria. These tokens must be necessary for the use of the application.

c. All changes must be approved by most of the network members. The application can adapt its response protocol to proposed improvements and market suggestions, but a majority consensus of its users must decide on all changes.

14.2.4.2 *Private Blockchain Technology*
The private blockchain technology is considered a centralized network because an organization controls the write permissions and verification and consensus process where not all nodes can participate in both processes, even if the nodes are from the same organization or group. On the other hand, the main difference with hybrid blockchain technology is that it comprises different groups (Zarrin *et al.*, 2021). Thus, Figure 14.3 shows three blockchains that can be differentiated by comparing their properties, distinguishing between public, hybrid, and private (Zarrin *et al.*, 2021).

14.2.4.3 *Hybrid Blockchain Technology*
Hybrid blockchain technology chooses nodes of a public or private web branch to handle its verification and consensus process. They are usually managed by a group of trusted people, entities, or authorities, thus becoming a hybrid between public and private blockchains. It is private since its logic only allows a few selected nodes the ability to read and write (Zarrin *et al.*, 2021). As a result, the consortium's blockchains have greater scalability, increasing the speed of transactions and providing greater privacy. An example of this type is seen most frequently used in the banking sector. Other examples are EWF (energy), B3l (insurance), and Corda.

14.2.5 Blockchain Technology Applications in the Food Industry
Many tests have been done for food chain blockchain applications, focusing mainly on traceability. However, a standard technology that can connect different blockchains is lacking due to the immaturity of this technology (Ciaian, 2018). In 2015, most of the existing blockchain systems for

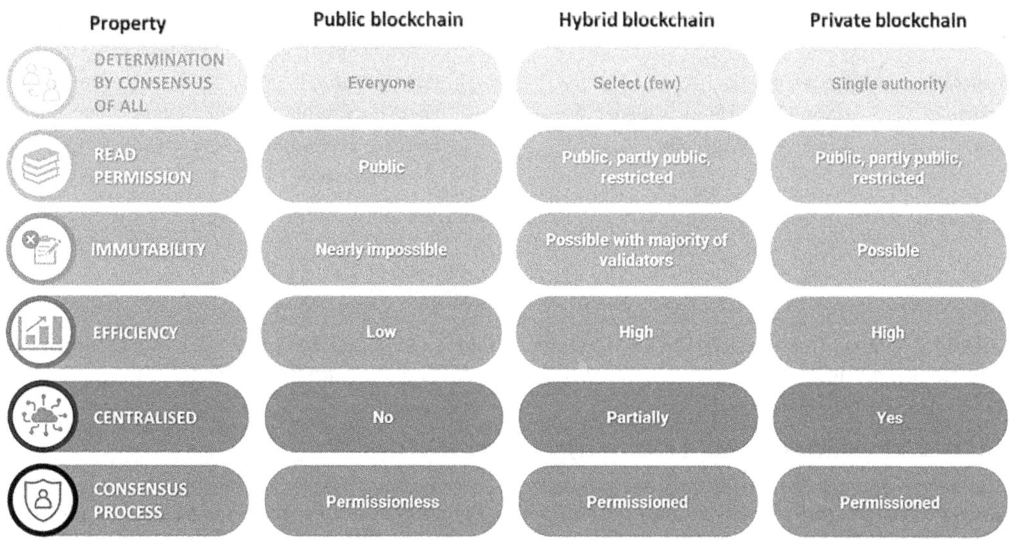

Property	Public blockchain	Hybrid blockchain	Private blockchain
DETERMINATION BY CONSENSUS OF ALL	Everyone	Select (few)	Single authority
READ PERMISSION	Public	Public, partly public, restricted	Public, partly public, restricted
IMMUTABILITY	Nearly impossible	Possible with majority of validators	Possible
EFFICIENCY	Low	High	High
CENTRALISED	No	Partially	Yes
CONSENSUS PROCESS	Permissionless	Permissioned	Permissioned

Figure 14.3 The figure presents four columns: property, public blockchain, hybrid blockchain, and private blockchain.

Table 14.1: Applications of Blockchain Technology in the Agricultural and Livestock Food Supply Chain

Goods/ Products	Initiative/Project/Company Involved	Objectives
Agri-food	AgriOpenData	Allow quality and digital identity to be certified
Agri-food	Supply Chain Traceability System for China Based on RFID & Blockchain Technology	Trusted information throughout the agri-food supply chain
Beef	"Paddock to plate" project, Beef Ledger; JD.com	Food traceability
Beer	Downstream	Food traceability
Chicken	Gogochicken; Grass Roots Farmers' Cooperative; OriginTrail ZhongAn	Food traceability, food safety concerns of urban consumers
Coffee	FairChain coffee: Bext360 in partnership with Moyee Coffee	Traceability, transparency of the value added
Fish	Provenance	Auditable system
Fresh food	Ripe	Enabling data transparency and transfer from farm to fork
Fruits	FruitChains	Public, immutable, ordered ledger of records
Grains	AgriDigital	Financial
Large enterprises	IBM	Food tracking project
Mangoes	Walmart, Kroger, IBM	Food traceability
Olive oil	OlivaCoin	Financial, small farmers support
Orange juice	Alber Heijn & Refresco	Show customers how and by whom products are made
Pork	Walmart, Kroger, IBM	Food traceability
Pork	Arc-net	Brand protection and security through transparency
Scotch Whisky	CaskCoin	Investing in maturing Scotch Whisky
Soybean	HSBC & Cargill; ING & Louis Dreyfus Co.	Help authenticate products as well as eliminate the "paper trail" of verification at every stage of the supply chain
Sugar cane	Coca-Cola	Humanistic
Turkeys	Cargill Inc., Hendrix Genetics	Food traceability, animal welfare
Wine	Chainvine Winecoin	Increase performance, revenue, accountability, and security

(Olsen *et al.*, 2019)

traceability management started to be developed (Galvez *et al.*, 2018). Table 14.1 is a summary of some blockchain technology projects in the agricultural food chain and the objectives they set out to achieve (Bermeo-Almeida *et al.*, 2018).

14.3 THE REVOLUTION OF TRADITIONAL TECHNOLOGY

Food safety and quality are important parameters to maintain or improve within the industry. Natural disasters and climatic changes affect both parameters because they have caused a decrease in food quality and production and, therefore, an increase in the demand by the population for minimally processed foods containing better physicochemical, nutritional, and sensory characteristics. To face the problems of food safety (short shelf life and low quality), new research has been carried out to improve the quality and preservation of foodstuffs. To this end, emerging technologies such as nanozymes, radiation, high hydrostatic pressure, encapsulation, cold plasma, and ultrasound are proposed; as a result, this second part of the chapter shows emerging low-temperature technologies that have improved food characteristics and safety.

14.3.1 Nanozymes

One of the technologies in which more has been invested for the early development of food safety is biosensors using enzymes, which help quickly and easily detect risks, ensuring food safety. The high sensitivity and specificity of enzymes allow this technology to be one of the most effective, with its adaptability to technologies that are commonly used in society, such as smartphones and 3D printing; in that war, we can talk about its advantages, such as short detection time, ease of use, and high sensitivity. It can also be used in labels that quickly visualize food insecurity (Huang *et al.*, 2019; Pan *et al.*, 2021). Nanozymes can be classified into two families.

14.3.1.1 Oxidoreductases

1. Oxidases: this reaction uses molecular oxygen as an electron acceptor to process the oxidation/reduction, producing H_2O or H_2O_2. In nanozymes, Au has been used as a catalyst for glucose oxidation; Au is a source of electrons and a bridge in the interaction between glucose and oxygen to produce H_2O_2 (Comotti *et al.*, 2006).

2. Peroxidases: like oxidases, these enzymes are redox, but use peroxide as an oxidant to oxidize the reducing substrate. In this field, horseradish peroxidase has been widely used and investigated in immunoassays because of its high stability, easy conjugation, and colorimetric detection (Montali *et al.*, 2020).

3. Catalases: act similarly to peroxidases since they oxidize peroxide, producing water and oxygen, but this enzyme does not require phenolic compounds like peroxidases (Castro *et al.*, 2006).

4. Superoxide dismutases: these enzymes can generate H_2O and O_2 from superoxide (O_2-) naturally produced in bacterial and animal cells. It is an important natural antioxidant for reducing free radicals (Ali *et al.*, 2004).

14.3.1.2 Hydrolases

This family of enzymes catalyzes hydrolysis reactions in the presence of water.

1. *Nucleases*: degrades nucleotide acids by the hydrolysis of phosphodiester groups.

2. *Esterases*: a hydrolase that acts on ester, amide, and thioester bonds; depending on the conditions, they can result in three reactions: hydrolysis, esterification, and transesterification (Huang *et al.*, 2019).

3. *Phosphatases*: catalyze the dephosphorylation reaction on a monophosphate ester to produce inorganic phosphate and alcohol (Huang *et al.*, 2019).

4. *Proteases*: cause hydrolysis of proteins at peptide bonds, producing free amino acids or peptides. These enzymes are highly important for degrading bacterial cells (Søltoft-Jensen and Hansen, 2005).

Due to the physicochemical and structural properties of nanozymes, there are different factors (pH, temperature, the presence of metal ions, irradiation) that influence their enzymatic activity; this advantage allows the reactions to be controlled according to the conditions in which they are found or substrate disposition, which will enable immunoassays and in situ reactions to be performed. Therefore, the factors in which enzymes present their maximum catalytic capacity continue to be studied (Castro *et al.*, 2006; Huang *et al.*, 2019). One of the most recent enzyme biosensor developments is the research of Montali *et al.*, (2020), with the presentation of a chemiluminescence folding paper for the detection of acetylcholinesterase inhibitors by enzymatic reactions sequenced by folding the paper, but without being mixed. The rationale for this work is based on three enzymatic reactions, each of the enzymes used in the reactions is on Whatman 1 paper; in the first reaction (Eq. 14.1), the sample (10 µL) is placed, and 5 µL of a 100 U/mL acetylcholinesterase solution at pH 7.0 is used for acetylcholine hydrolysis and choline production. In the second reaction (Eq. 14.2), 15 µL of 20 U/mL choline oxidase solution at pH 8 was used for choline oxidation and hydrogen peroxide production, which allows light emission. Finally, (Eq. 14.3), 15 µL of 108 U/mL horseradish peroxidase solution was used along with luminol for its oxidation and production of hydrogen (H_2), which allows the measurement of acetylcholinesterase inhibitory activity; the detection of light intensity can be performed with a smartphone.

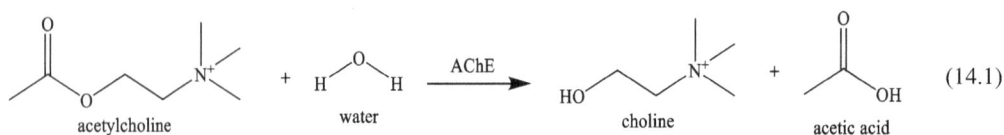

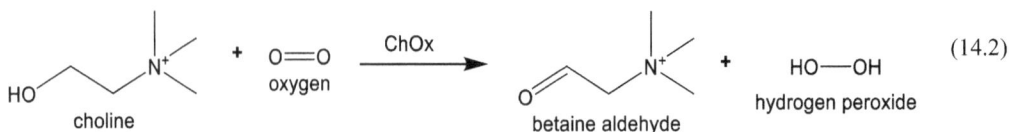

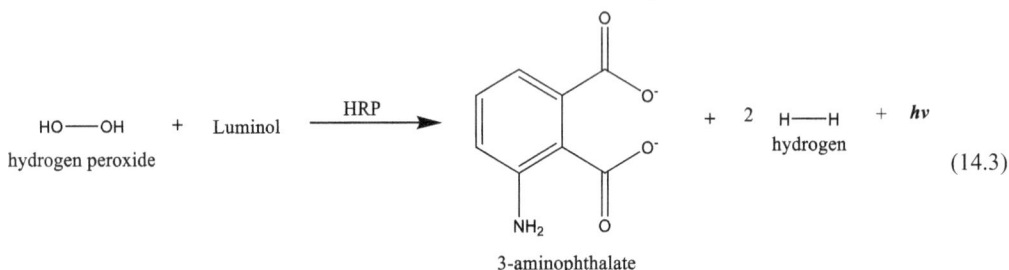

acetylcholine + water $\xrightarrow{\text{AChE}}$ choline + acetic acid (14.1)

choline + oxygen $\xrightarrow{\text{ChOx}}$ betaine aldehyde + hydrogen peroxide (14.2)

hydrogen peroxide + Luminol $\xrightarrow{\text{HRP}}$ 3-aminophthalate + 2 hydrogen + $h\nu$ (14.3)

The advantage of this innovative development is the small sample utilized for quantification, no need for complex apparatus, in situ, reduced detection time, can be used for organofluorine pesticides, nerve gases, and some drugs (Montali *et al.*, 2020; Zhang *et al.*, 2022). The challenges still (Tang *et al.*, 2022) facing nanozymes are as follows:

- Low catalytic activity
- Limitation of catalytic activity
- Low understanding of the catalytic mechanism
- Low understanding of toxicity and its mechanism
- Limitation of food applications.

Emerging enzyme technologies are not limited to applying detection and/or quantifying undesirable components. It has also been taken to the storage stage, waiting until it reaches the final consumer. Then, they put this technology on the labels to show the consumer and distributor the conditions of the product. The labels now feature nanozyme technology, maintaining standard enzymatic methods' physical and chemical characteristics, effectively detecting and analyzing food hazards. The nanozymes can be described as enzymes that use nanomaterials as support to catalyze their reactions, such as specific peroxidases, oxidoreductases, superoxide dismutase, and catalases. Therefore, nanozymes labels can be grouped into metal nanozymes, metallic oxides, carbon-based nanozymes, and others (Pan *et al.*, 2021; Shu *et al.*, 2022).

14.3.2 Radiation

This technology refers to the exposure of the food to ionizing energy for the inactivation of microbial cells and obtaining safe and additive-free food with higher sensory quality. It can be presented by five sources approved by the FDA as safe radiation: 60Co, 137Cs and 5 MeV, X-rays, UVC, and IR. These sources generate X-rays, gamma rays, electron beams, ultraviolet, and infrared light. The rationale is based on DNA damage production of hydroxyl and hydrogen radicals (OH- and H+). It´s a non-thermal technique in the food industry, which allows for prolonging the shelf life of products, being a safe method when the conditions are well defined to deactivate microbial cells (Caballero-Figueroa *et al.*, 2022; Panseri *et al.*, 2022). The main effects are described below (Caballero-Figueroa *et al.*, 2022):

- Lipids: these macromolecules can be the most sensitive, causing autooxidation, and due to the generation of free radicals (OH- and H+), it can induce oxidation, causing color changes, undesirable odor generation, deterioration, and rancid taste.

- Proteins: it was mainly believed that it could generate the breakdown of amino acids, but this technology is safe at high doses (>10 kGy).

- Bacterial cells: its effects are direct and indirect, but mainly on pathogens; direct damage refers to damage to carbohydrates, lipids, DNA, and RNA. Moreover, indirect damage occurs because of a reaction with free radicals.

To maintain food safety, regulations must be followed; the WHO has declared that there are no irrelevant changes in radiation less than or equal to 10 kGy; in addition, the European Union is governed by directives 1999/2/CE and 1999/3/CE. Nevertheless, despite the current regulations, markers are used to detect irradiation treatments (Panseri *et al.*, 2022):

1. Screening method based on microgel electrophoresis of suspicious DNA fragments by applying ionizing radiation.

2. Confirmation method based on microextraction, search, and quantification of butanones by mass spectrometry coupled to gas chromatography. Among its recent applications, we can describe the following in Table 14.2.

Table 14.2: Applications of UVC, Microwave-Assisted and Gamma Radiation in Food Matrix

Technology	Conditions and Results	Food Matrix	Reference
UVC	Use 2.93 kJ/L, 3.9 log reduction in aerobic colony count, and 2 log in coliforms. Without a significant change in nutritional and quality parameters.	Cold-pressed green juice blend (kale, romaine, celery, apple, and lemon).	Biancaniello *et al.*, 2018
Microwave-assisted	18 W/g for 48 s. In freeze-dried rice flour, no showed changes, and in 8, 20, and 30% of moisture microwaves exert a plasticizing effect. Microwave allows modulation of the techno-functional properties and rheological characteristics of the gels, with 8% of moisture being the most effective in these changes. These rice flours can be used in the production of food products for the celiac population.	Dry-heat and heat-moisture to modify techno-functional properties and gel viscoelasticity of rice flour.	Solaesa *et al.*, 2021
Gamma radiation	Evaluation of radiation with 250 and 500 Gy in Lulo, a better conservation of the fruit was observed at 500 Gy preserved at 4°C, increasing 7 days in comparison with the control.	Fresh Lulo fruit (*Solanum quitoense*).	Andrade *et al.*, 2019

Radiation is a non-thermal alternative technique to inactivate food spoilage microorganisms; without producing residues and/or toxic chemical by-products (X-rays, gamma rays, electron beams, ultraviolet light, and infrared light), low cost in installation and maintenance, green technology, and food quality maintenance. However, its limitations are low penetration, inactivation influenced by the size, angle, exposure (or surface area), usage may cause damage to skin and eyes, and overheating of food may occur (microwave) (Singh *et al.*, 2021).

14.3.3 High Hydrostatic Pressure (HHP)

It is one of the emerging non-thermal technologies that promote safe and minimally processed foods trying to maintain physicochemical properties without adding chemical preservatives. High hydrostatic pressure (HHP), also called high pressure (HP), ultra-high pressure (UHP), or high-pressure processing (HPP), is due to the application of temperatures ranging from –20 to 60°C and pressures in the range of 100 to 1,000 MPa for seconds or minutes. The U.S. Food and Drug Administration (FDA) and the U.S. Department of Agriculture (USDA) (Caballero-Figueroa *et al.*, 2022; Chiozzi *et al.*, 2022) have approved HHP as an alternative to traditional pasteurization. Eq. 14.4 can explain its principle, where the Gibbs free energy changes as a function of volume compression and temperature application. Pressurized water is used to inactivate the micro-organisms; this cold water envelops the food until the desired pressures are reached, usually 87,000 psi, 6,000, or 600 MPa (Aganovic *et al.*, 2021). Pressure applications can be made directly or indirectly. For example, a piston moves and causes a volume change inside the pressure vessel in direct application. In addition, in the indirect application, the pressure setting in the vessel varies depending on the fluid under pressure (Navarro-Baez *et al.*, 2022).

$$d\left(\Delta G\right) = \Delta 3Vdp - \Delta SdT \tag{14.4}$$

Due to the properties and behavior of high hydrostatic pressures, other uses in food have been given to this technology, mainly to provide changes in melting point, solubility, density, viscosity, dissociation of weak acids, ionization, and recently in the improvement of the extraction of phenolic compounds from fruits and vegetables by allowing the disruption of the cell wall, which enables the release of these compounds (Navarro-Baez *et al.*, 2022). Among the main positive effects of HHP are the following (Caballero-Figueroa *et al.*, 2022; Chiozzi *et al.*, 2022; Liu *et al.*, 2022):

- In physicochemical properties, the modifications are minimal since the conditions to which the food is subjected cannot break the covalent bonds of its chemical compounds, so it maintains the nutritional quality and commercial value (color, flavor, and aroma).

- In lipids, applying HHP inhibits lipid hydrolysis, delays the initial formation of radicals, and reduces peroxide formation; it has been reported that the effect is directly proportional to the pressure level applied.

- Proteins are not affected at pressures below 500 MPa; above this range, the quaternary structures dissociate, unfolding the proteins and allowing the sulfhydryl groups (thiol-SH) to be exposed.

Among the disadvantages of HHP are the following (Aganovic *et al.*, 2021; Chiozzi *et al.*, 2022): 1) the feed must contain a minimum air or gas content; 2) food packaging material because it cannot be contained in a rigid material such as glass and metal; 3) due to the low energy consumption, it is less expensive than pasteurization, and the water can be reused; 4) spores are resistant to high pressures; to ensure food safety, they must be combined with other applications; 5) high cost of equipment, configuration, and processing capacity; and 6) it cannot be used on high-pH foods, as they require pressures >800 MPa to inactivate their bacterial spores.

14.3.3.1 Applications

Argyri *et al.* (2018) inoculated *S. enteritidis* in chicken fillets at 3, 5, and 7 logs CFU/g and treated them with HHP at 500 MPa for 10 min; the storage at 4 and 12°C was investigated. They found that the population of *S. enteritidis* was reduced below 0.48 log CFU/g. Brochothrix thermosphacta was found to be the main spoilage microorganism surviving after HHP. In the sensory analysis, the chicken fillets showed clear pink color, and a whiter shade was observed in the raw chicken fillet. They concluded that the shelf life of chicken fillets increased by 6 and 2 days, at 4 and 12°C, respectively, compared to samples without HHP treatment. Taddei *et al.* (2020) evaluated the effect of high-pressure treatment on the viability of *Salmonella* spp. in traditional Italian dry-cured coppa. These bacteria were inoculated by immersion and then subjected to HHP treatment with a pressure of 593 MPa for 290 sec and a temperature of 14°C in water. The four samples found a 5-log CFU/g reduction of *Salmonella* spp. after HHP treatment.

14.3.4 Encapsulation

Encapsulation is a technique by which droplets are placed in a liquid state covering solid or gaseous particles with a porous polymeric film containing an active substance, allowing it to maintain its stability and viability. Microcapsules help food materials to be more resistant to processing and packaging conditions, improving flavor characteristics, aroma, compounds stability, nutritional value, and appearance (Parra-Huertas and Medina-Vargas, 2012). The microencapsulation method and coating material and how it can be used to enhance the performance of a particular ingredient condition the final characteristics of microencapsulated products. Coating materials are selected from a wide variety of natural or synthetic polymers, depending on the material to be coated and the characteristics of the microcapsules to be obtained (Poshadri and Aparna, 2010).

An ideal coating material should show the following characteristics (Desai and Jin Park, 2005): 1. Good rheological properties at high concentrations and easy to handle during encapsulation. 2. The ability to stabilize the emulsion and disperse or emulsify the active material. 3. No present reactivity with the material to be encapsulated during processing and storage. 4. Sealing and maintaining the active material within its structure. 5. The ability to completely release the solvent or other materials used during encapsulation (Poshadri and Aparna, 2010). The size, distribution, and morphology of the microcapsule depend on the core and shell material, and can be:

a. Mononuclear (nucleus/shell): the nucleus is wrapped with a continuous wall material.

b. Polynuclear: many nuclei coated with a shell material are generated.

c. Matrix encapsulation: the coating material is homogeneously distributed in the core.

d. Multilayer: a continuous core coated with a continuous multilayer sheath material (Giro-Paloma *et al.*, 2016).

14.3.4.1 Encapsulation Techniques

Spray drying is a method where a hot gas stream is applied to atomize the material to obtain the product powder. The gas used is air or nitrogen (being an inert gas rarely used). The encapsulation material should be selected for its high solubility, effective emulsification, efficient drying, and low viscosity, even in a high-concentration solution and film formation (Choudhury et al., 2021). Coacervation occurs when an active agent is distributed in a homogeneous polymer solution, and colloidal polymer aggregates (coacervates) are formed on the outer surface of a drop of the active agent when coacervation is triggered. Coacervation is initiated by varying some parameters in the system, such as temperature, pH, or the composition of the reaction mixture (addition of non-water miscible solvent or salt). There are two main coacervation techniques: simple coacervation and complex coacervation. They differ in the mechanism of phase separation. Simple coacervation occurs when the polymer is salted or dissolved, whereas complex coacervation is achieved by the complexation of two or more polyelectrolytes of opposite charge. The coacervation method is used in flavor preservation, thanks to the high payloads that can be achieved (up to 99%) and the possibilities of controlled release based on mechanical stress, temperature, or sustained release (Trojanowska *et al.*, 2017). In fluidized bed technology, the liquid coating is sprayed onto the particles; rapid evaporation occurs, which helps to form an outer layer on the particles. Coating thickness and formulations can be obtained as required. Different fluidized bed coatings include top spray, bottom spray, and tangential spray (Figure 14.4).

Pan coating is an effective physico-mechanical technique for the encapsulation of particles. The coating material solution is sprayed on; next, hot air is passed through to evaporate the solvent and obtain the microencapsulated sample (Jyothi *et al.*, 2010). The extrusion method is based on a polysaccharide gel that immobilizes the nucleus when it comes into contact with a multivalent ion, one of the drawbacks of this technique is the formation of larger particles (usually 500–1,000 mm). The principle on which this technique is based consists of incorporating the nucleus in a solution of sodium alginate, and the mixture obtained is subjected to a drop-by-drop extrusion by means of a

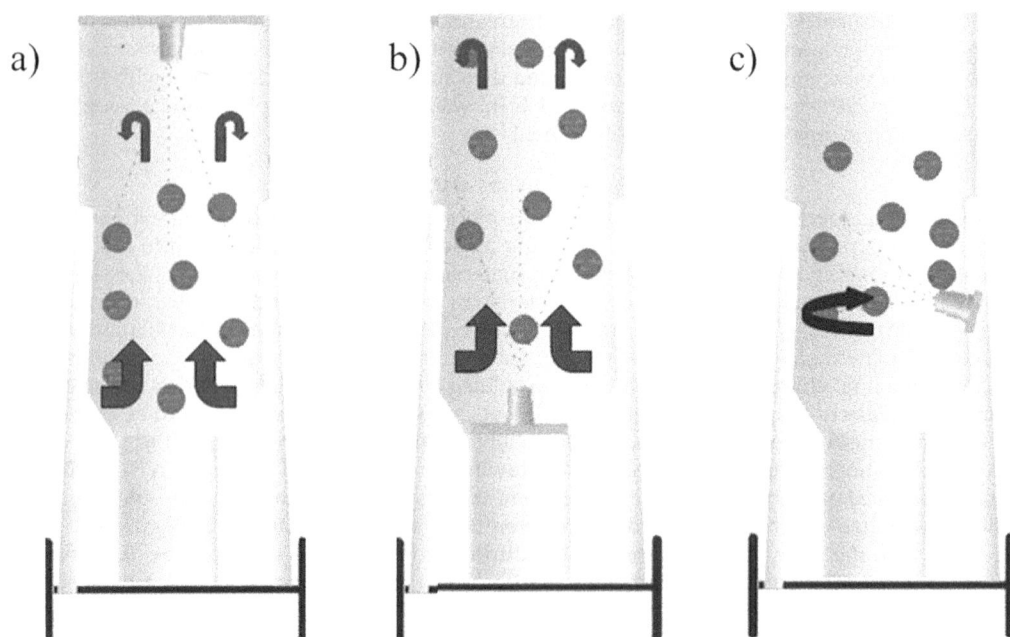

Figure 14.4 The picture shows three schemes of fluidizer bed coater (Redrawn from Ghosh, 2006).

small-caliber pipette or syringe in a solution that favors the formation of microcapsules, such as calcium chloride (Teixeira *et al.*, 2014). Solvent evaporation is widely used to produce solid and liquid core materials for water-soluble and water-insoluble materials. In this method, the coating material (polymer) that is immiscible in the liquid phase of the vehicle is dissolved in a volatile solvent. The core material (drug) that is to be microencapsulated is dispersed or dissolved in the coating polymer solution (Krishna and Jyothika, 2015).

Applications of microcapsules

1. Pharmaceutical industry: it helps with the controlled release of drugs, representing an advantage over traditional drugs.
2. Food industry: various substances can be encapsulated, including colorants, flavoring substances, vitamins, antioxidants, minerals, leavening agents, sweeteners, and enzymes (Calderón-Oliver and Ponce-Alquicira, 2022).

14.3.5 Cold Plasma

Cold plasma (CP) is of great importance in food technology. The novelty of this technology lies in its non-thermal, economical, versatile, and environmentally friendly nature (Pankaj et al., 2018). Cold plasma technology is a come-out treatment for food processing, most used for microbial decontamination effects, toxin removal, enzymatic inactivation, and food packaging modifications to ensure food safety and shelf life for consumers (Panka *et al.*, 2017). This method is particularly effective against major foodborne pathogenic microorganisms such as *Escherichia coli*, *Salmonella typhimurium*, *Staphylococcus aureus*, and *Listeria monocytogenes* (Panka *et al.*, 2017).

14.3.5.1 Plasma Chemistry: Process

In plasma, ionization of the process is considered the first important element, followed by other factors such as reaction rate, electron energy distribution, rate constants, and mean free path. The plasma chemical process is classified into two categories based on the reactions i) homogeneous gas phase reaction (e.g., generation of N3 from N2) and ii) heterogeneous reaction where the plasma meets the solid or liquid medium. The first reactions are further classified into three subcategories. In the first, the material is removed from the surface by etching or ablation; in the second, a process called chemical deposition takes place, where the material is added to the solid surface in the form of a thin film observed during plasma polymerization by a plasma-enhanced vapor phase. In the third, the substrate surface is physically and chemically modified during plasma exposure; in this case, no material is added or removed (Thirumdas *et al.*, 2014).

14.3.5.2 *Types of Cold Plasma Systems*

Cold plasma systems can operate at atmospheric pressure or in some degree of partial vacuum. The motive power can be electrical, microwave, or laser. The ionized gas can be something as common as air, nitrogen, or a mixture containing some proportion of noble gases, such as helium, argon, or neon (Niemira, 2012). Cold plasma systems intended for food treatment generally fall into one of three categories defined by where the food is placed. A cold plasma system is the first category where remote treatment is given. The plasma travels over the surface to be treated; this can be fed by a feed gas flow or (less commonly) manipulated using magnetic fields. This simplifies the design and operation of the device and increases flexibility in terms of the shapes and sizes of the objects to be treated. In addition, this type of system has the advantage of locating the surface to be treated at a physically separate point of generation (Niemira, 2012). In the second category, the generation equipment delivers active plasma directly to the object to be treated; this system is known as a direct treatment cold plasma system; as in the first category, the plasma is moved through the feed gas stream or a comparable medium. These systems can operate in pulsed mode, with plasma generated at pulse rates of hundreds or thousands of times per second. These systems provide higher concentrations of active agents because the target is relatively close to the cold plasma generation site and is exposed to the plasma before the dynamic species recombine and are lost (Niemira, 2012). In the last category (three), electrode contact systems, the surface to be sterilized is located between two electrodes or the neutral ground connection. The surface to be sterilized is physically located within the cold plasma generation field; the shape and composition of the electrodes must be carefully controlled to match the material to be treated to avoid discharge points and consequent heat build-up. The product is exposed to the broadest combination of active antimicrobial agents in these systems with the highest possible intensity of free electrons, radicals, ions, and UV radiation (Niemira, 2012).

14.3.5.3 *Limitations and Toxicology of Plasma Treatment*

As with any process, plasma processing has certain limitations, such as increased lipid oxidation, reduced color, decreased fruit firmness, increased acidity, etc. Therefore, it is essential to understand the interaction of reactive plasma species with food components, and it is necessary to investigate the effects of cold plasma on the physicochemical and sensory properties of food products at the molecular level (Thirumdas *et al.*, 2014).

14.3.6 Ultrasound

Ultrasound is a form of energy generated by sound waves (actually under pressure) of frequencies too high to be detected by the human ear, i.e., higher than 16 kHz. When these waves are propagated through a biological structure, they induce compressions and depressions of the particles in the medium, and a large amount of energy can be imparted. Depending on the frequency and amplitude of the applied sound wave, it allows various applications by observing chemical, physical, and biochemical effects (Teixeira *et al.*, 2014).

14.3.6.1 *Methods of Ultrasound*

Ultrasound can be used for food preservation and, along with other treatments, it improves its inactivation efficacy. For example, ultrasound combined with either pressure or temperature has been utilized (Ercan and Soysal, 2013).

1. Ultrasonication (US) can be used for heat-sensitive products because it applies ultrasound and low temperatures. However, it requires a long treatment time to inactivate enzymes and/or stable microorganisms, which can cause high energy consumption, and temperature increase can occur depending on the ultrasonic power and application time; in addition, it needs control to optimize the process (Ercan and Soysal, 2013).

2. Thermosonication (TS) is a method where moderate heat and ultrasound are combined and obtain a more significant inactivation effect of microorganisms than heat alone, especially when required to use processes of short time and low temperatures, achieving the same lethality values as with conventional methods (Ercan and Soysal, 2013).

3. Mannosonication (MS) is a combined method in which ultrasound and pressure are applied; this method allows the inactivation of enzymes and/or microorganisms by combining these parameters at low temperatures. Its inactivation efficiency is superior to ultrasound alone using same temperature (Ercan and Soysal, 2013).

4. Manothermosonication (MTS) is a combined method of heat, ultrasound, and pressure; these treatments inactivate various enzymes at lower temperatures and/or in a shorter time than thermal treatments at the same temperatures. The temperature and pressure applied to maximize the cavitation or implosion of bubbles in the medium increase the level of inactivation. One of the applications is in the inactivation of thermotolerant microorganisms, and too including some thermo-resistant enzymes, such as lipoxygenase, peroxidase, polyphenol oxidase, and thermostable lipases and proteases from Pseudomonas (Ercan and Soysal, 2013).

14.3.6.2 Ultrasound as a Food Preservation Tool

The use of ultrasound in food technology involves non-invasive analyses with particular reference to quality assessment; some examples of the use of these technologies are the analysis of droplet size in emulsions of edible fats and oils, in the localization of foreign bodies in food and the determination of the degree of crystallization in dispersed emulsion droplets (Mason 2005; Dolatowski 2007).

14.3.6.3 Filtration

In the food industry, the separation of solids and liquids is a process in which a solid-free liquid is produced, or a solid is isolated from its mother liquor. Unfortunately, the deposition of solid materials on the surface of the filtration membrane is one of the main problems. Applying ultrasonic energy improves the flux by breaking the polarization concentration and the support layer on the membrane surface while maintaining the intrinsic permeability of the membrane (Mason *et al.*, 2005). This principle is used to extract the pulp from apple juice, because the vibrational energy keeps the particles suspended, preventing agglomeration, which allows more space to separate the solvent (Mason *et al.*, 2005).

14.3.6.4 Microbial Growth

Alternative methods of food processing that have practically no effect on food quality have gained importance due to the growing consumer demand for minimally processed foods. With this ultrasound technology, high pressure, shear, and a temperature gradient are generated by high-power ultrasound (20 to 100 kHz), causing cell death by destroying cell membranes and DNA (Mason *et al.*, 2005).

14.3.6.5 Emulsification/Homogenization

An important means of introducing hydrophobic bioactive compounds into a range of food products is acoustic emulsification, which offers the following improvements over conventional methods: when producing the emulsion, where the range of size particles is at submicron values with an extremely narrow distribution, coupled with the stability of the emulsions; the addition of a surfactant to produce and stabilize the emulsion is not necessary; the energy required to produce an emulsion by acoustic waves is less than that required in other methods; and the energy required to produce an emulsion by acoustic waves is less than that required in other methods (Mason *et al.*, 2005).

14.3.6.6 Enzyme Inactivation

Another way to extend the shelf life of some foods is enzymatic inactivation, which can be easily achieved by heat treatment, which does not alter the properties and nutrients of the food. In some cases, it may not be easy to inactivate heat-resistant enzymes. The heat's magnitude may alter some food properties, so options are sought where these alterations are minimal. Inactivation can be more effective if ultrasound is combined with another inactivation method. Combining these three parameters: heat, sonication, and pressure have a synergistic effect of increasing enzyme inactivation, compared to ultrasound alone (Ünver, 2016).

14.3.6.7 *Advantages and Limitations of Ultrasonication*

Ultrasound applications offer numerous advantages in the food industry, some of which are enlisted as follows:

a. Ultrasonic waves are safe, non-toxic, and environmentally friendly.

b. It is considered an effective means of microbial inactivation combined with other non-thermal methods.

c. This method does not require sophisticated machinery or a wide range of technologies.

d. The use of ultrasound increases the extraction yield and speed compared to other conventional extraction methods.

e. There is minimal loss of flavor, superior consistency (viscosity, homogenization), and energy savings.

f. This method has acquired enormous applications in the food industry, such as processing, extraction, emulsification, conservation, homogenization, etc. (Majid and Nayik, 2015).

g. Despite having many advantages, the use of ultrasonication also has many disadvantages, such as:

h. Using ultrasonic methods requires a higher energy input, which makes it complicated to use this technique on a commercial scale.

i. Ultrasound can develop due to shear stress and shock wave eddies (mechanical effects), which cause inactivation of the released products.

j. These induce physicochemical effects, which can be responsible for the deterioration of the quality of food products by developing off-flavors, alterations of physical properties, and degradation of components.

k. The use of ultrasonication leads to forming radicals due to the critical conditions of temperature and pressure, which are responsible for changes in food compounds.

l. The radicals (-OH and H+) produced in the medium are deposited on the surface of a cavitation bubble, which stimulates radical chain reactions, leading to the formation of degradation products and thus leading to considerable quality defects in the products.

m. The frequency of ultrasonic waves can impose resistance to mass transfer.

n. The ultrasonic power modifies the characteristics of the medium. As a result, this power must be minimized in the food industry to achieve maximum results (Majid and Nayik, 2015).

14.4 CONCLUSIONS

The new technologies of food conserving have developed due to the need to reduce processing time, increase shelf life, maintain food safety, and improve nutritional and sensory qualities. Advantages of the technologies presented in this review include ease of handling, reduced equipment training, application time and energy consumption, and prevention of foodborne illness. These technologies have replaced heat treatments, which were affecting food, deteriorating its sensory characteristics, causing loss of nutrients, and excessive energy consumption to reach optimum temperatures that would guarantee the safety of the products. On the other hand, emerging technologies have high installation costs. However, research continues on technologies that allow lower equipment and installation costs, allowing scalability and adaptability in the food industry and generating more significant applications to improve the quality and safety of food products.

REFERENCES

Aganovic Kemal, Hertel Christian, Vogel Rudi F., Johne Reimar, Schlüter Oliver, Schwarzenbolz Uwe, Jäger Henry, Holzhauser Thomas, Bergmair Johannes, Roth Angelika, Sevenich Robert, Bandick Niels, Kulling Sabine E., Knorr Dietrich, EngelKarl Heinz, & Heinz Volker. "Aspects of high hydrostatic pressure food processing: Perspectives on technology and food safety". *Comprehensive Reviews in Food Science and Food Safety*, 20(2021): 3225–3266. 10.1111/1541-4337.12763

Ali S., Shameh Joshua I., Hardt Kevin L., Quick Kim-Han, Sook J., Erlanger Bernard F., Huang Ting T., Charles J., Epstein, & Dugan, L.L. "A biologically effective fullerene (C 60) derivative with superoxide dismutase mimetic properties". *Free Radical Biology and Medicine*, 37(2004): 1191–1202. 10.1016/j.freeradbiomed.2004.07.002

Andrade-Cuvi M.J., Valarezo L.E., Guijarro-Fuertes M., Lárraga-Zurita P., Alcívar C.D., Vasco, C., & Vargas-Jentzch, P. "Evaluación del uso de radiación gamma como tratamiento poscosecha en naranjilla (*Solanum quitoense*)". *Reportes Frutas*, 20(2019): 1–15. https://www.researchgate.net/publication/334836341_Evaluacion_del_uso_de_radiacion_gamma_como_tratamiento_poscosecha_en_naranjilla_Solanum_quitoense

Argyri A.A., Papadopoulou O.S., Nisiotou, A., Tassou C.C., & Chorianopoulos N. "Effect of high pressure processing on the survival of Salmonella Enteritidis and shelf-life of chicken fillets". *Food Microbiology*, 70(2018): 55–64. 10.1016/j.fm.2017.08.019

Bermeo-Almeida Oscar, Cardenas-Rodriguez Mario, Samaniego-Cobo Teresa, Ferruzola-Gómez Enrique, Cabezas-Cabezas Roberto, & Bazán-Vera William. "Blockchain in agriculture: A systematic literature review". *In International Conference on Technologies and Innovation, Springer, Cham*. (2018): 44–56. 10.1007/978-3-030-00940-3

Biancaniello Michael, Popović Vladimir, Fernandez-Avila Cristina, Ros-Polski, Valquiria, & Koutchma Tatiana. "Feasibility of a novel industrial-scale treatment of green cold-pressed juices by uv-c light exposure". *Beverages*, 4(2018): 1–15. 10.3390/beverages4020029

Blasetti Robert. "Blockchain for business, should you care?" (2017). https://blockgeeks.com/blockchain-for-business/

Caballero-Figueroa Esmeralda, Terrés Eduardo, Hernández-Hernández Hilda María, & Escamilla-García Monserrat. "Revisión sobre las tecnologías emergentes no térmicas para el procesamiento de alimentos". *TIP Revista Especializada En Ciencias Químico-Biológicas*, 25(2022): 1–14. 10.22201/fesz.23958723e.2022.459

Calderón-Oliver Mariel, & Ponce-Alquicira Edith. "The role of microencapsulation in food application". *Molecules*, 27(2022):1499. 10.3390/molecules27051499

Castro Jhon Alexander, Baquero Lucía Estrella, & Narváez Carlos Eduardo. "Catalasa, peroxidasa y polifenoloxidasa de pitaya amarilla (*Acanthocereus pitajaya*)". *Revista Colombiana de Química*, 35(2006): 91–101. https://www.redalyc.org/pdf/3090/309026667004.pdf

Chiozzi Viola, Agriopoulou Sofia, & Varzakas Theodoros. "Advances, applications, and comparison of thermal (pasteurization, sterilization, and aseptic packaging) against non-thermal (ultrasounds, UV radiation, ozonation, high hydrostatic pressure) technologies in food processing". *Applied Sciences*, 12(2022): 1–40. 10.3390/app12042202

Ciaian P. "Blockchain technology and market transparency". (2018). https://ec.europa.eu/info/sites/info/files/law/consultation/mt-workshop-blockchain-

Comotti Massimiliano, Della Pina Cristina, Falletta Ermelinda, & Rossi Michele. "Aerobic oxidation of glucose with gold catalyst: Hydrogen peroxide as intermediate and reagent". *Advanced Synthesis and Catalysis*, 348(2006): 313–316. 10.1002/adsc.200505389

Choudhury, Nitamani, Meghwal, Murlidhar, & Das, Kalyan (2021). Microencapsulation: An overview on concepts, methods, properties and applications in foods. *Food Frontiers*, 2, 426–442. 10.1002/fft2.94

Desai, Kashappa Goud H., & Jin Park, Hyun (2005). Recent Developments in Microencapsulation of Food Ingredients. *Drying Technology*, 23, 1361–139410.1081/drt-200063478.

Dolatowski Zbigniew, Stadnik Joanna, & Stasiak Dariusz. "Applications of ultrasound in food technology". *Acta Scientiarum Polorum, Technologia Alimentaria*, 6(2007): 89–99.

Dutta Pankaj, Choi Tsan-Ming, Somani Surabhi, & Butala Richa. "Blockchain technology in supply chain operations: Applications, challenges and research opportunities". *Transportation Research Part E: Logistics and Transportation Review*, 142(2020): 102067. 10.1016/j.tre.2020.102067

Ercan Songül, & Soysal Çiğdem. "Use of ultrasound in food preservation". *Natural Science*, 5(2013): 5–13. 10.4236/ns.2013.58A2002

Galvez Juan F., Mejuto J.C., & Simal-Gandara J. "Future challenges on the use of blockchain for food traceability analysis". *TrAC Trends in Analytical Chemistry*, 107(2018): 222–232. 10.1016/j.trac.2018.08.011

Giró-Paloma Jessica, Martínez Mònica, Cabeza Luisa, & Fernández A. Inés. "Types, methods, techniques, and applications for Microencapsulated Phase Change Materials (MPCM): A review". *Renewable and Sustainable Energy Reviews*, 53(2016): 1059–1075. 10.1016/j.rser.2015.09.040. https://ideas.repec.org/a/eee/rensus/v53y2016icp1059-1075.html

Ghosh, S.K., ed. *Functional Coatings*. Wiley, 2006. 10.1002/3527608478.

Huang Yanyan, Ren Jinsong, & Qu Xiaogang. "Nanozymes: Classification, catalytic mechanisms, activity regulation, and applications". *Chemical Reviews*, 119(2019): 4357–4412. 10.1021/acs.chemrev.8b00672

Ibarra-Sánchez Lidia Susana, Viveros-Ibarra Lidia Susana, González-Bernal Victor, & Hernández-Guerrero Felipe. "Transición Alimentaria en México. Mexico food transition". *Razón y Palabra*, 20(2017): 166–182. https://www.revistarazonypalabra.org/index.php/ryp/article/view/697

Johnston David, Yilmaz Sam Onat, Kandah Jeremy, Hashemi Farzad, Gross Ron, Wilkinson Shawn, & Mason Steven. "The general theory of decentralized applications". *GitHub*, (2014): 1–12. https://github.com/DavidJohnstonCEO/DecentralizedApplications

Jyothi N. Venkata Naga, Prasanna P. Muthu, Sakarkar Suhas, Prabha K. Surya, Ramaiah P. Seetha, & Srawan G. "Microencapsulation techniques, factors influencing encapsulation efficiency". *Journal of Microencapsulation* 27(2010): 187–197. 10.3109/02652040903131301

Kosba Ahmed, Miller Andrew, Shi Eelaine, Wen Zikai, & Papamanthou Charalampos. "Hawk: the blockchain model of cryptography and privacy-preserving smart contracts". *IEEE Access International Conference on Blockchain (Blockchain)*, 10(2016): 839–858. 10.1109/SP.2016.55, 2016.

Krishna Abbaraju, & Jyothika M. "A review on microcapsules". *Online International Journal*, 4(2015): 26–33. http://www.cibtech.org/cjps.htm

Lansiti Marco, & Lakhani Karim. "The truth about blockchain". *Harvard Business Review*, (2017). https://hbr.org/2017/01/the-truth-about-blockchain

Liu Lei, Deng Xi, Huang Lei, Li Yalin, Zhang Yu, Chen Xing, Guo Shuyu, Yao Yao, Yang Shuhui, Tu Mingxia, Li Hehe, & Rao Yu. "Comparative effects of high hydrostatic pressure, pasteurization and nisin processing treatments on the quality of pickled radish". *LWT - Food Science and Technology*, 167(2022): 113833. 10.1016/j.lwt.2022.113833

Majid Ishrat, & Nayik Gulzar. "Ultrasonication and food technology: A review". *Cogent Food & Agriculture*, (2015): 1–11. 10.1080/23311932.2015.1071022

Mason Timothy, Riera Enrique, Vercet Antonio, & Lopez-Buesa Pascual. "Application of Ultrasound" Emerging technologies for food processing. *Editor: Da-Wen Sun. Academic Press* (2005): 323–351. 10.1016/B978-012676757-5/50015-3

Meiklejohn Sarah, Pomarole Marjori, Jordan Grant, Levchenko Kirill, McCoy Damon, Voelker Geoffrey, & Savage Stefan. "A fistful of bitcoins: characterizing payments among men with no names". In: Proceedings of the 2013 Conference on Internet Measurement Conference (IMC '13). *Association for Computing Machinery*, (2013): 127–140. 10.1145/2504730.2504747

Montali Laura, Calabretta Maria Maddalena, Lopreside Antonia, D'Elia Marcello, Guardigli, Massimo, & Michelini Elisa. "Multienzyme chemiluminescent foldable biosensor for on-site detection of acetylcholinesterase inhibitors". *Biosensors and Bioelectronics*, 162(2020): 112232. 10.1016/j.bios.2020.112232

Navarro-Baez Jorge, Martínez Luz María, Welti-Chanes Jorge, Buitimea-Cantúa Génesis, & Escobedo-Avellaneda Zamantha. "High hydrostatic pressure to increase the biosynthesis and extraction of phenolic compounds in food: A review". *Molecules*, 27(2022): 1–15. 10.3390/molecules27051502

Niemira Brendan. "Cold plasma decontamination of foods". *Annual Review of Food Science and Technology*, 3(2012): 125–142. 10.1146/annurev-food-022811-101132

Olsen Petter, Borit Melania, & Syed Shadeen. "Applications, limitations, costs, and benefits related to the use of blockchain technology in the food industry". *Nofima AS*, 4(2019): 35.

Pan Ruiyuan, Li Guoliang, Liu Shucheng, Zhang Xianlong, Liu Jianghua, Su Zhuoqun, & Wu Yongning. "Emerging nanolabels-based immunoassays: Principle and applications in food safety". *TrAC -Trends in Analytical Chemistry*, 145(2021): 116462. 10.1016/j.trac.2021.116462

Pankaj Shashi, Wan Zifan, & Keener Kevin. "Effects of cold plasma on food quality: A review". *Multidisciplinary Digital Publishing Institute*, 7(2018): 4. 10.3390/foods7010004

Panseri Sara, Arioli Francesco, Pavlovic Radmila, Di Cesare Federica, Nobile Maria, Mosconi Giacomo, Villa Roberto, Chiesa Luca Maria, & Bonerba Elizabetta. "Impact of irradiation on metabolomics profile of ground meat and its implications toward food safety". *Lwt*, 161(2022): 113305. 10.1016/j.lwt.2022.113305

Parra-Huertas Ra, & Medina-Vargas Oscar. "Sobrevivencia y encapsulación de bacterias y su efecto en las propiedades sensoriales, fisicoquímicas y microbiológicas del yogurt". *Vitae*, 19(2012): S90–S92. http://www.redalyc.org/articulo.oa?id=169823914022

Poshadri Achina, & Kuna Aparna. "Microencapsulation technology: a review". *Angrau*, 38(2010): 86–102.

Shu Rui, Liu Sijie, Huang Lunjie, Li Yuechun, Sun Jing, Zhang Daohong, Zhu Ming-Qiang, & Wang Jianlong. "Enzyme-Mimetic nano-immunosensors for amplified detection of food hazards: Recent advances and future trends". *Biosensors and Bioelectronics*, 217(2022): 114577. 10.1016/j.bios.2022.114577

Singh Harpreet, Bhardwaj Sanjeev K., Khatri Madhu, Kim Ki-Hyun, & Bhardwaj Neha. "UVC radiation for food safety: An emerging technology for the microbial disinfection of food products". *Chemical Engineering Journal*, 417(2021): 128084. 10.1016/j.cej.2020.128084

Solaesa Ángela, Villanueva Marina, Muñoz José María, & Ronda Felicidad. "Dry-heat treatment vs. heat-moisture treatment assisted by microwave radiation: Techno-functional and rheological modifications of rice flour". *Lwt*, 141(2021): 110851. 10.1016/j.lwt.2021.110851

Søltoft-Jensen, J., & Hansen, F. "New chemical and biochemical hurdles". Editor, D.W. Sun, *Emerging Technologies for Food Processing: An Overview*, 401–403. Elsevier, 2005. 10.1016/B978-0-12-676757-5.50017-7

Taddei Roberta, Giacometti Federica, Bardasi Lia, Bonilauri Paolo, Ramini Mattia, Fontana Maria Cristina, Bassi Patrizia, Castagnini Sara, Ceredi Francesco, Pelliconi Maria Francesca, Serraino Andrea, Tomasello Federico, Piva Silvia, Mondo Elisabetta, & Merialdi Giuseppe. "Effect of production process and high-pressure processing on viability of *Salmonella spp.* in traditional Italian dry-cured coppa". *Italian Journal of Food Safety*, 9(2020): 141–145. 10.4081/ijfs.2020.8445

Tang Yinjun, Wu Yu, Xu Weiqing, Jiao Lei, Gu Wenling, Zhu Chengzhou, Du Da, & Lin Yuehe "Nanozymes enable sensitive food safety analysis". *Advanced Agrochem*, 1(2022): 12–21. 10.1016/j.aac.2022.07.001

Teixeira Pablo, Martins Leadir Lucy, Ragagnin Cristiano, Tasch Augusto, Schwan Carla Luisa, Francine Évelin, De Oliveira Juliana, & De Bona Cristiane. "Microencapsulation: concepts, mechanisms, methods and some applications in food technology". *Ciência Rural*, 44(2014): 1304–1311.

Thirumdas, R., Sarangapani, C., & Annapure, U. Cold plasma: A novel non-thermal technology for food processing. *Food Biophysics*, 10(2014): 1–11. 10.1007/s11483-014-9382

Trojanowska Anna, Giamberini Marta, Tsibranska Irene, Nowak Martyna, Marciniak Lukasz, Jatrzab Renata, & Tylkowski Bartosz. "Microencapsulation in food chemistry". *Journal of Membrane Science & Research*, 3(2017): 265–271. 10.22079/jmsr.2017.23652

Ünver Ahmet. "Applications of ultrasound in food processing". *Green Chemistry & Technology Letters*, 2-3(2016): 121–126.

Wang Jingzhong, Li Mengru, He Yunhu, Li Hong, Xiao Ke, & Wang Chao. "A blockchain based privacy-preserving incentive mechanism in crowdsensing applications". *IEEE Access*, 6(2018): 17545–17556. 10.1109/ACCESS.2018.2805837

Xu Jie, Guo Shuang, Xie David, & Yan Yan. "Blockchain: A new safeguard for agri-foods". *Artificial Intelligence in Agriculture*, 4(2021): 153–161. 10.1016/j.aiia.2020.08.002

Yang Xinting, Li Mengqi, Yu Huanjung, Wang Mingting, Xu Daming, & Sun Chuanheng. "A trusted blockchain-based traceability system for fruit and vegetable agricultural products". *IEEE Access*, 9(2021): 36282–36293. 10.1109/ACCESS.2021.3062845.

Zarrin Javad, Wen Phang Hao, Babu Saheer Lakshmi, & Zarrin Bahram. "Blockchain for decentralization of internet: prospects, trends, and challenges". *Cluster Computing*, 24(2021): 2841–2866. 10.1007/s10586-021-03301-8

Zhang Junjei, Huang Huang, Song Guangchun, Huang Kunlun, Luo Yunbo, Liu Qingliang, He Xiaoyun, & Cheng Nan. "Intelligent biosensing strategies for rapid detection in food safety: A review". *Biosensors and Bioelectronics*, 202(2022): 114003. 10.1016/j.bios.2022.114003

Zheng Zibin, Xie Shaoan, Dai Hongning, Chen Xiangping, & Wang Huaimin. "An overview of blockchain technology: architecture, consensus, and future trends". *IEEE 6th International Congress on Big Data, (BigData Congress) Honolulu* (2017). 10.1109/BigDataCongress.2017.85

15 Food Allergens

A Potential Health Hazard and Its Management

Deepshikha Buragohain, Mrinal Samtiya, Tejpal Dhewa, and Sanjeev Kumar

15.1 INTRODUCTION

A food allergen reacts with IgE antibodies, results in allergic sensitization, or produces allergic reactions, and a food allergy is an immune system reaction that happens soon after consuming a specific food (Aalberse, 1997). Any amount of allergenic food can result in stomach problems, hives, or enlarged airways. Food allergies (FAs), a serious public health concern affecting children and adults, have been more common over the past two to three decades. As a result, patients and their families must maintain constant vigilance, which is frequently unpleasant (Seth *et al.*, 2020). Food allergies are significantly influenced by the microbiota's makeup as well as lifestyle factors, including diet and maternal-fetal interactions (Anagnostou, 2018). The clinical manifestation of food allergy or hypersensitivity is typically the result of a complex interaction involving ingested food antigens, the digestive system, tissue mast cells, circulating basophils, and food antigen-specific IgE (Sicherer, 2011). Some persons with a food allergy may experience severe symptoms or even anaphylaxis, a potentially fatal reaction. An unwelcome immune reaction to dietary proteins causes a food allergy, which can cause a variety of symptoms (Tordesillas *et al.*, 2017). In a meta-analysis focusing on allergies to milk, eggs, peanuts, and shellfish, we discovered that 3.5% of people have food allergies. A recent study analysed numerous national health databases and health surveys; 3.9% of American children had a food allergy, and the frequency rose between 1997 and 2007, and it went up by 18%. Studies in the United States and the United Kingdom, in particular, have revealed that the prevalence of peanut allergy has risen to more than 1%, with the number of affected children has doubled over the past ten years (Luo *et al.*, 2022). It is simple to confuse a food allergy with an intolerance, a more typical reaction. Even though food intolerance doesn't affect the immune system, it is nevertheless a significant disorder that irritates (Valenta *et al.*, 2015). Various factors can cause adverse reactions to food, but a specific immunological reaction causes a food allergy. It is difficult to estimate the frequency of food allergies due to misclassification, skewed involvement, a lack of simple diagnostic testing, the rapid evolution of the disease, the large number of potential triggers, and the wide range of clinical presentations (Jones and Burks, 2017). IgE-associated food allergies affect about 3% of people and are prevalent not just in the gastrointestinal tract but also in other organ systems, impacting patients' day-to-day lives (Taylor *et al.*, 2007). Typically, symptoms of a food allergy start to manifest two hours or more after ingesting the allergen. Rarely, symptoms can not appear for several hours. The development and management of food allergies depend heavily on diet and nutrition. Diets during pregnancy may affect the probability that children will develop allergies. A mother's diet rich in fruits, vegetables, seafood, and foods strong in vitamin D has been related to a lower prevalence of allergic disease in their offspring, among other potentially aggravating factors. Even more surprisingly, the consumption of milk and butter has been shown to have a protective effect, especially in a farm environment (Neerven and Savelkoul, 2017). The infant's diet can be very important, not only for nursing but also for the variety of the diet, the timing of supplemental food introduction, and the effects of particular foods. More concrete proof of the value of food in prevention comes from newborn feeding practices that may maintain gut health by consuming considerable amounts of home-processed fruits and vegetables (Piccorossi *et al.*, 2020). Additionally, it has been proven that consuming fish within the first year of life is beneficial. The importance of dietary challenges in children and those with food allergies has been widely acknowledged in recent years. The most common foods that cause allergies in infants and children are milk, eggs, wheat, and soy. It may be challenging to avoid them nutritionally because milk, egg, wheat, and soy are the primary foods that trigger allergies in young children and newborns (Matthai *et al.*, 2020). As a result, avoidance might slow a child's growth who has a food allergy, especially before a diagnosis, when foods may be eliminated without seeing a nutritionist. Although avoiding the offending item remains the foundation of treatment for a food allergy, it is now understood that doing so can put children in particular, at nutritional risk if the rest of the diet is not closely regulated (Vandenplas, 2017). Many people will still avoid a variety of foods out of fear, even though it is generally considered that adults with food allergies do not need nutritional counseling. A nutrition assessment should be performed

DOI: 10.1201/9781003334859-15

on all individuals with food allergies, regardless of age, to ensure that they are consuming a healthy, balanced diet and are not purposely avoiding any food groups (Mennini *et al.*, 2020).

15.2 THE TYPICAL SIGNS AND SYMPTOMS OF A FOOD ALLERGY

The body's immune system defends against infections and other health risks. A food allergy reaction occurs when the immune system overreacts to a food or ingredient, interpreting it as a threat and mounting a defense mechanism (Hassan and Venkatesh, 2015). Even though food allergies typically run in families, it is difficult to predict if a child will inherit a parent's allergy or whether siblings will share the same sickness (Tuck *et al.*, 2019). It is possible for younger siblings of a child with a peanut allergy also to have the condition. Symptoms from food allergies can range from minor to serious. Anaphylaxis, a potentially fatal whole-body allergic reaction that can affect heart rate, blood pressure, and respiration, is the most severe allergic reaction. Anaphylaxis can manifest within minutes of exposure to the triggered food. It must be treated immediately with an epinephrine injection since it could be lethal (Gargano *et al.*, 2021). Hives, itching, tingling in the mouth, or eczema; wheezing, breathing issues, or nasal congestion; in addition to a swelling of the lips, face, tongue, throat, and other areas of the body vomiting, nausea, or diarrhea in the abdomen; fainting, feeling lightheaded, or dizzy are all symptoms. Currently, medical management focuses on the following: having an epinephrine auto-injector on hand, using it as soon as anaphylactic symptoms or signs develop, and being assessed right away at an emergency center for monitoring (Gendel, 2012).

15.3 FOOD ALLERGEN AND PUBLIC SAFETY

Food allergies are a significant threat to public health and safety. Food allergies are considered to be a serious public health concern that significantly lowers the quality of life for people who are allergic to or sensitive to particular foods (Gier and Verhoeckx, 2018). One of the main areas of food safety management in terms of public health (FSMSs) is food allergen management (FAM), which is required by standardized food safety management systems (Li *et al.*, 2018). In order to protect community members who have food allergies, food allergies must be handled properly. Food allergen labeling is critical for reducing exposure risk and preventing anaphylaxis for those with food allergies (Lee *et al.*, 2017). As the processed food industry expanded and the volume of international trade in food products increased, there was a greater demand for national and international regulatory organizations. It became evident that there is an urgent need to regulate food allergies (Bawa and Anilakumar, 2013). The safety of allergic consumers was discovered to depend on diets that eliminate certain allergens. The World Health Organization (WHO) and the Food and Agriculture Organization (FAO) recognized the relevance of food allergy as a worldwide issue for food regulation and sought expert assistance to identify which foods should always be labeled on food labels due to their allergenic qualities (Cardona *et al.*, 2020). A few countries and regulatory agencies have acknowledged the importance of providing this information, and regulatory risk management plans for consumers with allergies have focused on disclosing the existence of food allergens on labels and passing laws, regulations, or guidelines for the labeling of "priority allergens" in food products (Ortiz-Menéndez *et al.*, 2021). Individual susceptibility, food allergen type, processing effects on foods containing allergens, food ingredients, and their interactions are just a few of the variables that affect how clinical reactions to allergenic proteins are affected. In vulnerable individuals, side effects could range from minor to severe (such as anaphylaxis). Reactions may occasionally be lethal. According to estimates, children's food allergies cost the U.S. economy $24.8 billion annually. Food allergies' effects on the global economy have not been properly studied dates back years, but it wasn't until the 1990s that it was recognized as a food safety issue due to the rise in allergic disease cases. Consumer information laws that required the disclosure of allergen information were initially implemented in developed economies, but they now cover more than a third of the world's population (De Martinis *et al.*, 2020).

15.4 COMMON FOOD ALLERGENS

Food can cause an allergic reaction, accounting for more than 90% of all food allergic reactions. Out of the more than 170 foods known to cause allergic reactions, the top eight allergens are milk, eggs, soya beans, peanuts, tree nuts such as almonds, walnuts, pecans, cashews, Brazil nuts, hazelnuts, pistachios, and hickory nuts; seafood such as saltwater and freshwater finfish); crustaceans such as shrimp, prawns, crab, lobster, and crayfish; and molluscs such as snails, octopus, and clams (Matsuo *et al.*, 2015). Immunoglobulin E (IgE) antibodies that are specific to allergens are crucial in the emergence of food allergies. The effective analysis of food allergies has been made possible by recent developments in biological techniques. As a result, numerous food allergens have been discovered, along with their molecular makeup and IgE-binding epitopes (Bonneau, 1997).

In decreasing order of frequency, the most frequent causative foods are milk, wheat, eggs, fish, soybeans, tree nuts, shellfish, and peanuts. Despite these variations, there is a global tendency for the prevalence of food allergies to rise. Cow's milk has physicochemical characteristics. Milk is fractionated into two parts: casein and whey (Giannetti *et al.*, 2021). The allergens glutenins and gliadins can still cause baker's bronchial allergies and, in a few instances, an allergic reaction to wheat (Ricci *et al.*, 2019). An estimated 0.2%–0.3% and 0.6% of people are allergic to seafood. Allergies to shellfish, including shrimp, prawns, lobster, crab, cephalopods (squid, octopus), and various types of crustaceans, are most common in school-age children and adults. Shrimp allergen research is the most developed for shellfish allergy. The four edible shrimp most frequently consumed in Japan are crayfish, black tiger shrimp (*Panulirus japonicas*), Kuruma shrimp (*Marsupenaeus japonicus*), and white Pacific shrimp (*Litopenaeus vannamei*). Research on the allergens that cause allergies to shellfish is equally considerable. As allergens, myosin light chain, troponin C, tropomyosin, arginine kinase, and triose phosphate isomerase have all been found (Wai *et al.*, 2021). Allergic symptoms are primarily brought on by eating foods that cause plant food allergies, such as pollen food allergy syndrome and latex fruit syndrome. Pathogenic proteins such as profilins, seed storage proteins, and non-specific lipid transfer proteins are the allergen components of plant foods that have been most thoroughly studied (LTPs). These substances, which are collectively referred to as pan allergens, are found in large quantities in a wide range of plants and exhibit cross-reactivity with related plant species. Additionally, IgE-binding motifs for carbohydrates have been described as having a cross-reactivity with a number of plant allergens (Esser *et al.*, 2019).

15.5 ALLERGENS AND ANAPHYLACTIC SHOCK

Anaphylaxis happens when allergic swelling gets bad and obstructs the throat's ability to breathe. Blood pressure lowers and the pulse may become weak or weaker during anaphylaxis. If the enlargement prevents airflow long enough, the person may lose consciousness and die (Simons, 2010). The Gell-Combs classification classifies anaphylaxis as type 1 hypersensitivity or IgE-mediated hypersensitivity (Figure 15.1). A new, expanded definition of anaphylaxis refers to it as a serious, systemic hypersensitivity reaction that is life-threatening (Justiz Vaillant *et al.*, 2022). The reactions that were previously categorized as "anaphylactic reactions" are included in the current definition of anaphylaxis (Anagnostou, 2018). Only when something is eaten do food sensitivity symptoms manifest or are "triggered," or if the symptoms go away after coming into touch with sensitized food

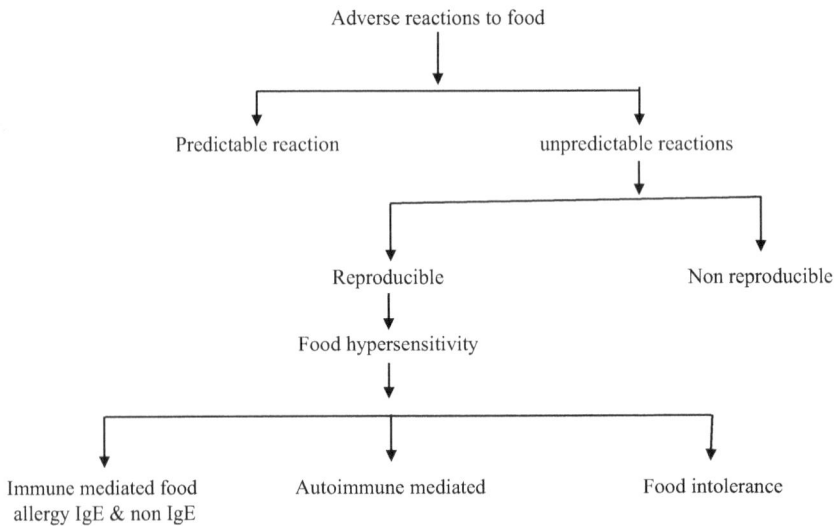

Figure 15.1 Specific dietary proteins trigger the development of the immune response. Since sensitization (production of allergen-specific IgE, sIgE) is conceivable even when there are no outward signs of an allergic reaction, sensitization and the induction of an unpleasant reaction after exposure to an allergen constitute the definition of food allergy.

(Cardona *et al.*, 2020). When a person stops eating, they may experience the initial subjective symptoms of itching and numbness, as well as a combination and series of aggravating objective symptoms like urticarial and eczema, swelling (of the lips, face, tongue, and throat), wheezing, chest tightness, stuffy nose, or difficulty breathing, as well as abdominal pain, diarrhea, nausea, or vomiting; chest pain; and dizziness/fainting. Anaphylaxis is a rare side effect of breathing difficulties and significant blood loss that could be fatal due to airway narrowing (Mills and Mackie, 2008).

15.6 MECHANISM OF FOOD ALLERGY

Type I hypersensitivity is the major mechanism of IgE-mediated food allergy. By comprehending the fundamental immune mechanisms, medications and other interventions can be targeted more effectively to prevent and decrease the effects of food allergies (Florsheim *et al.*, 2021). The primary coordinator of antibody production is cellular fingerprinting, which has been connected to peanut allergies, tolerance in children with IgE sensitivities, and immunological profiles of babies using mass cytometry. In contrast to sensitized and unsensitized individuals, who, respectively, experienced Th2 and Th1 altered peanut responses, patients with allergies and those without allergies have steady T regulatory responses. In contrast to unaffected controls, peanut-allergic individuals with disparity among effector and regulatory T-cells had less functional Tr1 cells. More responsive individuals had varied and diverged Th2 reactions (Justiz Vaillant *et al.*, 2022). Further research needs to focus on the function of antibodies in the allergy and understand the relationship between allergen-specific IgE and clinical response to food. For instance, the germinal center has been found to include a new subtype of T follicular helper cells (Tfh13). The Tfh13 cells stand out for their BCL6 and GATA3 transcription factor profiles and their capacity to produce IL4 and IL13. Tfh13 causes specific IgE to be produced, which can cause anaphylaxis. IgG and IgE rely on germinal centers and Tfh cells for expansion. However, IgA follows a different track by involving T-cells and CD40 ligands. It's interesting to note that research has indicated that somatic hyper-mutation and class switch recombination from IgG to IgE can occur in the guts of people who have peanut allergies. This emphasises how important gut-associated lymphoid tissue is for food allergy (Yu *et al.*, 2016). For overall health, it is essential that the immune system be able to distinguish between dangerous environmental antigens and beneficial pathogenic antigens. As a result, those who are not allergic to foods (i.e., those who are healthy or immune-tolerant) have an insensitivity to common food antigens. People with allergies to common foods experience inappropriate inflammatory immune reactions when they become sensitized to the allergens (Yu *et al.*, 2016). The allergens that cause food allergies are typically certain dietary proteins. IgE antibodies bind to the effector cells, mast cells in tissues, and basophils in blood. The cell membrane will be impaired because the food allergen will bind to mast cells or basophil-bound IgE and crosslink with other IgE antibodies upon repeated exposure (Kanagaratham *et al.*, 2020). This results in the release of histamine, neutral protease, and proteoglycan into the environment and the induction of typical allergy symptoms. Figure 15.2 depicts the primary pathway of allergy symptoms. A food allergy develops when the immune system overreacts to an allergen in food. Both the first phase of sensitization to a specific antigen (A) and the production of an allergic reaction after a later exposure to the same antigen are necessary for this occurrence (B).

15.7 LAWS AND REGULATIONS OF FOOD ALLERGENS

Different countries have different laws and rules regarding food allergies. There are stringent restrictions in some nations, the United States, requiring food makers to prominently label food containing known allergies (Roses, 2011). Other nations, Canada, allow for the voluntary labeling of food allergens. It is crucial for people with food allergies to be aware of any potential allergens present in the foods they eat, regardless of the rules in place. Millions of customers with food allergies now have better access to food label information thanks to the Food Allergen Labeling and Consumer Protection Act (FALCPA). Children who need to learn to identify allergens that they should avoid will benefit the most from this law (Shaker, 2017). An exhaustive study of regulatory databases, agency and governmental websites, literature citations, and references to other regulatory documents led to the discovery of the laws, directives, rules, regulations, and agency statements pertaining to food allergen labeling. Because the precise legal status of each relies on the type of government structure, these are referred to as regulatory frameworks. Standards set by the Codex are compared to country regulatory systems (Buhl *et al.*, 2008). Original sources (or direct translations of primary materials) have been incorporated into this analysis whenever possible. However, there have been instances where secondary sources were used instead of primary sources

(A)

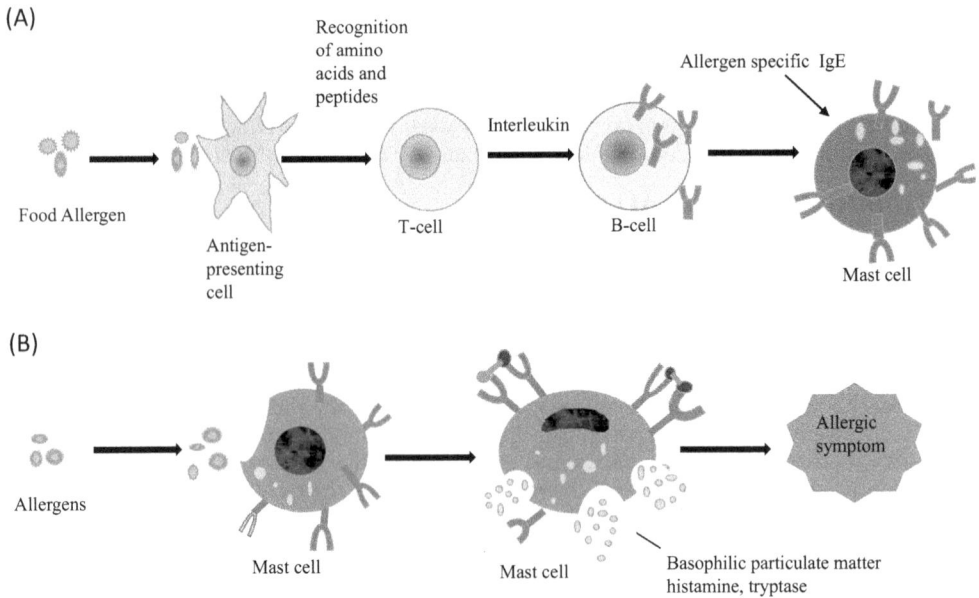

(B)

Figure 15.2 Mechanism of food allergy. (A) Food allergens are inducing dendritic cells and in turn activating B cells for antibody secretion and (B) the induction of an allergic response following subsequent exposure to the same antigen.

because they were unavailable (Olgin *et al.*, 2020). Table 15.1 provides a list of these supplementary sources. It was frequently impossible to review past data or illuminating documentation because it was only ever available in the most recent version. In total, 6 of the 19 recognized regulatory frameworks directly relate to or utilize Codex standards, while three directly refer to or use EU standards (Table 15.2). There is no need to explain these frameworks further. Worldwide regulatory frameworks for food allergy labeling differ significantly. Each jurisdiction identifies different priority allergens, and it is frequently unclear what standards were utilized to develop these priority allergen lists (Murdoch *et al.*, 2018).

15.8 PREVENTION AND PRECAUTION

The list of the most significant allergens and derived compounds mentioned in European Union rules is a crucial step. However, concealed allergens can manifest as conventional allergy symptoms. These ingredients either naturally occur in food or are added to it in the form of compound products, whose names rather than particular compositions are listed on labels (Turnbull *et al.*, 2015). Numerous foods can have allergies, and a single item may contain several allergens. Both naturally occurring and artificially introduced food components have allergenic characteristics. Similar to natural monosodium glutamate found in many foods like tomatoes, mushrooms, corn, peas, parmesan cheese, hydrolyzed vegetable proteins, yeast extracts, and caseinates, monosodium glutamate (Additive E 621) exhibits allergic qualities (Comberiati *et al.*, 2019). The majority of these research, including those on formulas with hydrolyzed proteins, indicated little to no benefit in avoiding allergy. Recent years have seen an unparalleled increase in interest in the challenging topic of allergies due to the rising prevalence of the disorders that underlie the phenomenon of allergic food sensitivity and improvements in knowledge in new fields, including immunology, molecular biology, and genetics. Avoiding food items that trigger allergic reactions is the cornerstone of treating food allergies. Thus, it is anticipated that this understanding will grow in the years to come (Ogulur *et al.*, 2021). This necessitates dedication to carefully selecting nutritional items and carefully reading the labeled information on product composition. The demands of allergy sufferers are taken into account on modern food labels. It is a crucial step to name the most significant allergens and derived components stated in the European Union Directives. However, hidden allergens can also be the source of common allergy symptoms (Herman *et al.*, 2020). These components are either found naturally in food or are added as compound products. Both naturally occurring and artificially introduced food

Table 15.1: Estimated Rates of Food Allergies with Different Foods Widely Used

Prevalence	Active Allergens	Infant/Child	Adult	Symptoms	Detection	Treatment	Reference
Milk	αs1-, αs2-, β-, and κ-casein from casein proteins and α-lactalbumin and β-lactoglobulin	2–3%	0.5%	Itchiness, rash, eczema, nausea, vomiting.	Medical history and standard allergy tests.	Epinephrine, Antihistamine (mild)	Muraro et al., 2014
Egg	Ovomucoid (Gal d 1), ovalbumin (Gal d 2), ovotransferrin (Gal d 3), lysozyme (Gal d 4)	1.8–2%	1.2–1.5%	Itchiness, rash, swelling of lips, tongue, or the whole face, wheezing, shortness of breath.	Medical history and standard allergy tests	Epinephrine, Antihistamine	Fineman et al., 2014
Wheat	Gliadin and glutenin	2–3%	1%	Swelling, itching, or irritation of the mouth, congestion.	Skin test, blood test, elimination diet.	Epinephrine, Antihistamine	Ogino et al., 2021
Seafood	Gelatin hydolysates, tropoyosin	3.5–5%	2–3%	Itchiness, rash, nausea, vomiting.	Skin prick test, blood test.	Epinephrine, avoid food.	Bonlokke et al., 2019
Peanut	Ara h 1,Ara h 2, Ara h 3	0.6%	1.5–3%	Itchiness, rash, eczema, nausea.	Allergy test, blood test.	Epinephrine, Anthistamines, avoid food	(Chen, Welch, and Laubach 2018)

Table 15.2: FALCPA Applies to Products Regulated by Food and Drug Administration in a Different Country

Jurisdiction	Source
United States	Food Allergen Labeling and Consumer Protection Act of 2004
Japan	Standards and Evaluation Division, Department of Food Sanitation, Pharmaceutical and Medical Safety Bureau, Ministry of Health, Labour and Welfare; FAQs on Labelling System for Foods Containing Allergens
Switzerland	Standard for Labelling of Pre-Packaged Foods, SVGNS 1 Part 3: 2000 (Rev. 2009) Ordinance of the Federal Department of Home Affairs on Food Labelling and Advertising of Foods (Information from WTO Notification G/TBT/N/CHE/106)
Korea	Korean Food and Drug Administration (Information from USDA ARS GAIN Report KS1102)
Philippines	Rules and Regulations Governing the Labelling of Pre-packaged Foods Products Distributed in the Philippines (Amending Administrative Order No. 88-B s. 1984) (Information from WTO Notification G/TBT/N/PHL/128)
China	General Rules for the Labelling of Pre-packaged Foods GB7718–2011 (Information from USDA GAIN Report CH110030)
Australia	Australia New Zealand Standard 1.2.3 – Mandatory Warning and Advisory Statements and Declarations
Canada	Regulations Amending the Food and Drug Regulations (1220 – Enhanced Labelling for Food Allergen and Gluten Sources and Added Sulfites) – Canada Gazette 145 (4) February 16, 2011

components have allergenic qualities. The characteristics of naturally occurring monosodium glutamate, which is found in many foods including tomatoes, mushrooms, corn, peas, parmesan cheese, hydrolyzed vegetable proteins, yeast extracts, and caseinates, are shared by monosodium glutamate (Additive E 621) (Hefle, 2001). Monosodium glutamate allergy symptoms are dose-dependent. Monosodium glutamate occurs naturally in several foods; thus, there should be no restrictions on consuming those foods. However, foods containing the additive E 621 should be avoided altogether. Interest in the challenging topic of allergies has never been higher due to the rising patient population and advancements in new domains, including immunology, molecular biology, and genetics (Skypala and McKenzie, 2019).

15.9 TREATMENT

The primary goal of treating food allergies is to completely remove the allergen that is the cause. When newborns are exclusively breastfed, they are more sensitive to allergens through breast milk, making maternal elimination diets successful. Supplements should not contain any food allergies (Heine, 2018). For infants who are formula-fed and have a cow's milk allergy (CMA), specialized hypoallergenic formulas are the best option. Although most babies (and those who need it) will be able to utilize a hypoallergenic formula without experiencing an allergic reaction, it is often derived from cow's milk. Parents should be aware that the American Academy of Pediatrics recommends breastfeeding as a child's first feeding option. It is also advised that the mother's diet be changed prior to introducing these formulas to infants who are at risk for allergies or who display intolerance symptoms (Aitoro *et al.*, 2017). A hypoallergenic elimination diet should be closely watched for proper nutrition. Despite efforts to systematically remove troublesome food allergies from the diet, unintended responses are rather prevalent. The labeling of preventative allergens is frequently still unclear or lacking. Thankfully, numerous nations have passed laws required for accurate allergen labeling (Anvari *et al.*, 2019). To treat food allergies, allergens must be eliminated from the diet. It has been proven that maternal elimination diets are beneficial for babies who are exclusively breastfed and react to allergies through breast milk. The offending item must not be included in the supplementary diet (Heine, 2018). The ideal method of care for formula-fed infants who are allergic to cow's milk is specialized hypoallergenic formulas. These therapeutic formulas fall into two main categories: EHF and formulae based on amino acids. It's important to regularly check nutritional sufficiency with hypoallergenic elimination diets. Despite efforts to completely eliminate aggravated food allergens from the diet, unintentional reactions are nonetheless rather common. The potential for unintended allergic reactions and anaphylaxis significantly influences patients and associated family members and society (Costa *et al.*, 2020).

15.10 CONCLUSION

Food hypersensitivity is becoming increasingly important to public health. It may be compulsory for the food industry to ensure risk management plans and allergy labeling. Consumers receive regular and clear risk advice from the food industry and regulatory enforcement bodies. This would necessitate hazard control procedures for all of the many stakeholders in the food chain. It is strongly advised to consistently classify foods according to their allergy danger status and quantitative reference doses. For effective food allergy prevention and treatment strategies, immediate attention is needed. Developing specific allergens T reg cells could be promising therapy in the coming time. The management of food allergies is still an active topic of research due to the significant progress that has been made and the availability of cutting-edge diagnostic and therapeutic approaches.

REFERENCES

Aalberse, R. C. 1997. "Food Allergens." *Environmental Toxicology and Pharmacology* 4 (1–2): 55–60. 10.1016/s1382-6689(97)10042-4.

Aitoro, Rosita, Lorella Paparo, Antonio Amoroso, Margherita Di Costanzo, Linda Cosenza, Viviana Granata, Carmen Di Scala, et al. 2017. "Gut Microbiota as a Target for Preventive and Therapeutic Intervention against Food Allergy." *Nutrients* 9 (7): 672. 10.3390/nu9070672.

Anagnostou, Katherine. 2018. "Anaphylaxis in Children: Epidemiology, Risk Factors and Management." *Current Pediatric Reviews* 14 (3): 180–186. 10.2174/1573396314666180507115115.

Anvari, Sara, Jennifer Miller, Chih-Yin Yeh, and Carla M. Davis. 2019. "IgE-Mediated Food Allergy." *Clinical Reviews in Allergy & Immunology* 57 (2): 244–260. 10.1007/s12016-018-8710-3.

Bawa, A. S., and K. R. Anilakumar. 2013. "Genetically Modified Foods: Safety, Risks and Public Concerns-A Review." *Journal of Food Science and Technology* 50 (6): 1035–1046. 10.1007/s13197-012-0899-1.

Bonlokke, Jakob H., Bang, Berit, Aasmoe, Lisbeth, Rahman, Anas M. Abdel, Syron, Laura N., Andersson, Eva, Dahlman-Höglund, Anna, Lopata, Andreas L., & Jeebhay, Mohamed (2019). Exposures and Health Effects of Bioaerosols in Seafood Processing Workers - a Position Statement. *Journal of Agromedicine* 24: 441–448. 10.1080/1059924x.2019.1646685

Bonneau, J. C. 1997. "[Food Allergens]." *Allergie Et Immunologie* 29 Spec No (July): 21–24.

Buhl, Timo, Hubert Kampmann, Jose Martinez, and Thomas Fuchs. 2008. "The European Labelling Law for Foodstuffs Contains Life-Threatening Exemptions for Food-Allergic Consumers." *International Archives of Allergy and Immunology* 146 (4): 334–337. 10.1159/000121467.

Cardona, Victoria, Ignacio J. Ansotegui, Motohiro Ebisawa, Yehia El-Gamal, Montserrat Fernandez Rivas, Stanley Fineman, Mario Geller, et al. 2020. "World Allergy Organization Anaphylaxis Guidance 2020." *The World Allergy Organization Journal* 13 (10): 100472. 10.1016/j.waojou.2020.100472.

Chen, Meng, Welch, Michael, & Laubach, Susan (2018). Preventing Peanut Allergy. *Pediatric Allergy, Immunology, and Pulmonology*, 31: 2–8. 10.1089/ped.2017.0826

Comberiati, Pasquale, Giorgio Costagliola, Sofia D'Elios, and Diego Peroni. 2019. "Prevention of Food Allergy: The Significance of Early Introduction." *Medicina (Kaunas, Lithuania)* 55 (7): 323. 10.3390/medicina55070323.

Costa, Célia, Alice Coimbra, Artur Vítor, Rita Aguiar, Ana Luísa Ferreira, and Ana Todo-Bom. 2020. "Food Allergy-From Food Avoidance to Active Treatment." *Scandinavian Journal of Immunology* 91 (1): e12824. 10.1111/sji.12824.

De Martinis, Massimo, Maria Maddalena Sirufo, Mariano Suppa, and Lia Ginaldi. 2020. "New Perspectives in Food Allergy." *International Journal of Molecular Sciences* 21 (4): 1474. 10.3390/ijms21 041474.

Esser, Philipp R., Sabine Mueller, and Stefan F. Martin. 2019. "Plant Allergen-Induced Contact Dermatitis." *Planta Medica* 85 (7): 528–534. 10.1055/a-0873-1494.

Fineman, Stanley M. (2014). Optimal Treatment of Anaphylaxis: Antihistamines Versus Epinephrine. *Postgraduate Medicine* 126: 73–8110.3810/pgm.2014.07.2785.

Florsheim, Esther B., Zuri A. Sullivan, William Khoury-Hanold, and Ruslan Medzhitov. 2021. "Food Allergy as a Biological Food Quality Control System." *Cell* 184 (6): 1440–1454. 10.1016/j.cell. 2020.12.007.

Gargano, Domenico, Ramapraba Appanna, Antonella Santonicola, Fabio De Bartolomeis, Cristiana Stellato, Antonella Cianferoni, Vincenzo Casolaro, and Paola Iovino. 2021. "Food Allergy and Intolerance: A Narrative Review on Nutritional Concerns." *Nutrients* 13 (5): 1638. 10.3390/nu13051638.

Gendel, Steven M. 2012. "Comparison of International Food Allergen Labeling Regulations." *Regulatory Toxicology and Pharmacology: RTP* 63 (2): 279–285. 10.1016/j.yrtph.2012.04.007.

Giannetti, Arianna, Gaia Toschi Vespasiani, Giampaolo Ricci, Angela Miniaci, Emanuela di Palmo, and Andrea Pession. 2021. "Cow's Milk Protein Allergy as a Model of Food Allergies." *Nutrients* 13 (5): 1525. 10.3390/nu13051525.

Gier, Steffie de, and Kitty Verhoeckx. 2018. "Insect (Food) Allergy and Allergens." *Molecular Immunology* 100 (August): 82–106. 10.1016/j.molimm.2018.03.015.

Hassan, A. K. G., and Y. P. Venkatesh. 2015. "An Overview of Fruit Allergy and the Causative Allergens." *European Annals of Allergy and Clinical Immunology* 47 (6): 180–187.

Hefle, S. L. 2001. "Hidden Food Allergens." *Current Opinion in Allergy and Clinical Immunology* 1 (3): 269–271. 10.1097/01.all.0000011025.21898.04.

Heine, Ralf G. 2018. "Food Allergy Prevention and Treatment by Targeted Nutrition." *Annals of Nutrition & Metabolism* 72 (Suppl 3): 33–45. 10.1159/000487380.

Herman, Rod A., Jason M. Roper, and John X. Q. Zhang. 2020. "Evidence Runs Contrary to Digestive Stability Predicting Protein Allergenicity." *Transgenic Research* 29 (1): 105–107. 10.1007/s11248-019-00182-x.

Jones, Stacie M., and A. Wesley Burks. 2017. "Food Allergy." *The New England Journal of Medicine* 377 (12): 1168–1176. 10.1056/NEJMcp1611971.

Justiz Vaillant, Angel A., Rishik Vashisht, and Patrick M. Zito. 2022. "Immediate Hypersensitivity Reactions." In *StatPearls*. Treasure Island, FL: StatPearls Publishing. http://www.ncbi.nlm.nih.gov/books/NBK513315/.

Kanagaratham, Cynthia, Yasmeen S. El Ansari, Owen L. Lewis, and Hans C. Oettgen. 2020. "IgE and IgG Antibodies as Regulators of Mast Cell and Basophil Functions in Food Allergy." *Frontiers in Immunology* 11: 603050. 10.3389/fimmu.2020.603050.

Lee, T. H., H. K. Ho, and T. F. Leung. 2017. "Genetically Modified Foods and Allergy." *Hong Kong Medical Journal = Xianggang Yi Xue Za Zhi* 23 (3): 291–295. 10.12809/hkmj166189.

Li, Yahong, Bing Zhou, Guoqiang Zheng, Xianhu Liu, Tingxi Li, Chao Yan, Chuanbing Cheng, et al. 2018. "Continuously Prepared Highly Conductive and Stretchable SWNT/MWNT Synergistically Composited Electrospun Thermoplastic Polyurethane Yarns for Wearable Sensing." *Journal of Materials Chemistry C* 6 (9): 2258–2269. 10.1039/C7TC04959E.

Luo, Jiangzuo, Qiuyu Zhang, Yanjun Gu, Junjuan Wang, Guirong Liu, Tao He, and Huilian Che. 2022. "Meta-Analysis: Prevalence of Food Allergy and Food Allergens - China, 2000-2021." *China CDC Weekly* 4 (34): 766–770. 10.46234/ccdcw2022.162.

Matsuo, Hiroaki, Tomoharu Yokooji, and Takanori Taogoshi. 2015. "Common Food Allergens and Their IgE-Binding Epitopes." *Allergology International: Official Journal of the Japanese Society of Allergology* 64 (4): 332–343. 10.1016/j.alit.2015.06.009.

Matthai, John, Malathi Sathiasekharan, Ujjal Poddar, Anupam Sibal, Anshu Srivastava, Yogesh Waikar, Rohan Malik, et al. 2020. "Guidelines on Diagnosis and Management of Cow's Milk Protein Allergy." *Indian Pediatrics* 57 (8): 723–729.

Mennini, Maurizio, Alessandro Giovanni Fiocchi, Arianna Cafarotti, Marilisa Montesano, Angela Mauro, Maria Pia Villa, and Giovanni Di Nardo. 2020. "Food Protein-Induced Allergic Proctocolitis in Infants: Literature Review and Proposal of a Management Protocol." *The World Allergy Organization Journal* 13 (10): 100471. 10.1016/j.waojou.2020.100471.

Mills, E. N. Clare, and Alan R. Mackie. 2008. "The Impact of Processing on Allergenicity of Food." *Current Opinion in Allergy and Clinical Immunology* 8 (3): 249–253. 10.1097/ACI.0b013e3282ffb123.

Muraro, A., Roberts, G., Worm, M., Bilò, M. B., Brockow, K., Fernández Rivas, M., Santos, A. F., Zolkipli, Z. Q., Bellou, A., Beyer, K., Bindslev-Jensen, C., Cardona, V., Clark, A. T., Demoly, P., Dubois, A. E. J., DunnGalvin, A., Eigenmann, P., Halken, S., Harada, L., Lack, G., Jutel, M., Niggemann, B., Ruëff, F., Timmermans, F., Vlieg-Boerstra, B. J., Werfel, T., Dhami, S., Panesar, S., Akdis, C. A., & Sheikh, A. (2014). Anaphylaxis: guidelines from the European Academy of Allergy and Clinical Immunology. *Allergy* 69: 1026–1045. 10.1111/all.12437

Murdoch, Blake, Eric M. Adams, and Timothy Caulfield. 2018. "The Law of Food Allergy and Accommodation in Canadian Schools." *Allergy, Asthma, and Clinical Immunology: Official Journal of the Canadian Society of Allergy and Clinical Immunology* 14: 67. 10.1186/s13223-018-0273-6.

Neerven, R. J. J. van, and Huub Savelkoul. 2017. "Nutrition and Allergic Diseases." *Nutrients* 9 (7): 762. 10.3390/nu9070762.

Ogino, Ryohei, Chinuki, Yuko, Yokooji, Tomoharu, Takizawa, Daigo, Matsuo, Hiroaki, & Morita, Eishin (2021). Identification of peroxidase-1 and beta-glucosidase as cross-reactive wheat allergens in grass pollen-related wheat allergy. *Allergology International* 70: 215–22210.1016/j.alit.2020.09.005.

Ogulur, Ismail, Yagiz Pat, Ozge Ardicli, Elena Barletta, Lacin Cevhertas, Ruben Fernandez-Santamaria, Mengting Huang, et al. 2021. "Advances and Highlights in Biomarkers of Allergic Diseases." *Allergy* 76 (12): 3659–3686. 10.1111/all.15089.

Olgin, Gabriella K., Annick Bórquez, Pieter Baker, Erika Clairgue, Mario Morales, Arnulfo Bañuelos, Jaime Arredondo, et al. 2020. "Preferences and Acceptability of Law Enforcement Initiated Referrals for People Who Inject Drugs: A Mixed Methods Analysis." *Substance Abuse Treatment, Prevention, and Policy* 15 (1): 75. 10.1186/s13011-020-00319-w.

Ortiz-Menéndez, Juan Carlos, Martha Cabrera, and Belén Garzón García. 2021. "Management of Food Allergens: Time to Prevent Food Allergic Reactions at School." *Pediatric Allergy and Immunology: Official Publication of the European Society of Pediatric Allergy and Immunology* 32 (5): 1106–1108. 10.1111/pai.13424.

Piccorossi, A., G. Liccioli, S. Barni, L. Sarti, M. Giovannini, A. Verrotti, E. Novembre, and F. Mori. 2020. "Epidemiology and Drug Allergy Results in Children Investigated in Allergy Unit of a Tertiary-Care Paediatric Hospital Setting." *Italian Journal of Pediatrics* 46 (1): 5. 10.1186/s13052-019-0753-4.

Ricci, Giampaolo, Laura Andreozzi, Francesca Cipriani, Arianna Giannetti, Marcella Gallucci, and Carlo Caffarelli. 2019. "Wheat Allergy in Children: A Comprehensive Update." *Medicina (Kaunas, Lithuania)* 55 (7): 400. 10.3390/medicina55070400.

Roses, Jonathan B. 2011. "Food Allergen Law and the Food Allergen Labeling and Consumer Protection Act of 2004: Falling Short of True Protection for Food Allergy Sufferers." *Food and Drug Law Journal* 66 (2): 225–242, ii.

Seth, Divya, Pavadee Poowutikul, Milind Pansare, and Deepak Kamat. 2020. "Food Allergy: A Review." *Pediatric Annals* 49 (1): e50–e58. 10.3928/19382359-20191206-01.

Shaker, Dana. 2017. "An Analysis of 'Natural' Food Litigation to Build a Sesame Allergy Consumer Class Action." *Food and Drug Law Journal* 72 (1): 103–140.

Sicherer, Scott H. 2011. "Epidemiology of Food Allergy." *The Journal of Allergy and Clinical Immunology* 127 (3): 594–602. 10.1016/j.jaci.2010.11.044.

Simons, F. Estelle R. 2010. "Anaphylaxis." *The Journal of Allergy and Clinical Immunology* 125 (2 Suppl 2): S161–S181. 10.1016/j.jaci.2009.12.981.

Skypala, Isabel J., and Rebecca McKenzie. 2019. "Nutritional Issues in Food Allergy." *Clinical Reviews in Allergy & Immunology* 57 (2): 166–178. 10.1007/s12016-018-8688-x.

Taylor, Steve L., Susan L. Hefle, Kevin Farnum, Steven W. Rizk, Jupiter Yeung, Michael E. Barnett, Francis Busta, et al. 2007. "Survey and Evaluation of Pre-FALCPA Labeling Practices Used by Food Manufacturers to Address Allergen Concerns." *Comprehensive Reviews in Food Science and Food Safety* 6 (2): 36–46. 10.1111/j.1541-4337.2007.00016.x.

Tordesillas, Leticia, M. Cecilia Berin, and Hugh A. Sampson. 2017. "Immunology of Food Allergy." *Immunity* 47 (1): 32–50. 10.1016/j.immuni.2017.07.004.

Tuck, Caroline J., Jessica R. Biesiekierski, Peter Schmid-Grendelmeier, and Daniel Pohl. 2019. "Food Intolerances." *Nutrients* 11 (7): 1684. 10.3390/nu11071684.

Turnbull, J. L., H. N. Adams, and D. A. Gorard. 2015. "Review Article: The Diagnosis and Management of Food Allergy and Food Intolerances." *Alimentary Pharmacology & Therapeutics* 41 (1): 3–25. 10.1111/apt.12984.

Valenta, Rudolf, Heidrun Hochwallner, Birgit Linhart, and Sandra Pahr. 2015. "Food Allergies: The Basics." *Gastroenterology* 148 (6): 1120–1131.e4. 10.1053/j.gastro.2015.02.006.

Vandenplas, Yvan. 2017. "Prevention and Management of Cow's Milk Allergy in Non-Exclusively Breastfed Infants." *Nutrients* 9 (7): 731. 10.3390/nu9070731.

Wai, Christine Y. Y., Nicki Y. H. Leung, Agnes S. Y. Leung, Gary W. K. Wong, and Ting F. Leung. 2021. "Seafood Allergy in Asia: Geographical Specificity and Beyond." *Frontiers in Allergy* 2: 676903. 10.3389/falgy.2021.676903.

Yu, Wong, Deborah M. Hussey Freeland, and Kari C. Nadeau. 2016. "Food Allergy: Immune Mechanisms, Diagnosis and Immunotherapy." *Nature Reviews. Immunology* 16 (12): 751–765. 10.103 8/nri.2016.111.

16 Assessment of Food Contaminants in Meat and Meat Products

Md. Shofiul Azam, Shafi Ahmed, Md. Wahiduzzaman, and Md. Abir Hossain

16.1 INTRODUCTION

In developing countries, meat and meat products are the main sources of food, which are susceptible to microbial growth. As a source of proteins, essential amino acids, fats, carbohydrates, vitamins, and minerals, meat contributes to cell maintenance and repair and provides energy for daily activities (Dharma *et al.*, 2022). Foodborne infections and disease are a major international health problem with consequent economic loss and deaths. Humans consume a great deal of protein from meat, and it is also one of the most perishable sources of protein. Since fresh red meat contains all the nutrients needed for bacterial growth, it poses a high risk of food poisoning (Mustefa, 2021). Meat refers to flesh, skeletal muscle, and any associated connective tissue or fat, excluding bones and bone marrow. Protein and essential fatty acids are found in meat, as well as minerals and vitamins, but meat is easily perishable because it provides a suitable environment for microorganisms to grow. In healthy animals, muscles contain few or no microorganisms, although meat can become contaminated during slaughtering and transportation. Contamination of raw meat can occur via knives, tools, clothes, hands, and air during bleeding, handling, and processing. Food hazards can occur from contaminated meat and meat products due to the presence of biological, chemical, physical, and in particular microbial contamination (Bantawa *et al.*, 2018). Increasingly, the public is becoming aware of the public health impact of zoonotic pathogens transmitted through animal products. *Salmonella* spp., *Listeria monocytogenes*, *Staphylococcus aureus*, *Escherichia coli*, *Clostridium perfringens*, *Campylobacter jejuni*, *Yersinia enterocolitica*, and *Aeromonas hydrophila* are the most common foodborne pathogens associated with meat. It is well known that *Salmonella* species, *Listeria monocytogenes*, *Campylobacter jejuni*, and verocytotoxin producing *E. coli* O157 are among the most serious public health threats (Bantawa *et al.*, 2018).

Foodborne diseases pose a significant public health threat. They are caused by bacteria, parasites, or viruses present in food or beverages. Symptoms include gastrointestinal discomfort and life-threatening situations. In 2011, sprouted seeds tainted with *E. coli* caused an outbreak of the disease. Chicken, pigs, cattle, and poultry are all common sources of food poisoning bacteria that live in their intestines. *Salmonella* and *Campylobacter* are particularly prevalent in poultry. On the farm, animals usually become infected through contact with other infected animals or their feces. Farm workers' shoes and clothes carry germs that can be transmitted to healthy animals. In addition, contamination may occur during transportation to the abattoir or in unhygienic slaughter environments (Aymerich *et al.*, 2008).

Meat products such as sausage, beef burgers, and luncheon are becoming increasingly popular since they are highly valued for their high biological value, reasonable price, agreeable flavor, and ease of preparation (Shaltout *et al.*, 2022).

16.2 SOURCE OF CONTAMINATION OF MEAT AND MEAT PRODUCTS

16.2.1 Veterinary Drug Residues in Meat-Related Edible Tissues

The use of veterinary drugs in poultry and livestock products is widespread. Human health is seriously threatened by the abuse of veterinary drugs. There are several aspects of risk assessment for veterinary drug residues in meat products explained in this chapter, including the principles and functions of risk assessment, a summary of the veterinary drug residue risk assessment process, and an outline of qualitative and quantitative risk assessment methods (Figure 16.1). A variety of veterinary drugs are used in farm animals as therapeutic and prophylactic measures. They include a variety of compounds that can be administered through feed or water. There is also a possibility that the residues come from contaminated animal feeds. Controlling veterinary drug residues is an important measure in ensuring consumer protection due to the presence of residues and their associated harm to humans. A variety of anabolic compounds and hormones may be present in animal products as residues (Kadim, 2016).

16.2.2 Toxic Elements

Contamination of meat products and fresh meat with potentially toxic elements (PTEs) is a serious health concern around the world. In general, PTEs fall into two categories: essential and non-essential elements (toxic). Several properties make them toxic, including their bioaccumulative,

DOI: 10.1201/9781003334859-16

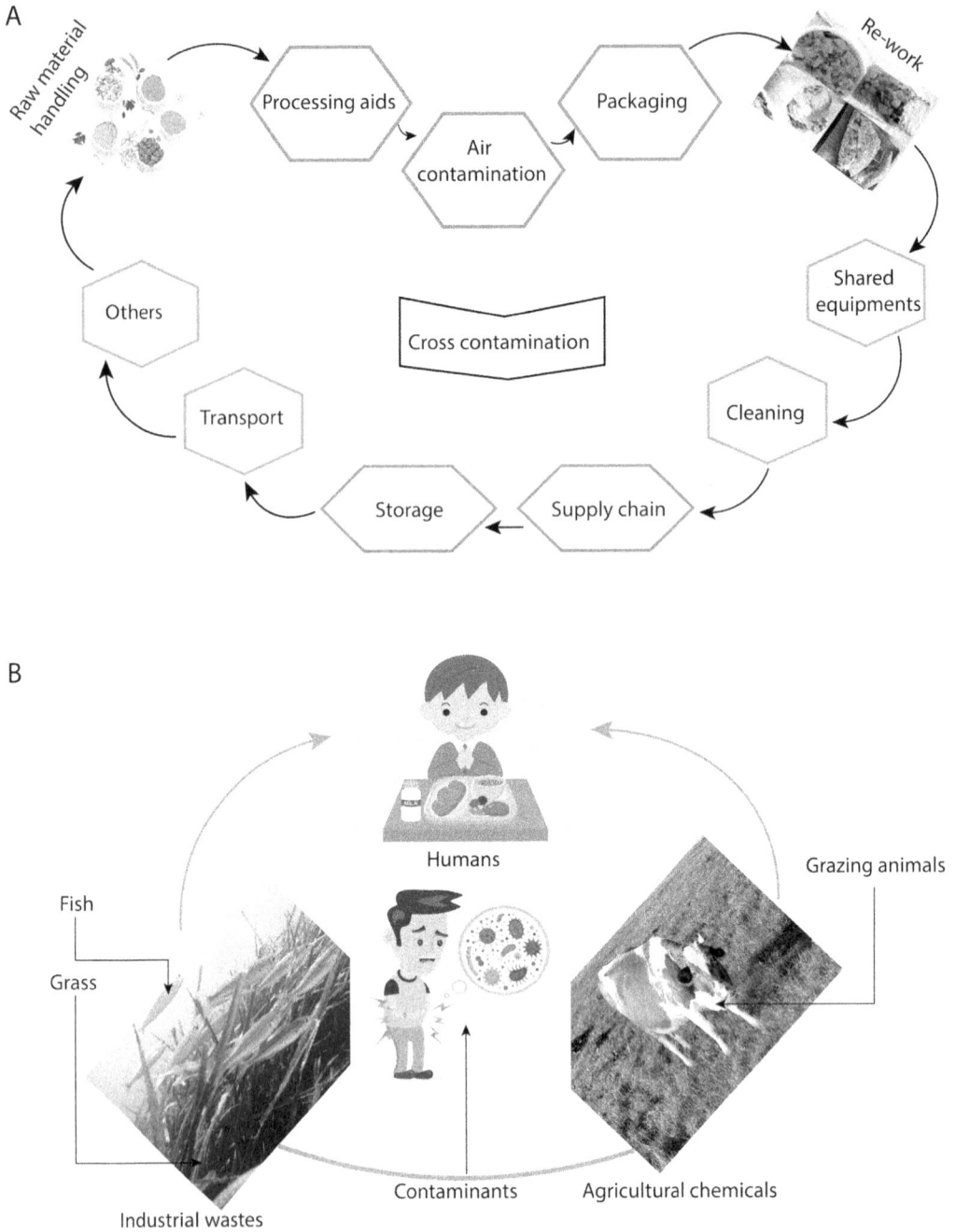

Figure 16.1 Contamination process of meat products.

biomagnifiability, non-degradability, and persistence in the environment and food chains. Exposure to PTE may cause DNA damage, apoptosis, and various types of cancer. Cadmium (Cd), lead (Pb), arsenic (As), and mercury (Hg) are the most critical heavy metals that cause health problems to several organs of humans, especially the kidney, lungs, heart, and brain, even in trace amounts when long-time exposure occurs. Moreover, Cd and Pb are linked to cardiovascular diseases, high blood pressure, and anemia (Han *et al.*, 2022). The central nervous system can also be affected by Pb, Cd, and Hg (Han *et al.*, 2022). Copper (Cu), chromium (Cr), iron (Fe), and zinc (Zn) are some of the essential elements that are necessary for human health in low concentration (Balali-Mood *et al.*, 2021). Even though they are micronutrients, these elements can become toxic at high concentrations. Cu and Zn are cofactors in several enzyme reactions and macronutrient metabolism; however, they can contribute to aging and heart diseases when consumed in high concentrations.

There have been reports of liver copper levels, muscle zinc levels, and kidney cadmium levels (in adult cattle) exceeding acceptable maximum levels adopted in some countries (Alonso *et al.*, 2002). The releasing of industrial wastes such as effluents, sludge, particles, etc., into the environment is a major contributor to heavy metal pollution. The metals, such as Cr, Ni, Pb, and As, are derived from tanneries and other industries. Metals in the environment pose a serious threat to the food chain, humans, animals, and ecosystems due to their toxicity. Consumer health, feed, and food safety are the main emerging issues in Bangladesh (Hossain *et al.*, 2022). It is therefore necessary to monitor the quality of beef available on the market in order to ensure food security, consumer safety, and public health. In order to produce beef cattle commercially, many animal husbandry practices are required, such as special care, management, feeding, grazing, breeding, housing, vaccination, medication, etc. (Jankeaw *et al.*, 2015). There are many human activities that take place within this industry, including farm operators, cattle integrators, feed millers, household owners, and others. In the preparation of cattle diets, various antibiotics, hormonal drugs, enzymes, feed additives, and medicines are often used, which can cause heavy metals to accumulate in animal tissues. As a result of the development of urban industries and human activities, one of the challenges to public health is the accumulation of heavy metals in food (Jankeaw *et al.*, 2015).

16.2.3 Microbial Contamination

The nutritional profile of meat provides the ideal conditions for spoilage microbes and foodborne pathogens to grow (Mohammed, 2004). Meat contamination often occurs as a result of microorganisms attaching to surfaces in which meat is stored and sold (Dharma *et al.*, 2022). *Listeria monocytogenes* is the major contamination in the RTE meat products (Figure 16.2). A study was conducted to investigate the prevalence of food infections and microbial contamination in meat products. A total of 42 samples of meat products, such as sausages, burgers, kebabs, and cutlets, were collected from 12 factories between 2011 and 2013. Microbial contamination of samples was examined according to Iranian national standards No. 5272, 9263, 2197, 10899–1 and 3, 1810, 6806–1 and 3 and 2946 for total count of microorganisms, coliform, *Clostridium*, mold and yeast, *Salmonella*, *Staphylococcus aureus,* and *Escherichia coli*, respectively (Kheyri *et al.*, 2014).

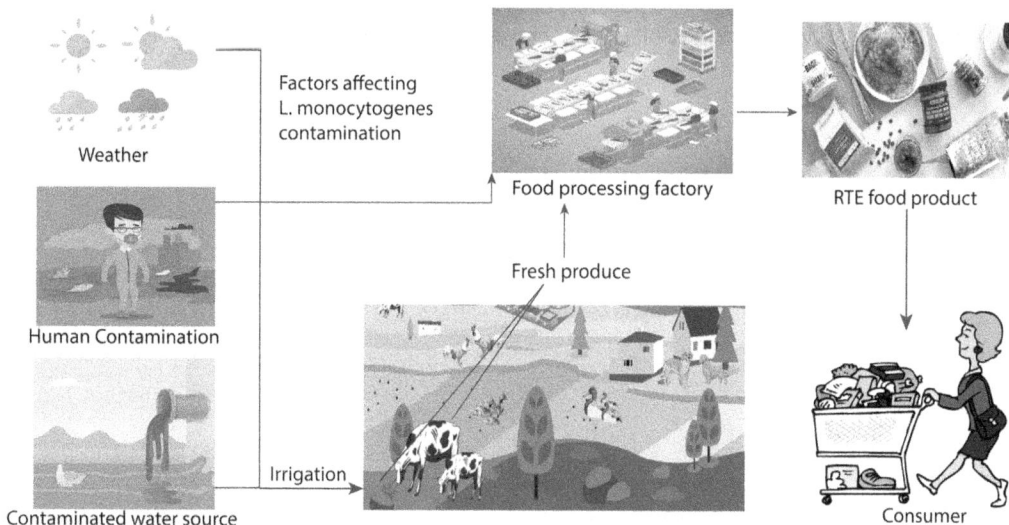

Figure 16.2 Steps involved in *Listeria monocytogenes* contamination in human RTE meat products.

In food processing plants, biofilms form primarily on damp surfaces where microorganisms can easily aggregate. Microorganisms such as *Staphylococcus, Pseudomonas,* and *Klebsiella* produce exopolymers that can fix additional microorganisms. The mixed biofilms formed by these organisms can be firmly attached to the surface. These bacterial communities contain both pathogenic and food spoilage microorganisms (Madoroba *et al.,* 2021). Furthermore, *Listeria monocytogenes* and *Enterobacter aerogenes* or bacteria from *Bacillus, Staphylococcus, Streptococcus, Escherichia, Shigella,* and *Klebsiella* survived cleaning and disinfection (Schlegelova *et al.,* 2010). Undercooked meat or chickens harboring bradyzoites may transmit *Toxoplasma gondii,* a zoonotic protozoan that forms meat cysts (Abd El-Razik *et al.,* 2014).

16.3 DETECTION METHOD MEAT CONTAMINATES

Evaluation of raw meat products for microbiological contamination by counting total aerobic spores, counting total bacteria by most probable number (MPN), testing for fecal coliforms and *E. coli,* and testing for *Salmonella* by polymerase chain reaction (Zafar *et al.,* 2016, Riffiandi *et al.,* 2022). In order to detect contaminants in meats products, it is crucial to develop rapid, sophisticated, reliable, and versatile screening methodologies. However, current detection methods may take several days to produce results. Establishing a national monitoring program and sampling will be recommended based on reliable and accurate procedures. It will reduce meat products' contamination levels in a sensible way. Effective control depends on the availability of simple and useful screening techniques (Figure 16.3). Various techniques can be used.

16.3.1 PCR-Based Method

Polymerase chain reaction (PCR) was used to detect pork and its derivatives based on pork-specific primers for the mitochondrial (mt) 12 S ribosomal RNA (rRNA) gene (Effendi *et al.,* 2020, Mohd-Hafidz *et al.,* 2020). Foodborne illness is often caused by *Staphylococcus aureus.* In light of the previous results, it can be concluded that meat products can serve as a suitable medium for the growth of *Staphylococcus* and the production of toxic substances (Hassan *et al.,* 2018). It has been strongly recommended that the highly sensitive, specific, and rapid M-PCR method be used for detecting adulteration in meat products and abuse of labeling requirements. This PCR method is useful for detecting adulteration in denaturized products (Abuelnaga *et al.,* 2021).

16.3.2 Spectroscopy

In order to detect and quantify spoilage as early as possible, several emerging techniques have been developed. Spectroscopy is an emerging detection technique based on the interaction between matter and electromagnetic radiation of varying wavelengths. The technique can be used to study, identify, and quantify materials and substances of varying biological complexity by measuring the light emitted, absorbed, or scattered by them. However, a number of notable applications are aided

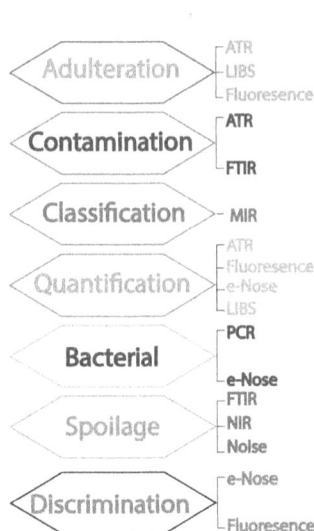

Figure 16.3 Summarized depiction of technologies for meat spoilage detection. Note: LIBS represents laser-induced breakdown spectroscopy.

by robust statistical and chemometric analysis, including laser-induced breakdown spectroscopy and Raman spectroscopy for food analysis, adulteration detection, authenticity testing, or even the pungency of spicy foods. There are many applications for mid-infrared (MIR) technology, such as food and beverage classification, shelf-life monitoring, and value-adding of foods. Attenuated total reflectance (ATR)-IR spectroscopy can also be used for detecting and identifying food adulterants, screening for food toxins, and quantifying food constituents, among other applications. These techniques are non-destructive, which makes them particularly useful for food applications (Fletcher *et al.*, 2018).

16.3.3 Odor Sensors and Electronic Nose Technology

Sensors of this type are based on human olfaction, the electronic nose (or e-nose) mimicking the chemical interactions between odor compounds and primary neurons found in the human nasal cavity. Electrochemical sensors and pattern classification algorithms are used in the e-nose to detect odors, similar to how primary neurons in the nasal cavity correspond with the chemical sensors of the e-nose. Sensor technology has advanced significantly, and chemosensors are readily available in many types. This section will briefly discuss metal oxide semiconductors (MOSs) and conducting organic polymers (CPs) (Fletcher *et al.*, 2018).

16.3.4 Metal Oxide Semiconductor

The most common semiconductor in e-noses is the metal oxide semiconductor MOS. Meat spoilage can be detected early using this type of gas sensor. An e-nose with MOS sensors was proposed by (Timsorn *et al.*, 2016) for evaluating the freshness of chicken meat and bacterial populations on chicken meat stored at 4.0°C and 30.0°C for up to 5 days. Principal component analysis (PCA) has been used by the authors to demonstrate the classification of chicken meat freshness as a function of storage days and temperatures in the study. In addition, the developed e-nose correlated well (0.94) with chicken bacterial populations, suggesting that it can serve as an alternative way of evaluating the bacterial population on meats in a rapid and alternative manner. The measurement is fast, portable, affordable, and non-destructive, with high relative accuracy, among other advantages. Kachele and colleagues used e-noses to assess microbial load, chemical changes, and sensory characteristics of silver carp fillets stored at 4°C for 14 days at two vacuum pressure levels (30 and 50 kPa). This system uses many metal oxide semiconductor sensors together with pattern recognition algorithms to construct an intelligent bionic olfactory e-nose. It provided valuable information on improving vacuum-packaging aspects, such as headspace and refrigeration conditions. Similarly, Zhang and collaborators designed an e-nose based on MOS sensors to detect microbial colonies that develop during vacuum-packaging of Yao meat. These findings offered new insights into packaging methods and improvements that can be made (Fletcher *et al.*, 2018).

16.3.5 Quantitative Microbial Risk Assessment Models

Beef is responsible for a high percentage of foodborne illness in Europe, with 2.3 million cases reported each year. These illnesses are caused by pathogenic bacteria that are contaminated and grown and/or not inactivated effectively along the whole farm-to-fork chain. A quantitative microbiological risk assessment (QMRA) plays an important role in ensuring consumer health in the area of food safety. QMRA has been applied in many areas over the past decades (Tesson *et al.*, 2020).

16.3.6 Quantitative Detection by ELISA

Many recent cases of pork adulteration of meats have reinforced the need for a method of detecting and quantifying pork contamination in other meats. A sandwich ELISA assay developed by Microbiologique, Inc. is able to detect pork quickly in cooked horse, beef, chicken, goat, and lamb meats. It is a global concern that pork is adulterated into other meats, and even small amounts of accidental contamination pose a serious religious issue for Muslim and Jewish consumers. Meat contamination levels have not been officially established in the United States by the U.S. Department of Agriculture. It is recommended by the United Kingdom Food Standards Agency (FSA) that contamination between 0.1 and 1% (w/w) should be investigated (Thienes *et al.*, 2018).

16.3.7 Smartphone-Based Biosensor

Technologies for rapid, non-destructive, and inexpensive detection of microbial contamination on meat are still in high demand. Smartphone-based biosensors have been developed to detect microbial spoilage on ground beef, without the use of antibodies or microbeads. This measurement can be used to screen meat products for microbial contamination, as well as to screen wounds for infection. Microbial spoilage was simulated by adding *Escherichia coli* K12 solutions to ground beef products. Ground beef was irradiated perpendicularly with an 880 nm near-infrared LED, and scatter signals at various angles were studied using the smartphone's gyro sensor and digital camera (Liang *et al.*, 2014). The development of biosensors has become an alternative approach to screening residues in animals in recent years (Khaniki, Ebrahim, and Parisa 2018).

16.3.8 Immunological Techniques

Based on the interaction between antigen and antibody, these methods are very specific for a specific residue. Most commonly, this type of test is done using enzyme-linked immunosorbent assays (ELISA). It is possible to find ELISA kits to test for a specific residue such as sulpametazine or a group of compounds including sulfonamides. It has been shown that ELISA kits are effective for analyzing antibiotic residues in meat (Khaniki, Ebrahim, and Parisa 2018).

16.3.9 High Performance Liquid Chromatography (HPLC)

The HPLC method is capable of detecting multiple residues in meat in both qualitative and quantitative ways (Khaniki, Ebrahim, and Parisa 2018). The accidental or fraudulent mixing of meat from different species is a highly relevant issue for quality control of food products, especially for consumers who are sensitive to species such as horse and pork. A sensitive mass spectrometric method was developed and demonstrated here for the detection of trace contamination in horse meat and pork, as well as for the specificity of the biomarker peptides against beef, lamb, and chicken (Gupta *et al.*, 2014). It was established that liquid chromatography-tandem mass spectrometry (HPLC-MS/MS) is an effective tool for identifying meat marker peptides and detecting exogenous meats in mutton (Gu *et al.*, 2018). The analysis and confirmation of chloramphenicol residues in edible animal tissues are performed using two high-performance liquid chromatography (HPLC) methods (Peris-Vicente *et al.*, 2022).

16.3.10 Charm II Technology

The Charm II technology is a new method of detection for residual compounds that is rapid, comprehensive, and semi-quantitative. Selectivity and sensitivity can be achieved with this technology. A combination of Charm II and HPLC separation provides an excellent method to detect and identify individual chemical and biological residues in animal tissue (Khaniki, Ebrahim, and Parisa 2018).

16.4 CONTROL OF THE CONTAMINANTS AT MEAT AND MEAT PRODUCTS

16.4.1 Physical Method for Decontamination

High pressure, UV-C, pulsed light, and cold plasma are the most effective in reducing bacteria, followed by gamma-ray and electron beam irradiation.

16.4.1.1 Steam Pasteurization

Steam pasteurization (SPS) uses potable water to produce superheated steam as a non-toxic and natural way of decontaminating the exposed surface of animal carcasses. The surface temperature of carcasses is instantly raised to 190°F (88°C) for 10 seconds by applying saturated steam. When the carcasses or trimmings enter the cooler, they are immediately chilled with cold water. There are two types of steam pasteurization systems on the market: the SPS 400 and the SPS 60. In the former, up to 400 carcasses can be treated per hour, while in the latter, up to 60 carcasses can be treated per hour (Cliver, 2007).

16.4.1.2 Irradiation Pasteurization

The FDA has approved irradiation pasteurization (IP) of red meat, but the USDA is currently developing appropriate regulations (AMI, 2000). The IP method emits pulses of intense energy that penetrate meat and destroy microorganisms using gamma radiation (e.g., Cobalt 60) or an electrical source (e.g., electron beam accelerators like SureBeam®). Microorganisms can be reduced on muscle foods using pulsed light (PureBright® TM), infrared irradiation, and ultraviolet light (SelectUV ®) (Cliver, 2007).

16.4.1.3 Ultrasound

The use of ultrasound treatment or ultrasonication (US) in food and non-food applications is an emerging technology. Sound waves that exceed the upper limit of human hearing (20 kHz) are considered ultrasounds, and they differ from audible and infrasonic sounds.

There are three types of ultrasound: power ultrasound (16–100 kHz), high-frequency ultrasound (100 kHz–1 MHz), and diagnostic ultrasound (1–10 MHz). The US has already been applied to a variety of applications, including distance measurement, cleaning, sonography, and waste-water treatment. In food processing, it is used for extraction, cleaning, emulsification, and homogenization. Since the US is an acoustic energy, its ionizing and invasive effects can be ignored. Moreover, this technology uses a non-polluting form of mechanical energy, making it a highly accepted method of food processing that does not affect food quality (Albert *et al.*, 2021).

16.4.1.4 Cold Atmospheric Plasma

Cold plasma, also called non-thermal atmospheric plasma, is another emerging food preservation technology that reduces foodborne pathogens without compromising product quality. A number of plasma-generating devices have been used to study the antibacterial properties of various food products in recent years. Cold plasma consists of electrons, excited atoms and molecules, ions, UV photons, free radicals, and reactive species (ozone, nitrogen oxides, hydroxyl radicals, atomic oxygen, superoxide, and singlet oxygen), which can kill bacteria, viruses, and fungi. Cells can be damaged by these compounds through lesions in the membrane, intracellular disorder, chemical bond breaks in the cell wall, loss of enzyme activity, damage to RNA and DNA, and denaturation of proteins (Albert *et al.*, 2021).

16.4.1.5 Packing Innovations

It is only possible to use these technologies for packaged meat, not for whole carcasses. These technologies include oxygen (O_2) adsorption, modified atmosphere packaging (MAP), and vacuum technology. A small pouch containing an oxygen scavenger (e.g., potassium permanganate) is enclosed with meat, so that the oxygen is captured by it instead of the meat. As a result, the meat color does not turn brown, and aerobic microorganisms have a difficult time multiplying under these conditions. The MAP technology flushes the red meat in a barrier package with a mixture of oxygen (80%) and carbon dioxide (20%). The oxygen in the package prevents microbial growth while the CO_2 delays it. The shelf life of red meat cuts can be extended by 7–12 days by modifying the atmosphere in the package. Last but not least, vacuum technology removes the atmosphere from a package by using a high barrier packaging material. Despite this, the meat does not bloom or turn red inside the vacuum package, which extends its shelf life to 21 days or longer. Packaging innovations under development include biosensors that detect chemicals, indicators of decomposition, and packaging that detects temperature changes (Cliver, 2007).

16.4.2 Chemical Method for Decontamination

16.4.2.1 Organic Acids

In the meat industry, organic acids are the most commonly used chemical decontaminants. By applying organic acids solutions to carcasses, microbial load and pathogen prevalence were reduced. The antimicrobial action of organic acids is due to their un-dissociated molecules, which accumulate in the cytoplasm and dissociate into protons at higher pH levels, thereby acidifying the cytoplasm. In addition to lactic acid and acetic acid, other organic acids such as citric acid, succinic acid, sodium hypochlorite, peroxy-acetic acid, etc. have also been used to decontaminate carcasses. On hot carcasses, organic acids are more effective (Bolder, 1997). As a means of reducing the microbial flora on animal carcasses and subprimals, several organic acids have been used. Acetic, lactic, citric, ascorbic, propionic, peracetic, and formic acids are some of them. USDA-FSIS approves citric, lactic, and acetic acid solutions at 1.5–2.5% for reducing carcass contamination. In general, both lactic and acetic acids are considered safe sanitizers. About 15% of beef processing plants mist carcasses with organic acids, according to Cliver (2007).

16.4.2.2 Ozonation

Ozone (O_3) is a soluble and unstable blue gas that is formed when ionizing radiation or electric charges pass through air or oxygen. The oxidizing properties of this substance make it highly effective at inactivating bacteria, and it was the first oxidizing agent used to disinfect water. Several muscle foods and fish were found to be effectively inactivated by ozonation (e.g., *E. coli* O157:H7, *Pseudomonas fluorescens*, *S. typhimurium*).

16.4.2.3 Hydrogen Peroxide

Hydrogen peroxide (H_2O_2) can rapidly kill microorganisms if used at adequate concentrations. In spite of its ability to inactivate microorganisms, H_2O_2 is not permitted as a food additive in many countries due to its bleaching and oxidizing effects. H_2O_2 (5%) and ozone (0.5%) are antimicrobial agents that decrease bacterial counts approximately 2.5 log10 CFU/cm^2 when used in meat as a decontaminants (Cliver, 2007).

16.4.2.4 Sodium Chloride

Sodium chloride (NaCl) is a generally recognized as safe (GRAS) antimicrobial and non-intentional food additive. Numerous studies have demonstrated NaCl's effectiveness as an antimicrobial agent. Inhibition of the majority of foodborne pathogenic bacteria can be achieved by using 13% (w/v) NaCl (equivalent to water activity of 0.92 or less), except for *S. aureus*. NaCl inhibits the growth of spoilage microorganisms in fermented meat products and promotes the growth of lactic acid bacteria. Foodborne pathogens such as *Staphylococci* in fermented meats can be inhibited by the growth of lactic acid bacteria (Figure 16.4). Salt tolerance varies among bacteria based on their intrinsic properties and extrinsic and intrinsic growth factors such as water activity, nutrients, pH, temperature, and oxygen availability. Generally, *Salmonella* spp. can survive in salty environments with NaCl concentrations as high as 3.25.3%, *L. monocytogenes* 8.0%–12%, *C. perfringens* 8.0%, *S. aureus* 18%–20%, and *C. botulinum* 11%–12%, respectively (Stein, 2000). There are several factors that contribute to the mode of preservation of NaCl as an antimicrobial agent, including (i) dehydration, (ii) effect of chloride ion, (iii) oxygen removal, and (iv) the interference with proteolytic enzyme (Cliver, 2007).

16.4.2.5 Acidified Sodium Chlorite

It has been revealed that chlorite stabilized in acid (e.g., acidified sodium chlorite, $NaClO_2$, or ASC) has effective effects of decontaminating carcasses, since it has a combination of antimicrobial effects caused by low pH due to the acid content of the spray, along with chlorine's antimicrobial properties. ASC is a powerful oxidant formed by reacting sodium chlorite with

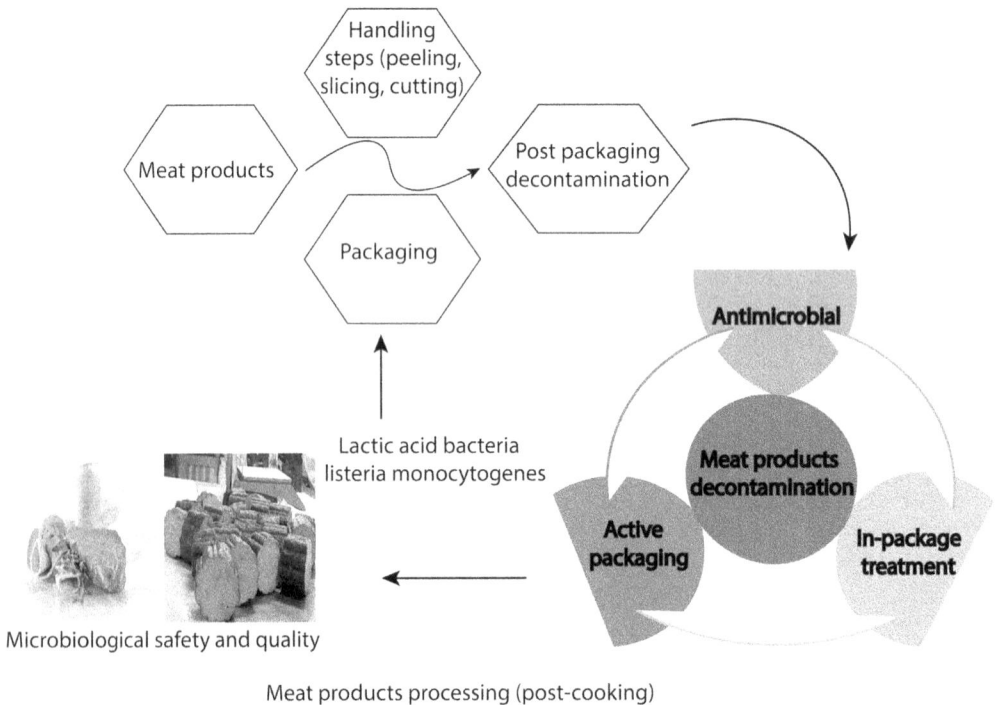

Figure 16.4 Microbiological safety and quality of meat products.

citric acid. The Food and Drug Administration approved ASC solutions as a direct food additive for decontaminating poultry and red meat carcasses. Throughout the dairy industry, it has been used to prevent and reduce intramammary infections by applying sodium chlorite to the udder and teats. Other applications include decontaminating chicken skin, altering *Salmonella* colonization in broiler chickens, and bleaching fruits and vegetables. *S. typhimurium* and *E. coli* O157:H7 were found to be susceptible to ASC when applied to beef surfaces (Cliver, 2007).

16.4.2.6 Chlorine Washes

During the chilling process of carcasses, rinsing with chlorine-containing water has been used to reduce or prevent the proliferation of bacteria. Water treated with chlorine at 200 parts per million (ppm) was found to reduce aerobic bacteria (APC) and the reduction was higher after 24 hours of treatment. Both chlorine dioxide (ClO_2) and calcium hypochlorite ($CaCl_2O_2$) reduced contamination on beef forequarters. Sprouting seeds have also been disinfected with chlorine water. The rapid inactivation of chlorine in organic systems like meat is one of the drawbacks of this treatment. Chlorine is the most commonly used sanitizer in the food industry. Chlorinated chemicals sanitize microbes by disrupting chemical bonds. In meat processing plants, chlorinated water is used to disinfect beef and poultry carcasses and prevent cross-contamination because of its ease of application, stability, quick effect, and low cost. During slaughtering, washing poultry carcasses with chlorinated water reduces total aerobic counts, total coliforms, *Salmonella*, and *E. coli* (Bolder, 1997).

16.4.2.7 Trisodium Phosphate

The USDA-FSIS (USDA, 1996a) has approved the application of trisodium phosphate (TSP) as an antimicrobial treatment before chilling (24 to 48 hours before fabrication) of beef carcasses. There has been extensive research on the capacity of TSP to deactivate pathogenic and nonpathogenic bacteria, as well as minimize the adhesion of microorganisms to carcasses and meat trimmings. *Salmonella* was reduced by 1.6–1.8 log10 when post-chilled chicken was dipped for 15 seconds in 10% TSP at 50°C. Intentionally inoculated pathogens on lean and adipose tissue were reduced by TSP, where greater reductions were observed for Gram-negative pathogens (e.g., *Salmonella* species and *E. coli*) than Gram-positive pathogens (e.g., *L. monocytogenes*).

16.4.2.8 Lactates

It has been reported that sodium and potassium lactates are effective in limiting the growth and development of aerobes and anaerobes in meat, as well as in antibotulinal and antilisterial actions. In cook-in-bag turkey products, sodium lactate delayed *Clostridium botulinum* toxin production; sodium, potassium, and calcium lactates exhibited antilisterial activity; calcium salt showed greater effect than sodium or potassium salt (Cliver, 2007).

16.5 RISK ASSESSMENT

Muscle, which is the main component of meat-producing animals, contains very low levels of microbial load. Microorganisms are present in their gastrointestinal tracts and excrement, in their mouths, and on the exterior of their bodies. HACCP has long established that missteps in handling, slaughter, dressing, and dissection (hide removal, evisceration, etc.) of these ruminants can have a major impact on food safety. It is not intended to minimize the importance of maintaining strong sanitation practices in the receiving and holding areas prior to abattoirs. The importance of assessing the prevalence of mud and feces on incoming animals, defining procedures for older or non-ambulatory animals, restricting employee movement from dirty to clean areas, and more cannot be overstated. Nevertheless, after animals are humanely stunned and exsanguinated, much attention must be paid to the safe science of removing the hide and splitting edible parts from those that shouldn't be eaten. According to previous study, more than 6% of cattle hides contain the dangerous *E. coli* O157 bacterium, so it is imperative that processors follow hygienic hide and pelt removal practices at this stage, and also strategically sample carcasses according to regulatory performance standards in order to determine and improve process control based on the results. As carcasses are eviscerated, viscera can release a number of unwanted microorganisms (gram-positive bacteria such as *Listeria*, gram-negative bacteria such as *E. coli*, *Salmonella*, and *Campylobacter*, as well as numerous yeasts and molds), which can spread, attach, and adhere to other parts (Raspor and

Jevšnik, 2009). Together, with the following information, a chemical risk profile of meat and meat products was conducted:

1. It is important to identify the chemicals that may potentially affect public health and safety in the Australian meat supply chain.

2. Assess the potential risks of these chemicals to public health and safety, within the current regulatory framework.

3. Identify any areas of the current regulatory system that need further attention in order to address potential public health and safety risks associated with chemicals in meat.

From meat production to retail of meat products, the chemical risk profile was identified and examined where chemicals might enter the meat supply chain. Furthermore, it considered the inputs into the primary production and processing chain of meat. According to the chemical risk profile, the following factors were considered:

- Chemicals used in primary production of agriculture and veterinary medicine;

- Heavy metals, organic contaminants, and other environmental contaminants;

- Natural chemicals found in plants, bacteria, or fungi associated with plants; and

- By-products of food processing.

Chemicals migrate from packaging, such as food additives and processing aids. Among the key findings from the risk profile regarding chemical hazards are:

- Regulations and non-regulatory measures are in place along the meat industry's primary production chain, resulting in minimal public health and safety concerns.

- Chemical residues in meat have been monitored extensively over many years, demonstrating high compliance.

- The meat industry will maintain a high standard of public health and safety by maintaining current management practices, especially chemical monitoring programs along the primary production chain.

- Further research or monitoring of potential chemical hazards would help provide further assurance that the public health and safety risk is low (Pointon *et al.*, 2006).

16.6 CONCLUSION

Several non-thermal technologies have been examined in this chapter, including irradiation, UV-C light, pulsed light, high pressure, cold plasma, and ultrasound, which can be used to reduce microorganisms on raw meat surfaces while maintaining the quality of food. These treatments are free of chemicals and leave no residues, environmentally friendly, and in most cases, one of these treatments is sufficient to significantly decrease the pathogenic load on meat. In the case of pulsed light, or UV-C microorganisms on the product surface are inactivated within seconds, while ultrasound requires a longer exposure time. However, Gram-negative pathogens and spores tend to react more strongly than Gram-positive pathogens and spores. Different compatible methods can be applied combined to achieve sterility effects. In addition, food safety laws vary from country to country, so some techniques may not yet be permitted. Therefore, it is very difficult to determine which method will prove most effective. The application of the decontamination methods described in this chapter should be considered as a supporting measure in the fight against food-relevant infectious agents. The primary focus should remain on good hygiene practices: they should never replace hygiene measures.

16.7 FUTURE TRENDS

In order to ensure that chemicals used or present in meat and meat products present a very low risk to public health and safety, extensive regulatory and non-regulatory measures have been put in place. Regulations and control measures in place along the meat primary production chain have resulted in minimal public health and safety concerns regarding chemicals in meat and meat products. Over many years, extensive monitoring of chemical residues in meat has demonstrated high compliance with regulations. Additionally, the chemical risk profile identified a number of areas in which further research or monitoring would provide further assurance that the public health and safety risk is low. As long as the current management practices are continued, particularly monitoring programs for chemicals along the primary production chain, the meat industry will maintain a high level of public health and safety.

REFERENCES

Abd El-Razik, K.A., H.A. El Fadaly, A.M.A. Barakat, and A.S.M. Abu Elnaga. 2014. "Zoonotic hazards T. gondii viable cysts in ready to eat Egyptian meat-meals." *World Journal of Medical Sciences* 11 (4):510–517.

Abuelnaga, Azza Sayed Mohammed, Khaled Abd El-Hamid Abd El, Mona Mohamed Hassan Soliman Razik, Hala Sultan Ibrahim, Mona Momtaz Mohamed Abd-Elaziz, Amany Hamed Elgohary, Riham Hassan Hedia, and Elgabry Abd-Elalim Elgabry. 2021. "Microbial contamination and adulteration detection of meat products in Egypt." *World* 11 (4):735–744.

Albert, Thiemo, Peggy G. Braun, Jasem Saffaf, and Claudia Wiacek. 2021. "Physical methods for the decontamination of meat surfaces." *Current Clinical Microbiology Reports* 8 (2):9–20.

Alonso, M. López, J.L. Benedito, M. Miranda, C. Castillo, J. Hernández, and R.F. Shore. 2002. "Contribution of cattle products to dietary intake of trace and toxic elements in Galicia, Spain." *Food Additives & Contaminants* 19 (6):533–541. doi: 10.1080/02652030110113744.

Aymerich, Teresa, Pierre A. Picouet, and Joseph M. Monfort. 2008. "Decontamination technologies for meat products." *Meat Science* 78 (1–2):114–129.

Balali-Mood, Mahdi, Kobra Naseri, Zoya Tahergorabi, Mohammad Reza Khazdair, and Mahmood Sadeghi. 2021. "Toxic mechanisms of five heavy metals: Mercury, Lead, Chromium, Cadmium, and Arsenic." *Frontiers in Pharmacology* 12. doi: 10.3389/fphar.2021.643972.

Bantawa, Kamana, Kalyan Rai, Dhiren Subba Limbu, and Hemanta Khanal. 2018. "Food-borne bacterial pathogens in marketed raw meat of Dharan, eastern Nepal." *BMC Research Notes* 11 (1):1–5.

Bolder, N.M. 1997. "Decontamination of meat and poultry carcasses." *Trends in Food Science & Technology* 8 (7):221–227.

Cliver, Dean O. 2007. "Microbial decontamination, food safety and antimicrobial interventions." *ErişimAdresi*: http://www.vetmed.ucdavis.edu/PHR/phr250/27.

Dharma, Edy, Haryono Haryono, Aldi Salman, Pangesti Rahayu, and Widagdo Sri Nugroho. 2022. "Impact of hygiene and sanitation in ruminant slaughterhouses on the bacterial contamination of meat in Central Java Province, Indonesia." *Veterinary World* 15 (9):2348–2356

Effendi, Mustofa Helmi, Shelma Warda Afdilah, Dhandy Koesoemo Wardhana, Fredy Kurniawan, and Rurini Retnowati. 2020. "The identification of pork contamination on beef by polymerase chain reaction (PCR)." *Systematic Reviews in Pharmacy* 11:634–637.

Fletcher, Bridget, Keegan Mullane, Phoebe Platts, Ethan Todd, Aoife Power, Jessica Roberts, James Chapman, Daniel Cozzolino, and Shaneel Chandra. 2018. "Advances in meat spoilage detection: A short focus on rapid methods and technologies." *CyTA-Journal of Food* 16 (1):1037–1044.

Gu, Shuqing, Lina Zhan, Chaomin Zhao, Jiang Zheng, Yicun Cai, and Xiaojun Deng. 2018. "Identification of meat marker peptides and detection of adulteration by liquid chromatography-tandem mass spectrometry." *Se pu= Chinese Journal of Chromatography* 36 (12):1269–1278.

Gupta, Vikas, Sushma Singh, Bhavesh Patel, Om Shankar, Kuldeep Kumar, Sanjeev Shukla, Shubhra Shukla, Navneet Kaur, Manisha Dubey, and Lakshya Singh. 2014. "High-performance liquid chromatography method validation for determination of tetracycline residues in poultry meat." *Chronicles of Young Scientists* 5:72–74. doi: 10.4103/2229-5186.129344.

Han, Jian Long, Xiao Dong Pan, and Qing Chen. 2022. "Distribution and safety assessment of heavy metals in fresh meat from Zhejiang, China." *Scientific Reports* 12 (1):1–8.

Hassan, Mohamed A., Reham A. Amin, Nesreen Z. Eleiwa, and Hala W. Gaafar. 2018. "Detection of *Staphylococcus aureus* in some meat products using PCR technique." *Benha Veterinary Medical Journal* 34 (1):392–403.

Hossain, Md Mosharaf, Asma Hannan, Md Mostofa Kamal, and Mohammad Abul Hossain. 2022. "Detection of heavy metals and evaluation of beef procured from the different market of Dhaka in Bangladesh." *European Journal of Food Science and Technology* 10 (2):1–10.

Jankeaw, Montip, Natagarn Tongphanpharn, Rattanapon Khomrat, Chuleemas Boonthai Iwai, and Nisa Pakvilai. 2015. "Heavy metal contamination in meat and Crustaceans products from Thailand local markets." *International Journal of Environmental and Rural Development* 6 (2):153–158.

Khaniki, Gholamreza Jahed, Aghaee Ebrahim Moglaee, Sadighara Parisha. 2018. "Chemicals and drugs residue in meat and meat products and human health concerns." *Journal of Food Safety and Hygiene* 4(1–2):1–7

Kheyri, Aghakhan, Maryam Fakhernia, Nasrin Haghighat-Afshar, Hassan Hassanzadazar, Behzad Kazemi-Ghoshchi, Farah Zeynali, Jafar Asadzadeh, Farzaneh Rahmanpour, and Mahmoud Bahmani. 2014. "Microbial contamination of meat products produced in the factories of West Azerbaijan Province, North West of Iran." *Global Veterinaria* 12 (6):796–802.

Liang, Pei-Shih, Tu San Park, and Jeong-Yeol Yoon. 2014. "Rapid and reagentless detection of microbial contamination within meat utilizing a smartphone-based biosensor." *Scientific Reports* 4 (1):1–8.

Madoroba, Evelyn, Kudakwashe Magwedere, Nyaradzo Stella Chaora, Itumeleng Matle, Farai Muchadeyi, Masenyabu Aletta Mathole, and Rian Pierneef. 2021. "Microbial communities of meat and meat products: An exploratory analysis of the product quality and safety at selected enterprises in South Africa." *Microorganisms* 9 (3):507.

Mohammed, Arafa Mohammedeen. 2004. "Contamination of ready to eat vended food of meat origin with aerobic bacteria in Khartoum State." *University of Khartoum*.

Mohd-Hafidz, M.M., W.H. Makatar, H. Adilan, and T. Nawawee. 2020. "Detection of pork in processed meat products by species-specific PCR for halal verification: food fraud cases in Hat Yai, Thailand." *Food Research* 4 (S1):244–249.

Mustefa, Wazir Shafi. 2021. "Microbiological evaluation of meat sold in butcheries shop of Cheleleka town in Anchar woreda, West Harerge, Oromia, Ethiopia Western Ethiopia." *Journal of Food Science and Nutrition Therapy* 7 (1):033–039.

Peris-Vicente, J., Ester Peris-García, Jaume Albiol-Chiva, Abhilasha Durgbanshi, Enrique Ochoa-Aranda, Samuel Carda-Broch, Devasish Bose, and Josep Esteve-Romero. 2022. "Liquid chromatography, a valuable tool in the determination of antibiotics in biological, food and environmental samples." *Microchemical Journal* 177:107309.

Pointon, Andrew, Ian Jenson, David Jordan, Paul Vanderlinde, Jo Slade, and John Sumner. 2006. "A risk profile of the Australian red meat industry: Approach and management." *Food Control* 17 (9):712–718.

Raspor, Peter, and Mojca Jevšnik. 2009. "Novel food safety concepts for safe food: case meat processing industry." *Scientific Journal" Meat Technology"* 50 (1–2):1–10.

Riffiandi, N., G.G. Maradon, and V.R. Pertiwi. 2022. "The detection of microbial contamination on goat meat from traditional market in bandar lampung city." *IOP Conference Series: Earth and Environmental Science.*

Schlegelova, Jarmila, Vladimir Babak, Martina Holasova, Lucie Konstantinova, Lenka Necidova, F. Šišák, Hana Vlkova, Petr Roubal, and Zoran Jaglic. 2010. "Microbial contamination after sanitation of food contact surfaces in dairy and meat processing plants." *Czech Journal of Food Sciences* 28 (5):450–461.

Shaltout, Fahim Aziz Eldin, Abdelazez Ahmed Helmy Barr, and Mohamed Elsayed Abdelaziz. 2022. "Pathogenic microorganisms in meat products." *Biomedical Journal of Scientific & Technical Research* 41 (4):32836–32843.

Tesson, Vincent, Federighi Cummins, Juliana De Oliveira Mota, Sandrine Guillou, and Géraldine Boué. 2020. "A systematic review of beef meat quantitative microbial risk assessment models." *International Journal of Environmental Research and Public Health* 17:688. doi: 10.3390/ijerph17030688.

Thienes, Cortlandt P., Jongkit Masiri, Lora A. Benoit, Brianda Barrios-Lopez, Santosh A. Samuel, David P. Cox, Anatoly P. Dobritsa, Cesar Nadala, and Mansour Samadpour. 2018. "Quantitative detection of pork contamination in cooked meat products by ELISA." *Journal of AOAC International* 101 (3):810–816.

Timsorn, Kriengkri, Theeraphop Thoopboochagorn, Noppon Lertwattanasakul, and Chatchawal Wongchoosuk. 2016. "Evaluation of bacterial population on chicken meats using a briefcase electronic nose." *Biosystems Engineering* 151:116–125. doi: 10.1016/j.biosystemseng.2016.09.005.

Zafar, Ayesha, Erum Ahmed, Hafiza Wajiha, and Abdul Basit Khan. 2016. "Microbiological evaluation of raw meat products available in local markets of Karachi, Pakistan: Microbial evaluation of raw meat products." *Proceedings of the Pakistan Academy of Sciences: B. Life and Environmental Sciences* 53 (2):103–106.

Index

For Product Safety Concerns and Information please contact our EU
representative GPSR@taylorandfrancis.com
Taylor & Francis Verlag GmbH, Kaufingerstraße 24, 80331 München, Germany

www.ingramcontent.com/pod-product-compliance
Lightning Source LLC
Chambersburg PA
CBHW061346210326
41598CB00035B/5891